Lecture Notes in Mechanical Engineering

Lecture Notes in Mechanical Engineering (LNME) publishes the latest developments in Mechanical Engineering—quickly, informally and with high quality. Original research reported in proceedings and post-proceedings represents the core of LNME. Volumes published in LNME embrace all aspects, subfields and new challenges of mechanical engineering. Topics in the series include:

- Engineering Design
- Machinery and Machine Elements
- Mechanical Structures and Stress Analysis
- Automotive Engineering
- Engine Technology
- Aerospace Technology and Astronautics
- Nanotechnology and Microengineering
- Control, Robotics, Mechatronics
- MEMS
- Theoretical and Applied Mechanics
- Dynamical Systems, Control
- Fluid Mechanics
- Engineering Thermodynamics, Heat and Mass Transfer
- Manufacturing
- Precision Engineering, Instrumentation, Measurement
- Materials Engineering
- Tribology and Surface Technology

To submit a proposal or request further information, please contact the Springer Editor of your location:

China: Ms. Ella Zhang at ella.zhang@springer.com
India: Priya Vyas at priya.vyas@springer.com
Rest of Asia, Australia, New Zealand: Swati Meherishi at swati.meherishi@springer.com
All other countries: Dr. Leontina Di Cecco at Leontina.dicecco@springer.com

To submit a proposal for a monograph, please check our Springer Tracts in Mechanical Engineering at https://link.springer.com/bookseries/11693 or contact Leontina.dicecco@springer.com

Indexed by SCOPUS. All books published in the series are submitted for consideration in Web of Science.

Oleg Anatolyevich Gorbachev · Xiaoguang Gao ·
Bo Li
Editors

Proceedings of 10th International Conference on Recent Advances in Civil Aviation

Springer

Editors
Oleg Anatolyevich Gorbachev
Department of Computer Science
Moscow State Technical University of Civil Aviation, Irkutsk Branch
Irkutsk, Russia

Xiaoguang Gao
Department of System Engineering
Northwestern Polytechnical University
Xi'an, China

Bo Li
Department of Computer Science
Northwestern Polytechnical University
Xi'an, China

ISSN 2195-4356 ISSN 2195-4364 (electronic)
Lecture Notes in Mechanical Engineering
ISBN 978-981-19-3787-3 ISBN 978-981-19-3788-0 (eBook)
https://doi.org/10.1007/978-981-19-3788-0

This Springer imprint is published by the registered company Springer Nature Singapore Pte Ltd.
The registered company address is: 152 Beach Road, #21-01/04 Gateway East, Singapore 189721, Singapore

Preface

Dear readers,

We are pleased to present you the Proceedings of the 10th International Scientific and Practical Conference "Current problems and prospects for development of civil aviation" which covers the issues of air traffic management and control, quality control and increasing the reliability of radio equipment and avionics, designing and testing aircraft units and mechanisms, increasing the reliability of aircraft control systems, managing aircraft enterprises, etc. The book is aimed at specialists dealing with problems of development and operation of aircraft systems, air transport economics as well as at a wide range of persons interested in current problems of civil aviation.

The conference proceedings will allow you to see and estimate new ideas and ways of solving the problems faced by developers and operators of aircraft and aircraft equipment, determine promising areas of its development.

Among the authors of the papers contained in the proceedings, you can find the names of the scientists from Republic of Belarus (Belarusian State Academy of Aviation), Vietnam (Vietnam Aviation Academy), as well as representatives of Russian higher educational establishments and enterprises such as Moscow State Technical University of Civil Aviation (Moscow), Moscow Aviation Institute (National Research University) (Moscow), Saint Petersburg State University of Civil Aviation (Saint Petersburg), Saint Petersburg State University of Aerospace Instrumentation (Saint Petersburg), Irkutsk National Research Technical University (Irkutsk), Irkutsk State University (Irkutsk), Irkutsk Branch of Moscow State Technical University of Civil Aviation (Irkutsk), Arkhangelsk ATM Center of the North-West Air Navigation Branch of the Federal State Unitary Enterprise State ATM Corporation in the Russian Federation, (Mezen), Papanin Institute for Biology of Inland Waters, Russian Academy of Sciences (Borok, Yaroslavl Region).

The texts of papers are published in the authors' edition.

Kind regards,
Irkutsk, Russia Oleg Anatolyevich Gorbachev
Xi'an, China Xiaoguang Gao
Xi'an, China Bo Li

Acknowledgements

The program committee of the conference expresses its deepest gratitude to everyone who contributed to the success of the proceedings:

Shamsiev, Zair—Doctor of Technical Sciences, Full Professor, Professor, Chair of Air Navigation Systems, Faculty of Aerospace Technologies, Tashkent State Technical University named after Islam Karimov, Tashkent, Uzbekistan

Eshmuradov, Dilshod—Candidate of Technical Sciences, Academician of the Academy of Sciences "Turon," Associate Professor of the Department of Metrology, Standardization and Certification, Tashkent State Technical University named after Islam Karimov, Tashkent, Uzbekistan

Prof. Xiaoguang Gao—Professor, Northwest Polytechnical University, China

Dr. Bo Li—Associate Professor, Northwest Polytechnical University, China

Markus Hennig—Paderborn University, Germany, Scientist

Dr. Daqing Chen—Deputy Head of Division: Computer Science and Informatics, School of Engineering, London South Bank University

Abdukayumov A.—Doctor of Technical Sciences, Professor, Department of Electronic Devices and Systems, Tashkent State Transport University, Tashkent, Uzbekistan

Dmitry Shomankov—Ph.D. (Engineering), Head of Cycle, Minsk, Republic of Belarus

Dr. Katharina Dück—Doctor of Philosophy, Scientific Assistant at the Leibniz–Institute of German Language (IDS), Mannheim, University Lecturer in German Philology at the University of Mannheim, Germany

Khamida Khusnutdinova—Tashkent Institute of Railway Transport Engineers, Tashkent, Uzbekistan

Lilia Eberleir—jekleitung für AMS-Maßnahmen und ÖSD, Salzburg, Austria

Martin Ŝikýř—Ph.D. Assistant Professor, MIAS Czech Technical University in Prague, Czech

MudherAl-Rubaye—Ph.D. (Engineering), Iraqi Ministry of Defence/Airforce Command/Air Operations/Air Control Section, Iraq

Naing Linn Soe—Candidate of Technical Sciences, Assistant Lecture, Department of Chemistry, Defence Services Academy, Pyin Oo Lwin, Republic of Union of Myanmar
Oleg S. Ablyalimov—Candidate of Technical Sciences, Senior Staff Scientist, Tashkent Institute of Railway Transport Engineers, Tashkent, Uzbekistan
Paulushkin Eryk—Associate Professor of the Department of Technical of Aviation Radio-Electronic Equipment of the Educational Intuitional of the Belarusian State Aviation Academy, Minsk, Republic of Belarus
Petr A. Khmarski—Ph.D. (Engineering), Senior Researcher Scientist of the Institute of Applied Physics of the National Academy of Sciences of Belarus, Minsk, Republic of Belarus
Saw Ulin Myint—Ph.D. of Industrial Ecology, Republic of Union of Myanmar, Sc. Candidate MUSTK named after D. I. Mendeleev
Veronika Fujimoto—Ph.D. of Education and Linguistic, Russian, Japanese and English Interpreter, Japan
Vladimir Vashkevich—Ph.D. (Engineering), Docent, Minsk, Republic of Belarus
Xikui Li—Professor, School of Foreign Studies, University of International Business and Economics, Beijing, China

About This Book

The book presents the Proceedings of the 10th International Scientific and Practical Conference "Current Problems and Prospects for Development of Civil Aviation." It covers the issues of air traffic management and control, quality control and increasing reliability of radio equipment and avionics, designing and testing aircraft units and mechanisms as well as transport specialists' training.

The volume is aimed at specialists dealing with problems of development and operation of aircraft systems, radio communication and navigation, aircraft and aircraft engines, teachers at technical higher education institutions as well as at a wide range of persons interested in current problems of civil aviation.

Contents

About the Editors

Oleg Anatolyevich Gorbachev Doctor of Technical Sciences, full professor, Director of Irkutsk branch of Moscow State Technical University of Civil Aviation (MSTU CA). In 2009 he defended his doctorate thesis on "Navigational support of civil aircraft under disturbed ionospheric conditions". He has published more than 100 scientific works, 19 of them being indexed in WoS, Scopus. Currently, he is an editor of several journals, proceedings of scientific conferences, including International informational and analytical journal *Crede Experto: Transport, Society, Education, Language*. He is an IOP reviewer (IOP Conference Series: Materials Science and Engineering).

Xiaoguang Gao Professor at Northwest Polytechnical University, China, is a Subject Director for System Engineering, Deputy Head of Aviation Science and Technology Key Laboratory of Avionic Systems. She also is a Chinese Head of the Sino-Russian teaching and scientific research center for Effectiveness Evaluation of Avionic Systems, a Head of Science and Technology Innovation Team. She received her Ph.D. in Control, Guidance and Simulation of Vehicle in 1989. Her major research interests include advanced control theory and its applications in complex systems, integrated aviation and aerospace electronics system and its warfare effectiveness evaluation, airborne robotic control system and its effectiveness evaluation.

Bo Li is currently working as an Associate Professor in System Engineering, Northwest Polytechnical University, China. He received his Ph.D. in 2008 and his PostDoc in 2010. His major research interests include advanced airborne fire control technology, combat simulation and efficiency analysis of weapon system, large-scale simulation optimization and distributed parallel computing. He is a member of Institute of China Command and Control.

Aircraft, Aircraft Engines and Their Operating Practice

The presented papers consider: increasing reliability of an aircraft and its power plant; strength, ruggedness and damage of materials and structures; motion of objects and apparatuses in different environments; gas and hydrodynamics of technical and natural systems; development of new construction materials and structural coatings; thermodynamic processes in engineering systems; transport power engineering (of land, water, air, space transport); improving methods of operation and technical state control; modeling of engineering systems; reliability and fault-tolerance of engineering systems; diagnostics of technical state and testing; complexing and processing information in engineering systems; increasing the level of technical personnel's training.

Keywords Aircraft · Aircraft engine · Hydrodynamics · Transport power engineering · Modeling of engineering systems · Air transport · Technical personnel's training

Thermodynamic Calculation of the Parameters of a Gas Medium in a Pulsed Combustion Chamber

Andrey Safarbakov and Sergey Kuznetsov

Abstract The paper presents a method of thermodynamic calculation of the parameters of the gas medium in a single combustion cycle in a pulse combustion chamber. The combustion process of the fuel–air mixture is considered as a single pulse, which includes the process of filling the combustion chamber with the fuel–air mixture and its combustion. The parameters of the gas medium in the combustion process of the fuel–air mixture can be determined using the equations of equilibrium thermodynamics. The parameters are calculated in the quasi-stationary formulation. This means that the entire unsteady combustion process can be considered as a continuous sequence of each single state of the parameters at each moment of time.

Keywords Pulsed combustion chamber · Thermodynamic calculation of parameters · Gas medium · Ball combustion layer · Flame front · Combustion pressure · Gas density · Adiabatic index

1 Introduction

The pulse combustion chamber is a cylindrical flame tube with a gas collection device made in the form of a conically tapered nozzle. The front device is a check valve, the plate of which lets air into the combustion chamber [1]. When the pressure in the flame tube volume rises, the valve disc is pressed against its seat, and the air supply stops. In turn, the nozzle is a throttling device, and its diameter at the cutoff determines the structure of the gas flow and its parameters in the volume of the combustion chamber. A pulse fuel nozzle for pulse fuel delivery and a spark plug are placed in the flame tube. Thus, the combustion process of the fuel–air mixture in such a combustion chamber occurs in pulses [1, 2]. The pulse repetition rate depends on the time of filling the flame tube with fuel–air mixture and the time the hot gases flow through the nozzle [3]. The article considers a single impulse which includes the process of filling the combustion chamber with a fuel–air mixture and its combustion. It is

A. Safarbakov (✉) · S. Kuznetsov
Irkutsk Branch of Moscow State, Technical University of Civil Aviation, Irkutsk, Russia
e-mail: safarbakov@yandex.ru

O. A. Gorbachev et al. (eds.), *Proceedings of 10th International Conference on Recent Advances in Civil Aviation*, Lecture Notes in Mechanical Engineering, https://doi.org/10.1007/978-981-19-3788-0_1

possible to determine the parameters of the gas medium in the process of combustion of the fuel–air mixture using the equations of equilibrium thermodynamics [4]. We calculate the parameters in the quasi-stationary formulation. This means that the entire unsteady combustion process can be considered as a continuous sequence of every single state of the parameters at every moment of time [5].

According to the theory of combustion and detonation of gases, the combustion of the air–fuel mixture in the closed volume of the pulse combustion chamber proceeds with a very high speed [6, 7].

In a small volume of fuel–air mixture from the spark plug, an electrical discharge occurs, and the oxidation reaction begins. When the rate of the oxidation reaction in this volume reaches a maximum, then combustion will occur. A flame front appears at the boundary of this volume of the fuel mixture with the medium at the initial moment the flame front is a sphere (see Fig. 1) [6, 7].

The temperature of the combustion front can be very high. At the surface of the combustion front, this temperature is equal to the autoignition temperature of the fuel. The temperature of the fuel–air mixture that is not yet burning also increases due to thermal conductivity. The expanding flame front will capture more and more volumes of the fuel–air mixture, and the combustion temperature will increase. The expanding flame front will displace the unburned mixture towards the front device and the pulse combustion chamber nozzle [2, 8].

The combustion sphere of the flame will be observed until it comes into contact with the walls of the combustion chamber. Due to the increase in pressure during combustion, the check valve of the front device will close, and the flame front which is a truncated sphere will begin to spread in the combustion chamber tube only towards the nozzle, forcing the unburned fuel–air mixture into the surrounding space where it will burn out (see Fig. 2).

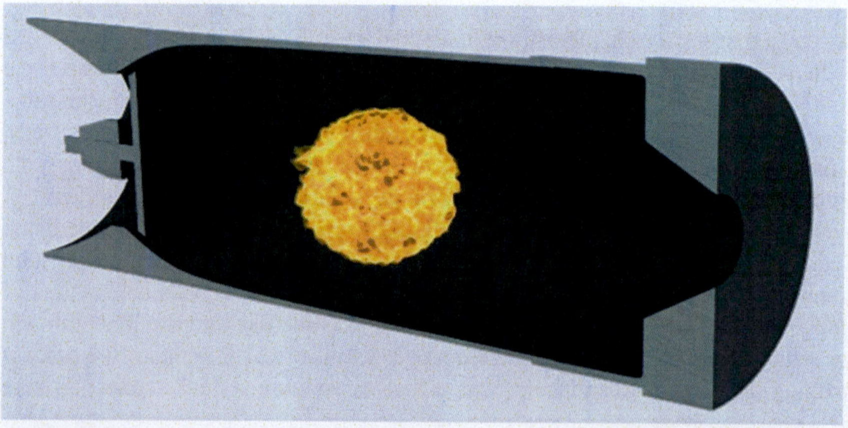

Fig. 1 Sphere of combustion of the air–fuel mixture

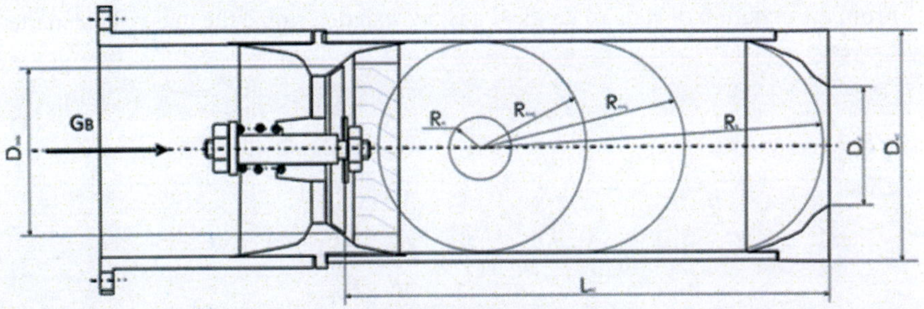

Fig. 2 Flame front spreading in an impulse combustion chamber

The length of the combustion chamber nozzle influences on the combustion process. Lengthening the nozzle leads to the occurrence of flame vibrations and detonation waves. The higher the combustion pressure, the more intense the detonation wave will be in a long nozzle [9].

2 Thermodynamic Calculation

To calculate the parameters of the gas medium in the combustion process, it is first necessary to calculate the volume of the combustion chamber $V_{c.c.}$, then the volume of the increasing combustion sphere and the volume of the ball layer $V_{b.l.}$.

For the convenience of using formulas, we introduce a list of abbreviations:

- R—specific gas constant $R \approx 287$ J/(kg K);
- $T_{ini.}$—air temperature at the initial moment before combustion, °K;
- $p_{ini.}$—air pressure at the initial moment before combustion, Pa;
- $m_{a.b.l.}$—mass of air in the ball layer, kg;
- α—excess air ratio;
- L_0—required amount of air to burn 1 kg of kerosene;
- $V_{h.g.}$—hot gas volume in the flame front, m^3;
- $V_{c.c.}$—combustion chamber volume, m^3;
- $V_{b.l.}$—ball layer volume, m^3;
- $V_{c.g.}$—cold gas volume, m^3;
- $m_{h.g.}$—mass of hot gas in the flame front, kg;
- $T_{com.}$—kerosene combustion temperature, °K;
- $p_{com.}$—combustion pressure, Pa;
- k—adiabatic index, $k = 1.25$.

Since the ignition of the air–fuel mixture occurs at a distance of 0.035 m from the check valve, the ball segment that has reached the frontal device is not considered further. The front of the flame rushes towards the exit from the nozzle of the combustion chamber.

From the equation of state of an ideal gas, we can determine the mass of air in the ball layer:

$$p_{\text{ini.}} \cdot V_{\text{b.l.}} = m_{\text{a.b.l.}} \cdot R \cdot T_{\text{ini.}} \tag{1}$$

Then:

$$m_{\text{a.b.l.}} = \rho \cdot V_{\text{b.l.}} = \frac{p_{\text{ini.}} \cdot V_{\text{b.l.}}}{R \cdot T_{\text{ini.}}} \tag{2}$$

where ρ is the density of air before the combustion process $\rho = \frac{p_{\text{ini.}}}{R \cdot T_{\text{ini.}}}$, kg/m^3.

Knowing the mass of air involved in combustion, the mass of fuel m_T [10] in the ball layer is determined:

$$m_T = \frac{m_{\text{a.b.l.}}}{L_0 \cdot \alpha} \tag{3}$$

The combustion temperature of kerosene depends on its mass composition and is approximately equal to [11]:

$$T_{\text{com.}} \approx 2162\,^{\circ}\text{K}$$

The calculation of the parameters of the gas medium will be carried out for the combustion of the fuel–air mixture in a single pulse. For this the combustion process is divided into iterations. Each iteration will correspond to a single time moment $\boldsymbol{\tau}$.

Knowing the kerosene combustion temperature, we can determine the combustion pressure at the initial time of the first iteration $p_{\text{com.1}}$:

$$p_{\text{com.1}} = \frac{T_{\text{com.}} \cdot p_{\text{ini.}}}{T_{\text{ini.}}} \tag{4}$$

During pulsed combustion of the fuel–air mixture in the combustion chamber, there will be hot gas in the ball layer $\rho_{\text{h.g.b.l.1}}$, hot gas in the flame front $\rho_{\text{f.f.1}}$, hot gas in the combustion chamber $\rho_{\text{h.g.c.c.1}}$, and cold gas $\rho_{\text{c.g.1}}$, for which their densities can be determined in the first iteration.

$$\rho_{\text{h.g.b.l.1}} = \frac{p_{\text{com.1}}}{R \cdot T_{\text{com.}}}$$

$$\rho_{\text{f.f.1}} = \frac{p_{\text{com.1}}}{R \cdot T_{\text{com.}}}$$

$$\rho_{\text{h.g.c.c.1}} = \frac{p_{\text{ini.}}}{R \cdot T_{\text{ini.}}}$$

$$\rho_{\text{c.g.1}} = \frac{p_{\text{ini.}}}{R \cdot T_{\text{ini.}}}$$

During pulsed combustion of the fuel–air mixture, there will be a hot and a cold component of the fuel–air mixture in the combustion chamber. In this case, the total pressure of the gas mixture $p_{\text{g.m.}}$ at the first iteration is the sum of the pressures of the cold $p_{\text{g.m.c.1}}$ and hot $p_{\text{g.m.h.1}}$ components of the mixture:

$$p_{\text{g.m.1}} = p_{\text{g.m.h.1}} + p_{\text{g.m.c.1}} \tag{5}$$

To calculate the pressures of the cold and hot components of the mixture, we turn to their specific volumes, since in subsequent iterations, it is more convenient to calculate the densities of the components of the gas mixture through specific volumes.

Since the combustion sphere expands with each moment of time and captures a new volume, the pressure of the hot mixture component $p_{\text{g.m.h.1}}$ in the new volume can be determined by the formula that describes the adiabatic process (adiabatic equation) [12]:

$$\frac{p_2}{p_1} = \left(\frac{V_1}{V_2}\right)^k$$

or:

$$\frac{p_{\text{g.m.h.1}}}{p_{\text{com.1}}} = \left(\frac{V_{\text{h.g.1}}}{V_{\text{c.c.}}}\right)^k$$

Then, the pressure of the hot component of the mixture $p_{\text{g.m.h.1}}$:

$$p_{\text{g.m.h.1}} = p_{\text{com.1}} \cdot \left(\frac{V_{\text{h.g.1}}/m_{\text{h.g.1}}}{V_{\text{c.c.}}/m_{\text{h.g.1}}}\right)^k \tag{6}$$

In formula (6) in the numerator, there is a specific volume of hot gas in the flame front $\vartheta_{\text{f.f.1}}$. In the denominator, there is a specific volume of hot gas in the combustion chamber $\vartheta_{\text{h.g.c.c.1}}$.

The mass of hot gas in the flame front is determined by the formula:

$$m_{\text{h.g.1}} = \rho_{\text{f.f.1}} \cdot V_{\text{h.g.1}}$$

The pressure of the cold component of the mixture can also be expressed through the adiabatic equation.

In the first iteration:

$$p_{\text{g.m.c.1}} = p_{\text{ini.}} \cdot \left(\frac{\vartheta_{\text{c.g.1}}}{\vartheta_{\text{c.c.1}}}\right)^k \tag{7}$$

where: $\vartheta_{\text{c.g.1}} = \frac{V_{\text{c.g.1}}}{m_{\text{c.g.1}}}$—specific volume of cold gas in the combustion chamber; $\vartheta_{\text{c.c.}} = \frac{V_{\text{c.c.}}}{V_{\text{c.c.}} \cdot \rho_{\text{c.g.1}}}$—specific gas volume in the combustion chamber. Here, $(V_{\text{c.c.}} \cdot \rho_{\text{c.g.1}})$—mass of cold gas in the combustion chamber $m_{\text{c.g.}}$; $m_{\text{c.g.1}} = \left(V_{\text{c.c}} \cdot \rho_{\text{h.g.c.c.1}}\right) - m_{\text{a. b.l.1}}$—mass of cold gas in the combustion chamber.

Having determined the pressures of the cold (7) and hot (6) components of the gas mixtures, we determine the mixture pressure $p_{\text{g.m.1}}$ by formula (5) and proceed to calculate the gas temperature $T_{\text{g.m.1}}$ in the combustion chamber.

Temperature is a thermodynamic parameter of a gas, which is understood as a measure of the average kinetic energy of the forward motion of gas molecules. Consequently, it is necessary to know the temperatures of the cold components of the system $T_{\text{g.m.c.}}$ and hot $T_{\text{g.m.h.}}$. From Poisson's equation for adiabatic compression [12]:

$$\frac{T_2}{T_1} = \left(\frac{p_2}{p_1}\right)^{\frac{k-1}{k}}$$

or in the first iteration for hot gas:

$$T_{\text{g.m.h.1}} = T_{\text{com.}} \left(\frac{p_{\text{g.m.1}}}{p_{\text{com.1}}}\right)^{\frac{k-1}{k}} \tag{8}$$

for cold gas:

$$T_{\text{g.m.c.1}} = T_{\text{ini.}} \cdot \left(\frac{p_{\text{g.m.1}}}{p_{\text{g.m.c.1}}}\right)^{\frac{k-1}{k}} \tag{9}$$

To determine the temperature of the gas mixture in the combustion chamber volume, it is necessary to use the law of energy conservation [13]. Since the volume of the combustion chamber is insulated, the total energy of the mixture of hot and cold gas cannot change during mixing [14]. The internal energy n moles of an ideal monatomic gas are equal to the kinetic energy of chaotic motion of gas molecules:

$$U = \frac{3}{2} n R T_{\text{g.m.1}} \tag{10}$$

where $T_{\text{g.m.1}}$—temperature of the mixture of cold and hot gas in the first iteration and R—universal gas constant.

Taking into account expressions (8) and (9), the internal energy of the mixture of cold and hot gas (10) will be equal to

$$U = \frac{3}{2} n_1 R T_{\text{g.m.h.1}} + \frac{3}{2} n_2 R T_{\text{g.m.c.1}} \tag{11}$$

where n_1—number of hot gas moles and n_2—number of cold gas moles.

When mixing cold and hot gas, the temperature of the mixture became $T_{CM.1}$, and the internal energies of the gases became

$$U = \frac{3}{2} n_1 R T_{g.m.1}, \text{ and, } U = \frac{3}{2} n_2 R T_{g.m.1}$$

According to the law of energy conservation (11):

$$U = \frac{3}{2} n_1 R T_{g.m.h.1} + \frac{3}{2} n_2 R T_{g.m.c.1} - \frac{3}{2} n_1 R T_{g.m.1} + \frac{3}{2} n_2 R T_{g.m.1}$$

Then, the temperature of the whole mixture is determined from this expression $T_{g.m.1}$:

$$T_{g.m.1} = \frac{n_1 T_{g.m.h.1}}{n_1} + \frac{n_2 T_{g.m.c.1}}{n_2} = \frac{\frac{n_1}{n_2} T_{g.m.h.1} + T_{g.m.c.1}}{\frac{n_1}{n_2} + 1} \tag{12}$$

Formula (12) includes the ratio of moles of hot and cold gas. It can be found from the equation of gas state for both gases before mixing:

$$p_{g.m.h.1} \cdot V = n_1 \cdot R \cdot T_{g.m.h.1}$$
$$p_{g.m.c.1} \cdot V = n_2 \cdot R \cdot T_{g.m.c.1}$$

By dividing the first equality by the second one, we can determine the ratio of moles of hot and cold gas:

$$\frac{n_1}{n_2} = \frac{p_{g.m.h.1}}{p_{g.m.c.1}} \cdot \frac{T_{g.m.c.1}}{T_{g.m.h.1}}$$

We substitute this expression for the temperature of the mixture (12) and obtain:

$$T_{g.m.1} = \frac{T_{g.m.h.1} \cdot T_{g.m.c.1} \cdot \left(p_{g.m.h.1} + p_{g.m.c.1}\right)}{p_{g.m.h.1} \cdot T_{g.m.c.1} + p_{g.m.c.1} \cdot T_{g.m.h.1}} \tag{13}$$

In the second iteration, the calculation of the parameters of the gas medium in the pulse combustion chamber will change slightly. In this case formulas (6) and (7) for calculating the pressures of the hot $p_{g.m.h.2}$ and cold $p_{g.m.c.2}$ components of the mixture will be:

$$p_{g.m.h.2} = p_{com.2} \cdot \left(\frac{V_{h.g\,2}/m_{h.g.2}}{V_{c.c.}/m_{h.g.2}} \right)^k \tag{14}$$

where: $p_{com.2} = \frac{T_{com.} \cdot p_{g.m.1}}{T_{g.m.c.1}}$—combustion pressure in the second iteration;

$V_{h.g.2} = V_{b.l.2} - V_{b.l.1}$—the volume of the flame front in the second iteration;

$m_{h.g.2} = \rho_{f.f.2} \cdot V_{h.g.2}$—mass of hot air in the expanded flame front.

For the cold component in the second iteration:

$$p_{g.m.c.2} = p_{g.m.c.1} \cdot \left(\frac{\vartheta_{c.g.2}}{\vartheta_{c.g.r.2}} \right)^k \quad (15)$$

where:

- $\vartheta_{c.\,g.2} = \frac{V_{c.g.2}}{m_{c.g.2}}$—specific volume of cold air in the combustion chamber in the second iteration;
- $V_{c.g.2} = V_{c.c.} - V_{b.l.2}$—reduced volume of cold air;
- $m_{c.g.2} = (V_{c.c.} \cdot \rho_{h.g.b.l.2}) - m_{a.b.l.2}$—mass of cold gas in the second iteration;
- $\vartheta_{c.g.r.2} = \frac{V_{c.g.2}}{m_{c.g.2} - m_{c.g.r.1}}$—specific volume of cold air, taking into account the mass of cold gas released from the combustion chamber nozzle;
- $m_{c.g.r.1} = G_{g.1} \cdot \tau$—the mass of cold gas that came out of the nozzle in the first iteration,
- τ—expiration time,
- $G_{g.1}$—gas flow from the nozzle, kg/s.

The gas flow rate for a tapered nozzle is calculated by the formula:

$$G_{g.1} = m_g \cdot \frac{p_{g.m.1}}{\sqrt{T_{g.m.1}}} \cdot F_n \cdot q(\lambda)_1$$

where:

- $m_g = \sqrt{\frac{k}{R} \cdot \left(\frac{2}{k+1}\right)^{\frac{k+1}{k-1}}}$—gas type ratio;
- F_n—nozzle area of the impulse combustion chamber;
- $q(\lambda)_1 = \left(\frac{k+1}{2}\right)^{\frac{1}{k-1}} \cdot \lambda_1 \left(1 - \frac{k-1}{k+1} \cdot \lambda_1^2\right)^{\frac{1}{k-1}}$—relative flux density in the first iteration;
- $\lambda_1 = \sqrt{\frac{k+1}{k-1}\left[1 - \Pi(\lambda_B)_1^{\frac{k-1}{k}}\right]}$—speed ratio;
- $\Pi(\lambda_B)_1 = \frac{p_{g.m.1}}{p_{21}^*}$—relative pressure in the combustion chamber;
- $p_{21}^* = p_{g.m.1} \cdot \left(1 + \left(\frac{k-1}{2}\right) \cdot M^2\right)^{\frac{k}{k-1}}$—total gas pressure in the combustion chamber;
- $M = \frac{c_{com.}}{a}$—Mach number in the combustion chamber;
- $a = \sqrt{k \cdot R \cdot T_{g.m.1}}$—local sound speed;
- $c_{com.}$—flame propagation speed in the combustion chamber.

Having the pressures of hot (14) and cold (15) mixture in the second iteration, the total pressure of the mixture is determined by formula (5):

$$p_{g.m.2} = p_{g.m.h.2} + p_{g.m.c.2}$$

After determining the pressure of the gas medium in the second iteration by formulas (8) and (9), the temperature of the hot and cold mixture is determined:

$$T_{g.m.h.2} = T_{com.}\left(\frac{p_{g.m.2}}{p_{com.2}}\right)^{\frac{k-1}{k}}$$

where: $p_{com.2} = \frac{T_{com.} \cdot p_{g.m.1}}{T_{g.m.c.1}}$—combustion pressure in the second iteration.

$$T_{g.m.c.2} = T_{g.m.c.1} \cdot \left(\frac{p_{g.m.2}}{p_{g.m.1}}\right)^{\frac{k-1}{k}}$$

The total temperature of the gas mixture in the second iteration is determined by the formula (13):

$$T_{g.m.2} = \frac{T_{g.m.h.2} \cdot T_{g.m.c.2} \cdot \left(p_{g.m.h.2} + p_{g.m.c.2}\right)}{p_{g.m.h.2} \cdot T_{g.m.c.2} + p_{g.m.c.2} \cdot T_{g.m.h.2}}$$

Further calculation of the gas mixture parameters in the pulsed combustion chamber in subsequent iterations will proceed in the same way as described above. The parameters of the gas medium in the pulsed combustion chamber were studied using the proposed method of thermodynamic calculation.

Initial data for calculation:

Geometric characteristics of the impulse combustion chamber	
Impulse combustion chamber length, $L_{c.c}$ (m)	0.13
Conical nozzle length, L_n (m)	0.01
Combustion chamber radius, $r_{c.c}$ (m)	0.035
Nozzle cutting radius, r_n (m)	0.01
Initial parameters of the working body	
Atmospheric pressure, p (PA)	95,849
Initial temperature, $T_{ini.}$ (°K)	313
Pressure generated by the blower, p_{blower} (Pa)	2045
Initial pressure in the combustion chamber, $p_{ini.}$ (Pa)	97,894
Thermodynamic parameters	
Thermodynamic parameters, R/M (J/(kg K))	287
Universal gas constant, R (J/(mole*K))	8.314
Required amount of air to burn 1 kg of kerosene, L_0 (kg)	14.8
Excess air ratio, α	1
Kerosene combustion temperature, T_{com} (°K)	2162
Adiabatic index, k	1.25

(continued)

(continued)

Flame propagation speed in the combustion chamber, $c_{com.}$ (m/s)	45
The amount of heat released by the combustion of kerosene, Q (kJ/kg)	43,050

The thermodynamic calculation of the gas parameters was performed for a single pulse. In turn, the pulse duration was divided into iterations with a step of 0.001 s, with a combustion duration of 0.022 s.

As a result of thermodynamic calculation, graphs (see Figs. 3, 4, 5, 6 and 7) showing the change in the main parameters of the gas mixture during pulse combustion were plotted.

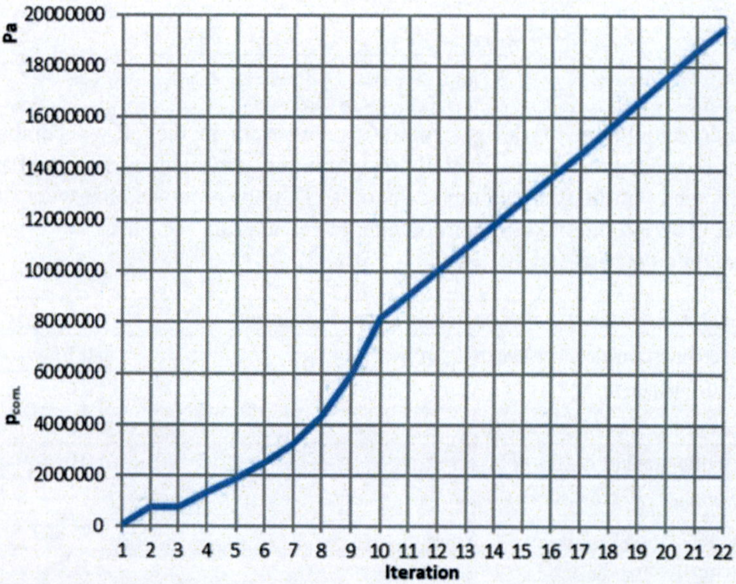

Fig. 3 Pressure changes in the combustion front

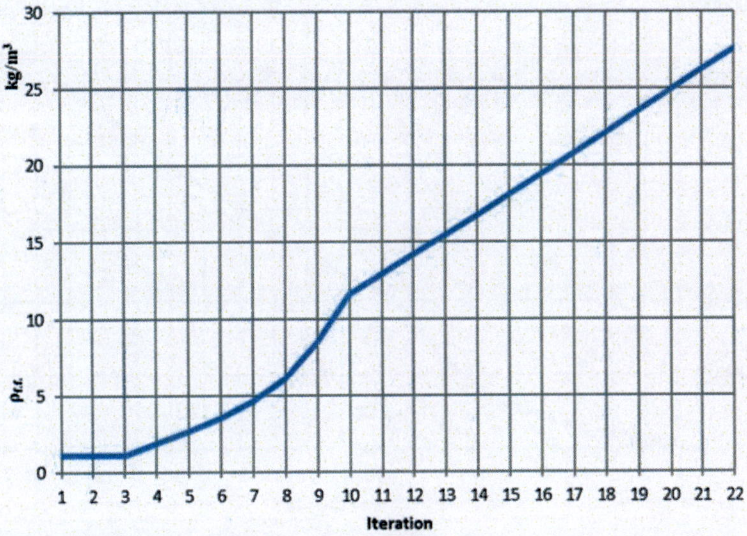

Fig. 4 Variation of gas density in the combustion front

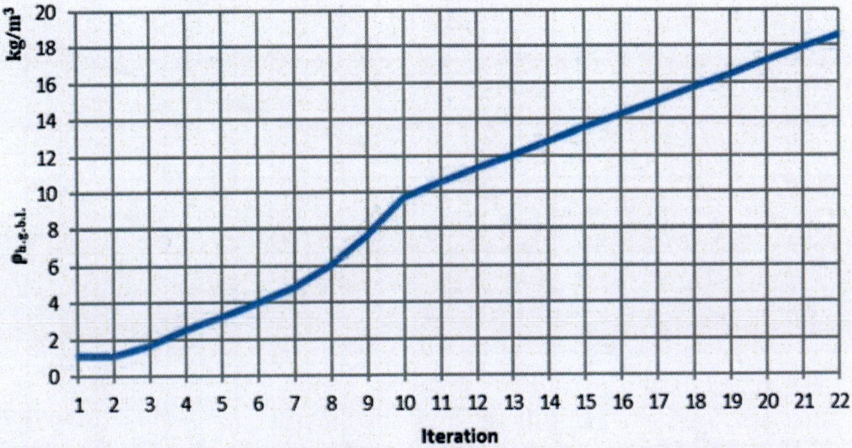

Fig. 5 Change of gas density in the ball layer

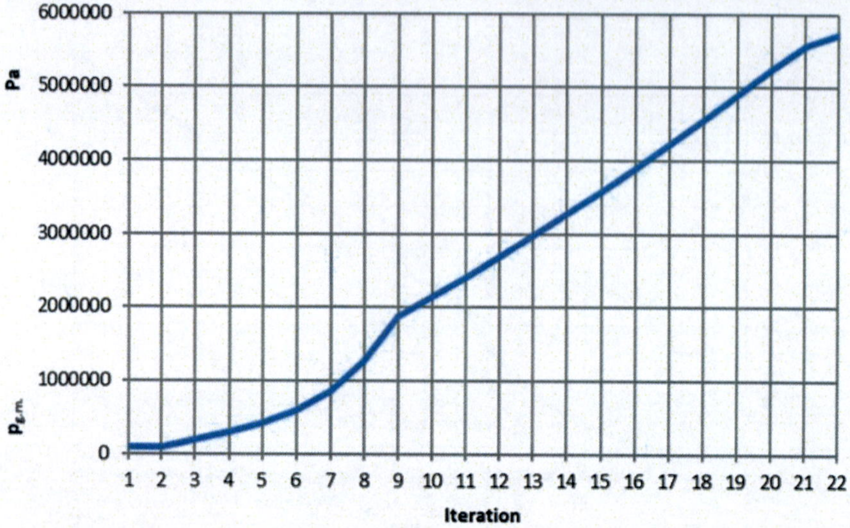

Fig. 6 Changing the pressure of the cold and hot gas mixture

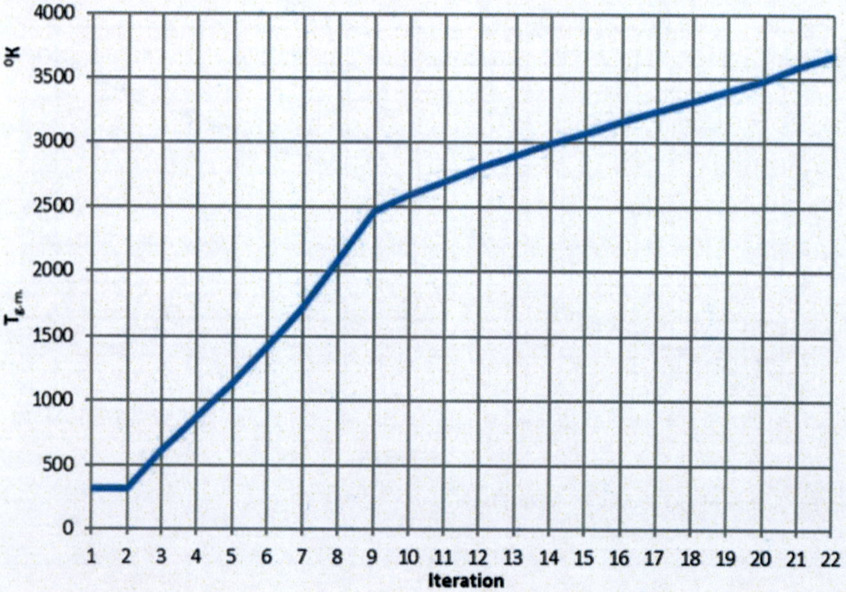

Fig. 7 Changing the temperature of the cold and hot gas mixture

3 Conclusion

The values of the calculated parameters can vary with changes in the flame propagation velocity in the pulsed combustion chamber. In addition, the mixture temperature and pressure values will be significantly affected by the nozzle diameter and shape. The calculation was performed for a tapered nozzle. It is impossible to accelerate the flow to supersonic velocities in such a nozzle, although it is possible to reach the critical mode of gas flow through the nozzle at the third iteration. Therefore, the Laval nozzle can be used.

References

1. Isaev AI, Safarbakov AM, Bogdanovich DV, Mayrovich YI (2012) Design of a pulse combustion chamber for a gas turbine engine: high technology, economics, industry. T. 2. In: Kudinov AP (ed) Part 2: collection of articles of the thirteenth international scientific and practical conference "fundamental and applied research, development and application of high technologies in industry and economy, 24–26 May 2012, Polytechnical University Publisher, St. Petersburg, Russia, 212 p
2. Kailasanath K (2000) Review of propulsion applications of detonation waves. AIAA J 39(9):1698–1708
3. Schauer FR, Miser CL, Tucker KC, Bradley RP, Hoke JL (2005) Detonation initiation of hydrocarbon–air mixtures in a PDE. In: AIAA paper, no 2005-1343
4. Bazarov IP, Gevorkyan EV, Nikolaev PN (1989) Nonequilibrium thermodynamics and physical kinetics. Publishing House of Moscow State University, Moscow, p 240
5. Moiseev MG, Tsirkunov YM (2006) Fundamentals of aerogasdynamics: tutorial. Baltic State Technical University, SPb., p 144
6. Ting JM, Bussing TRA, Hinkey JB (1995) Experimental characterization of the detonation properties of hydrocarbon fuels for the development of a pulse detonation engine. In: AIAA paper, no 95-3154
7. Govorov VI, Plotnikov VM, Karatay EV (2007) Theoretical foundations of combustion and explosion. Course of lectures for specialties: "work and life safety", "safety of life", "protection in emergency situations", "safety and environmental protection". KSTIU, Temirtau, 89 p
8. Isaev AI, Safarbakov AM, Khodatsky SA, Mayrovich YI (2016) Study of fuel-air mixture formation in the pulse combustion chamber and thermodynamic calculation of pulse combustion. In: [Electronic resource] proceedings of MAI. MAI, Moscow, p 91
9. Zeldovich, YB (1984) Selected works. chemical physics and hydrodynamics. Nauka, Moscow, 374 p
10. Schauer F, Stutrud J, Bradley R (2001) Detonation initiation studies and performance results for pulsed detonation engine applications. In: AIAA paper, no 2001-1129
11. Portola VA, Lugovtsova NY, Torosyan ES (2001) Calculation of combustion and explosion processes: manual. Yurginsky Technological Institute, Publishing House of Tomsk Polytechnic University, Tomsk, p 108
12. Glagolev KV, Morozov AN (2007) Physical thermodynamics, 2nd edn. Publishing House of Bauman Moscow State Technical University, 270 p. ISBN 978-5-7038-3026-0
13. Barilovich BA, Smirnov YA (2014) Fundamentals of technical thermodynamics and theory of heat and mass transfer. In: INFRA-M, 432 p. ISBN 978-5-16-005771-2
14. Lightman AP (2000) Great ideas in physics: the conservation of energy, the second law of thermodynamics, the theory of relativity, and quantum mechanics, 3rd edn. McGraw-Hill Professional, 300 p

Environmental Hazards of Aviation Events in Air Transport

Nikolay Nikolaykin, Evgeniy Starkov, Igor Merzlikin, and Natalia Korepina

Abstract The article offers an analysis of geotechnical systems formed in anthropogenic emergency zones during aviation events. The authors provide examples of transport facilities where aviation events lead to the formation of geotechnical systems. The emerging systems levels are presented with a hierarchy of ecological extreme zone levels. The study carries out an assessment of the aviation event's harm to the environment. It also substantiates that improving flight safety is the most appropriate and effective option for reducing the negative impact on the environment in air transport.

Keywords Aviation event · Environment · Geotechnical system · Environmental hazard · Air transport

1 Introduction

Air transportation of passengers and cargo is very significant in the Russian Federation and abroad. Covering significant distances and connecting remote areas and regions into a single whole is necessary. At the beginning of the third millennium, the constant growth of air transportation volume and the development of numerous air-related geographical points on the country's map (as well as worldwide) contributed to an increase in aviation popularity as the most mobile and fastest means of transport communications. At the same time, the concerns about the safety level of services [1] provided by civil aviation are constantly growing. It primarily relates to the flight safety level [2], aviation security, and environmental safety [3, 4].

Today, the modern concept of acceptable risk is globally recognized, according to which there is no absolute safety. The probability of an aviation accident always

---

N. Nikolaykin (✉) · E. Starkov · I. Merzlikin
Moscow State Technical University of Civil Aviation, Moscow, Russia
e-mail: nikols_n@mail.ru

N. Korepina
Irkutsk National Research Technical University, Irkutsk, Russia

O. A. Gorbachev et al. (eds.), *Proceedings of 10th International Conference on Recent Advances in Civil Aviation*, Lecture Notes in Mechanical Engineering, https://doi.org/10.1007/978-981-19-3788-0_2

remains, but the risk to flight safety should not exceed an acceptable level in all operations. Flights are operated with varying degrees of difficulty, often associated with the "human factor" [5]. Therefore, the accident risk in such flights has a different level, which is not always acceptable. So, following the civil aviation statistics, aircraft maintenance and repair errors cause 7–10% of accidents and 20–25% of aviation incidents [2].

2 Relevance Rationale

Nowadays, understanding the traditional word "flight safety" has significantly transformed. It means eliminating unacceptable impact on nature [4, 6], excluding loss of life and material losses [7, 8].

There are currently various attempts to develop risk management methods in air transportation [9]. However, the aviation community should be prepared for improbable but possible accidents, i.e., be ready to eliminate possible severe environmental consequences since the acceptable level of civil aviation flights safety is not quantified. Therefore, it is crucial to reduce the environmental hazard in civil aircraft accidents.

In aircraft accidents, the main property of the impact on nature is the unexpected moment in time of an event and the exclusive territory uncertainty that will be affected by the crashing equipment [10, 11]. The chances of predicting the amount and rate of impact increase are meager [12]. The impact itself is multi-vector and multi-profile, depending on heterogeneous factors of variable intensity. The impact on the natural objects biocenosis is primarily differentiated into the direct damage and the damage from subsequent indirect events [13, 14].

To date, there is no data on the form, types, and volumes of the adverse aviation events' environmental impact in the regularly ICAO Environmental Protection Report (every three years) [15].

Considering the relevance of maintaining global ecological balance and preventing disproportionate negative impact on nature, practical guidelines for emergencies and actions analysis to reduce the negative environmental impact are being developed [6, 12].

The nature damage from aviation emergency impact is summarized from the damage to natural resources and the cost of the lost biocenosis compensation (restoration). It is not possible to plan preliminary expenses for losses compensation in the event of an accident [16].

A proper understanding of the risk magnitude and the possible damage is the basis for further work on the biocenoses restoration affected by accidents with damaged aircraft and helicopters. A comprehensive assessment of the probable risks of natural ecological systems disruption should be based on heterogeneous data expressed in the same measurement units. That is, a specific uniform assessment scale should be used. It is proposed to borrow such a normalized scale in [17], provided and convincingly substantiated.

Let us consider the risk as to the probability of an adverse event occurring and assume that there may be several hazardous events in the reporting period. In each event, the damage must be taken into account from the events dependent on each other and occurring in parallel over a single timespan on one territory. Hence, it is advisable [8] to use the expression:

$$R = \sum_{i=1}^{n} 1 \sum_{j=1}^{m} PijUi \tag{1}$$

where *Pij* is the probability of damage *Ui* upon event *j*; *n, m* is the number of biocenosis damage variants in potential negative cases.

It is important to note that in the current ICAO documents regulating the international aviation organizations' activities, there is no mention of the negative contribution of incidents and accidents to the "carbon footprint" left by the international civil airlines operating air transportation on our planet. For example, the UNDP report on the aviation sector transformation to reduce carbon dioxide emissions from air travel, funded by The Global Environment Facility (GEF), does not take into account the environmental costs of the flight delays and impossibility of flying to some airports, one way or another, affected by aviation accidents [18].

3 Methods

In case of an adverse aviation event, a powerful simultaneous impact occurs both on the biotope and on the territory biocenosis where the event took place. As a result, the traditional ecosystem is transformed into a zone of technogenically caused catastrophic changes under the aircraft event factors influence with the intensity exceeding the permitted level.

The subsequent study of the aviation event environmental results was carried out as the development of the theory of physical and chemical systems (from here on—PhChS) and the theory of geotechnical systems (from here on—GTS) [19], necessary to improve the set of environmental protection methods of the transport communications hubs [4].

In [19], the authors proposed the following definitions of the concepts of PhChS and GTS:

> PhChS is an m-phase and n-phase continuous medium distributed in space and time variables. In the presence of a source or sink, matter or energy is transferred at each point of the homogeneous medium and at the phase boundary.
>
> GTS is an open system in which an industrial (transport or any other production) facility exchanges energy with the environment. Mass, energy, and information transfer processes in the GTS obey the same general laws as in artificially created physicochemical systems (PhChS).

It is these definitions that we use in our research further.

Another study [4] reflects that, in the last quarter of the twentieth century, the subordination of PhChS and GTS was made for the case of industrial organizations biocenoses impact. Further, at the turn of the XX and XXI centuries, an adaptive restructuring of this theory was implemented for the network of transport communications hubs, differentiated by the size of objects and the amount of cargo transported. As a result, it became possible to take into account the interrelationships and mutual influence of negative impact factors on nature, arising from civil aircraft and the corresponding infrastructure, from the point of view of the integration taking place in the entire resource cycle, according to ISO 14040:2006 requirements, of the activities of enterprises and organizations of the country's civil air fleet [10]. It could also be probable to give first estimates of harmful aviation event hazards for the accident sites ecosystems [13].

The fractionation of the corresponding hazard factors into groups and subgroups according to clearly defined classification criteria is an initial act to generate an integral system to eliminate the negative impact on biocenoses.

The authors [4] managed to develop and use a fractionation variation of traditional production and communications PhChS and GTS in practice, reasonable to employ for the fractionation of the corresponding systems' hierarchy that arises during critical aviation events.

The assessment of the unexpected impact of different factors on a murky territory is possible, as noted above, only with a unified system for measuring values on a normalized scale. In the literature, it is proposed to borrow the total criterion of the environmental impact of air transportation on the environment—I_{CA} [17] for the analysis of the considered emergencies.

Such a criterion, as, for example, it is shown in [7], makes it possible to single out specific fractions of harmful action sources and their consequences for the natural environment during "flight debriefing" under stable conditions of the system regulation of air transport enterprises and organizations operation. Therefore, in unconventional conditions after adverse aviation events, the I_{CA} criterion should be revised, expanded, and extended to new conditions with the designation of the criterion—IAC.

$$I_{\text{AC}} = \sum_{k} N I_k \tag{2}$$

where NI is an adverse effect on nature; k is the type of adverse action or component of pollution.

In the new interpretation of the I_{AC} criterion, it is recommended to leave the measurement unchanged, i.e., in units of the number of adverse effects on the environment, which are set equal to such a volume of action on the atmospheric air (and then on other components of the natural environment), which corresponds to the damage caused by the emission of carbon monoxide (CO) weighing 1 ton.

4 Experiment

The sequence of what is happening in a typical situation developing during an accident, taking into account the proposed classification, will be depicted in the form of a diagram shown in Fig. 1.

Group complexes of impact are the forms of adverse impact on nature, NI_{gen}, such as the inflow of pollution into the air basin, natural surface water bodies, various solid and concentrated liquid components entering the lithosphere, as well as complexes of physical factors (acoustic, radiation, and optical):

$$NI_{gen} = NI_{air} + NI_{w} + NI_{lith} + NI_{phys} \tag{3}$$

For each complex of impact unfavorable to nature caused by a catastrophic aviation event on a particular territory, the impact measure NI is determined by the formula:

$$NI = M \times H \times K_{t.b.b.} \times K_{unfav.f.} \times K_{aux.prot} \tag{4}$$

where M is the measure of the substance (energy, information) introduced into the biotope and biocenosis during an adverse aviation event, as well as from related phenomena;

H is the conditional unfavorableness level, defined as the value of CO, equal to the damage caused by carbon monoxide emission with a mass of 1 ton;

$k_{t.b.b.}$ is the indicator of the territory biocenosis balance;

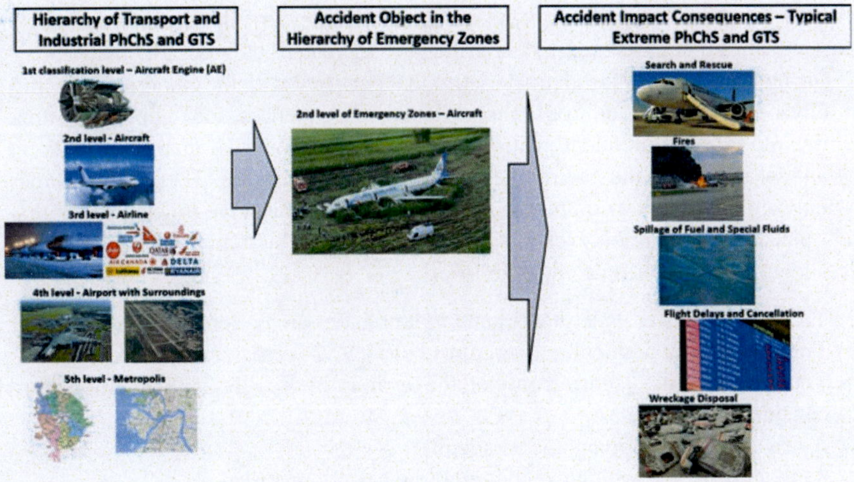

Fig. 1 Physicochemical and geotechnical systems formed on the accident site

$k_{\mathrm{unfav.f.}}$ is the indicator of unfavorableness of the surrounding atmospheric environment of the residential area;

$k_{\mathrm{aux.prot.}}$ is the auxiliary indicator of the protected area (equal to one or 1.2).

5 Results

According to the proposed method, we obtain the amount of natural systems loss due to an adverse aviation event with an emergency aircraft as the output of the research algorithm. If generalized qualimetry is acceptable, a less detailed assessment technique is relevant, as it is usually used for road traffic events. Then, the levels of pollution of soils ($\mathrm{Lp_s}$), surface waters (Lp_{w}), atmosphere ($\mathrm{Lp_a}$), and phytocommunities ($\mathrm{Lp_{phc}}$) are considered. The following dependence is used to calculate the damage to nature:

$$\mathrm{Lp}_n = \mathrm{Lp_s} + \mathrm{Lp_w} + \mathrm{Lp_a} + \mathrm{Lp_{phc}} \tag{5}$$

Using the above developments and applying them to measure the aviation accident impact level on a specific ecosystem, it is possible to introduce the analysis of the negative impact on the ecosystem as a simulation model. This model estimates direct and indirect impacts. In addition, the mutual influence of factors of various aviation accident impact types on nature among themselves and on the total amount of environmental pollution is taken into account.

The data on actual aviation accidents from the materials of the Interstate Aviation Committee on their investigation was taken as a basis. The model parameters were selected through expert assessments and generalizations of the aviation accidents characteristics that have occurred over the past 15 years (2006–2020).

The heterogeneity of the characteristics that occurred with the emergency aircraft by types of aircraft, conditions, the scale of consequences, and numerous other features does not allow identification and take into account all direct and indirect variants of impact on the environment and their characteristics. Therefore, in order to analyze the process of mutual influence of different negative impacts types that took place in the aviation events, some limitations and assumptions to the model were identified and formulated, as follows:

1. The accidents over water bodies and mountainous areas were not considered.
2. The accidents at ambient temperatures below 5 °C were not considered.
3. The accidents that occurred only on the territory of the Russian Federation were considered.
4. A fire did not accompany the accidents.
5. At the time of the accidents, precipitation was not recorded.

The model distinguishes between two fundamentally different cases: the absence or presence of protective systems (and, accordingly, measures to protect the environment).

The simulation model of the accident impact on the environment makes it possible to formulate and qualitatively solve the variational problem of minimizing the functional of the economic damage value. It also completes the functional of ecosystem restoration costs, where the target function Φ, as well as the complex index of harmful effects *I*, depends on three groups of functions: *X*—linear in time, *Y*—quadratic in time, and *Z*—more difficult time dependences.

6 Discussion

The optimization was carried out using the system of Euler–Lagrange equations, including critical, protective, and preventive adjustments. The system of equations for the target function Φ has the form:

$$\frac{\partial \Phi}{\partial x_n} - \frac{\mathrm{d}}{\mathrm{d}t}\left(\frac{\partial \Phi}{\partial x'_{nt}}\right) = 0; \ \frac{\partial \Phi}{\partial y_m} - \frac{\mathrm{d}}{\mathrm{d}t}\left(\frac{\partial \Phi}{\partial y'_{mt}}\right) = 0; \ \frac{\partial \Phi}{\partial z_k} - \frac{\mathrm{d}}{\mathrm{d}t}\left(\frac{\partial \Phi}{\partial z'_{kt}}\right) = 0 \tag{6}$$

Further transformation makes it possible to obtain a differential equation for a generalized quantitative indicator θ, which generally characterizes the system's intake of chemicals, energy, organisms, and information.

$$A\theta\frac{\mathrm{d}\theta}{\mathrm{d}t} + B\frac{\mathrm{d}\theta}{\mathrm{d}t} + C\theta^2 + D\theta + E = 0 \tag{7}$$

The coefficients of the equation *A*, *B*, *C*, *D*, and *E* are expressed through the parameters of the ecosystem and all types of adjustments (critical, preventive, and protective). They are time-constant values for a specific ecosystem of the territory of the aircraft accident site.

These parameters are primarily:

- the maximum capacity of the ecosystem for the entry of a pollutant into it, that is, the maximum amount of a substance whose entry into the ecosystem leads to an unacceptable shift in the ecological balance; after that, natural self-regulations can no longer prevent the destruction of the ecosystem;
- the effectiveness of the protective system that implements preventive and protective adjustments;
- coefficient of specific suppression of the negative impact caused by a unit of intensity of one or another adjustment;
- coefficient of completeness of the actual use of the intensity unit of a particular adjustment;
- safety barrier for the respective pollutant.

Equation (7) can be represented as an equation with separable variables and is generally solvable. The general solution has a complex form for physical interpretation. It contains the terms competing with each other for dominance, depending on the ratio of the equation coefficients.

While analyzing the reliability of the simulation model, a numerical experiment for some limiting cases was carried out. The limiting cases can be compared with an actual situation. The model parameters are obtained from expert assessments and analysis of the accident investigation materials for the last 15 years (2006–2020).

In the absence of any adjustments, the coefficients A and C in Eq. (7) are equal to zero. Then, the solution takes the form:

$$\theta = \alpha e^{\beta t} - \gamma \tag{8}$$

The exponent β in expression (8) is positive, so the amount of harmful substances or information (causing indirect chemical effects) increases with time. This solution is similar to the Malthus model [20, 21] for populations without limiting the size of the ecological niche. In real life, such behavior does not occur since, within the mechanism of ecosystems' natural homeostasis, the processes of natural self-regulation proceed in the form of natural successional processes of biological replacement and development. Natural dissipation of the introduced pollution and complexes formed in the biotope also occurs.

However, an increase in the harmful effects indicator, close to exponential, can be observed in a limited time interval before the start of protective and preventive measures.

It is possible to obtain a logistic function concerning the natural self-regulation processes. If protective and preventive measures have not begun by the time corresponding to the function inflection point, then the curve reaches a particular constant value that does not decrease before adjustments start.

The presence of such a limiting case confirms the developed simulation model reliability.

7 Conclusion

The use of a set of theoretical provisions of the existing theories of physicochemical and geotechnical systems made it possible to propose a classification of the harmful impact types on the environment caused by accident, which differs from the known ones, as it additionally takes into account the results of environmental (chemical) pollution from the information impact of the event in question.

In the future, this allows starting the problem solving of reducing an emergency aircraft impact on the environment, which will require the development of methods and algorithms for the comprehensive protection of ecological systems when organizing work at the accident site.

References

1. Ministry of Transport of the Russian Federation. Federal Agency of Air Transport. Rosaviatsiya homepage. https://favt.gov.ru/dejatelnost-vozdushnye-perevozki-osnovnye-proizvodstvennye-pokazateli-ga/. Accessed 2021/10/09
2. IATA Safety Report 2018 (2019) 55th edn. IATA, pp 56–61
3. Daley B (2016) Air transport and the environment. Cambridge University Press, Cambridge
4. Nikolaykin N, Nikolaykina N (2016): Environmental safety. Industrial, transport and energy hubs. LAP LAMBERT Academic Publishing, Deutschland
5. Nikolaykin NI, Tsetlin VV, Stepanova GP, Koroleva EK, Starkov EY (2020) Cosmophysical disturbances and the human factor in aviation as the cause of negative events. Scientific Bulletin UI CA 12:56–61
6. Environmental Impacts Assesments of Projects. European Union (2017). https://ec.europa.eu/environment/eia/pdf/EIA_guidance_EIA_report_final.pdf. Accessed 2021/10/08
7. Sharov VD, Vorob'ev VV, Nikolaikin NI, Kuznetsova VL, Tolstykh SA (2020) Methodology for estimating the safety and quality of the aviation service provider activities using the principal component analysis. Russ Aeronaut 63:575–585. https://doi.org/10.3103/S1068799820040030
8. Sharov VD, Vorobyov VV, Zatuchny DA (2021) Probabilistic-statistical methods for risk assessment in civil aviation. Springer Aerospace Technology. Springer Nature
9. Sharov VD, Vorobyov VV, Zatuchny DA (2021) Risk management methods in the aviation enterprise. Springer, Singapore
10. Nikolaykin NI (2006) The ecological estimation of the whole living cycle of civil aviation enterprise activity. Civil Aviation High Technol 108:73–79
11. Kurakina IN, Ivlichev IA (2015) Environmental risk assessment methods based on heterogeneous data. Izvestiya SPbGETU "LETI" 2:46–51
12. Wisner B, Adams J (2002) Environmental health in emergencies and disasters. A practical guide. World Health Organization, Geneva
13. Nikolaykin NI, Nikolaykina NE, Sekerin VD, Gorokhova AE (2017) Environmental and economic model of an aircraft accident evaluation. J Environ Manage Tourism 8, 5(21):1128–1135
14. Ritchie H (2021) Climate change and flying: what share of global CO_2 emissions come from aviation? https://ourworldindata.org/co2-emissions-from-aviation. Accessed 2021/10/08
15. ICAO 2019 environmental report. https://www.icao.int/environmental-protection/Pages/envrep2019.aspx. Accessed 2021/10/09
16. Should we give up flying for the sake of the climate? https://www.bbc.com/future/article/20200218-climate-change-how-to-cut-your-carbon-emissions-when-flying. Accessed 2021/10/08
17. Nikolaykin NI, Matyagina AM, Smirnova YuV (2007) A method of ecological estimation for man-made chemical and greenhouse gas pollution. Chem Pet Eng 43(9):612–616
18. ICAO (2018) Transforming the global aviation sector: emission from international aviation. Emission reductions brochure final. https://www.icao.int/environmental-protection/Documents/EmissionReductions_Brochure-web-1up.pdf. Accessed 2021/10/09
19. Balabekov OS, Vorobiev OG, Shakirov BS (1993) Genesis, classification and ecological optimization of physicochemical. Bullet Nat Acad Sci Repub Kazakhstan 3:40–43
20. Gloveli GD (2021) Malthusianism. https://bigenc.ru/economics/text/2171438. Accessed 2021/10/14
21. Agarwal P (2021) Malthusian theory of population. Intelligent economist. https://www.intelligenteconomist.com/malthusian-theory/. Accessed 2021/10/14

Vortex Combustion Chamber with Angular Flame Stabilizer: Design and Experimental Investigations

Sergey Skorobogatov

Abstract The combustion chamber has a significant impact on the gas turbine engines efficiency. Special attention is paid to the improvement of combustion chamber design. The design of the vortex combustion chamber was performed as part of the ongoing research aimed at ensuring a minimum level of temperature non-uniformity at the gas turbine inlet. After determining the inlet geometry of the vortex combustion chamber with angular flame stabilizer, the discrete cannular-type outlet geometry was designed. Numerical simulations showed that discrete outlet configuration provide a 33% reduction in radial non-uniformity was observed relative to a conventional gas turbine engine combustion chamber. The computational methodology was verified by comparing the results of numerical simulation with the results of hydrodynamic studies and with the results of bench tests on a quantitative and qualitative levels. As a further development of the vortex combustion chamber with angular flame stabilizer, a continuous outlet concept was proposed. The analysis of the temperature field showed a further decrease in the level of temperature non-uniformity with a continuous outlet.

Keywords Vortex combustion chamber · Gas turbine engine · Uneven temperature field

1 Introduction

Most of the existing gas turbine engines combusting chambers have an uneven temperature distribution in the outlet section. Uneven temperature distribution is an unfavorable factor provoking formation and growth of cracks in the combustion chamber and turbine structural elements. As a result, engine overhaul period decreases and the frequency of borescope inspections increases, leading to higher costs for airlines. The intensity of the temperature irregularity depends on various factors, one of which is widespread using flow swirlers. Despite all the advantages

S. Skorobogatov (✉)
Irkutsk Branch of Moscow State Technical University of Civil Aviation, Irkutsk, Russia
e-mail: maestro.ru@mail.ru

O. A. Gorbachev et al. (eds.), *Proceedings of 10th International Conference on Recent Advances in Civil Aviation*, Lecture Notes in Mechanical Engineering, https://doi.org/10.1007/978-981-19-3788-0_3

[1], this type of flame stabilizer includes many separate inlet channels through which fuel–air mixture goes into the combustion chamber. This circumstance causes the formation of many individual vortices in the annular duct. These vortices interfere with each other and introduce heterogeneity into the gas flow parameters.

Turbine blades defects are one of the main reasons for engine removal and repair. Intense temperature irregularity worsens the cooling degree of the turbine blades and accelerates fracture process in the protection cover.

Another important requirement is to ensure high fuel efficiency of the gas turbine engine. This requirement can be achieved by increasing gas temperature in front of the turbine and reducing the cooling air flow rate for the hot engine elements. Increasing the gas temperature exacerbates the problem of temperature irregularity.

However, in addition to fuel efficiency and overhaul period, an important requirement for gas turbine engines is to ensure their environmental friendliness. Among the substances formed in the combustion chamber, special attention is paid to the level of nitrogen oxides and carbon monoxide, as well as unburned hydrocarbons. At the same time, in order to reduce the volume of NO_x emissions it is necessary to reduce the combustion temperature below a certain level.

Thus, several contradictory requirements are imposed on the combustion chamber: on the one hand, it is necessary to increase the combustion chamber outlet temperature as much as possible to increase overall efficiency index and fuel efficiency; on the other hand, the temperature cannot be increased to reduce the volume of nitrogen oxide emissions. It is also necessary to ensure a low temperature irregularity at the turbine inlet in all engine thrust rating. To ensure all of the above requirements, the engineers compromise.

The low emission level combined with high fuel efficiency is achieved by dividing the combustion zone into several sections. This can be realized in a single swirler design, like the Twin Annular Premixing Swirler [2], or by installing multiple individual swirlers around the annular duct, like the NASA multipoint LDI combustion chamber [3]. In this approach, a single combustion zone is divided into a pilot zone and a main zone. In the pilot zone, the combustion process is continuous, while the main zone is activated during heavy engine operation [4, 5].

Yet, a significant increase in the number of individual flame stabilizers entails the problem of ensuring quality fuel atomization due to extremely low fuel-flow rate. In addition, the discrete inlet of an annular combustion chamber increases non-uniformity in the temperature field in the outlet section.

To eliminate the above drawbacks, a vortex combustion chamber is proposed, in which the combustion zone is formed by an angular flame stabilizer. The potential for using vortex combustion chambers is actively investigated and this research direction is considered prospective [6–12]. However, most of the ongoing studies of vortex combustion chambers assume a discrete inlet. The proposed concept makes it possible to ensure air supply, fuel atomization and combustion by a single inlet along the entire perimeter of the annular duct.

2 Research Tools and Methods

The research was conducted using numerical and natural experiments. Ansys Fluent was used as a computer simulation environment. Verification and validation of calculations were carried out on the basis of the hydrodynamic experiment and bench test results under laboratory conditions (Fig. 1).

Shear-Stress Transport (SST) k-ω Reynolds-Averaged Navier–Stokes (RANS) model was used during computer modeling in the Ansys Fluent. During the hydrodynamic studies it was found that the SST k-ω RANS model provides the greatest agreement with the experiment results on a quantitative and qualitative level. This model also showed high stability and fast convergence during calculation [13] (Fig. 2).

The non-premixed combustion model was chosen to calculate combustion. This model allows the fuel to be fed separately from the oxidizer, which is necessary in

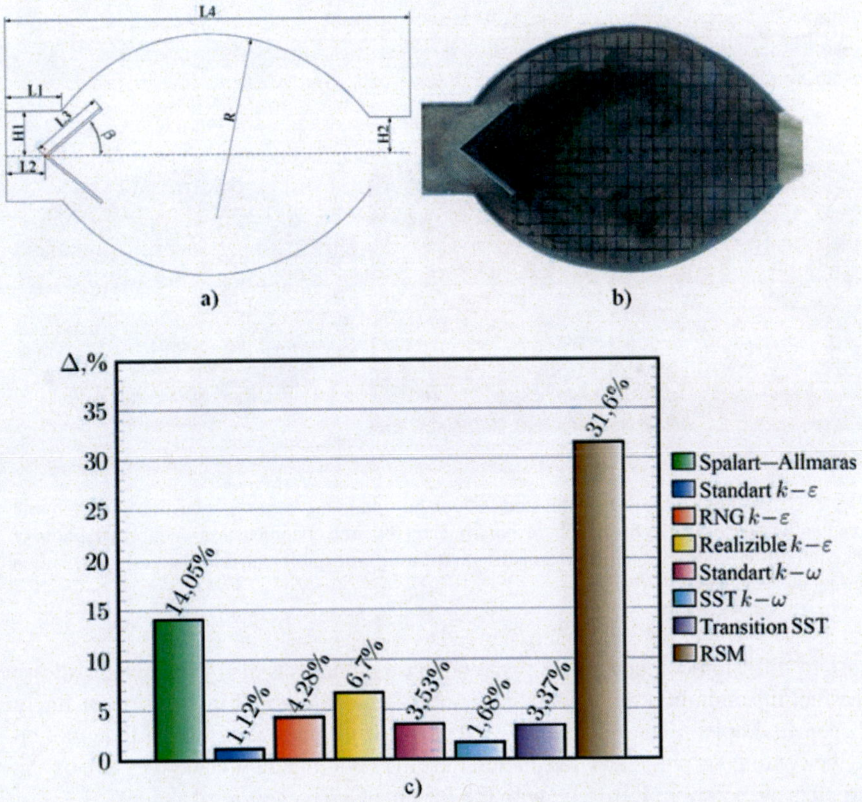

Fig. 1 The comparison results of the experimental and computational flow structure for a simplified vortex combustion chamber with angular flame stabilizer: **a** scheme of a simplified vortex combustion chamber with angular flame stabilizer; **b** flow structure during the hydrodynamic experiment; **c** relative deviation from the experiment of the return flow zone area determined by different turbulence models

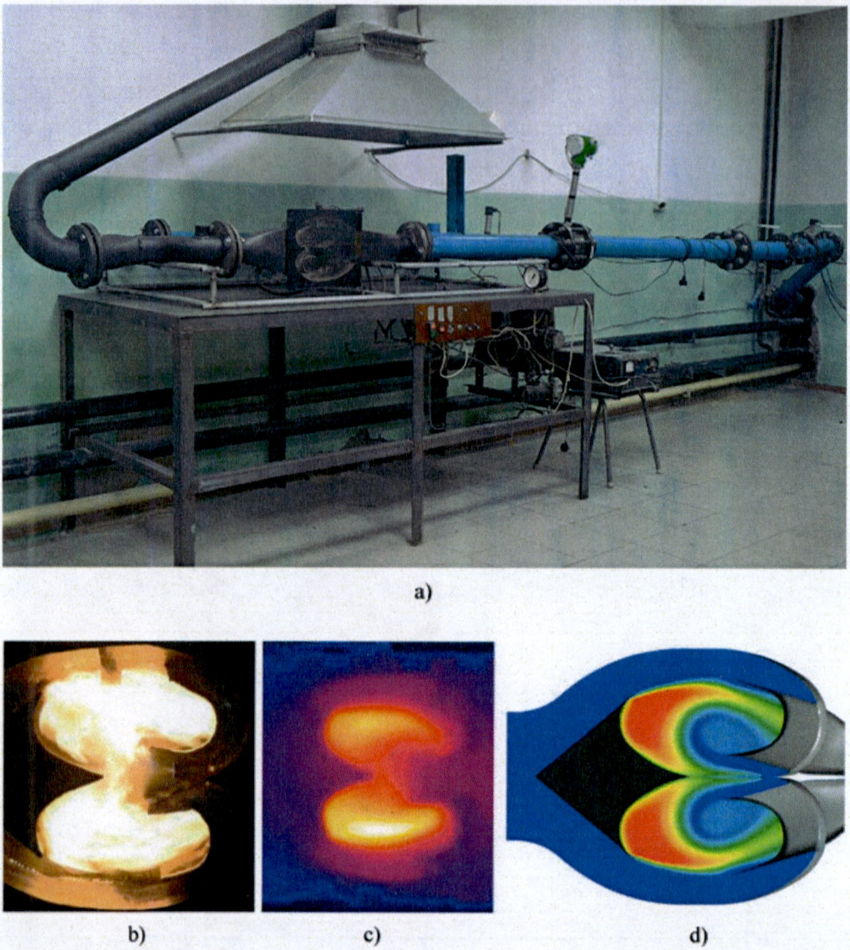

Fig. 2 Calculation and experimental study of combustion in the vortex combustion chamber with angular flame stabilizer: **a** fire bench for experimental research of combustion chambers; **b** photo of the combustion process; **c** temperature distribution during the experiment; **d** temperature distribution during the calculation

case of diffusion combustion. A discrete phase model was also used, which allows the fuel injection into the calculation region to be configured in the form of finely dispersed droplets. The calculation used a two-way coupling of fuel particle motion to the continuous phase and considered the effect of temperature and pressure on the droplets evaporation. Kerosene with the formula $C_{12}H_{23}$ was used as fuel.

The combustion calculation approach used was verified during tests on the fire bench [14].

3 The Investigation Results and Their Discussion

According to Fig. 3, the operation process in the vortex combustion chamber with angular flame stabilizer is as follows. The compressor air flows around the corner flame stabilizer, then part of this air goes into the gas manifold, in which the gas flow reverses and forms return flow (denoted by the letter "*B*"). The return flow interacts with the inner wall of the corner flame stabilizer and the flow reverses toward the outlet. The combustion area is marked with the letter "*F*". The combustion products enter the exhaust manifold channels, mixing in with some of the compressor air. The combustion products leave the exhaust manifold channels, combining in the transition liner and then entering the turbine [15] (Figs. 4 and 5).

The operation of the vortex combustion chamber was simulated in the Ansys Fluent software environment. The parameters of temperature, velocity and pressure of the air flow at the combustion chamber inlet corresponded to the cruising mode of the PS-90A prototype engine. The fuel was sprayed toward the inner wall of the angular flame stabilizer [16, 17].

As a calculation result, temperature distribution pattern in the outlet section was obtained. The turbine blade, as it rotates, takes the average temperature of the combustion chamber outlet gas. Thus, to estimate the temperature non-uniformity, it is necessary to have averaged temperature values at some radius *i*. For this purpose, a radial

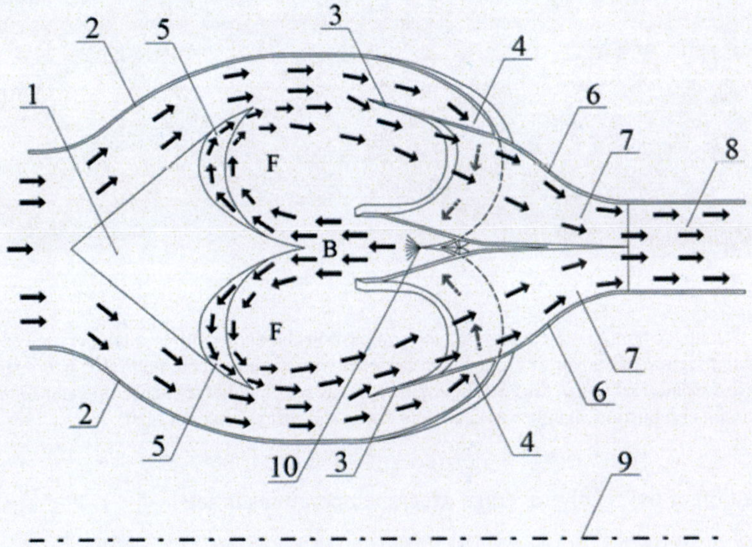

Fig. 3 Scheme of the vortex combustion chamber with continuous inlet and discrete outlet: 1—angular flame stabilizer; 2—combustion chamber walls; 3—separating surface of the gas manifold; 4—gas manifold channels; 5—inner wall of the angular flame stabilizer; 6—exhaust manifold walls; 7—exhaust manifold channels; 8—combustion chamber transition liner; 9—engine longitudinal axis; 10—fuel injector

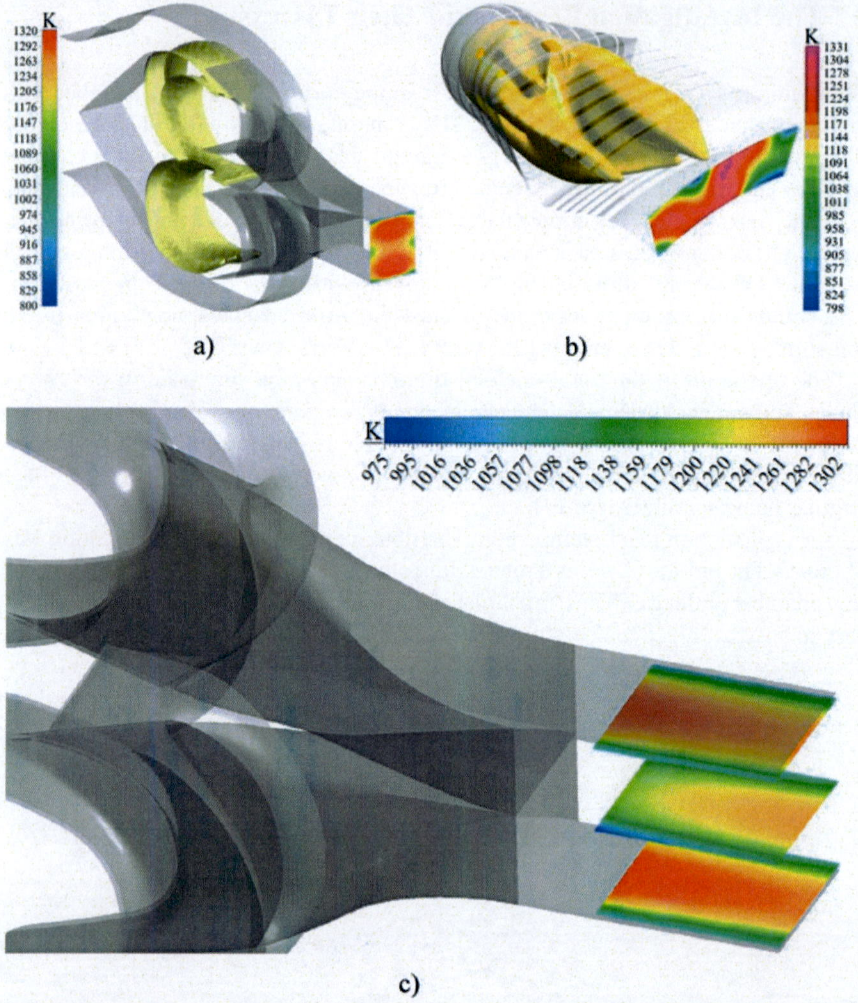

Fig. 4 Calculated temperature contours at the combustion chambers outlets: **a** temperature contour in the outlet section of the vortex combustion chamber with angular flame stabilizer; **b** temperature contour in the outlet section of the PS-90A engine combustion chamber; **c** slice temperature contours of the vortex combustion chamber with angular flame stabilizer transition liner

distribution of the relative average excess temperatures is set:

$$\theta_{i_{\text{mid}}} = \frac{T^*_{\text{CC}_i} - T^*_{\text{Com}}}{T^*_{\text{CC}} - T^*_{\text{Com}}} \tag{1}$$

where $\theta_{i_{\text{mid}}}$—is the relative average excess gas temperature at radius i of the combustion chamber outlet section; $T^*_{\text{CC}_i}$—is the average gas temperature at radius i; T^*_{CC}—is

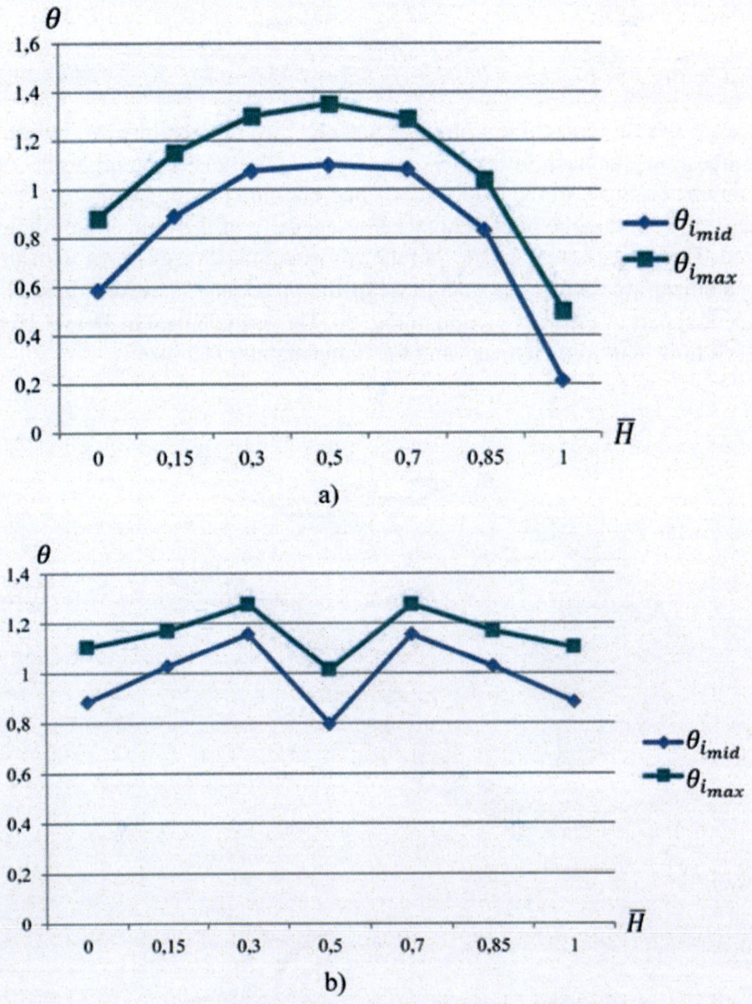

Fig. 5 Radial distribution of gas temperature in the combustion chambers outlet sections: **a** radial distribution of the relative average excess temperatures and the maximum relative excess temperatures in the combustion chamber outlet section of the PS-90A engine; **b** radial distribution of the relative average excess temperatures and the maximum relative excess temperatures in the outlet section of the vortex combustion chamber with angular flame stabilizer

the average gas temperature at the combustion chamber outlet section; T^*_{Com}—is the air temperature at the combustion chamber inlet section.

In addition, to ensure the operation of the turbine nozzle blades, the radial distribution of the maximum relative excess gas temperatures at the combustion chamber outlet is set:

$$\theta_{i_{\max}} = \frac{T^*_{CC_{i_{\max}}} - T^*_{Com}}{T^*_{CC} - T^*_{Com}} \tag{2}$$

where $\theta_{i_{\max}}$—is the maximum relative excess gas temperature at radius i of the combustion chamber outlet section; $T^*_{CC_{i_{\max}}}$—is the maximum value of gas temperature at radius i of the combustion chamber outlet section (Fig. 6).

Due to the continuous inlet, a higher homogeneity of the temperature field was achieved. The values of $\theta_{i_{mid}}$ and $\theta_{i_{\max}}$ of the vortex combustion chamber with angular flame stabilizer are 33% lower than those of the prototype. However, there is a dip in the central part of the inlet section in Fig. 5b. This is explained by the interference of the gas flow manifold elements and the flame angular stabilizer.

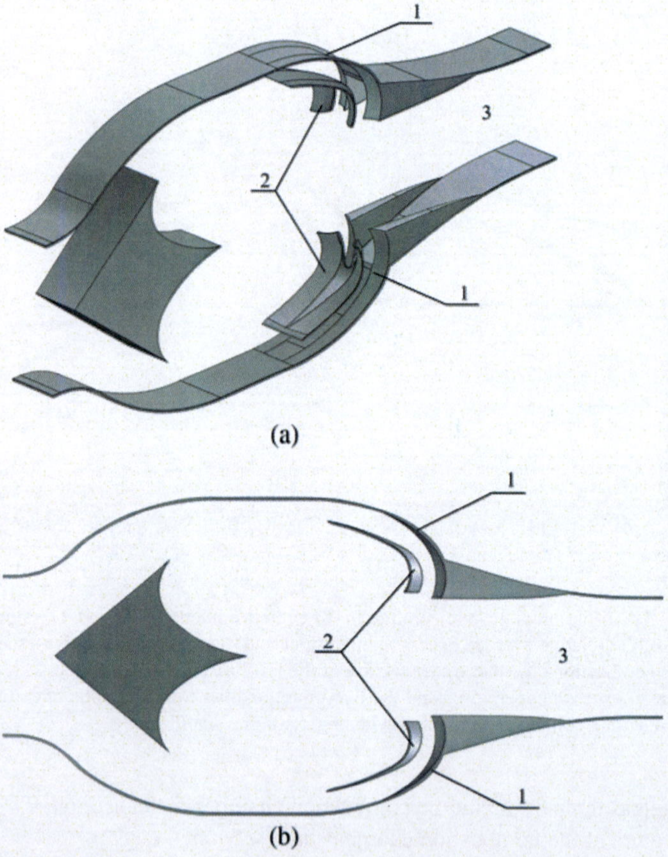

Fig. 6 Vortex combustion chamber with corner flame stabilizer and continuous outlet: **a** axonometric projection; **b** side projection; 1—combustion chamber walls; 2—separating surface of the gas manifold; 3—outlet line

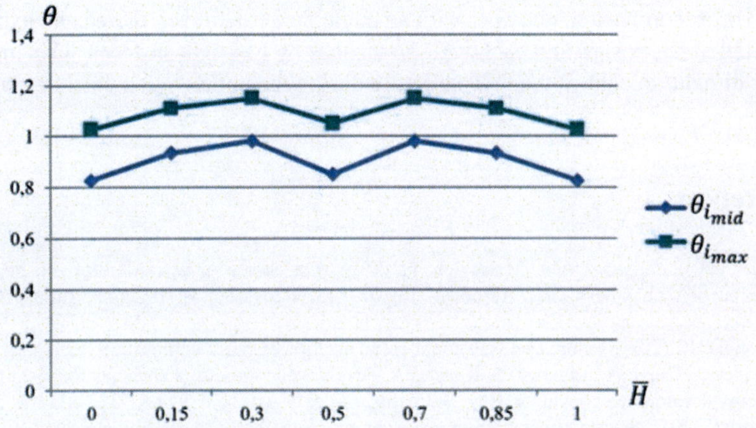

Fig. 7 Radial distribution of the relative average excess temperatures and the maximum relative excess temperatures in the outlet section of the vortex combustion chamber with angular flame stabilizer and continuous outlet

To make the temperature parameters even more homogeneous, a vortex combustion chamber variant with a continuous inlet and continuous outlet was studied. According to Fig. 5, in this case, the separation of the forward flow and the return flow is carried out by a gas-dynamic way, and there are no walls between these flows. This approach to flow control makes it possible to considerably simplify the vortex combustion chamber design.

As a calculation result, it was found that the elimination of the discreteness in the vortex combustion chamber outlet section provides an additional reduction of the temperature irregularity level by 7%. The dip in the central part of Fig. 7 is due to the ejection of primary airflow in the gas manifold area. This is related to suboptimal geometry of the manifold and gas collector (Fig. 6a, b). However, even in the first approximation, the lack of discreteness is favorable for the temperature uniformity.

4 Conclusions

The conducted studies allowed us to establish that the continuous inlet in the vortex combustion chamber with an angle flame stabilizer in case of a discrete outlet makes it possible to reduce temperature irregularity in the outlet section by 33% in comparison with a serial cannular combustion chamber of a gas turbine engine. The combination of a continuous inlet and continuous outlet in the vortex combustion chamber with angular flame stabilizer allows an additional 7% reduction in the temperature irregularity level in the outlet section. In addition, the removal of mechanical flow separation at the inlet and outlet allowed us to considerably simplify the vortex combustion chamber design. However, with a continuous inlet and continuous outlet in the vortex combustion chamber with an angle flame stabilizer, there is no explicit delimitation of forward and backward flow zones by gas-tight surfaces. This makes it very difficult to control the flow structure during the optimization calculations.

References

1. Isaev AI, Safarbakov AM, Mairovich YI (2013) On choice of a swirler for the impulse combustion chamber. Russ Aeronaut 56:407–413. https://doi.org/10.3103/S1068799813040144
2. Mongia HC (2014) Future trends in commercial aviation engines' combustion. In: Agarwal A, Pandey A, Gupta A, Aggarwal S, Kushari A (eds) Novel combustion concepts for sustainable energy development. Springer, New Delhi. https://doi.org/10.1007/978-81-322-2211-8_7
3. Heath C (2016) Parametric modeling investigation of a radially-staged low-emission aviation combustor. In: 54th AIAA aerospace sciences meeting, p 1394
4. Magerramova LA, Nozhnitsky YA, Volkov SA, Volkov ME, Chepurnov VZ, Belov SV, Verbanov IS, Zaikin SV (2019) Prospects of application of additive technologies to develop parts and components of gas turbine engines and ramjets. Vestnik Samara Univ Aerosp Mech Eng 18(3):81–98. https://doi.org/10.18287/2541-7533-2019-18-3-81-98
5. Aleksandrov YB, Vafin II, Volkov SA, Mingazov BG (2018) Investigating the formation of temperature patterns in gas turbine engine combustion chambers. Eng J Sci Innov 1(73). https://doi.org/10.18698/2308-6033-2018-1-1716
6. Bruno C, Losurdo M (2007) The trapped vortex combustor: an advanced combustion technology for aerospace and gas turbine applications. In: Syred N, Khalatov A (eds) Advanced combustion and aerothermal technologies. NATO science for peace and security series C: environmental security. Springer, Dordrecht. https://doi.org/10.1007/978-1-4020-6515-6_28
7. Sun H, Yan P, Tian L et al (2021) Numerical simulation of inverse diffusion combustion and flow characteristics in a trapped vortex combustor. Int J Aeronaut Space Sci 22:625–637. https://doi.org/10.1007/s42405-020-00335-x
8. Wu Z, He X (2020) Investigations on emission characteristics of a liquid-fueled trapped vortex combustor. J Therm Sci 29:69–80. https://doi.org/10.1007/s11630-019-1232-3
9. Yi JIN, Xiaomin HE, Jiang B, Zejun WU, Guoyu DING (2012) Design and performance of an improved trapped vortex combustor. Chin J Aeronaut 25(6):864–870
10. Wu Z, Jin Y, He X, Xue C, Hong L (2015) Experimental and numerical studies on a trapped vortex combustor with different struts width. Appl Therm Eng 91:91–104
11. Jingyu Z, Xiaomin H, Lu W, Yi J (2015) Experimental and numerical investigations on liner cooling characteristics of a trapped vortex combustor. Appl Therm Eng 80:66–75
12. Zhao D, Gutmark E, de Goey P (2018) A review of cavity-based trapped vortex, ultra-compact, high-g, inter-turbine combustors. Prog Energy Combust Sci 66:42–82
13. Isaev AI, Skorobogatov SV (2017) Hydrodynamic verification and validation of numerical methods of the flow calculation in combustion chamber of a gas turbine engine. Trudy MAI 97:28 (in Russian)
14. Isaev AI, Skorobogatov SV (2018) Methodological aspects of burning process experimental research in combustion chambers of gas-turbine engines. Trudy MAI 98:14 (in Russian)
15. Safarbakov AM, Skorobogatov SV, Isaev AI (2020) Annular combustion chamber of gas turbine engine and method of arrangement of working process therein. RU Patent 2,716,992, C2 Mar 2020

16. Isaev AI, Skorobogatov SV (2019) Assuring of operational requirements when designing the flame head of the combustion chamber with a transverse vortex system. Scientific Bullet State Sci ResInst Civ Aviation 25:35–42
17. Isaev AI, Skorobogatov SV (2019) Operational aspects at the stage of formation of the appearance manifold of the gas flow in the combustion chamber. Scientific Bullet State Sci Res Inst Civ Aviation 25:32–40

Influence of Coriolis Force on the Peculiarities of Aircraft Operation at the Aerodrome

Nikolai Danilenko and Anton Kirenchev

Abstract The article discusses one of the main problems faced by engineering and technical personnel during the operation of aircraft with gas turbine engines—vortex suction of foreign objects into the engine air path. The analysis of works on the topic under study was carried out and the main results were identified. The article presents the basics of the vortex field of the Earth's rotation relative to the polar axis, its altitudinal and latitudinal vortex characteristics, and their consequences. The working process of physical manifestation of the investigated vortex field on the character of the atmospheric environment in the locations of airfield nodes of the Northern Hemisphere of the Earth and the direction of the impact of atmospheric vortex activity on the features of operation of gas turbine power plants of aircraft at the airfield is proved. The factor of the Coriolis force, which influences the formation of vortices under the air intakes of aircraft engines, but was not taken into account in the analyzed works, is highlighted. The dependence of the influence of this factor on the geographic location of the aerodrome for the operation of the aircraft on the Earth's surface has been investigated. A way of practical application of the revealed feature when adjusting the values of hazardous zones of aircraft engines for different geographic latitudes is proposed.

Keywords Air intake vortices · Gas turbine engine · Hazardous areas · Coriolis force

1 Introduction

Aircraft are among the most efficient means of transportation, enabling passengers and cargo to be transported over vast distances in the shortest possible time. For this reason, increasing the efficiency and cost-effectiveness of air passenger and cargo transportation has been one of the main objectives of the transport industry for many years.

N. Danilenko (✉) · A. Kirenchev
Irkutsk Branch of Moscow State, Technical University of Civil Aviation, Irkutsk, Russia
e-mail: danko_irk@mail.ru

O. A. Gorbachev et al. (eds.), *Proceedings of 10th International Conference on Recent Advances in Civil Aviation*, Lecture Notes in Mechanical Engineering, https://doi.org/10.1007/978-981-19-3788-0_4

The main way to increase the efficiency of air transportation is to increase the airflow rate through the air path of the gas turbine engine (GTE) in order to improve its performance characteristics [1]. If we look at the graph of second mass air flow rate (G_a) through the engine depending on the year of its release (Fig. 1), we can see a tendency for growth of this parameter—for the last decades the value of G_a has grown more than three times.

A significant increase in the value of the second mass flow rate through the air duct of the GTE allowed for a higher efficiency of its operation. However, an increase in the airflow rate at the engine air intake led to the formation of vortices under it (Fig. 2a), which gave rise to a number of problems. The main and most significant of which was the vortex suction of foreign objects into the air intake of the aircraft engine (Fig. 2b). The consequence of this problem was damage to the elements of the flowing part of the engine air-gas path (Fig. 2c), maintenance and repair of which are expensive operations, and also require temporary removal of the GTE from the aircraft, and sometimes its complete replacement [2–4].

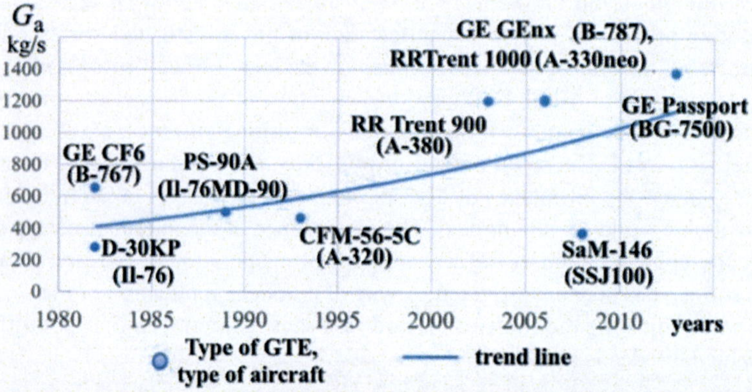

Fig. 1 Diagram of second air mass flow rate through the engine depending on the year of its production

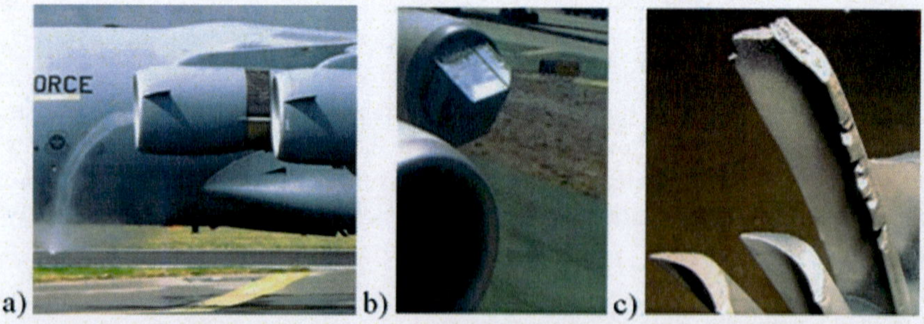

Fig. 2 Vortex under the GTE air intake (**a**), foreign object sucked into the GTE (**b**) and compressor blades damage (**c**) [www.airliners.net]

2 Problem Statement

The problem of vortex suction of foreign objects over the years of its existence has been studied by many scientists, based on their studies the following regularities and features of vortex behavior under the air intakes of aircraft engines have been revealed:

- gaseodynamically, the process of vortex formation under the air intake of an aircraft engine is similar to the process of formation of large atmospheric vortices—tornadoes [1, 5–8];
- the vortex intensity depends not only on the value of the second mass flow rate through the GTE but also on the height of its location above the airfield surface and the shape of its air intake [1, 9].

The conducted studies served as a basis for creating various protective devices [1, 10], both protecting GTE elements from foreign objects sucked in by the vortex (protective nets) and directly influencing the vortex formation process and its intensity (sliding shields, jet protection systems, make-up windows, etc.). Many of these devices have found application in military aviation.

In order to improve aviation safety and reduce the likelihood of foreign objects getting into the air intake of the aircraft gas turbine power plant (GTPP) during its operation, such a concept as " hazardous areas" was introduced—these are areas in front of and behind the engine, within which during its operation there should be no foreign objects and personnel. These areas are individual for each aircraft, depending on the type of engine installed on it and its height above the surface of the airfield. The size of these areas is indicated both in the relevant documentation and on the engine air intake cladding (Fig. 3).

Obviously, the problem of vortex suction of foreign objects is well studied, but it is actual nowadays, and considering a tendency in development of GTE to increase

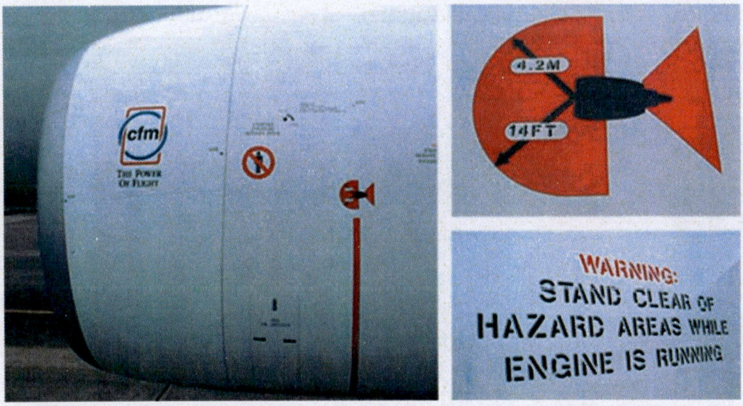

Fig. 3 Hazardous areas of gas turbine engine [tacomaworld.com]

the second mass flow rate, we can assert about strengthening of its importance in the future.

3 Factors of Vortex Activity

It is appropriate to recall that to the above factors of vortex activity at the inlet to the GTPP air intakes we should add the natural whirling of the Earth's atmosphere in the field of its rotation around the polar axis NS. It is this vortex activity of our planet that is the working environment for tornadoes, cyclones, and their small-sized analogs—vortices of air intakes (Fig. 4) during working of gas turbine engines on the ground at higher operating modes.

A number of studies have established that the vortex field of the Earth's environment during its diurnal rotation serves as a constant source of the vortex activity of the GTE's air intake. Due to the constancy of the angular velocity Ω of the rotation of our planet relative to the NS polar axis, this field is stable and uninterrupted. This vortex activity is the basis for the formation of not only vertical but also horizontal vortices (see Fig. 4).

The foundations of the Earth's vortex field were first outlined in the article "Tornado" [6] and substantiated in the monograph [1]. The vortex field is represented by the Earth profile of a simplified vortex model based on the Bio-Savara law. The author managed to obtain the altitudinal and latitudinal vortex characteristics in the form of patterns of the natural circulation density γ of the Earth's environments depending on the relative altitude $\overline{H}$ and geographic latitude λ in its range of 0–90°.

A simplified model of the Earth and the kinematics of motion of the studied point with the required geometrical parameters is presented in Fig. 5. In the left part of the figure is given the scheme of the projection γ_n and γ_τ on their direction "n" and "τ".

The obtained patterns of the Earth's media velocity circulation are functions of two parameters of geographical latitude λ and relative altitude $\overline{H}$ [6]. Consequently, it is possible to obtain latitudinal and altitudinal characteristics of the Earth's vortex field, viz:

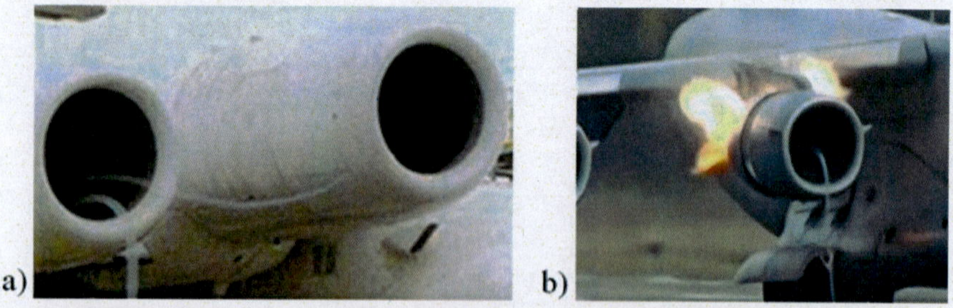

Fig. 4 Small-size man-made analogs of natural air vortices: **a** vertical and horizontal vortices; **b** GTE's surge in the field of counter-curling vortex [www.airliners.net]

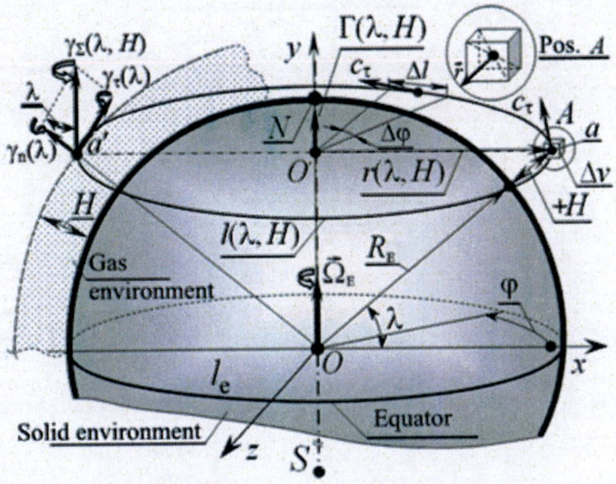

Fig. 5 Simplified kinematic model of the earth and its geometrical parameters

- latitudinal characteristics of the circulation rate density $\overline{\gamma}(\lambda)$:

$$\overline{\gamma}_n(\lambda) = \frac{\gamma_n}{\Omega_E R_E} = \left(1 \pm \overline{H}\right)\cos\lambda \cdot \sin\lambda = f\left(\overline{H} = \text{const}\right); \tag{1}$$

$$\overline{\gamma}_\tau(\lambda) = \frac{\gamma_\tau}{\Omega_E R_E} = \left(1 \pm \overline{H}\right)\cos^2\lambda = f\left(\overline{H} = \text{const}\right); \tag{2}$$

- altitudinal characteristics of the circulation density $\overline{\gamma}(\overline{H})$:

$$\overline{\gamma}_n\left(\bar{H}\right) = \frac{\gamma_n}{\Omega_E R_E} = \left(1 \pm \bar{H}\right)\cos\,\lambda \cdot \sin\,\lambda = f(\lambda = \text{const}); \tag{3}$$

$$\overline{\gamma}_\tau\left(\bar{H}\right) = \frac{\gamma_\tau}{\Omega_E R_E} = \left(1 \pm \bar{H}\right)\cos^2\lambda = f(\lambda = \text{const}); \tag{4}$$

Here lies the possibility of representing the characteristics (1–4) in normal parameters, for example, the altitude characteristic of the circulation Γ of the Earth's vortex field in relative altitude "$\pm\overline{H}$" by means of a dimensionless parameter $(1 \pm \overline{H})$, that is, in the bowels of the Earth's body.

$$\Gamma\left(\bar{H}\right) = 2\pi\Omega_E R_E^2 (1 \pm \bar{H})^2 \cos^2\lambda = f(\lambda = \text{const}). \tag{5}$$

The graphical representation of the latitudinal and altitudinal characteristics of the Earth's vortex field in the criterion parameters is presented in Fig. 6.

The analysis of vortex characteristics allows us to establish their consequences and applied factors of influence on ground operation of aircraft GTE, and on the zones of special danger in the area of input devices.

1. According to the characteristic $\overline{\gamma}_n = f(\lambda)$, the atmospheric environment of middle latitudes contains the maximum linear density of the circulations of the cyclonic type normally oriented to the Earth $\overline{\gamma}_n = \overline{\gamma}_{n\,\max}$. This circulation is the

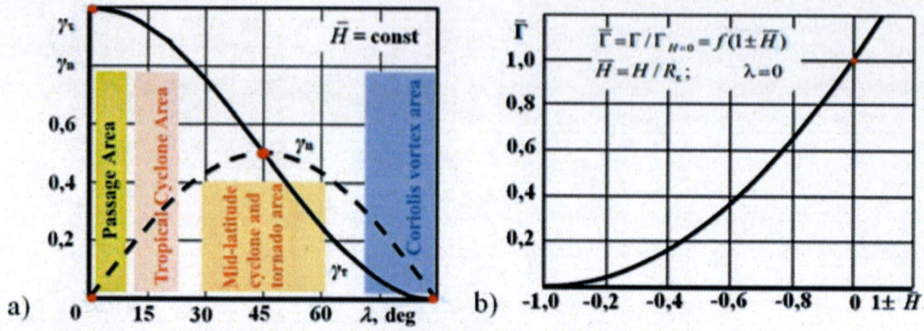

Fig. 6 Characteristics of the earth's vortex field: **a** latitudinal; **b** altitudinal

"building" material for accumulation of cyclones, effluent tornadoes, and their small-sized counterparts—vertical technogenic vortices of GTE's input devices.

2. The flat maximum of the function $\overline{\gamma}_n(\lambda)$ (Fig. 6) falls on the geographic latitude $\lambda = 45°$, and the range of this danger zone is within $\Delta\lambda \approx \pm\, 15°$. It is at these latitudes that cyclones and their extremely active vortex product tornadoes.
3. At latitude $\lambda = 45°$, the densities of normal and flow circulation ($\gamma_n(\lambda)$ and $\gamma_\tau(\lambda)$) are equal to each other. This equality gives grounds to assert that not only vertical but also horizontally oriented vortices should appear and arise in front of GTPP air intakes during ground operation of the engine (see Fig. 4). Their presence is capable of generating surge of one of two adjacent and simultaneously operating engines (Fig. Ibid.). Because the starting mechanism of axial compressor (engine) surge is a counter vortex swirl of the effluent flow at the compressor inlet, which provokes flow stall.

4 Influence of the Coriolis Force on the Aircraft Operation

Despite the fact that the similarity of the working process of air intake vortices and atmospheric vortices was proved in the above works [1, 6, 11–13], the influence of Coriolis force, which is one of the generators of the process of formation of large atmospheric vortices, has not been studied in the works of other scientists on this subject. As a consequence, this feature of the vortex formation process has not found practical application in solving the problem of vortex suction.

The study of the Coriolis force's influence on the vortex formation process was carried out by the authors [14, 15], namely, it was found that this force affects the tangential flow velocity near the ground, increasing it by 11.26%, this score is correct only for the territory of Irkutsk (the experimental site), where the geographical latitude $\lambda = 52°$. Using the Coriolis force formula from theoretical mechanics [14, 15], we can calculate the percentage of flow velocity increment Δc at the airfield surface in the vicinity of the vortex for all geographic latitudes of the Earth's surface:

$$P_{\text{cor}} = m 2\omega_E u \sin\lambda, \tag{6}$$

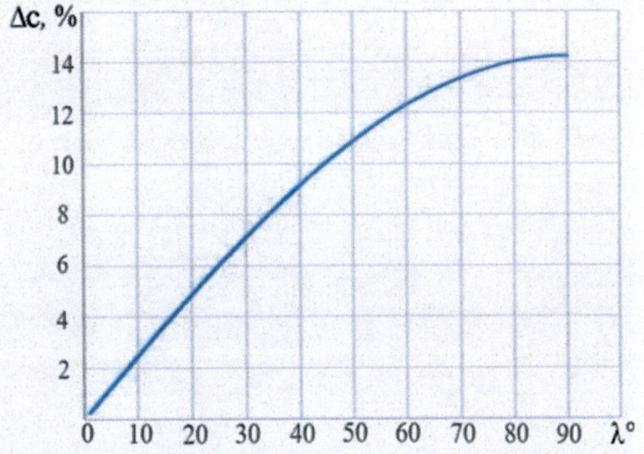

Fig. 7 Dependence of tangential velocity increment of GTE vortex flow on geographical latitude of airfield location

where *m*—the mass of an object or particle; ω_E—angular velocity of the Earth's daily rotation; *u*—the velocity of the object or particle relative to the Earth's surface; λ – geographic latitude.

On the basis of the obtained results, it is possible to plot the dependence of the increment of tangential velocity of the vortex flow of GTE on the geographic latitude of the airfield where the aircraft with this engine is operated (Fig. 7).

Analysis of Fig. 8 allows us to argue that Coriolis force has a significant impact on the process of vortex formation under GTE air intakes, increasing the tangential flow rate from 0 at the equator to 14, 23% at the poles. But for more correct assessment of influence of the investigated factor, it is necessary to take into account at what latitudes of the Earth are most often operated by aircraft. For this purpose, we will use the air traffic statistics for 2019 for different zones of the globe (Fig. 9) and flight intensity data per day, taken from the Flightradar24 website (Fig. 10). The statistics show that most of the air traffic occurs in the Northern Hemisphere, namely in its middle latitudes (the value of λ from 20 to 50°). In these latitudes, the value of two increments of tangential flow velocity from the Coriolis force varies on average from 5 to 11%, such a deviation certainly requires taking into account when operating the aircraft GTE at the aerodrome.

The results obtained can be taken into account when correcting the hazardous zones of engines, which were described above (see Fig. 2), because their size depends on the type of GTE and its height above the airfield surface, and hence on the value of the tangential velocity in front of the air intake of this engine, which is directly proportional to the vortex intensity [1]. If we make the assumption that the size of the danger zone of the engine directly depends on the flow velocity under its air intake, then the increment of the linear size of the zone will be equal to the increment of this velocity $\Delta l \approx \Delta c$. In this case, the graph in Fig. 7 can be represented as a dependence of Δl on λ (Fig. 10). Consequently, when operating the aircraft at various airfields, it should be taken into account that, depending on the degree of influence of the Coriolis force, the value of dangerous zones can change by the increment Δl, which

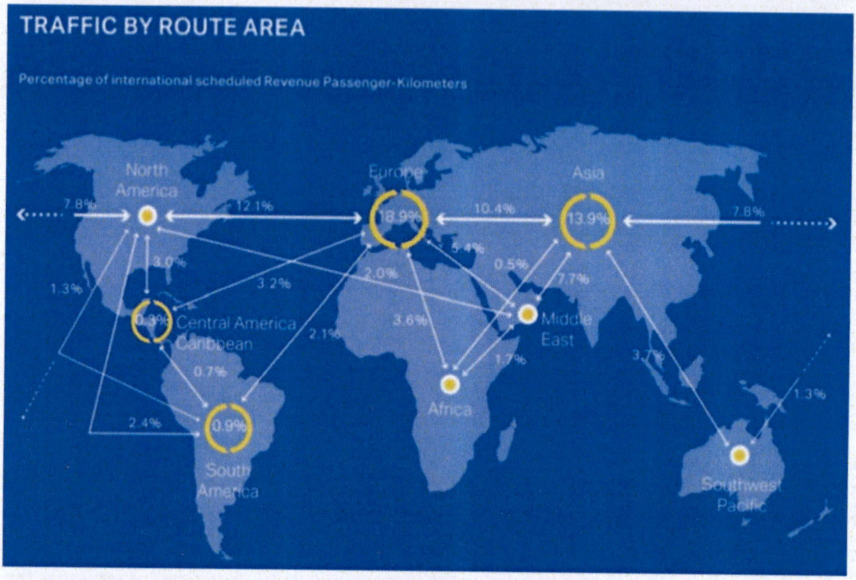

Fig. 8 Traffic percentages in commercial passenger-kilometers by zone for 2019 [16]

Fig. 9 Global flight intensity on 10.10.2021 [17]

is equal to zero at the equator, where the influence of Coriolis forces is minimal. At the Earth's poles, it reaches its maximum value of 14 23%.

The change in the magnitude of the increment of hazardous areas in various Russian cities is shown in Table 1.

Analysis of the table data allows us to quantify the possible inaccuracies in determining hazardous zones at various airfields. It should be noted that for Russian

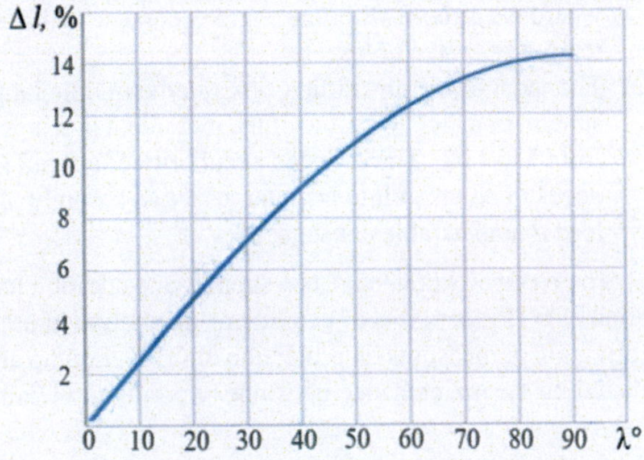

Fig. 10 Dependence of geometric increment of hazardous areas value on the latitude of aerodrome location

Table 1 Distribution of "hazardous areas" increments in Russian cities

Name of the city of Russia	Latitude of location λ, °	Δl, %
Makhachkala	42.984913	9.7
Sochi	43.581509	9.8
Rostov-on-Don	47.227151	10.4
Khabarovsk	48.472584	10.6
Voronezh	51.661535	11.2
Irkutsk	52.28638	11.2
Novosibirsk	55.028739	11.7
Moscow	55.755773	11.8
Perm	58.004785	12.1
St. Petersburg	59.938806	12.3
Yakutsk	62.027833	12.6
Arkhangelsk	64.539304	12.9
Murmansk	68.9695	13.3
Norilsk	69.349039	13.3

cities, the currently unaccounted increment of hazardous zone values varies from 9.7 to 13.3%, which may greatly affect the safety of aircraft operations in our region.

5 Conclusions

Based on the given material of this article, followed conclusions can be made.

1. The value of hazardous zones of aviation GTE depends on the influence of Coriolis force and changes depending on the geographical location of the aerodrome

where the aircraft with this engine operates, on the value from 0% at the equator to 14.23% at the poles.

2. The most of the air traffic occurs in the middle latitudes of the Northern Hemisphere, in which the value of the increment of dangerous zones varies on average from 5 to 11%. The revealed peculiarity of aircraft GTE operation, undoubtedly, needs to be taken into account, as it can strongly affect the safety of flights and lead to undesirable consequences.

The received knowledge is a support and addition to the competent ground operation of GTPP in places of extreme manifestation of atmospheric vortex activity and formation of measures on protection of GTE of pomp and suction of foreign objects, including the maintenance personnel working in vicinity of hazardous areas.

References

1. Danilenko NV, Krivel' PM, Pahomov SV (2011) Teoriya vihrej pered vozduhozabornikami samolyotov pri rabote gazoturbinnyh dvigatelej na aerodrome (The theory of vortices in front of the air intakes of aircraft during the operation of gas turbine engines on the airfield). Monografiya. IrGTU, Irkutsk (in Russian)
2. Cui G, Huang W, Lin M, Xu C, Zhang Z (2017) Numerical study of aircraft wake vortex evolution near ground in stable atmospheric boundary layer. Chin J Aeronaut 30(6):1866–1876. https://doi.org/10.1016/j.cja.2017.08.012
3. Bednarz A, Witek L, Stachowicz F (2015) Fatigue analysis of compressor blade with simulated foreign object damage. Eng Failure Anal 58(1):229–237. https://doi.org/10.1016/j.engfailanal.2015.09.002
4. Cheng L, Chen X, Li P, Shu C, Xu Y (2020) Foreign object damage performance and constitutive modeling of titanium alloy blade. Int J Aerosp Eng. https://doi.org/10.1155/2020/2739131
5. Arsen'yev SA (2011) Mathematical modeling of tornadoes and squall storms. Geosci Front 2(2):215–221. https://doi.org/10.1016/j.gsf.2011.03.007
6. Danilenko NV (2004) Tornado (Tornado). Vestnik Irkutskogo gosudarstvennogo tekhnicheskogo universiteta 2(18):64–72 (in Russian)
7. Ashrafi A, Romanic D, Kassab A, Hangan H, Ezami N (2021) Experimental investigation of large-scale tornado-like vortices. J Wind Eng Ind Aerodyn 208:104449. https://doi.org/10.1016/j.jweia.2020.104449
8. Makarieva AM, Gorshkov VG (2009) Condensation-induced kinematics and dynamics of cyclones, hurricanes and tornadoes. Phys Lett A 373(46):4201–4205. https://doi.org/10.1016/j.physleta.2009.09.023
9. Evdokimov AI (1987) Separaciya chastic vo vneshnij kontur. Metod zashchity TRDD ot povrezhdeniya lopatok kompressora postoronnimi predmetami (Separation of particles into the external circuit. Method of GTE protection against compressor blades damage by foreign objects). Materialy 4-j mezhvedomstvennoj NTK po problemam zashchity GTD ot povrezhdeniya ih postoronnimi predmetami, vodoj, peskom i pyl'yu. ZHukovskij, pp 188–202 (in Russian)
10. Komov AA, Yurin SP (2014) Uroven' zashchishchennosti aviacionnyh dvigatelej otechestvennyh vozdushnyh sudov ot povrezhdenij postoronnimi predmetami (Level of protection of aircraft engines of domestic aircraft from damage by foreign objects). Nauchnyj vestnik GosNII GA 4(315):42–48 (in Russian)
11. Archer CL, Nouri R, Vasel-Be-Hagh A (2020) The Coriolis force and the direction of rotation of the blades significantly affect the wake of wind turbines. Appl Energy 277:115511. https://doi.org/10.1016/j.apenergy.2020.115511

12. Baker C, Sterling M (2019) Are tornado vortex generators fit for purpose? J Wind Eng Ind Aerodyn 190:287–292. https://doi.org/10.1016/j.jweia.2019.05.011
13. Hangan H, Parvu D, Refan M, Romanic D (2018) Wind and tornado climatologies and wind resource modelling for a modern development situated in "tornado alley." Renewab Energy 115:97–112. https://doi.org/10.1016/j.renene.2017.08.026
14. Danilenko NV, Kirenchev AG (2016) Problemy eksperimental'nogo modelirova-niya stokovyh vihrej koriolisovoj sily i puti ih preodoleniya (Problems of experimental modeling of Coriolis force of runoff vortices and ways to overcome them). Sbornik trudov V Vserossijskoj nauchno-tekhnicheskoj konferencii «Aktual'nye problemy i perspek-tivy razvitiya grazhdanskoj aviacii» 22 – 24 marta 2016. Irkutsk: Irkutskij filial MGTU GA, pp 301–307 (in Russian)
15. Danilenko NV, Kirenchev AG (2020) Problemy processov tekhnicheskogo obsluzhivaniya vozdushnyh sudov grazhdanskoj aviacii v ozhidaemyh usloviyah ekspluatacii (Problems of civil aviation aircraft maintenance processes in expected operating conditions). Crede Experto: transport, obshchestvo, obrazovanie, yazyk 1:59–73
16. World air transport statistics 2019 edition. https://www.iata.org/contentassets/a686ff624550453e8bf0c9b3f7f0ab26/wats-2019-mediakit.pdf. Accessed 10 Oct 2021
17. Flightradar24: live flight tracker. https://www.flightradar24.com/. Accessed 10 Oct 2021

Estimating the Operability of the Gas Turbine Engine Inter-Shaft Bearing

Sergei Khodatsky

Abstract Operational experience shows that a quite large number of aircraft engines are taken out from service due to inter-shaft bearing failures. The main causes of their failure can be a violation of lubrication and cooling conditions under various operating conditions. The article presents the methodology of estimating the durability of the inter-shaft bearing taking into account the rotor bending vibrations and the experimental studies' results. Due to the imbalance of the rotor, its shaft makes an oscillating process. Since the oil drive through which the oil is supplied to lubricate the inter-shaft bearing is located inside the shaft, it is also involved in the oscillating process. This increases the rotor vibration level and reduces the oil flow through the oil drive. In addition, the durability of the bearing is affected by its temperature level. Vibration and increased bearing temperature generally have a negative effect on bearing durability.

Keywords Aircraft engine · Inter-shaft bearing · Vibration velocity · Durability · Oil drive · Experimental unit

1 Introduction

The D-30KP bypass aircraft engines are widely used in the civil aviation in Russia. They have included in the power plant of the Il-76 aircraft and its modifications as well as the Il-62M passenger aircraft.

This engine belongs to the third-generation engines and has sufficiently high parameters of the working process. The engine is reliable in operation, technologically advanced, and has an inter-repair service life of 9000 h which is a rather high indicator at the present stage.

The D-30KP is a dual-rotor and dual-circuit engine. The high-pressure rotor is three-supported and the low-pressure rotor is four-supported. One of the low-pressure rotors supports is an inter-shaft roller bearing.

S. Khodatsky (✉)
Irkutsk Branch of Moscow State, Technical University of Civil Aviation, Irkutsk, Russia
e-mail: sergeixodatski.kafedra@mail.ru

O. A. Gorbachev et al. (eds.), *Proceedings of 10th International Conference on Recent Advances in Civil Aviation*, Lecture Notes in Mechanical Engineering, https://doi.org/10.1007/978-981-19-3788-0_5

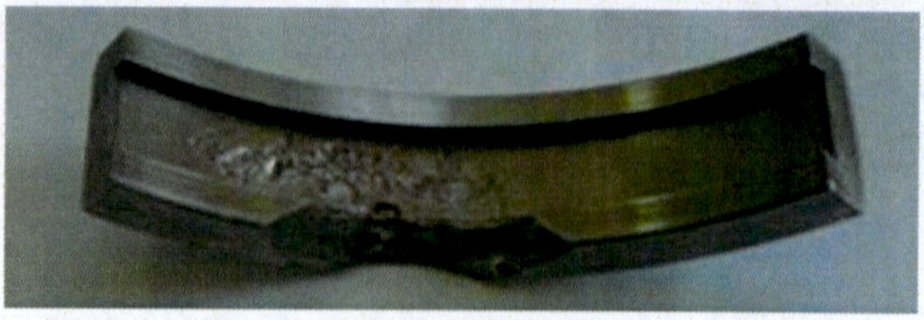

Fig. 1 Fragments of the inter-shaft bearing

It should be noted that a significant design disadvantage of the D-30KP engine is insufficiently high operational reliability of the inter-shaft bearing. Due to the significance of the consequences in case of failure of the inter-shaft bearing during flight, its malfunction with a high probability can lead to a catastrophic situation. This circumstance explains the accidents with the IL-76 (1989) and IL-62M (1986).

An analysis of the total number of failures shows that 45–81% of all failures and malfunctions are in the inter-shaft bearing.

This is due to the fact that these bearings operate in difficult conditions characterized by insufficient oil pumping, vibration, and temperature heating.

The most characteristic damages of bearings are pinching of bearing due to reduction of internal clearances, wear of raceways and rolling elements, corrosion of surfaces of parts, fatigue pitting of metal on working surfaces of rings and rolling elements [1]. In addition to the above-mentioned causes of premature bearing, damage can be design flaws or unsuccessful selection of the bearing type for operation in given conditions, manufacturing and technological shortcomings, and violation of operating and maintenance conditions of the engine (Fig. 1).

It follows from the above that the failure of the inter-shaft bearing is a serious factor that reduces the operational reliability of the engine. Further development of aviation double-circuit engines, complication of their schemes, increase of parameters of working process assumes presence of inter-shaft (inter-rotor) bearings in power systems of the rotor. This fact confirms the relevance of the presented research.

2 Evaluation of Bearing Durability

The characteristics of GTE bearings are studied in the works of many authors [2–6].

For a number of years, the Department of Aircraft and Engines of the Irkutsk branch of MSTU CA has been studying the operational characteristics of the inter-shaft bearing [1, 7–11]. Using the developed experimental units, various operational characteristics of the bearing and its lubrication system are taken.

The high level of vibration of a motor's rotor largely depends on its unbalance. Rotor unbalance is defined as the product of the rotor weight force by the eccentricity. Rotor imbalance can be corrected by static and dynamic balancing. Static balancing balances all forces acting on the rotor supports, while dynamic balancing balances the acting moments.

If the rotor imbalance increases during operating conditions, the radial and axial loads increase, these loads are transferred to the rotor bearings of the motor [12].

A calculated relationship to estimate the effect of rotor unbalance on bearing life is given in [12]:

$$L_H = a_1 \cdot a_{23} \cdot \frac{10^6}{60 \cdot n} \cdot \left(\frac{c_{\text{dyn}}}{P + P_{\text{add}}} \right)^p \tag{1}$$

where a_1—safety coefficient; a_{23}—safety coefficient of lubricant and technology; c_{dyn}—dynamic load capacity; P—equivalent dynamic load; P_{add}—additional radial load.

It should be noted that the above methodology for calculating rolling bearing life does not take into account the fact that the two bearing rings rotate at different angular velocities during operation of inter-shaft bearings. In many calculations, assumptions are made to account for this circumstance. In addition to this, the rings may be misaligned during operation of the inter-shaft bearings, which is also difficult to take into account in the calculation.

It was stated in [13] that when the D-30KP engine operates in all operating modes from flight idle to takeoff mode, as well as when the engine is throttled, the relative speed of the inter-shaft bearing rings remains practically unchanged.

From the above it follows that assuming that there is no misalignment of the rings of the inter-shaft bearing, the calculated ratio can be used to estimate the bearing life, taking into account the vibration state of the rotor (1).

Additional radial load due to rotor vibration can be determined by the ratio:

$$P_{\text{add}} = P_j = m \cdot j = m \cdot A \cdot \omega^2 = m \cdot v_j \cdot \omega \tag{2}$$

where P_j—inertial force of rotor; m—inter-shaft bearing reaction force from the mass of the rotor; A—rotor amplitude; v_j—vibration speed; ω—angle rate of the rotor.

The results of calculating the durability of the inter-shaft bearing of the D-30KP engine taking into account vibration are shown in Fig. 2.

Analysis of the information presented in Fig. 1 allows us to formulate the following conclusions:

- the durability of the inter-shaft bearing changes insignificantly (by 10–14%), in the entire range of rotational speeds from flight idle to takeoff mode with a constant vibration level;
- with an increase in the level of vibration in a stationary mode of engine operation, the durability of the inter-shaft bearing is significantly reduced (by 74–85%).

Therefore, when assessing the operational reliability of the D-30KP engine and the durability of the inter-shaft bearing, vibration loads should be taken into account.

Since the low-pressure rotor of the engine has a fairly large length, and the oil supply line to the inter-shaft bearing runs inside the rotor, this imposes particularities on the oil supply process.

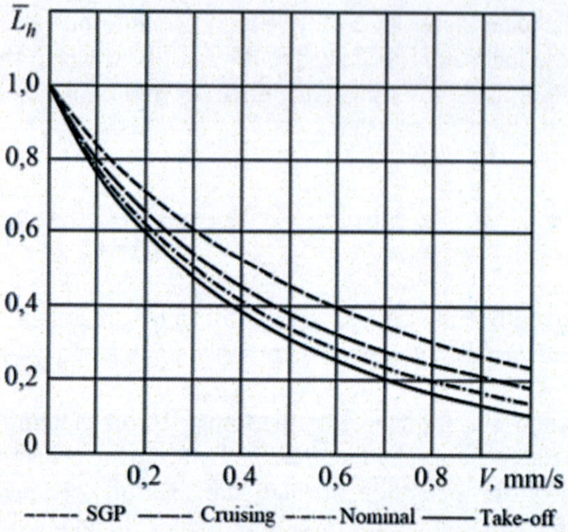

Fig. 2 Evaluating the durability of the inter-shaft bearing when changing engine operating modes and vibration levels

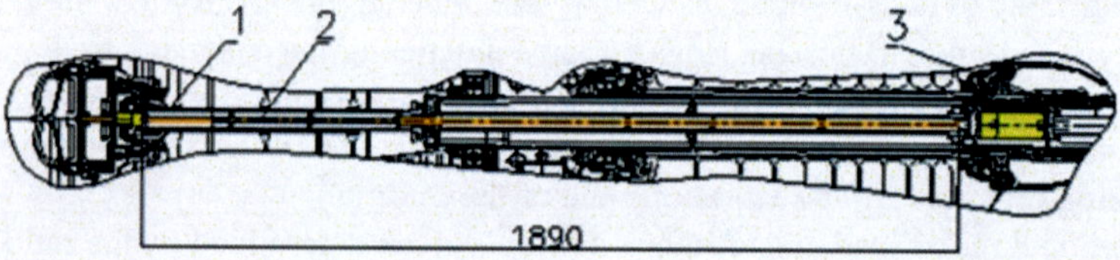

Fig. 3 Placement of the oil pipe in the low-pressure rotor: 1—tube; 2—oil pipe; 3—inter-shaft bearing

Figure 3 shows a structural diagram of the oil pipe location [9].

The oil pipe is located inside the low-pressure rotor. The oil is supplied to the oil pipe through ducts inside the rotor front bearing. The oil pipe itself is stationary, but its end part in the area of the inter-shaft bearing is in the form of a rotating sleeve with oil inlet holes in the cavity of the inter-shaft bearing.

Since the oil pipe is placed inside the low-pressure rotor, its vibration state is similar to the vibration state of the rotor itself. Thus from [7] follows that the level of vibration (vibration velocity) in area of location of engine supports reaches the value $V_{\text{vibration}} = 60\text{–}90$ mm/s, which corresponds to overload factor $Kj = 4\text{–}5$.

3 Experimental Studies

In order to assess the change in the performance of the oil drive when changing the vibration level of the engine rotor, an experimental setup "oil pipe" was developed (see Fig. 4).

The oil from the oil tank enters the oil pump and then enters the oil distribution sleeve through the inlet channel.

Since the main part of the oil supply line to the inter-shaft bearing is located stationary inside the rotor, and the closing part—the oil distribution sleeve, which directly supplies oil to the inter-shaft bearing, is movable, it rotates together with the rotor.

The oil distribution sleeve contains a certain volume of oil which is subject to centrifugal forces as the rotor rotates. The magnitude of the centrifugal force is proportional to the square of angular velocity, the mass of oil in the oil-distributing sleeve, and the radius. It will vary according to the operating conditions of the engine.

Figure 5 shows a diagram of the oil distribution sleeve [9].

If during motor operation the rotor rotates with circular frequency ω and simultaneously performs oscillatory movements with circular frequency p, it performs a complex motion called precession.

Fig. 4 Experimental setup for investigating the characteristics of the oil pipe: 1—oil distribution sleeve; 2—protective casing; 3—oil pump electric motor; 4—oil pump; 5—electric drive of oil distribution sleeve

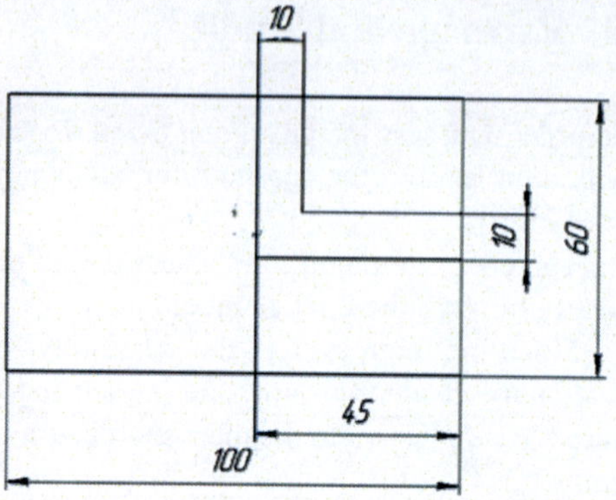

Fig. 5 Diagram of the oil distribution sleeve

If the direction of rotor rotation coincides with the direction of circular motion of the shaft, the precession is called direct, otherwise, it is called inverse. If p and ω are equal in absolute value, the precession motion is called synchronous.

In the general case ω may differ from p both in magnitude and direction, since the value of ω is determined by the mode of motor operation, and the value of p is determined by the intrinsic bending properties of the rotor. But the most important cases are direct synchronous precession ($\omega = p$) and reverse synchronous precession ($\omega = p$).

With bending vibrations of the rotor and precessional movement of the shaft, inertial forces will act on the volume of oil in the oil distributor sleeve. These forces will impede the supply of oil to the inter-shaft bearing. This effect will be especially great in the case of direct synchronous precession. In this case, the direction of the radial oil supply channel of the oil distributor sleeve will coincide with the direction of the shaft deflection (see point 1, Fig. 6).

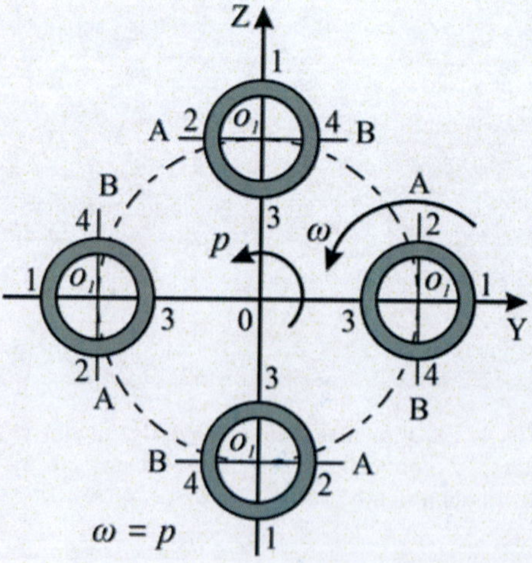

Fig. 6 Scheme of the precessional movement of the shaft

Thus, the maximum inertial force will be at point 1 [9]. In this case, as already mentioned, the oil supply to the inter-shaft bearing through the oil distribution sleeve will be disturbed. This will complicate working conditions of the inter-shaft bearing and affect its reliability and durability.

In order to investigate the characteristics of the oil pipeline taking into account the above factors, a ring weight was placed on the distribution sleeve near the outlet section of the radial channel to simulate different levels of imbalance.

In view of these features, experimental studies were carried out to estimate oil flow rate through the oil duct. Experimental studies [14] have shown that oil consumption reduction at a vibration velocity of 90 mm/s due to the influence of inertial forces and shaft precession motion was 35.4%.

In order to study the characteristics of the oil pipeline on the temperature condition of the cages of the inter-shaft bearing, an experimental setup was created [9]. The object of the research was an inter-shaft bearing of the engine D-30KP (Fig. 7).

The inter-shaft bearing 3 is located in the housing fixed on frame 1.

The inter-shaft bearing under study is shown in Fig. 7 [7].

Frame 1 has a rectangular shape. In the upper part of the frame, there is an oil tank 2. Laptop 4 was provided for recording the results of bearing cage temperature measurements. The licensed software package ZetLab with digital measuring modules ZET7121 was installed on it. Four thermoelectric converters ChEMT-50M were placed in slots of the driven shaft (Fig. 8).

The ChEMT is wired to the ZET 7121 module via the drive shaft.

ZET 7121 and ZET 7176 modules provide conversion of the ChEMT signal into the temperature value of the bearing cage.

Oil supply was carried out by pressure pump 6, and oil pumping—by pump 7. Oil temperature and pressure were controlled by sensors and indicators 8. MK-8P oil was used as oil (Fig. 9).

Figure 10 shows the main unit of the experimental setup.

During the assembly, the inter-shaft bearing was pressed by an inner ring onto the driven shaft. The outer bearing race was secured to the drive shaft (Fig. 11).

An oil cooler was used to cool the oil passing through the inter-shaft bearing. The radiator was installed behind the oil feed pump.

Fig. 7 Inter-shaft roller bearing

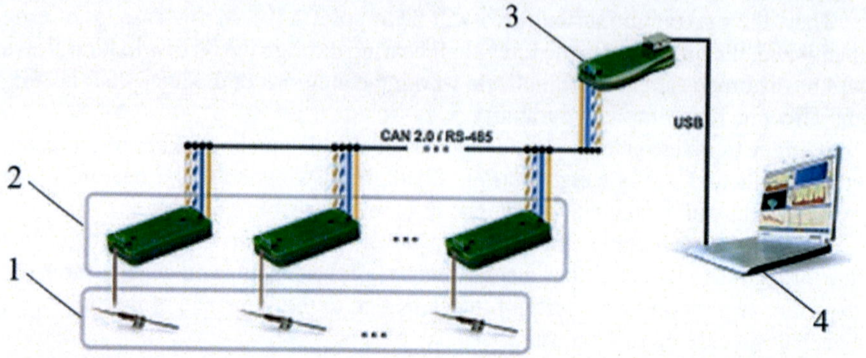

Fig. 8 ZetLab measurement system: 1—primary converters, 2—measuring modules, 3—converter of in-interfaces, 4—computer

Fig. 9 Experimental setup for studying the characteristics of an inter-shaft bearing: 1—barrel; 2—tank; 3—bearing; 4—computer; 5—interface converter; 6—motor; 7—motor drive; 8—manometers

4 Results of the Experiment

During experimental studies, the temperature of the oil passing through the inter-shaft bearing was measured, as well as the temperature of the bearing inner ring Tb. The measurement was carried out at constant speed of the input shaft of the inter-shaft bearing *n* and constant oil pumping through the bearing *q* (Fig. 12).

As the oil pumping decreases, the degree of heating of the bearing Δt increases (see Fig. 13) [9].

Fig. 10 Research object: 1—supports; 2—driven shaft; 3—drive shaft; 4—housing; 5—cover; 6—supporting bearing; 7—gland; 8—retaining bearing; 9—supports; 10—belt drive pulley; 11—oil supply connection; 12—oil outlet connection

Fig. 11 Driven and drive shaft

Since the radial load on the inter-shaft bearing on the experimental setup is 70 H, and on the D-30KP engine—2650 H, the method of linear interpolation was used in order to ensure that the results of the experimental studies correspond to real operating conditions (see Fig. 13) [9].

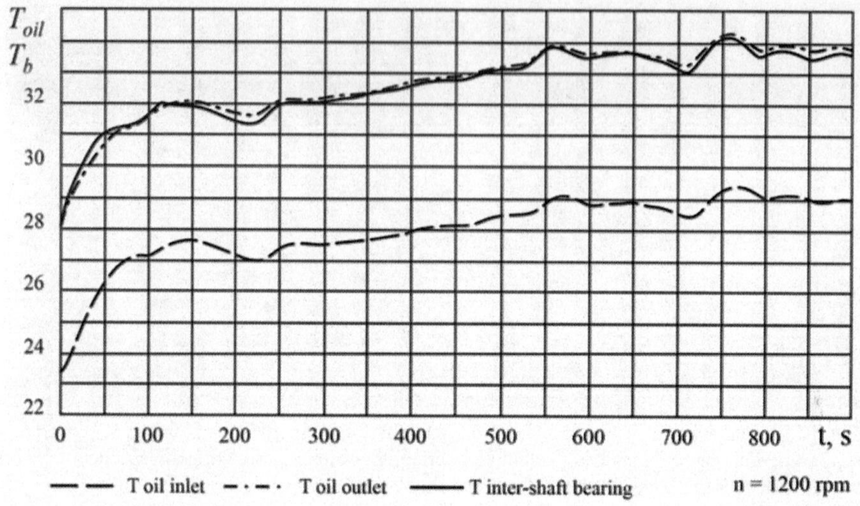

Fig. 12 Temperature measurement results

Fig. 13 Experiment results

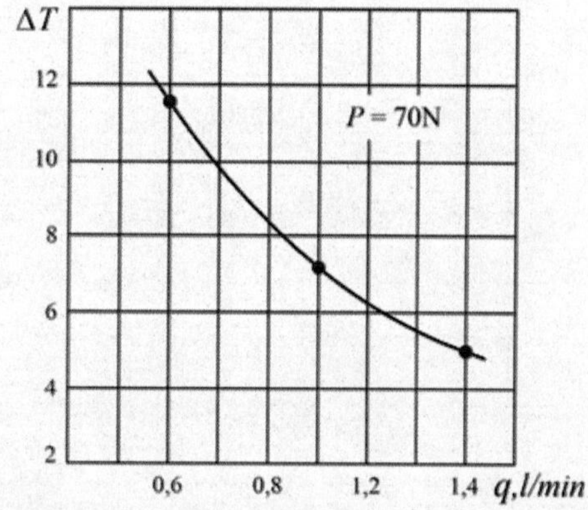

5 Thermal Condition of the Bearing

The results of studies of the thermal condition of the engine oil system show that the heat input to the oil in the bearing units is approximately 90% of the total heat input to the oil in the engine. If there is insufficient heat transfer in the bearing, a significant temperature gradient between the bearing rings can occur, which leads to radial clearance sampling in the bearing and its jamming. Sources of oil heating in bearing units are: friction heat in bearings and contact seals; heat dissipation from air penetrating into oil cavities; heat transfer through the walls of the oil cavity and heat transfer by heat conduction from engine elements mating with the bearing units. Depending on the design of bearing assemblies of motor rotor bearings, heat transfer due to friction in the bearing can reach 50% of the total heat transfer to the oil in the bearing.

In general, the causes of significant heat loss in bearings are:

- friction between lattice layers of the bearing material during mutual deformation of the rolling body and the raceway in the contact patch;
- rolling element slippage in the contact patch;
- friction of the rolling elements in the cage seats, and in roller bearings—friction of the roller ends against the guide edges of the rings, etc.

In the process of "oil starvation" during the bearing operation, the degree of its heating increases. An additional factor reducing the reliability of the bearing operation is an increased level of vibration of the motor rotor. The following assumptions were made to evaluate the bearing performance under these conditions:

- does not take into account heat input during bearing operation from bearing parts;
- the heat generated by the bearing during operation is directly proportional to the friction torque generated in the bearing;
- under these assumptions, the temperature of the oil passing through the bearing and the temperature of the bearing itself is the same.

In order to assess the frictional moment arising in the inter-shaft bearing, the relation was used [15]:

$$M_{\text{fric}} = d \cdot \mu \cdot P \tag{3}$$

where M—friction moment in the bearing; d—inside diameter of the bearing; μ—friction coefficient; P—equivalent dynamic load.

The heat released during friction can be estimated using the dependence [7] and taking into account the results of experimental studies.

$$B_{\text{fric}} = q \cdot c \cdot \rho \cdot \Delta t, \tag{4}$$

where q—oil pumping through the bearing; c—oil heat-sink capacity; ρ—oil density; $\Delta t = t_{\text{out}} - t_{\text{in}}$—increase in oil temperature in the bearing.

Given the assumption that the B_{fric} heat generated by friction in the bearing is proportional to the friction torque, the amount of B_{fric} heat generated in the motor bearing can be determined.

$$B_{\text{fric}\,b} = \frac{B_{\text{fric}} \cdot M_{\text{fric}\,b}}{M_{\text{fric}}}, \tag{5}$$

where $B_{\text{fric}\,b}$—heat, distinguished in bearing at the stand; $M_{\text{fric}\,b}$—engine bearing friction moment.

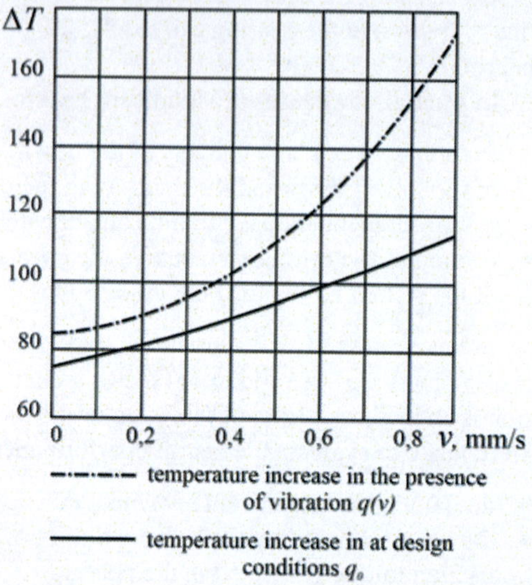

Fig. 14 Change of bearing temperature

Taking into account Eq. (4), we determine the heating value of the inter-shaft bearing installed on the engine (see Fig. 14) [9].

The steeper rise of the bearing heating line $q = q(v)$ is explained by a decrease in the oil line performance when the vibration velocity increases.

The influence of the inter-shaft bearing temperature on its durability was evaluated according to the generally accepted methodology by introducing a temperature coefficient into the calculation relation (1). The calculation results are shown in Fig. 15.

It is known that in a number of cases the service life of aircraft bearings is less than their calculated service durability, which is calculated on the basis of bearing load capacity and operating conditions. It is explained by a number of reasons. First of all, the durability of aircraft bearings is determined by calculation for fatigue failure, while analysis of statistics of failure of gas turbine engine roller bearings shows that in most cases, bearing failure is caused by wear of raceways and rolling elements. In this case, especially often observed seizure with subsequent progressive wear of the working surfaces, caused by slippage in the contacts of the rollers with the raceways of the rings.

To improve the reliability of aircraft bearings, some technological measures are used, such as heat treatment of rings for size stabilization, special selection of metal for rings and rolling elements, manufacturing of bearing inner rings from tubular billets with roll-out chute, the use of vacuum smelting steel and electroslag remelting.

Fig. 15 Influence of oil line performance on the durability of the inter-shaft bearing

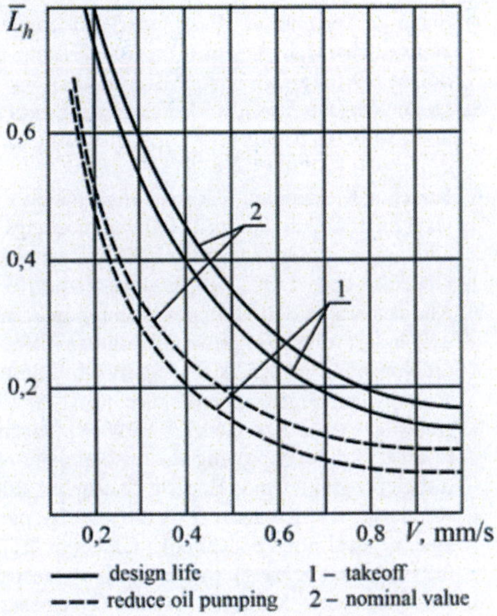

6 Conclusion

As a result of the research work, the following conclusions can be made:

1. The influence of rotor bending vibrations on the durability of the inter-shaft bearing with intra-rotor oil supply was evaluated.
2. Reduction of oil pumping through the inter-shaft bearing, due to increase of vibration velocity of the turbofan rotor, is accompanied by reduction of its durability by 45–53%.
3. The revealed significant influence of oil pumping value on durability of inter-shaft bearing requires development of measures increasing reliability of bearing operation in operational conditions.

References

1. Khodatsky SA, Karavaev YuA, Safarbakov AM (2015) Ocenka napryazhenno - deformirovannogo i teplovogo sostoyaniya mezhval'nyh podshipnikov GTD (Assessment of the stress-strain and thermal state of inter-shaft bearings of GTEs). Mod Technol Syst Anal Model 1(45):41–48 (in Russian)
2. Gao P, Hou L, Chen Y (2021) Dynamic load and thermal coupled analysis for the inter-shaft bearing in a dual-rotor system. Meccanica 56(11):2691–2706. https://doi.org/10.1007/s11012-021-01410-7
3. Hao X, Zheng Y, Lu L, Pan H (2021). Research on intelligent fault diagnosis of rolling bearing based on improved deep residual network. Appl Sci 11(22):10889.https://doi.org/10.3390/app112210889

4. Shuai Z, Yongcun C, Zhonghui H, Xiaomin Y, Yan L, Sier D (2021) Thermal-stress-wear coupled characteristics of oil seal in airframe rod end-bearing. Tribol Int 163:107132.https://doi.org/10.1016/j.triboint.2021.107132
5. Lv D, Wang H, Che C (2021) Fault diagnosis of rolling bearing based on multimodal data fusion and deep belief network. Proc Inst Mech Eng Part C J Mech Eng Sci 235(22):6577–6585. https://doi.org/10.1177/09544062211008464
6. Kudelina K, Vaimann T, Rassolkin A, Kallaste A (2021) Impact of bearing faults on vibration level of BLDC motor. In: IECON proceedings (industrial electronics conference) book 2021. 47th annual conference of the IEEE industrial electronics society, IECON 2021 Toronto13–16 Oct 2021, code 173927.https://doi.org/10.1109/IECON48115.2021.9589268
7. Khodatsky SA (2020) Puti povysheniya nadezhnosti mezhval'nyh podshipnikov gazoturbinnyh dvigatelej (Ways to improve reliability of intershaft bearings of gas-turbine engines). In: Actual problems and prospects of civil aviation development. Proceedings of the IX international scientific and practical conference, pp 81–88 (in Russian)
8. Karavaev YuA, Khodatsky SA (2019) Sovershenstvovanie metodiki ocenki ostatochnogo resursa TRDD (Improving the methodology of estimating the residual life of double-circuit turbojet engine). Crede Experto Transp Soc Educ Lang 2:10–24 (in Russian)
9. Khodatsky SA, Karavaev YuA (2018) Vliyanie vibracij rotora GTD na tekhnicheskoe sostoyanie mezhval'nogo podshipnika (Effect of GTE rotor vibrations on the technical condition of the inter-shaft bearing). Crede Experto Transp Soc Educ Lang 4:31–41 (in Russian)
10. Khodatsky SA, Karavaev YA (2017) Ocenka raskhodnyh harakteristik masloprovoda pri vnutrirotornom podvode masla k mezhval'nomu podshipniku aviacionnogo GTD (Estimation of flow characteristics of an oil-pipeline for int-rotor oil supply to the inter-shaft bearing of an aircraft GTE). In: Topical problems and prospects for the development of civil aviation in Russia. Proceedings of the all-Russian scientific and technical conference dedicated to the 50th anniversary of Irkutsk branch of MSTU CA, pp 136–143 (in Russian)
11. Karavaev YA, Khodatsky SA, Matvienko AS (2016) Ocenka vliyaniya maslorapredelitel'nyh ob"emov pri vnutrirotornom masloprovode aviacionnogo dvigatelya (Assessment of the influence of oil-distribution volumes with an intra-rotor oil line of an aircraft engine). Transp Infrastruct Siberian Region 2:395–399 (in Russian)
12. Sidorov VA, Sotnikov AL, Sushko AE (2009) Metodika ocenki ekonomicheskoj effektivnosti balansirovki rotorov v promyshlennyh usloviyah (Methodology of evaluation of economic efficiency of rotors balancing in industrial conditions). Mach Vibr Measur Reduction Protection Int Sci Techn Ind J DonGTU Technopark UNITECH 2(7):38–43 (in Russian)
13. Ivko SY, Karavaev A, Khodatsky SA (2013) Povyshenie nadezhnosti mezhval'nogo podshipnika (Improvement of reliability of inter-shaft bearing). In: VI regional scientific-practical conference of students and graduate students. Irkutsk State Technical University of CA, Irkutsk, pp 195–199 (in Russian)
14. Karavaev YuA, Khodatsky SA, Shusharin VA (2014) Ocenka parametrov potoka masla pri vnutrirotornom podvode k mezhval'nomu podshipniku aviacionnogo GTD (Assessment of oil flow parameters at intra-rotor supply to the inter-shaft bearing of aviation GTE). Mod Technol Syst Anal Model Sci J 42:50–54 (in Russian)
15. Using the friction model as an engineering tool (2006) [Electronic resource]. http://www.skf.com/ru/products/bearings-units-housings/principles/bearing-selection-process/operating-temperature-and-speed/friction-powerloss-startingtorque/index.html. Accessed 10 Oct 2021

Aircraft Electrical Systems and Flight and Navigation Equipment

The chapter discusses the issues and promising areas of developing the aircraft equipment: aircraft electrical systems, flight and navigation equipment, implementation of artificial intelligence technologies in aircraft systems; control, diagnostics and prediction of technical state of aircraft equipment; organization and support of aircraft equipment operation.

Keywords Aircraft equipment · Aircraft electrical systems · Flight and navigation equipment Operation of aircraft equipment

Development of an Algorithm for Determining the Sleepiness of a Crew

Vladislav Karelin and Vladimir Popov

Abstract The paper discusses the development of an algorithm for an intelligent system for monitoring the sleepiness of a crew in real time using a neural network model. The following signs of sleepiness were investigated: the frequency of yawning, the direction of gaze, the frequency and speed of blinking of the eyelids, and the duration of eye closure.

Keywords Computer vision · Recurrent neural network · Deep learning · Object detection · Mediapipe · Python

1 Introduction

Fatigue and sleepiness of pilots are among the main risks in the field of civil aviation and also affect their performance and health.

Pilots' fatigue leads to deterioration in their well-being, which increases the risk of accidents and incidents. Let us consider the consequences of fatigue, which determine the violation of mental and psychological parameters of a crew.

Violation and deterioration of pilots' mental parameters consist of slowing down the speed and reducing the accuracy of reproducing control actions, scattering attention. Violation of cognitive parameters affects memory, alertness, and speed of decision-making. These factors lead to missing or insufficient response to weak signals, forgetting or deviating from a specified algorithm of actions during the flight, inadequate perception of the information received.

Fatigue causes illusions, false perceptions of the spatial position of the aircraft. In these cases, the pilot begins to fall asleep involuntarily ("nodding off"). The crew even forgot to release the landing gear when landing.

V. Karelin (✉) · V. Popov (✉)
Irkutsk Branch of Moscow State Technical University of Civil Aviation, Irkutsk, Russia
e-mail: fl1ckje@mail.ru

V. Popov
e-mail: povlamix@mail.ru

O. A. Gorbachev et al. (eds.), *Proceedings of 10th International Conference on Recent Advances in Civil Aviation*, Lecture Notes in Mechanical Engineering, https://doi.org/10.1007/978-981-19-3788-0_6

During long flights, due to the automation of aircraft control, the crew control piloting and navigation parameters only, which often causes calmness and boredom. So, in the case of an emergency, they will need more time to make a decision.

Some airlines have begun to appoint two crews or an additional crew member on long-haul flights to improve flight safety. Also, fake situations prepared on the ground arise in flight to increase the attention and vigilance of the crew.

It was found that the level of fatigue of the pilot when landing at 5 a.m. is the same as with 0.08% alcohol in the blood.

The flight operation manual contains regulations for organizing a rest in flight for a two-member crew, a reinforced crew, and a double crew. The International Civil Aviation Organization (ICAO) has also developed a manual for operators on the implementation of a fatigue risk management system (FRMS), allowing them to work both safer and more efficiently [1]. The document states that sleep and its lack reduce the amount of attention, memory, and concentration during the flight.

For this purpose, for example, the "Volga-Dnepr" airline is introducing a procedure of controlled rest for 30–40 min during a level flight in the cockpit of a two-member crew as a means of combating fatigue during a long flight. This procedure reduces the risk of micro-sleep and raises the activity of the crew at the most crucial stage of the flight during landing. To exclude the co-pilot's sleep and increase the reliability of the proposed procedure, it is proposed to use a third person to check the procedure.

But the pandemic has made its adjustments to the actions of airline executives. Due to the decrease in the number of flights, pilots began to be reduced, while increasing the load on the remaining ones. As a result, the number of flight accidents and incidents related to pilot errors due to excessive load, fatigue, and, as a result, sleepiness has increased.

Therefore, systems have been developed to determine the drowsy state of the crew.

The first system measures biological signals from the pilot's body. The most common are the following: automatic electroencephalogram analysis, electroencephalography, electrocardiogram, electrooculography, surface electromyogram, galvanic skin reaction, and respiration [2, 3]. Such devices create physical discomfort for the crew when controlling fatigue.

There are also systems based on a scale that measures fatigue. The most common scale is the "Karolinska Sleepiness Scale" [4]. The fatigue measurement is implemented with the aid of an observer who asks a crew member about his condition at some time interval. This approach requires the pilot to report his condition frequently. It is a rather intrusive measure that cannot be used in the real system.

The system that determines the state of sleepiness of the pilot based on the analysis of data from the signals mentioned above requires an individual approach to each person; therefore, it will not be applicable for a new crew member.

The most promising is an intelligent system based on the control of the pilot's face using a camera and image processing methods to obtain physical indicators of sleepiness, such as:

- frequency of yawning;

- direction of view;
- slow eye rotation;
- blinking speed and duration of eye closure.

This approach is quite effective since the pilot's sleepiness manifests itself in signs of fatigue on the face, and secondly, this approach does not irritate the pilot.

Consider the physical indicators of sleepiness.

1. The frequency of yawning. The main reason for yawning is fatigue and overwork overload of the body. Yawning, a person supplies oxygen to the brain, maintaining its activity. Fatigue also makes you yawn. When a person gets tired, the work of the organ system slows down, breathing becomes shallow, the content of carbon dioxide in the blood increases, so yawning activates the blood flow, contributing to the removal of carbon dioxide from the body. Mental overstrain is also the cause of yawning, which activates tired brain. The yawn lasts an average of 6 s. After yawning for the second time, a person usually yawns not earlier than after one and a half minutes. A sign of sleepiness is 2–4 yawns per minute.
2. Direction of gaze, head nodding. One of the noticeable signs of sleepiness is the moment when the pilot lowers his head.
3. The frequency and speed of eyelid blinking. Under normal conditions, there are about 10–15 movements of the eyelids within one minute. The average rate is 2/5 s, but with the development of sleepiness, the blinking of the eyes becomes more extended. When blinking, the eyelids begin to fall with a small amplitude. A sign of sleepiness is the time of the closed state of the eyelids—the blinking speed is 3–5 s, and the frequency of a blinking of the eyelids is 3–5 times per minute.
4. Duration of eye closure. The criterion of sleepiness is a condition when the eyes are closed more than 80% per minute. If the percentage of closing the eyes exceeds 80% within a minute, then the person is considered to be dozing.

2 The Main Part

2.1 Problem Solution

The solution consists in monitoring the pilot's face using a camera and methods of digital analysis and image processing to obtain some information, namely physical indicators of the pilot's condition, such as:

- frequency of yawning;
- direction of view;
- frequency and speed of blinking of the eyelids;
- duration of eye closure.

To obtain this data, we first need to get information about the face of the pilot and specifically about the key points of the face. To do this, we used the MediaPipe project from Google [5]. It is an open-source framework that provides readymade, customizable, and cross-platform solutions in machine learning and computer vision for streaming media information, which mainly enters the computer in real time. It supports many modern operating systems and a wide range of central and graphics processors (CPUs and GPUs). In addition, it has a Face Mesh algorithm that evaluates 468 key points of the face in three-dimensional space in real time. This algorithm is not demanding on equipment and requires only the presence of a digital camera.

2.2 *The Algorithm Development*

The algorithm consists of two neural network models: the BlazeFace detector [6], which calculates the location of faces in the frame, and a model for determining face points, which is based on the regression method. Due to the lightweight architecture of this model, it is possible to achieve good performance and acceptable accuracy even when only computing by means of CPU (without acceleration using GPU). An example of the result of the model is shown below (see Fig. 1).

This model is the first component of the overall cascade of models, of which the algorithm for determining the sleep state consists.

Next, we need to filter out the entire polygonal grid and leave only those points that form the contours of the lips at the mouth and eyelids of the eyes (see Fig. 2).

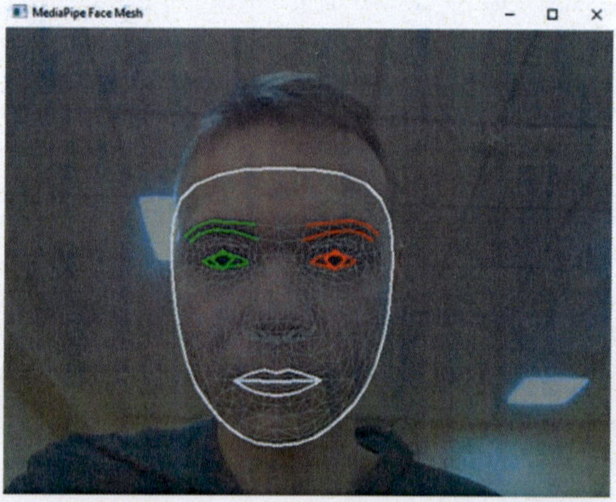

Fig. 1 Example of the face mesh algorithm

Fig. 2 Eye contour selection by points

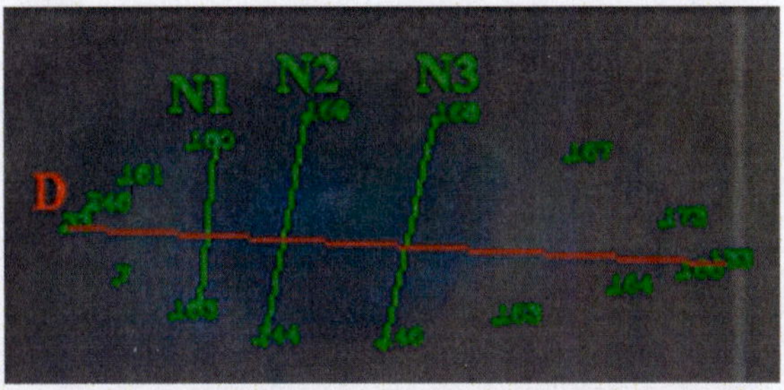

Fig. 3 Visualization of distances between eye points

Each point is characterized by the x and y coordinates of the pixel in which the point is located in the image. This is enough to determine the position of the head.

To determine open/closed eyes and yawning, the eye aspect ratio (EAR—eyes aspect ratio) and the mouth aspect ratio (MAR—mouth aspect ratio) are used [7, 8]. For a more precise definition of whether a person yawns or not, the relation of MAR to EAR (MOE—mouth over eyes) is also introduced.

EAR is calculated using the formula:

$$\frac{N1 + N2 + N3}{3 \cdot D} \quad (1)$$

where $N1, N2$, and $N3$ are the distance between the points vertically; D is the distance between the points horizontally (see Fig. 3).

Similarly, MAR is calculated, and MOE = MAR/EAR. An issue may arise during the first experiments: This algorithm is tied to the threshold values at which the determination of closed/open eyes and mouth will occur correctly. To fix this issue, before the full operation of the software, a preliminary calibration is performed, during which arithmetic averages and standard deviations for EAR, MAR, and MOE are calculated in the sum of 25 frames.

The dependence of EAR, MAR, and MOE when recording during 600 frames (600/27 ~ 22 s) is shown below (see Fig. 4).

The bursts on the graph, highlighted by segments *a*, *b*, and *c*, correspond to changes in the physical information about the pilot's face: closing the eyes, yawning, and blinking (once).

Knowing the shape of these sections (approximate duration and amplitude) and considering the normalization of EAR, MAR, and MOE, it is possible to select the conditions and threshold values where determining a performed action (closed his eyes, yawned, or yawned blinked) becomes possible. Accordingly, since there are frames along the x-axis, knowing the frame rate of the image analysis, it is possible to find the duration of a change in one or another aspect ratio. Having EAR, MAR, and MOE, it is possible to classify a pilot's condition, but the values of these characteristics from one frame are not enough. There is a need to divide the time into groups of frames (e.g., 5), during which it will be possible to judge more accurately about the presence or absence of the pilot's sleepiness. To do this, we decided to

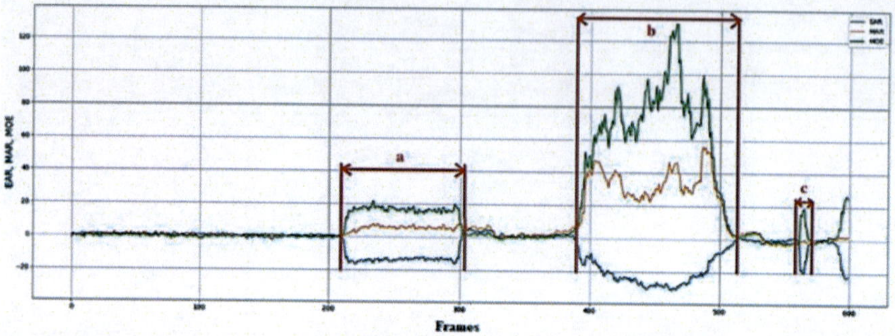

Fig. 4 Dependencies of EAR, MAR, and MOE after calibration: *a*—the length of time at which the eyes were closed; *b*—the length of time at which the yawn was made with the mouth open (the eyes are closed due to the yawn); *c*—the length of time at which the short-term blinking occurred

create and train a recurrent neural network with a long-term short memory due to the lack of ability of neural networks to remember previous information, extrapolate it, and make predictions. In other words, neural networks imply independent inputs and outputs due to missing feedback loop. Recurrent neural networks (RNNs) have it, so RNNs can be considered as several copies of the same network, each of which transmits information from the previous one to the subsequent copy. Figure 5 shows general structure of the RNN [9].

$X(t)$ is the input signal at the t-step, and $S(t)$ is the hidden state at the t-step. This is essentially network memory that depends (as a function) on the previous states and the current input:

$$S(t) = f(U_x(t) + W_S(t-1)) \tag{2}$$

As a rule, the nonlinear (ReLU or tanh) function is chosen [10]. $O(t)$ is the output signal at the t-step. In our case, we had to choose a particular type of the RNN-LTSM (long short-term memory), i.e., long short-term memory networks. When compared with the RNN, it better stores long-term dependencies and is more suitable for solving this problem. The structure of the classifier model was compiled empirically but taking into account the prevention of overfitting (the model works well on training datasets, but poorly on real data for testing) using dropout blocks, which discards (sets zero) with a given probability some parts of the input signal before it is transmitted to

Fig. 5 Generalized structure of the RNN

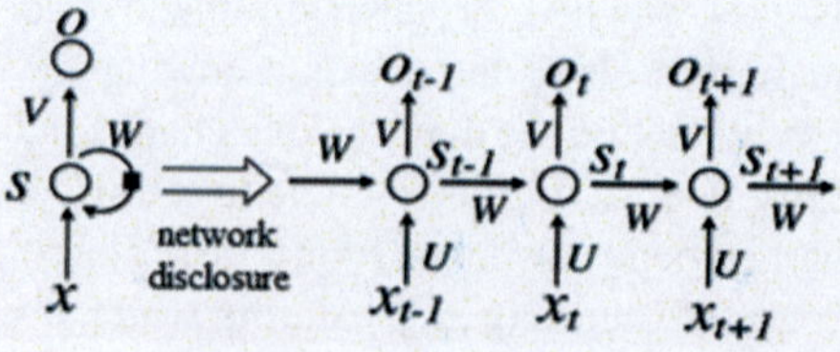

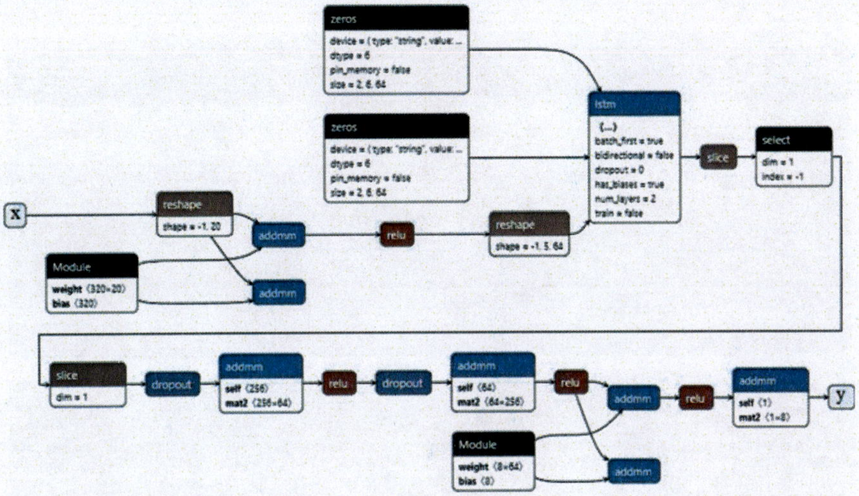

Fig. 6 Structure of the pilot status classifier model by EAR, MAR, and MOE values using the LSTM network

the next layer. Figure 6 shows the structure of the classifier model with an integrated LSTM network [11, 12].

The neural network model is implemented in the PyTorch library [13–15]. ReLU are used as activation functions. "Addmm" blocks are responsible for matrix multiplication of output signals with weights taking into account their shifts. Reshape layers are responsible for changing the transformation of output signals from previous layers. There are linear layers for classification before the output. As a result, the prediction becomes the output signal of the model, namely an integer from 0 to 1. 0 indicates that the pilot is not sleeping, and 1 indicates that he has already fallen asleep.

Figure 7 shows the final generalized block diagram of the algorithm for determining the sleep state.

Figure 8 shows the result of the algorithm.

The FPS designation (frames per second) is the number of frames per second. Sleepy/alert is the signature of the classifier prediction result, denoting the sleeping state/awake state, respectively.

Fig. 7 Generalized block diagram of the sleep state algorithm

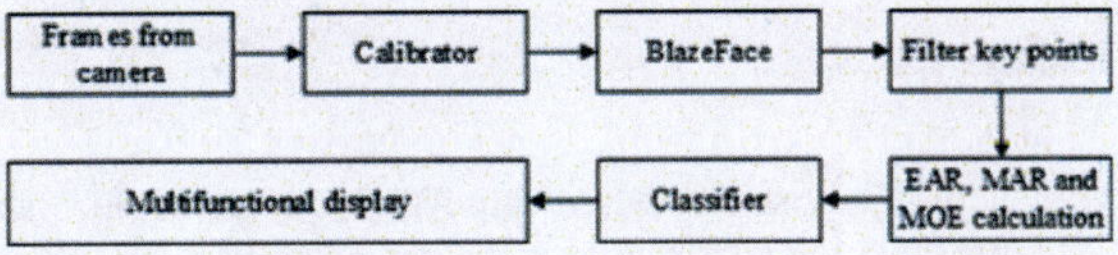

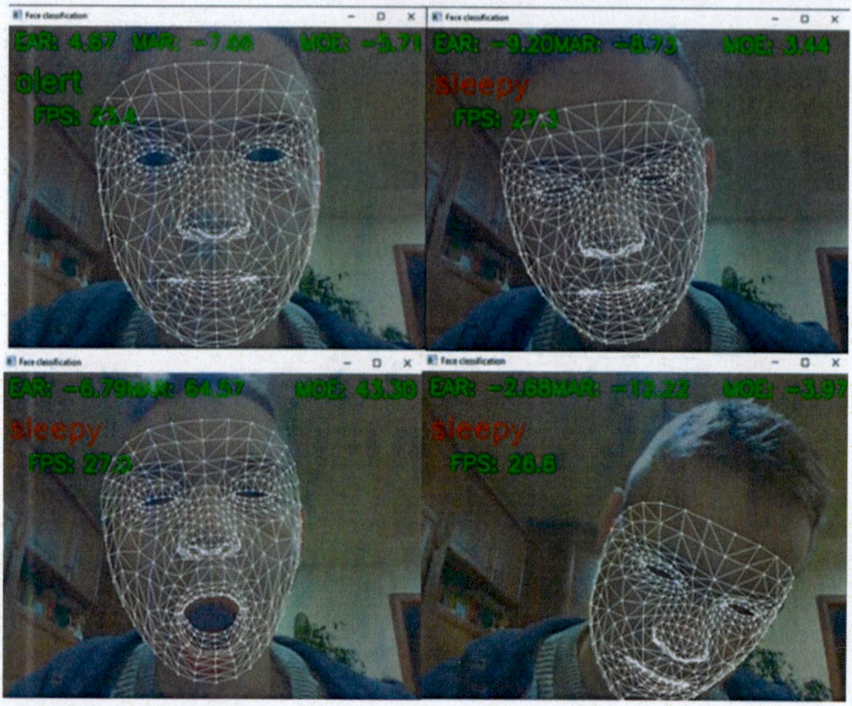

Fig. 8 Demonstration of the developed software

3 Conclusion

The study shows that the developed algorithm for determining the drowsy state allows one to adapt the EAR and MAR parameters to the physiological characteristics of the pilot's face, thereby ensuring its versatility in practical application. It can also identify such signs of sleepiness of the aircraft crew as eye closure, yawning, and blinking. To improve the accuracy of detecting faces and their key points, one should train the model based on a more voluminous and diverse set of EAR, MAR, and MOE values obtained from real videos with people in alert and sleepy states.

References

1. Guziy AG, Lushkin AM, Mishin AV (2016) Utomlenie kak neblagopriyatnoe funkcional'noe sostoyanie letchika bezopasnost' zhiznedeyatel'nosti (Fatigue as an unfavorable functional state of the pilot) Life safety: science, education, practice, pp 210–214 (in Russian)
2. Turan G, Gupta S (2013) Road accidents prevention system using driver's drowsiness detection. Int J Adv Res Comput Eng Technol (IJARCET) 2

3. Berka C et al (2007) EEG correlates of task engagement and mental workload in vigilance, learning, and memory tasks. Aviation Space Environ Med 78(5):B231–B244
4. De Rosario H et al (2010) Controlled inducement and measurement of drowsiness in a driving simulator. IET Intell Transp Syst 4(4):280–288
5. Lugares C et al (2019) Mediapipe: a framework for building perception pipelines. arXiv:1906.08172
6. Bazarevsky V et al (2019) Blazeface: sub-millisecond neural face detection on mobile gpus. arXiv:1907.05047
7. Mehta S et al (2019) Real-time driver drowsiness detection system using eye aspect ratio and eye closure ratio. Proceedings of international conference on sustainable computing in science, technology and management (SUSCOM), Amity University Rajasthan, Jaipur, India
8. Maior CBS et al (2020) Real-time classification for autonomous drowsiness detection using eye aspect ratio. Expert Syst Appl 158:113505
9. Medsker LR, Jain LC (2001) Recurrent neural networks. Des Appl 5:64–67
10. Hara K, Saito D, Shouno H (2015) Analysis of function of rectified linear unit used in deep learning. 2015 international joint conference on neural networks (IJCNN). IEEE, pp 1–8
11. Sherstinsky A (2020) Fundamentals of recurrent neural network (RNN) and long short-term memory (LSTM) network. Phys D Nonlin Phenomena 404:132306
12. Xingjian SHI et al (2015) Convolutional LSTM network: a machine learning approach for precipitation nowcasting. Adv Neural Inf Process Syst:802–810
13. Ketkar N (2017) Introduction to PyTorch. Deep learning with python. Apress, Berkeley, CA. pp 195–208
14. Subramanian V (2018) Deep learning with PyTorch: a practical approach to building neural network models using PyTorch. Packt Publishing Ltd.
15. Paszke A et al (2019) Pytorch: an imperative style, high-performance deep learning library. Adv Neural Inf Process Syst 32:8026–8037

Towards an Intelligent Decision Support System for Aircraft Troubleshooting

Aleksandr Yurin, Yuri Kotlov, Vladimir Popov, and Sergey Mishin

Abstract Currently, troubleshooting an aircraft remains a promising area for automation and intellectualization. At the same time, existing solutions in this area in the form of electronic technical manuals do not always meet the requirements of technical personnel when searching and troubleshooting aircraft. In this regard, the development of another class of systems based on artificial intelligence methods is relevant. These systems can provide not only the search and elimination of failures and malfunctions but also self-learning by accumulating the experience. This paper proposes the basic principles of such an intelligent system, called the AirTech Assistant, designed for use by technical personnel engaged in the maintenance and repair of the power supply system of the Sukhoi Superjet (RRJ-95) aircraft. In particular, a fragment of the conceptual model of the domain is given, functional, operational, and quality requirements are formulated, architecture, as well as fundamental algorithms and a stack of implementation technologies, are defined. The main components of the designed system will be expert systems implementing case-based and rule-based reasoning. An additional study was conducted in terms of testing the formalism of event trees for knowledge base engineering.

Keywords Intelligent decision support system · Conception · Troubleshooting · Aircraft · RRJ-95

A. Yurin (✉)
Matrosov Institute for System Dynamics and Control Theory, Siberian Branch of Russian Academy of Sciences (ISDCT SB RAS), Irkutsk, Russia
e-mail: iskander@icc.ru

A. Yurin · Y. Kotlov (✉) · V. Popov (✉) · S. Mishin (✉)
Irkutsk Branch of Moscow State Technical University of Civil Aviation, Irkutsk, Russia
e-mail: yukotlov@rambler.ru

V. Popov
e-mail: povlamix@mail.ru

S. Mishin
e-mail: sv_mishin@mail.ru

O. A. Gorbachev et al. (eds.), *Proceedings of 10th International Conference on Recent Advances in Civil Aviation*, Lecture Notes in Mechanical Engineering, https://doi.org/10.1007/978-981-19-3788-0_7

1 Introduction

Troubleshooting an aircraft remains a promising area for automation and intellectualization, especially in the context of big data processing. In practice, each aircraft is equipped with a large volume of documentation, the volume of which can reach up to 8 tons of paper and up to 50,000 pages (A-320 aircraft), and specialized methods and means are required for its operational use. At the same time, digitization of this documentation and its representation in the form of interactive electronic technical manuals supporting keyword search become insufficiently effective. The effectiveness of such systems can be increased through training and self-learning; accumulating and reusing new experience in the form of malfunctions and failures, not yet reflected in the manuals, but recorded by technical specialists; narrowing and ranking the list of possible systems which are the candidate for failure or malfunction.

In this paper, we present the basic principles of an intelligent decision support system for troubleshooting an aircraft, called the AirTech Assistant, designed for use by technical personnel engaged in the maintenance and repair of the power supply system of the Sukhoi Superjet (RRJ-95) aircraft. The basic principles include functional, operational, and quality requirements, architecture, basic algorithms, and data structures. We carried out an analysis of the domain when developing these principles; the results of this analysis are also given. At the same time, the main components of the designed system will be expert systems implementing case-based [1] and rule-based [2] reasoning. The first one is designed for accumulating and reusing information about failures and malfunctions that are not accounted in the current version of the documentation. The second one is designed for codifying electronic manuals and providing support in the case of the formation of plans for the search, confirmation, troubleshooting, and confirmation of troubleshooting.

The AirTech Assistant will reduce the time effort of technical personnel during troubleshooting, and one of its features is the configurability with the account of a certain type of aircraft.

The paper is organized as follows. Section 2 presents the state of the art. Section 3 contains the problem statement. Section 4 presents the results of domain analysis and conceptual modeling. The description of the basic concepts of the intelligence system, including the requirements for the functions, operating conditions and quality, the structure of software, and the implementation technologies are presented in Sect. 5, while Sect. 6 presents some concluding remarks.

2 The State of the Art

2.1 Aircraft Troubleshooting

Existing solutions for decision support in the task of finding and troubleshooting failures and malfunctions of an aircraft can be divided into two main groups:

- Information systems in the form of electronic technical manuals, representing digitized documentation supplemented by the ability to search by keywords. The most well-known system of this type is the AirNav Maintenance [3] developed by the Airbus Company. Further development of this class of systems is their integration with the onboard maintenance system (OBMS), an example of such work is [4].
- Intelligent decision support systems involving the use of certain models and methods of artificial intelligence. An example of such works is [5], where the author considers an aircraft troubleshooting system, which includes a mathematical model of equipment. Using this model, a correspondence is established between the equipment failure and its causes.

From our point of view, the second group is more promising. Since the solutions of the first group have practically exhausted the potential for their perfection, having reached their maximum efficiency, which, nevertheless, is insufficient from the point of view of technical personnel and the tasks they solve. At the same time, the works of the second group are mainly theoretical and methodological, do not specify the type of aircraft and the diagnostic models`used. For this reason, they are difficult to reproduce.

Next, we will consider the basic techniques that can be used to solve this task following the principles of the second group: case-based reasoning, rule-based reasoning, and end-user development.

2.2 *Case-Based Reasoning*

Onc of thc tcchniques of decision support used for solving tasks of troubleshooting is reasoning by cases [1]. This approach is based on the implementation of the principle of reasoning "by analogy".

Case-Based Reasoning (CBR) [6–9] is a methodology for solving tasks by reusing and adapting (if necessary) previously obtained solutions of similar tasks. An important concept of this methodology is a case [9] as a form of structured representation of accumulated experience in the form of data and knowledge intended for subsequent automated processing with the aid of specialized software.

The main characteristics of cases are the following: context-bound; clear identification of functional parts (a task and a solution); a variety of forms for representation (textual, structured/templated, question–answer, logical rules).

To extract the cases, the procedure of their vectorization is used with a further calculation of the distances between the vectors using various metrics. The most widely used [1, 6, 8] are Euclidean, Hamming, probabilistic similarity measure, Rogers-Tanimoto, Manhattan, Chebyshev, Mahalanobis, Zhuravlev, Bray–Curtis, Chekanovsky, Jacquard, etc.

The possibility of reuse of cases is achieved by adapting (transforming) the extracted solutions. Currently, there are several adaptation methods [1, 6, 7, 10] based on either transformational or generative approaches. The transformational adaptation uses a set of adaptation rules or operators that describe how differences in the description of the task affect changes in the solution. Methods of generative adaptation for their effective application require the presence of a solver (inference machine) capable of solving the task based on generalized knowledge (without using cases). In this case, no solutions are used, but a sequence of steps (order, path) to solve the task described in the cases.

2.3 Rule-Based Reasoning

Rule-based expert systems [11, 12] are a classic and the most methodologically developed method of the "black box" artificial intelligence. They provide a representation of knowledge in the form of logical rules of the "IF (condition) THEN (action)" type.

Logical rules can be formalized by the following:

$$(i)\colon Q;\ P;\ A \rightarrow B;\ N$$

where i is a logical rule name; Q is the scope of a logical rule application; P is the condition of the logical rule applicability; $A{\rightarrow}B$ is the core of the logical rule (IF … THEN …); N is the postcondition of the logical rule.

Many algorithms have been developed for making decisions or reasoning based on rules, and the RETE [13] algorithm is the most famous one. Currently, there are languages and software that provide the construction of rules, including integrated expert systems. The most popular are the following: CLIPS, JESS, DROOLS.

2.4 End-User Development

One of the complex tasks connected with expert systems engineering is the task of extracting knowledge from experts, consequent to its codification and filling knowledge bases. The way to increase the efficiency of this process is the use of methods based on End-User Development (EUD) [14] principles. EUD can be defined as a set of methods, techniques, and tools for non-professional software developers to create, modify or extend software artifacts. The following EUD approaches are the most popular [15]: component-based, wizard-based, programming by example, natural language processing, visual programming, etc. At the same time, wizard-based approach and visual programming are most widely used in the field of artificial intelligence systems engineering and they are implemented, for example, in the PESoT (Prototyping Expert Systems Based on Transformations) technology [16].

3 Problem Statement

The purpose of the study is to provide decision support for technical personnel during the maintenance and repair of aviation equipment, in particular, during searching failures and malfunctions, and troubleshooting.

The creation of an experimental (pilot) version of an intelligent system, namely AirTech Assistant (Aviation Technical Assistant) is considered as a solution.

It is necessary to solve the following tasks:

1. Domain analysis with the construction of conceptual models describing main concepts and relationships.
2. Development of a conception with the formulation of basic principles, namely: requirements for the software, architecture, basic algorithms, and principles of intermodule interaction, as well as conducting exploratory research in terms of selecting the best techniques for the implementation of methods and functions.
3. Detailed design of the system, including its separate modules and subsystems.
4. Step by step software implementation with prototyping and testing of separate subsystems.
5. Debugging, taking into account testing and solving applied tasks.

This paper presents the results of the first two tasks.

4 Domain Analysis and Conceptual Modeling

4.1 Technical Documentation Preview

The analysis of technical documentation is the first step in analyzing the domain and building a conceptual model.

In particular, we analyzed technical operation manuals (OM) and troubleshooting manuals (TM). At the same time, the OMs describe information on maintenance, replacement, adjustment, inspection, and verification of the equipment of aircraft systems. The TMs also contain information about inspections and maintenance of the airframe structure and contains a description of the procedures for scheduled maintenance of the aircraft. In turn, the TM lists possible failures and malfunctions, the sequence of operations, and works to identify and troubleshoot each aircraft system.

In this study, section no. 24 related to the power supply system of RRJ-95 was analyzed, where failures and malfunctions of seven subsystems are considered.

4.2 Basic Domain Concepts and Relationships

Technical diagnostics (diagnosis of technical conditions) is one of the main concepts of the domain. This concept can be defined as the process of determining the technical condition of a product with a certain accuracy, the result of which is a conclusion on the technical condition of the object, indicating, if necessary, the location, type, and causes of defects [17, 18].

Technical diagnostics, in turn, defines such concepts as the object of diagnostics, where the processes due to functioning take place, and the process of technical diagnostics, which is a sequence of operations (works) aimed at solving particular problems of the process (Fig. 1).

We used the ontological template [19] to clarify the object of diagnosis. This template reflects the hierarchical structure: "system—subsystem—sub-subsystem", and after its adaptation allows us to describe the mechanical system in the form of the following structure: "aircraft—aircraft systems—aircraft subsystems—aircraft sub-subsystems". A hierarchy of the "general—specific" type was used to systematize information about the object.

Based on the "general—specific" hierarchy, the "part—whole" hierarchy for the structure of the aircraft was described (Fig. 2), a similar description was later carried out for operations (works) by various classifications. In particular, Fig. 3 is an example of a description of operations (works) according to the classification by a functional code.

The issue of detailing the properties of operations (works) and their relationships for separate systems was investigated too. In particular, troubleshooting work includes preparatory work, confirming the malfunction, etc. (Fig. 4).

Currently, the conceptual model in the form of ontology contains 355 concepts and 1066 axioms.

In the future, we plan to detail the conceptual model by covering other aircraft systems and subsystems. The conceptual model will be used as an initial model for the automated creation of case-based and rule-based knowledge bases for the

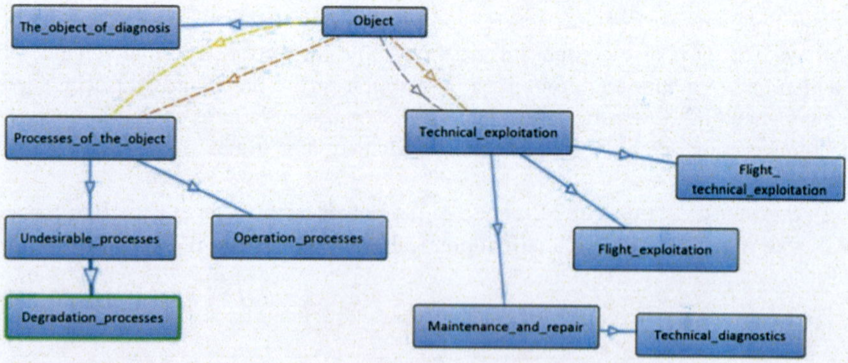

Fig. 1 The relationship between the concepts of the object and the process of technical diagnostics

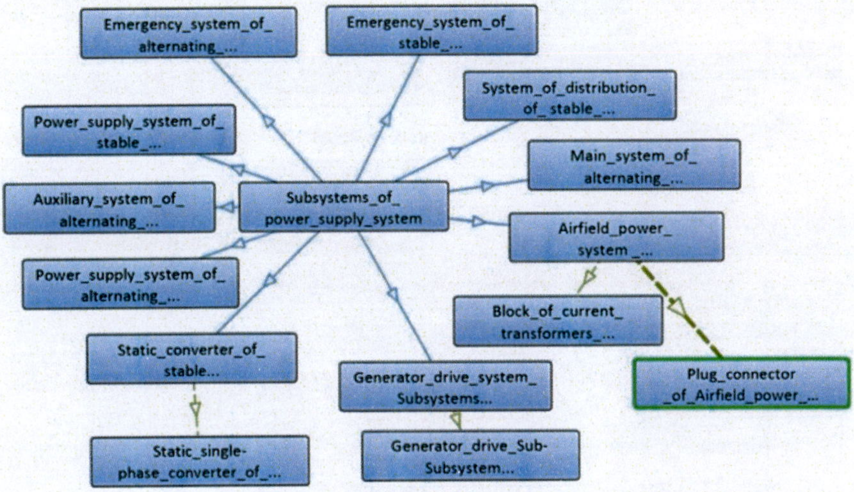

Fig. 2 A fragment of a conceptual model for the RRJ-95 power supply system

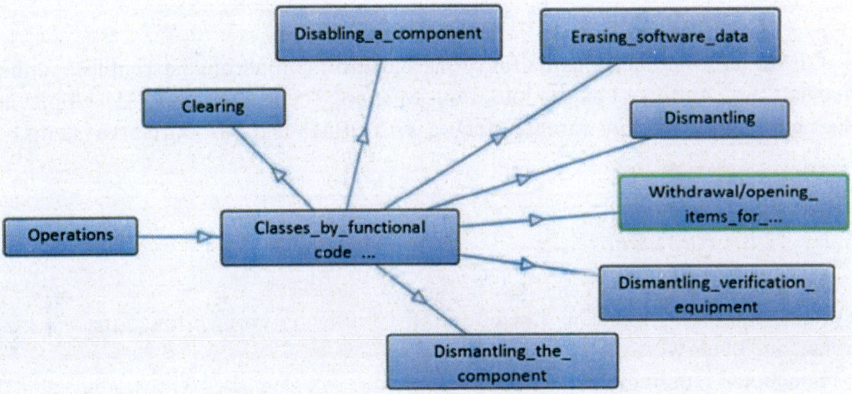

Fig. 3 Classification of operations (works) by a functional code

implementation of decision-making support in the search and troubleshooting of aircraft.

5 Basic Principles of Intelligent System

The intelligent system is designed to support decision-making technical personnel during maintenance and repair of aviation equipment, in particular when troubleshooting.

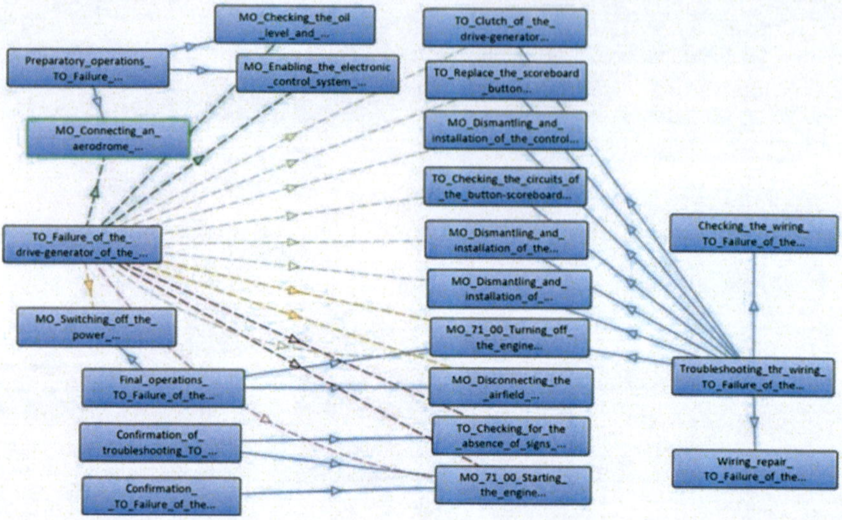

Fig. 4 A detailed structure of the work "failure of the drive-generator of the left engine or the drive-generator and the drive box of the left engine are disconnected"

OBMS information about malfunctions and information from the troubleshooting manuals are considered as the initial information. The system under development does not provide for hardware interfacing with OBMS and other aircraft systems.

5.1 Main Requirements

System requirements can be divided into the following groups: functional, operational, and quality.

Functional requirements:

- Entering, editing, and storing information about aircraft (and its systems), technical operations, malfunctions (failures), and troubleshooting;
- Searching and ranking the possible failed systems on the OBMS information about failures and malfunctions;
- Accumulating the information about failures and malfunctions that are not accounted for by the current version of the documentation (to support the case-based expert system);
- Entering data to the knowledge base of a rule-based expert system;
- Forming planes (schedules) of operations for troubleshooting;
- Providing access for main functions through domain-specific graphic user interface.
- Operational requirements are determined by the client-server architecture of the intelligent system:

- The client part should be multiplatform, i.e. operate under different operating systems;
- The client part should implement the principle of a "thin" client, i.e. it should not require the installation of any additional software;
- The server part must operate under Unix family operating systems;
- The implementation of the backend should provide the ability to scale the application and load balancing;
- A DBMS should be used to store information about new malfunctions and failures in the form of cases;
- Quality requirements determine the completeness, efficiency, and accuracy of the main functions:
- Maximum volume of the base of cases: 10,000 records;
- The time of formation of recommendations based on cases: < 30 s;
- The volume of the base of rules: 10,000 rules;
- The time of forming the logical conclusion of the rule-based expert system: < 30 s;
- The time of forming plans for troubleshooting: < 30 s.

5.2 *The Architecture*

To implement the considered requirements, the following architecture is proposed (Fig. 5), which includes the following main subsystems:

- Storage (database) containing information from operating manuals, maintenance manuals, troubleshooting manuals, and information about an aircraft.
- A case-based expert system for accumulating, searching, and storing information about failures and malfunctions, including:
  - a database (or a case-based knowledge base) with information about failures and malfunctions;

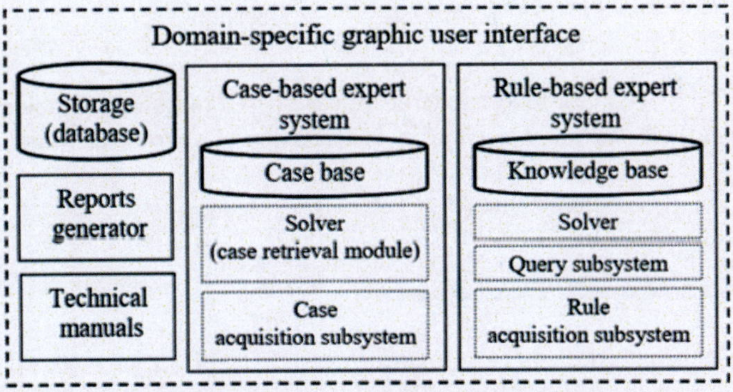

Fig. 5 The principal architecture of an intelligent decision support system

  - a cases retrieval subsystem (solver);
  - a subsystem for the acquisition (accumulation) of knowledge, which provides an extension of the case base.

- A rule-based expert system for defining potentially failed systems and the formation of plans of work on search, verify and troubleshoot failures, including:
  - a knowledge base with information about the failures and task cards are required to maintain, repair, replacement, setup, adjustment, inspection, and control of equipment and systems on the aircraft;
  - a logical inference engine (solver);
  - a query subsystem;
  - a knowledge acquisition subsystem that provides the expansion of the rule base through visual programming and formalisms of event trees, decision tables, and state transition diagrams describing work plans for finding, confirming failures and malfunctions, and troubleshooting.
- The subsystem for generating reports.
- The domain-specific graphic user interface.

5.3 *Basic Algorithms*

The implementation of separate blocks of an intelligent system requires the creation of appropriate algorithms and models. Let's consider the most important of them.

A case model. For their effective processing, cases should be described by a specific model. In most cases the model of cases includes two main parts:

- an identifying (characterizing) part that describes the experience in a way that allows one to assess the possibility of its reuse in a certain task;
- a learning part that describes the solution (decision) of the task or part of it.

Therefore, the case model can be formalized as follows [20]:

$$M^{\text{Task\_CBR}}\colon \text{Problem}^{\text{CBR}} \rightarrow \text{Decision}^{\text{CBR}},$$

where $M^{\text{Task\_CBR}}$ is a task model in terms of case-based reasoning; $\text{Problem}^{\text{CBR}}$ is a task (problem) description, $\text{Decision}^{\text{CBR}}$ is a decision of a problem, while:

$$\text{Problem}^{\text{CBR}} = \langle c^*, C\rangle, C = \{c_1 \ldots c_K\}, c^* \notin C,$$

where c^* is a new case, C is a case base.

$$\text{DecisionCBR} = \{d1, \ldots, dR\}, di = (ci, si), ci \in C, s \in [0; 1]$$

where $\text{Decision}^{\text{CBR}}$ is a task decision in the form of a set of retrieved cases with similarities s_i.

Let's formalize a case description:

$$ci = \left\{\text{Prob}_i^{\text{Problem}}, \text{Prop}_i^{\text{Decision}}\right\},$$

where $\text{Prob}_i^{\text{Problem}}$ is an identifying (characterizing) part of a case, $\text{Prob}_i^{\text{Decision}}$ is a learning part of a case. In addition, each of these parts contains task properties and the composition of the parts has a problem-specific character:

$$\begin{aligned}
&\text{Prob}_i^{\text{Problem}} = \{p_1, \ldots, p_m\}, \text{Prop}_\text{i}^{\text{Decision}} = \{p_{m+1}, \ldots, p_N\},\\
&\text{Prob}_i^{\text{Problem}}, \cup \text{Prob}_i^{\text{Decision}} = \text{Prob}, \text{Prob}_j^{\text{Problem}} \cap \text{Prob}_j^{\text{Decision}} = \emptyset,\\
&p_i \in \text{Prob}, \text{Prob} = U_i p_i, i = [1, N].
\end{aligned}$$

where p_i are task properties (significant characteristics), Prop is a set of task properties.

Thus, we developed the conceptual model mentioned above to define the task properties. Concepts and relationships from this model will be used to fill parts of the case model.

Retrieving cases. The Zhuravlev metric [21] with the normalization will be used for the implementation of a case retrieval procedure:

$$d_G(\overline{x}, \overline{y}) = \sum_{i=1}^{N} w_i h_G(x_i, y_i)/N,$$

$$h_G(x_i, y_i) = \begin{cases} \text{for quantitative} \begin{cases} 1, \text{ if} |x_i - y_i| < \xi \\ 0, \text{ otherwise} \end{cases}, \\ \text{for qualitative} \begin{cases} 1, \ x_i = y_i \\ 0, \ x_i \neq y_i \end{cases} \end{cases}$$

where w_i is the information weight, and ξ is the constraint on the difference between the values of properties. This metric supports processing both quantitative and qualitative values of task properties.

Reusing case decisions. When solving this problem, it is not necessary to change the solutions obtained by analogy, thus we propose to use the so-called "zero adaptation", based on copying parts of the retrieved cases decisions without any modifications.

Logical rules representation. The representation of logical rules and their structure will also be determined based on the developed conceptual model, which can be represented in the form of a UML class diagram (Fig. 6).

Prototypes of the rules were built in [22] and they provide automated formation of operation (work) plans (Fig. 6, "Operation" class) as some scenarios based on information about the signs of a malfunction (Fig. 6, "Malfunction" and "Signs of malfunction" classes).

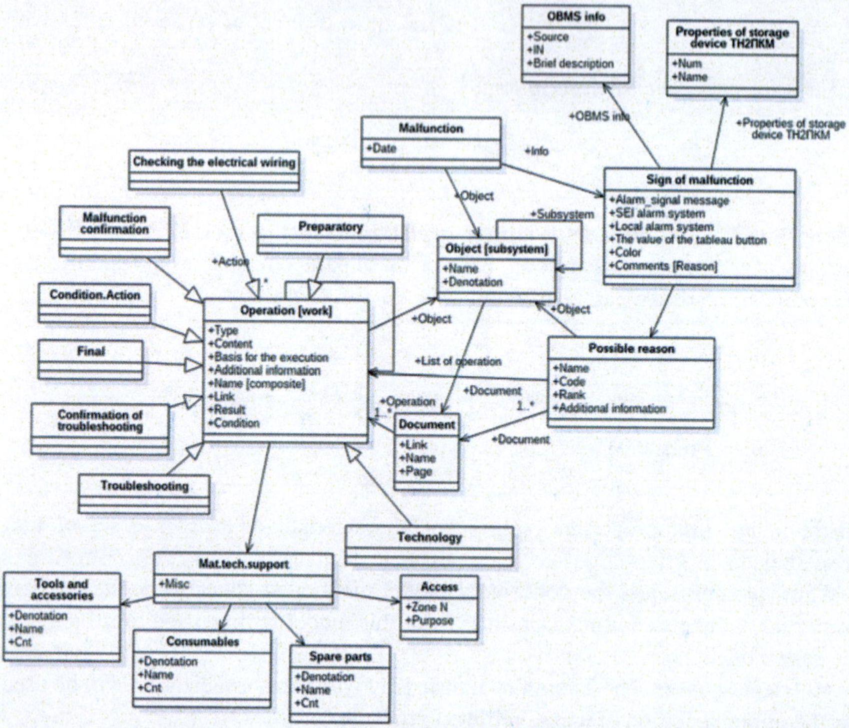

Fig. 6 The conceptual model in the form of a UML class diagram

Knowledge acquisition. Extracting knowledge from experts and filling knowledge bases remains the most difficult tasks in the development of intelligent systems. In case-based reasoning this task is solved by automatically saving information about new failures and malfunctions in the form of cases according to a certain model, in rule-based reasoning, some methods, including those based on EDU, should be used.

As already noted, in the context of the task being solved, among the EDU methods, the most promising is the use of visual programming. In this project, we tested the use of event trees formalism (Fig. 7) [22] for this purpose. This test showed:

- event trees are a handy tool for visualizing scenarios or plans;
- software that provides not only the construction of such models but also the synthesis of codes for expert systems and knowledge bases are not available;
- some difficulties are associated with the inability to create cycles;
- it is promising to test state transition diagrams for this task.

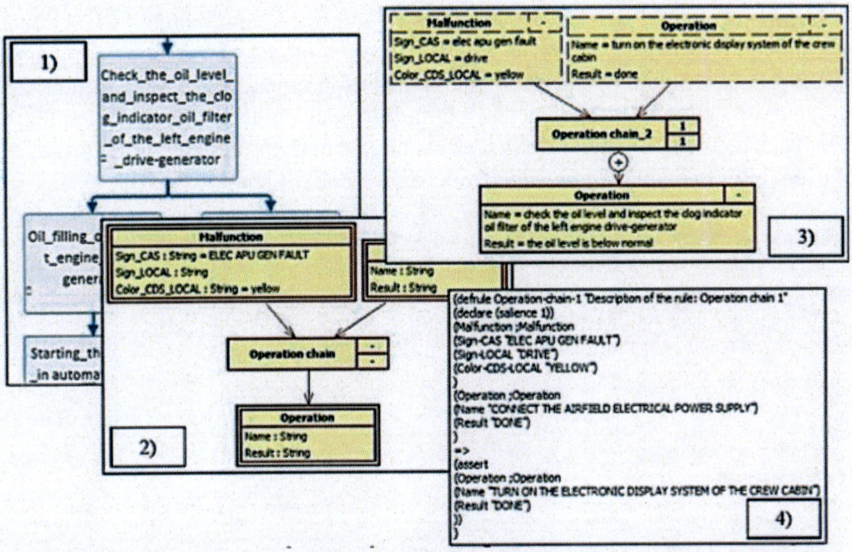

Fig. 7 An example of the use of event trees for the synthesis of knowledge bases: (1) a designed event tree fragment; (2) a logical rule template corresponding to the event tree fragment; (3) a specific rule corresponding to the rule template; (4) obtained CLIPS code corresponding to the specific rule

5.4 Implementation Techniques

Based on the analysis of the requirements for functionality, operating conditions, and quality, the following stack of technologies was formed for the software implementation of an intelligent system: the Yii Framework, PHP, MySQL. The prototypes of knowledge bases will be designed with the aid of the Personal Knowledge Base Designer [23].

6 Conclusion

Troubleshooting an aircraft remains a promising area for automation and intellectualization, especially in the context of big data processing. However, the currently existing systems are mainly electronic technical manuals, the capabilities of which are limited and do not always meet the requirements of technical personnel in the search and troubleshooting of aircraft.

In this regard, the development of another class of systems based on artificial intelligence methods is urgent. These systems can self-learn in the process of their operation.

In this paper, the basic principles of an intelligent system, namely the AirTech Assistant, are proposed. In particular, a conceptual model of the domain has been developed, functional, operational and quality requirements have been formulated, architecture as well as fundamental algorithms and implementation technologies have been defined. A special study was conducted in terms of testing the formalism of event trees to extract knowledge from experts and fill knowledge bases.

Acknowledgements The present study was supported by the Ministry of Education and Science of the Russian Federation (Project no. 121030500071-2 "Methods and technologies of a cloud-based service-oriented platform for collecting, storing and processing large volumes of multi-format interdisciplinary data and knowledge based upon the use of artificial intelligence, model-driven approach and machine learning").

References

1. Aamodt A, Plaza E (1994) Case-based reasoning: foundational issues, methodological variations, and system approaches. AI Commun 7(1):39–59. https://doi.org/10.3233/AIC-1994-7104
2. Jackson P (1998) Introduction to expert systems. Addison-Wesley
3. AirNav-Maintenance. https://www.airnav.com/maintenance/index.html. Accessed 23 Oct 2021
4. Savvina AM (2019) Predlozhenie po modernizacii bortovoj sistemy tekhnicheskogo obsluzhivaniya samoleta SSJ 100 (Proposal for the modernization of the onboard maintenance system of the SSJ 100 aircraft). Crede Experto Transp Soc Educ Lang 3(22):27–35 (in Russian)
5. Perfiliev OV, Ryzhakov S, Dolzhikov VA (2018) Intellektual'naya sistema poiska neispravnosti na samolete (Intelligent system for finding mal-functions an aircraft). Proc Samara Sci Center Russian Acad Sci 4(3):326–331 (in Russian)
6. Bergmann R (2002) Experience management: foundations, development methodology, and internet-based applications. In: Lecture notes in artificial intelligence, p 2432.https://doi.org/10.1007/3-540-45759-3
7. Watson I (1999) Case-based reasoning is a methodology not a technology. Knowl-Based Syst 12:303–308. https://doi.org/10.1016/S0950-7051(99)00020-9
8. De Mantaras LR, Mcsherry D, Bridge D, Leake D, Smyth B, Craw S, Faltings B, Maher ML, Cox MT, Forbus K, Keane M, Aamodt A, Watson I (2005) Retrieval, reuse, revision and retention in case-based reasoning. Knowl Eng Rev 20(3):215–240. https://doi.org/10.1017/S0269888906000646
9. Nikolaychuk OA, Yurin AY (2006) Automating the identification of mechanical systems' technical state using case-based reasoning. In: IEEE intelligent systems. Processing of the 3rd international IEEE conference intelligent systems, pp 30–35. https://doi.org/10.1109/IS.2006.348389
10. Yurin AY (2015) Group decision-making methods for adapting solutions derived from case-based reasoning. Sci Tech Inf Process 42(5):375–381. https://doi.org/10.3103/S014768821505010X
11. Chastikov AP, Gavrilova TA, Belov DL (2003) Razrabotka ekspertnyh sistem. Sreda CLIPS (Development of expert systems. CLIPS). BHV-Petersburg, 608 p (in Russian)
12. Giarratano J, Riley G (2004) Expert systems: principles and programming. https://doi.org/10.5860/choice.27-4542
13. Forgy C (1982) Rete: a fast algorithm for the many pattern/many object pattern match problem. Artif Intell 19:17–37. https://doi.org/10.1016/B978-0-934613-53-8.50041-8

14. Lieberman H, Paternò F, Klann M, Wulf V (2006) End-user development: an emerging paradigm. Human-Comput Interaction Ser 9:1–7. https://doi.org/10.1007/1-4020-5386-X_1
15. Barricelli BR, Cassano F, Fogli D, Piccinno A (2019) End-user development, end-user programming and end-user software engineering: a systematic mapping study. J Syst Softw 149:101–137. https://doi.org/10.1016/J.JSS.2018.11.041
16. Yurin AY (2020) Technology for prototyping expert systems based on transformations (PESoT): a method. CEUR Workshop Proc 2677:36–50
17. GOST 20911-89 (2009) Tekhnicheskaya diagnostika. Terminy i opredeleniya (Technical diagnostics. Terms and definitions). Terms and definitions. Standartinform. Moscow (in Russian)
18. Berman AF, Nikolaychuk OA, Yurin AY, Pavlov AI (2014) A methodology for the investigation of the reliability and safety of unique technical systems. Proc Inst Mech Eng Part O J Risk Reliab 228:29–38. https://doi.org/10.1177/1748006X13494820
19. Zagorulko YA, Borovikova OI (2018) Podhod k realizacii patternov soderzhaniya pri razrabotke ontologij nauchnyh predmetnyh oblastej. Sistemnaya informatika (Approach to the implementation of content pat-terns for the development of ontology of scientific subject domains). Syst Inform 12:27–40. https://doi.org/10.31144/si.2307-6410.2018.n12 (in Russian)
20. Maltugueva GS, Yurin AY (2019) Improving case-based reasoning with the aid of multi-criteria and group decision-making methods. In: Proceedings of the 42nd international convention on information and communication technology, electronics and microelectronics (MIPRO), pp 1031–1036. https://doi.org/10.23919/MIPRO.2019.8756874
21. Zhuravlev IY, Gurevitch IB (1989) Raspoznavanie, klassifikaciya, prognoz. Matematicheskie metody i ih primenenie (Pattern recognition, classification, forecasting: mathematical techniques and their application) 2:5–72 (in Russian)
22. Dorodnykh NO, Kotov YV, Nikolaychuk OA, Popov VM, Yurin AY (2021) End-user development of knowledge bases for semi-automated formation of task cards. CEUR Workshop Proc 2913:60–73. https://doi.org/10.47350/ICCS-DE.2021.05
23. Yurin AY, Dorodnykh NO (2020) Personal knowledge base designer: software for expert systems prototyping. SoftwareX 11:100411. https://doi.org/10.1016/j.softx.2020.100411

Aircraft Electrical Power Systems from the Viewpoint of Requirements of Modern Regulatory Documents

Sergey Mishin and Vladimir Popov

Abstract It is shown that the currently operating generation systems based on the drive-generator units of various types have achieved the best technical characteristics. Systems of the "variable speed—constant frequency" type are promising, but their implementation is strongly limited by the level of development of the domestic semiconductor element base. The implementation of the concept of an all-electric aircraft has now become important. This required not only a significant increase in the installed capacity of the aviation power supply system but also changes in the requirements of regulatory documents for voltage level and electric energy quality. In fact, these documents lay down the requirements for the structure of the onboard equipment complex. The emphasis in the work is made on the structural features of building an electrical power system for aircraft, taking into account the requirements of the modern regulatory documents.

Keywords Generating system · Electrical power system · Constant electric frequency · Constant speed drive · Electromechanical generation system · Variable electric frequency · All-electric aircraft · Modern regulatory framework

1 The Problem Statement

On board most modern aircraft, the generation of electrical energy of alternating current of constant frequency is carried out by a generation system. In such a system, a special device is located between the aircraft engine (ENG) and the synchronous generator (G). This device converts the variable speed of ENG rotation into a constant frequency of rotation of the generator. Such a device is called a constant speed drive (CSD). It provides a constant electrical frequency at the output of the generator. The

S. Mishin (✉) · V. Popov (✉)
Irkutsk Branch of Moscow State Technical University of Civil Aviation, Irkutsk, Russia
e-mail: sv_mishin@mail.ru

V. Popov
e-mail: povlamix@mail.ru

O. A. Gorbachev et al. (eds.), *Proceedings of 10th International Conference on Recent Advances in Civil Aviation*, Lecture Notes in Mechanical Engineering, https://doi.org/10.1007/978-981-19-3788-0_8

generator and the constant speed drive in the modern version are known as integrated drive-generator (IDG) (see Fig. 1) [1].

Most modern aircraft and helicopters use a 400 Hz alternating current electrical power system, based on pneumomechanical and hydromechanical IDG types. As an example, the structure of the electrical power system of the Airbus A320 aircraft is presented below (see Fig. 2). The electrical power systems of Boeing concern aircraft are built in the same way.

The new domestic Sukhoi Superjet 100 airliner (the international abbreviation SSJ-100, a type certificate was obtained for an aircraft with the RRJ-95 abbreviation which stands for Russian Regional Jet) was no exception.

However, the technical characteristics of various IDG types have reached their best values and this makes them unpromising. In this regard, the problem of developing fundamentally new ways of generating alternating current electrical energy of a constant frequency becomes urgent.

The most promising way to solve this problem is the development of "variable speed—constant frequency" systems with static frequency converters of the cycloconverter type or electromechanical generation systems based on cascade

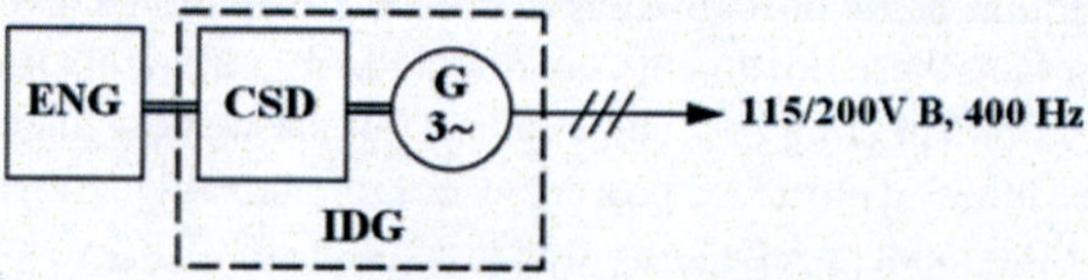

Fig. 1 A block diagram of a generation system with the integrated drive-generator

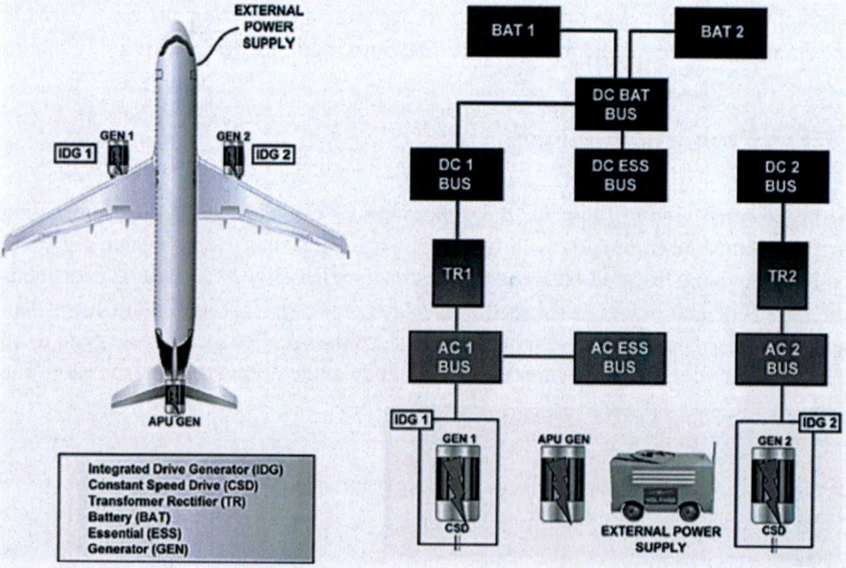

Fig. 2 A power supply system of the Airbus A320 aircraft

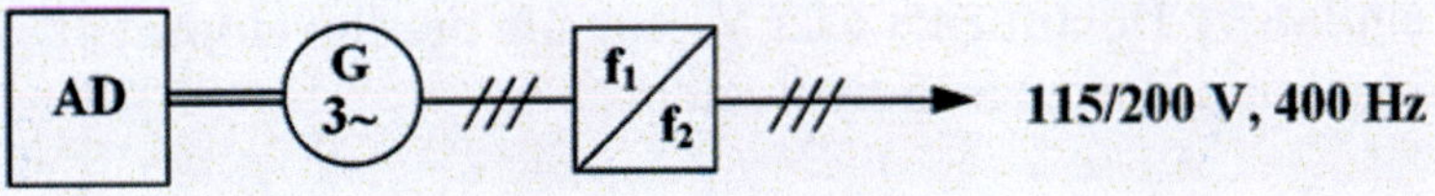

Fig. 3 A block diagram of the "variable speed—constant frequency system"

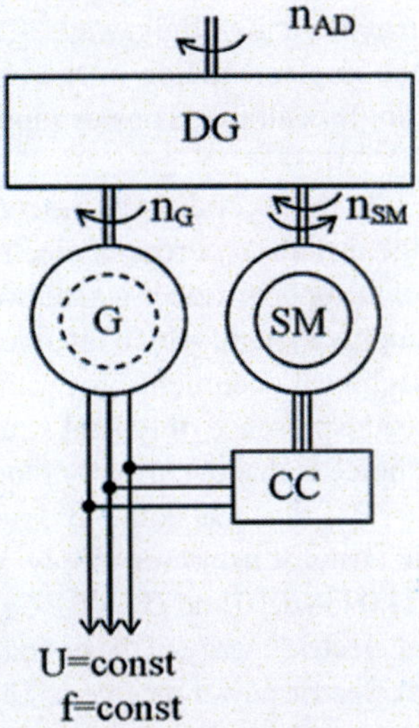

Fig. 4 A block diagram of an electromechanical generating system: DG—differential gear; SM—support machine; CC—control device

connections of synchronous and asynchronous electric machines (see Figs. 3 and 4) [2–4].

Electromechanical systems' technical perfection, relatively low cost of production and operation, and are distinguished by very high reliability. However, such systems implement the already well-known principle of electric frequency stabilization by stabilizing the rotation speed of a synchronous generator.

There is no doubt that the future belongs to systems of the "variable speed—constant frequency" type. They contain a minimum number of mechanical moving parts and allow for phase-by-phase regulation of the output voltage. However, mass production of such high-power systems is difficult. The reason is that the power conductors must simultaneously meet the following requirements: operating voltages up to 400 V, currents up to 100 A. The switching frequency should be up to 10^{-6} s [4–8].

2 Regulatory Documents as a Vector for the Development of Aviation Power Supply Systems

All of the above is completely true for the structure of the construction of the power supply system of a modern aircraft. However, the reality is that the power of the onboard complex of electrified equipment is steadily growing and already reaches units of gigawatts. This has become especially relevant in connection with the implementation of the concept of a so-called all-electric aircraft. It assumes a single centralized power supply system that provides the energy needs of the aircraft [9–15].

In these conditions, the requirements increase, both in terms of the total power of the installed sources of electrical energy and an increase in the supply voltage level of the onboard electrical network. In addition, the design of the structure of the power supply system, which involves the designer's knowledge of dozens of normative and technical documents governing the design stages of power supply systems, as well as the presence of typical structures of power supply systems, from which a specific choice is made, currently requires its maximum automation.

The time has come to optimize the structure of the onboard equipment complex in terms of its power supply. So before the introduction of the new Russian GOST P 54,073–2010 and GOST P 54,073–2017, the regulatory requirements for the quality of electric energy of the onboard power supply system were based on the requirements of electric power receivers. Therefore, to determine the range of permissible electrical frequencies in the steady-state operation of the onboard power supply system, it was customary to divide all receivers into three groups:

- Receivers, the operation of which does not depend on the frequency of the current of their power sources (electric heating and anti-icing equipment, lighting and light-signaling equipment on incandescent lamps);
- Receivers, allowing deviations of the supply current frequency within $\pm$ 5% (asynchronous motors, transformer-rectifiers, etc.);
- Receivers, requiring precision accuracy of stabilization of the power frequency, the error of which does not exceed $\pm$ 0.05% (calculating devices, navigation systems, etc.).

Since most alternating current receivers operate satisfactorily with a frequency deviation of $\pm$ 5% from the nominal level of 400 Hz, this range is adopted in regulatory documents as the standard for a steady-state frequency value at the nominal or partial operation of the power supply system.

At present, it is understood that it is the regulatory documents that should establish the values and permissible ranges of deviation of the parameters of electrical energy in the onboard network from the nominal values.

In addition, the time has come to optimize the structure of the onboard equipment complex in terms of its power supply. The approach, when the needs of receivers of electrical energy were decisive and formed the basis of regulatory documents, must go away. It is the regulatory framework that should establish the values and permissible ranges of deviation of the parameters of electrical energy in the onboard network from the nominal values/

According to the new regulatory documents, the following types of power supply systems are allowed on modern aircraft:

(a) Alternating three-phase current with a rated voltage of 115/200 V constant rated frequency of 400 Hz;
(b) Alternating three-phase current with a rated voltage of 230/400 V constant rated frequency of 400 Hz;
(c) Alternating three-phase current with a rated voltage 115/200 V variable frequency 360–800 Hz;
(d) Alternating three-phase current with a rated voltage 230/400 V variable frequency 360–800 Hz;
(e) Direct current with a nominal voltage of 27 V;
(f) Direct current with a nominal voltage of 270 V.

Only half (a, c, e) of presented types of electrical energy were used earlier. Moreover, the use of a low DC voltage of 27 V causes a high mass of electrical machines and distribution systems. Therefore, this type of electrical energy should be considered unpromising.

Three-phase AC voltage of 36 V of constant nominal frequency of 400 Hz has already become unacceptable on newly developed types of aircraft.

In (b, d, f) items, new types of electric energy are indicated. They are the future and there are several reasons for this.

The first one is an increase relative to the operating voltage level (230/400 V versus 115/200 V for alternating current and 270 V versus 27 V for direct current). Therefore, there is a proportional gain in the estimated mass of electric machines and distribution systems.

The second reason is that the high level of DC voltage of 270 V is a breakthrough from the point of view of the possibility of introducing a power electric drive on board an aircraft based on contactless direct current electric motors. This type of drive is characterized by good adjustment characteristics and the relative simplicity of its digital control.

A separate important category should include the (d) item a three-phase alternating current power supply system with a rated voltage of 230/400 V of variable frequency 360–800 Hz. This is due to a radical restructuring of the structure not only of the SES itself but also of the entire complex of onboard equipment. Let's consider this approach on the example of a promising Russian domestic aircraft MS-21.

It is assumed that the main sources of electrical energy on the MS-21 will be contactless generators L GEN and R GEN (see Fig. 5). They are driven directly by the main aircraft engines. Taking into account the wide range of operating speeds of rotation of these motors, the range of electric frequency variation will be 360–800 Hz.

The 27 V direct current power supply system is secondary. The main sources in it are two transformer-rectifiers L TRU and R TRU.

The MS-21 power supply system fully meets the requirements of the modern regulatory documents. The power supply system, both for alternating and direct

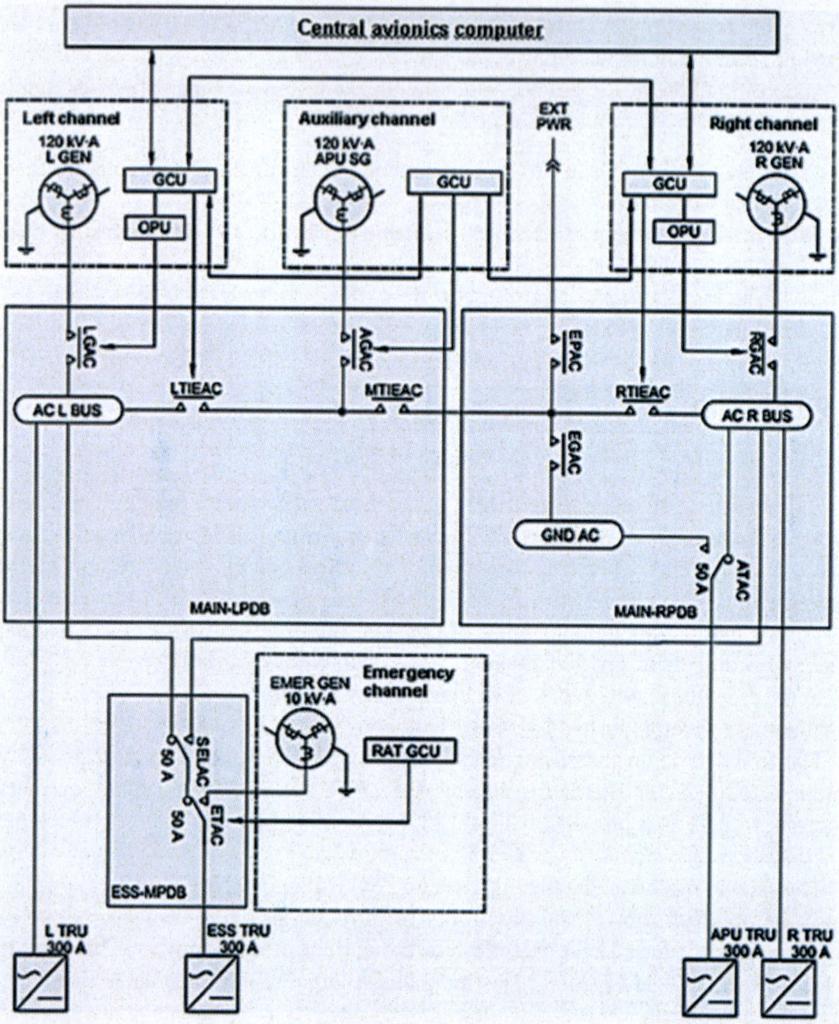

Fig. 5 A scheme of the MS-21 aircraft power supply system

current, is two-channel with a separate operation of the sources of each of the channels for its group of electric energy receivers. When disconnecting a faulty source, it is provided that its receivers are connected to the source of a working channel.

In the case of a failure of all main sources of electricity, the power supply to the receivers of the first category can be provided in the process of lowering to the starting height of the auxiliary power plant and starting it by batteries or a generator driven by the EMER GEN turbine released into the airflow.

3 Conclusion

More recently, manufacturers of aviation equipment not only set requirements for the quality of electrical energy but also, in fact, for the very structural construction of the onboard power supply system. Currently, there has been a radical change in the role of the regulatory documents. They began to determine the prospects for the development of the onboard equipment complex. It is the regulatory documentation that specifies the values of voltage and electrical frequency, their permissible ranges of deviation from the nominal values. The old ones go down in history and new, promising voltage and frequency levels come.

References

1. Khalyutin SP, Zhmurov BV, Tyulyaev ML (2010) Sistemy elektrosnabzheniya letatel'nyh apparatov. (Aircraft power supply systems). Moscow: VUNTS VVS "VVA", pp 428 (In Russian)
2. Krasnoshapka MM (1974) Generators of stable and adjustable frequency. Kiev: Technika, pp 168
3. Mishin SV, Mishina IV (2012) Features of work processes in electromechanical systems for generating alternating current of constant frequency. Scientific bulletin of the Moscow state technical university of civil aviation. 185:138–145
4. Mishin SV, Golovanov IG (2007) Problems of generating AC electric power of constant frequency on modern aircraft. Scientific bulletin of the Moscow state technical university of civil aviation. 115:132-135.
5. Levin AV, Khalyutin SP, Davidov AO, Zhmurov BV, Kharitonov SA, Zharkov MA, Kharitonov AS (2017) Starter-generatornaya sistema dlya vspomogatel'noj silovoj ustanovki. (Starter-generator system for auxiliary power unit). Civil aviation high technologies. 20(5):50–66. (In Russian). https://doi.org/10.26467/2079-0619-2017-20-5-50-66
6. Xiuxian X (2011) Dynamic power distribution management for all electric aircraft. Cranfield University, pp 114
7. Shenai K, Scott RS, Baliga BJ (1989) Optimum semiconductors for high power electronics. IEEE Trans Electron Devices 36(9):1811–1823
8. Makarov DV, Khlebnikov AS, Geist AV, Bachurin PA (2011) Generation system with variable frequency and constant amplitude. Energetics (IYCE). In: Proceedings of the 2011 3rd international youth conference, 7–9 July 2011, pp 1–9
9. Levin AV, Khalyutin SP, Zhmurov BV (2015) Trends and prospects of development of aviation electrical equipment. Scientific bulletin of the Moscow state technical university of civil aviation. 213:50-57
10. Levin AV, Musin SM, Kharitonov SA, Kovalev KL, Gerasin AA, Khalyutin SP (2014) Electric aircraft: concept and technologies. Ufa: UGATU, pp 388
11. Voronovich S, Kargopoltsev V, Kutakhov VP (2009) Fully electric airplane. Aviapanorama 2:14–17
12. Masson PJ, Brown GV, Soban DS, Luongo CA (2007) HTS mashines as anabling technology for all-electric airborne vechicles. Supercond Sci Technol 20:748–756
13. Ganev E (2006) High performance electric generators for aerospace more electric architectures. Aerospace engineering & technology. Honeywell international

14. Kivokurtsev AL (2019) (Problems and prospects of using integrated modular avionics in onboard equipment complexes. In collection: Actual problems and prospects for the development of civil aviation. In: Proceedings of the VIII All-Russian scientific and practical conference with international participation. October 14–16, 2019—Irkutsk: IF MSTU CA, pp 62–67
15. Zhmurov BV (2018) Process proektirovaniya sistem elektrosnabzheniya vozdushnyh sudov kak ob"ekt avtomatizacii. (Aircraft power supply system design process as an automation object). Civil aviation high technologies; 21(1): 88–103. https://doi.org/10.26467/2079-0619-2018-21-1-88-103. (In Russian)

Development and Adaptation to the Application of Methodological Tools for Assessing the Quality Level of Experts System

Ljudmila Bolshedvorskaya and Nikolay Koryagin

Abstract The criterion for the quality level of the formation and development of professional competencies of aviation specialists is a set of characteristics that reflect their professionalism, the ability to acquire knowledge, apply the acquired knowledge in the profession. In the scientific literature, there are various interpretations of the concepts of "competence," "competency." In almost every interpretation of domestic and foreign authors, it can be noted that, unlike the term "qualification," competencies reflect not only the level of professional knowledge and skills but also such qualities as initiative, the ability to work in a team, communication skills, the ability to learn, evaluate and to think logically, select and use information, etc. Previous studies have emphasized the difficulty of obtaining quantitative values of the level of development of individual competencies since the applicability of specific statistical data is very limited. This creates objective difficulties for assessing the professional preparedness of personnel through the use of the achievements of scientific and technological progress in the field of artificial intelligence. This determines the direction of the research, the main purpose of which is the adaptation and application of methodological tools for assessing the quality level of expert systems. *Publication was prepared within the framework of project No. 19-08-00028, supported by a grant from the Russian Foundation for Fundamental Research (RFFI).

Keywords Competencies · Competence · Expert systems · Safety inspectors

1 Introduction

In recent years, "quality of education" and "expert systems" are mentioned in parallel with such concepts as "competence and competency," which have become a complex

L. Bolshedvorskaya (✉) · N. Koryagin (✉)
Moscow State Technical University of Civil Aviation, Moscow, Russia
e-mail: l.bolshedvorskaya@mstuca.aero

N. Koryagin
e-mail: n.korygin@mstuca.aero

O. A. Gorbachev et al. (eds.), *Proceedings of 10th International Conference on Recent Advances in Civil Aviation*, Lecture Notes in Mechanical Engineering, https://doi.org/10.1007/978-981-19-3788-0_9

concept to include three groups of characteristics: the quality of the potential for achieving educational goals; the quality of the process of forming professional skills and abilities; the quality of educational outcomes.

Competence means a professional area in which a specialist is well aware and shows preparedness for professional activity.

Competency means knowledge, skills, and abilities that reflect personal qualities in combination with the outcomes of professional personnel training in the chosen specialty and field of activity.

As part of the study and previously published works, individual issues related to the use of expert systems were formulated, and solutions to them were proposed [1]. Nevertheless, the issue of assessing the quality level of expert systems remains insufficiently studied, which is very relevant within the framework of the Russian Foundation for Basic Research (RFBR) grant "Development of the Concept for Building the Architecture and Composition of Algorithms for an Expert System Aimed at Improving the Efficiency of Training for Civil Aviation Safety Supervisors."

In this regard, the goal of this study is to develop recommendations for the use of mathematical tools for assessing the quality level of expert systems.

2 Research Tools and Methods

To achieve this goal, a scheme has been developed to reflect the interaction of the main components of the characteristic processes of training specialists for civil aviation—the formation of basic professional competencies and their development in work environment, including the solution of the following issues:

- development of a list of competencies, indicating the stages of their formation, which is one of the main components of the formation of a set of common cultural and professional competencies;
- development of training modules for assessing professional competencies, which are involved in the second important component associated with assessing competency, depending on the formation and development of professional qualities using expert systems;
- development of models and methods for assessing the quality level of expert systems in order to monitor the level of development of professional competencies (Fig. 1).

Competency as an educational goal has gained particular relevance since its inclusion in the Program for International Student Assessment (PISA). [2, 3]. The authors of the study emphasize that against the backdrop of the increased interest of the scientific community in the problem of assessing competency, there is a number of poorly studied issues that require additional research. For this purpose, the study analyzes the prerequisites, approaches, and theoretical aspects of concept building with a special emphasis on the intercultural educational model of competency. At

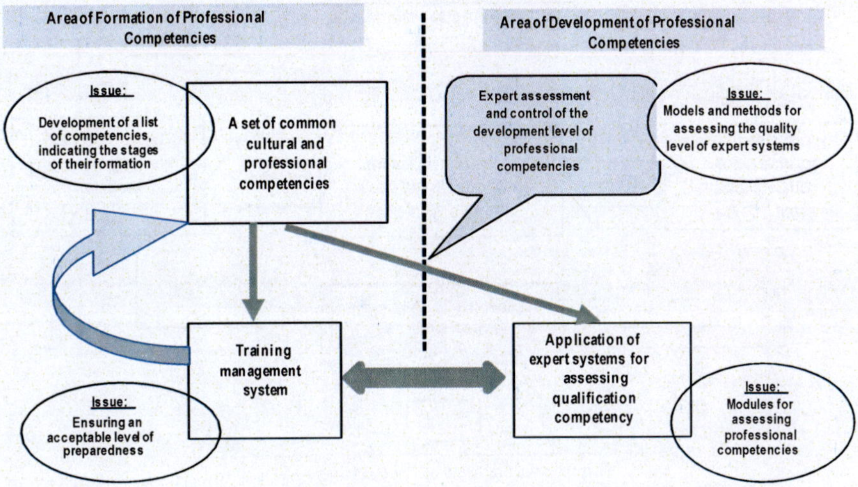

Fig. 1 Place and role of the quality of expert systems in the formation and development of professional competencies

the same time, it emphasizes the complexity of solving this problem, due to the lack of effective tools for assessing competencies at the international level.

Researchers from the University of Gothenburg have demonstrated an approach to solve the problem of assessing the professional competency of students through tests using simulators [4, 5]. The reliability of the data obtained is quite high since the methodological research is based on video materials and the results of the experiment. Using the method of recorded conversation, the study involved an in-depth analysis of the interactions within the investigated dialogues between students and trainers. As part of the study, interest in the results obtained is contingent on the fact that interactive sequences are introduced into assessment methods, including both training and grading.

The importance of an integrated approach to the formation of competency is emphasized in hazards and disaster researchers [6]. The four-step procedure set out in the study helps researchers move from cultural awareness to cultural knowledge, cultural sensitivity, and, ultimately, cultural competence. The time and effort spent on developing cultural competencies can improve the quality of creation and mobilization of knowledge.

In order to create a system for assessing the quality of expert systems in the process of managing personnel training, it is necessary to take a fresh look at the formation of the necessary competencies and the assessment of their development, which, in addition to their main purpose, namely, modeling the management system of the educational process, would ensure:

- increased efficiency in the process of recruiting and selecting new employees;
- increased control over the development of professional competencies;
- diagnosis of the problem areas in the training of aviation personnel;

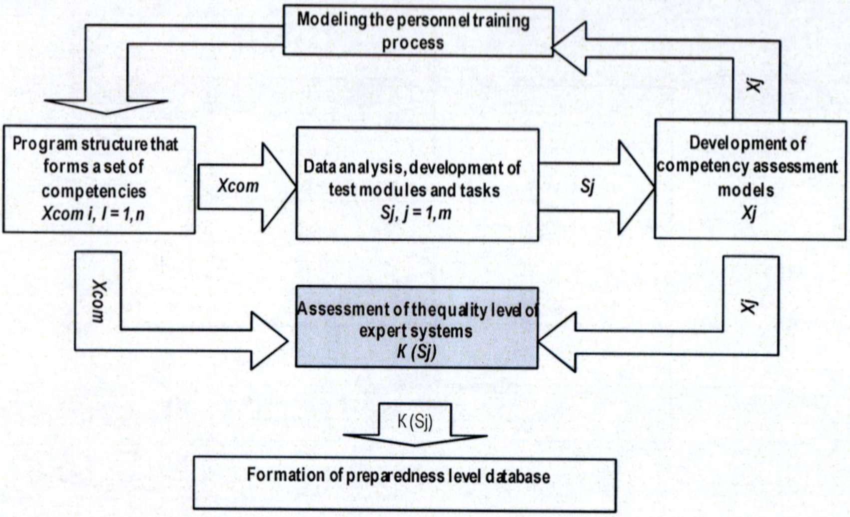

Fig. 2 Algorithm for assessing the quality level of expert systems

- increased efficiency of training and professional development;
- succession planning, etc.

In this case, by studying, accumulating, and combining information about competencies and assessing their development, it is possible to create a new approach to modeling the personnel training process using expert systems, eliminating existing contradictions and difficulties (Fig. 2).

In this regard, it is suggested to present the set of competencies for the training of flight safety supervisors in the form of generalized groups and subgroups. Table 1 illustrates this idea, providing an example of groups of basic professional competencies of flight safety supervisors, as well as subgroups with detailed description of achievement index. Achievements in each subgroup are assessed by the number of points depending on test and case study results. To assess an achievement index, such as the level of mastering working skills in the field of regulatory control supervision, the relative score between the number of $AI1_{(PC1)}$ points and the total number of points in the group of competencies (PC_1) should be calculated. To assess the level of mastering a specific competence, for example (PC_2), the ratio of the total number of points in this group to the total number of points received for all competencies is calculated. This example shows the total number of points (PC_1 and PC_2).

This approach allows calculating a score of the level of development for each competence, taking into account the index of mastering necessary skills and use the results obtained to assess the level of quality of expert systems.

To solve the key issue of this study—the development of a mechanism for assessing the quality level of expert systems based on the adaptation of existing mathematical approaches—the analysis of existing methods and scientific achievements has been carried out.

Table 1 Example of grouping basic professional competencies for flight safety supervisors

Groups of competencies		Subgroups of competencies with a description of achievement index	
(PC_1)	Proficiency in flight safety management methods	$AI1_{(PC1)}$	has mastered the skills of work in the field of regulatory control supervision
		$AI2_{(PC1)}$	is able to plan, conduct, and analyze inspections using an approach based on international standards and recommended practices
		$AI3_{(PC1)}$	is able to identify significant deficiencies depending on the operator's flight safety risk factors
		$AI4_{(PC1)}$	understands the difference between compliance-based supervision and performance-based supervision
(PC_2)	Decision-making skills needed to make judgments based on the information available	$AI1_{(PC2)}$	is able to critically and accurately analyze trends, problem situations, and issues
		$AI2_{(PC2)}$	is able to use logic and analysis to draw appropriate inferences based on relevant information and assumptions
		$AI3_{(PC2)}$	is able to exercise judgment, intelligence, and discretion when making decisions
		$AI4_{(PC2)}$	has skills to identify alternative solutions and their consequences
		$AI5_{(PC2)}$	has the skills to manage emotions and perception problems to ensure objectivity in decision-making in stressful situations

3 The Investigation Results and Their Discussion

3.1 Selection and Justification of a Model for Assessing Professional Competencies of Flight Safety Supervisors

Ranking methods are common methods for assessing training outcomes. One of the methods that has proven its practical application in the field of assessing the quality of training is the Kendall method [7].

The essence of this method is to calculate the rank correlation coefficient, which characterizes the increase or decrease in the effective feature with the increase in the factorial one. For example, using this approach, it is possible to calculate a coefficient that characterizes the degree of relationship between test results. Another method, which is similar to the previous one in its functional purpose, is the calculation of Spearman's rank correlation coefficient [7]. The complexity of justification of the factors affecting the quality of expert systems and expression of the desired outcomes limits the use of such approaches within the framework of the issue at hand.

As an alternative, a method of integral assessment of the quality level of technical products can be considered [8–11].

The applicability of this method is explained by the fact that the final indicator of the quality level can be not only an integral indicator but also a complex one, summarizing several different indicators. Since the expert system relates to intelligent computing systems that include the knowledge of specialists, it is characterized by the following:

- the availability of the results of theoretical and practical experience in a specific area;
- the ability to perceive opinion and reasoning with limited initial data;
- the ability to logically construct the results of judgments;
- the ability to adapt to changes in scientific and technological progress;
- the ability to adapt the user's thinking mechanism to solving specific issues;
- the ability to draw certain conclusions based on the knowledge possessed;
- the ability to formulate specific answers or recommendations;
- the ability to show high performance level.

Therefore, one part of the indicators characterizing the quality of the expert system can have clear quantitative values based on the principles of their measurement or statistical data. Another part of the indicators implies some uncertainty and can be estimated through the use of probabilistic and expert assessments [12].

In recent years, researchers have paid increased attention to the problem of collective decision-making associated with uncertain assessments of preferences [13]. To solve this problem, the authors of the study propose to use the matrix of group preferences based on the ranking of individual preferences. The relevance of the results obtained and the practical applicability of ranking alternative assessments by using the Bradley–Terry–Luce model is confirmed by experimental results and illustrated through a specific numerical example. This study stands out in that it opens up new possibilities for solving the problems of aggregation of preferences when making group decisions using elements of the theory of probability, thereby providing a solid theoretical basis, as well as reliable statistical interpretation for the generalized assessment of expert opinions.

In addition, the method for assessing the quality of expert systems should take into account the importance of certain indicators in the time or the field of application of these expert systems. On these grounds, the following approach is proposed to be a methodological tool for assessing the quality level of expert systems:

$$\kappa(S_j) = A_x(y_i) * B_x(y_i)/B_x(y_i) \tag{1}$$

where $K(S_j)$ is an integral indicator of the quality of expert systems; $A_x(y_i)$ is the degree of availability of the j-th characteristic of the expert system; $B_x(y_i)$ is the degree of importance of the j-th characteristic of the expert system for the development of the necessary competencies.

An important aspect of this approach is obtaining reliable information, especially when assessing the degree of importance of the characteristics of the expert system in specific conditions. If there are several experts, the assessment of the degree of importance of the j-th characteristic can be subjective. Therefore, the opinion of each expert should be taken into account with a certain weighting factor, reflecting the weight of every participant of the expert group. In this case, the sum of the weights must be equal to 1. Then, $B_x(y_i)$ will have the following form:

$$M_j^0 = 1/n * K_{i\ Bx(yi)} \tag{2}$$

where $M^0{}_j$ is an average assessment of the degree of importance of the characteristics of expert systems; n is the number of experts.

3.2 *Application of Methodological Tools for Assessing the Quality Level of Expert Systems*

As the evidence base of the proposed method, the study considered its application to assess the professional suitability of aerodrome operators. The choice is conditioned by the importance of the human factor in the technically dominant field of flight safety, which requires considering issues of flight safety and aviation security using a system approach [14, 15]. The high level of relevance of study results obtained by the authors is the identification of a clear relationship between the practice of austerity in terms of human resource management, the state of working conditions and risks to flight safety and aviation security in commercial aviation. The conclusion was made on the basis of the results of a survey of a group of aviation experts and the development of a model based on the least squares method, which was tested to confirm that the formation of qualification requirements for the quality of training and human resource management indirectly increase the risks in flight safety.

In this regard, the study developed and carried out the specification of the qualification competencies of the operating personnel ensuring flight safety in the aerodrome area in Table 2.

The operating personnel includes specialists that have qualification competencies in the field of operational, rescue, firefighting, and technical procedures.

The formation of professional competencies of operating personnel is aimed at developing responsibility for ensuring FS, as well as identifying HF and voluntarily submitting reports about them.

The term "professional competency" means the suitability of the specialists, managers, and responsible employees of the aerodrome for the performance of their

Table 2 Specification of the qualification competencies of the operating personnel ensuring flight safety in the aerodrome area

Topics for Developing Qualification Competencies	Categories of Specialists			
	Responsible Manager	Top Management	Middle-Ranking Managers	Operating Personnel
Flight safety (FS) policy and objectives	■	■	■	■
Decomposition of duties and responsibilities	■	■		
Identification of hazard factors (HF)		■	■	■
HF consequences and risks for FS		■	■	■
FS risk management process		■	■	■
Analysis of FS and AS (aviation security) data		■	■	
Establishing an acceptable level of FS		■		
Systems providing information about FS		■	■	■
Popularization of FS issues		■		

duties. This can be achieved by fulfilling the necessary conditions, such as completing the required training; obtaining a diploma or degree; gaining relevant experience. In addition, professional suitability reflects the abilities, aptitude, knowledge, or skills that correspond to the position held or can qualify the job seeker as a suitable candidate for the performance of certain duties, holding of the position.

Certain positions may in nature be associated with a specific skill in a specific field (e.g., rescue and firefighting, civil, mechanical or electrical aerodrome engineering, fauna biology, etc.).

In such cases, the person holding the position is expected to have the required qualifications corresponding to the applicable national legislation, training programs, and testing of professional suitability.

The training program for aerodrome flight safety personnel should involve all personnel, which can be divided into two classification groups. The first group consists of people directly involved in the operation, maintenance, and management of the aerodrome, such as inspectors, managers, senior managers, and responsible managers. The second group includes personnel working unaccompanied in the traffic area and other operational areas of the aerodrome, regardless of the position they hold in the company.

The training program for the first group specialists should include specifications for knowledge, skills, and abilities that correspond to the responsibilities of this specialist and the level of its participation in ensuring FS. For this personnel category, tasks that develop human and organizational factors are of great importance during training. A distinctive feature of training programs for personnel included in the second group, i.e., personnel operating or providing services at the aerodrome, is the inclusion of the relevant rules for the flight safety reporting system, the aerodrome flight safety program, etc. The general approach to program development is reflecting the following procedures:

- establishing training standards, including the schedule and frequency for each type of training;
- assessment of training efficiency;
- conducting initial training for each specific job;
- consolidation of theoretical knowledge by testing them in the working environment;
- carrying out recurrent training and advanced training.

Testing for each training course should be carried out in a manner consistent with that training element. Moreover, training elements that require personal participation can be combined with practical tests.

Testing of professional suitability should confirm that personnel are familiar with the policies and procedures that are relevant to their duties.

Given the dynamic processes in place in the industry, recurrent training and advanced training should be carried out at intervals not exceeding 12 months from the date of the initial completion of training program. In the event that the recurrent training is carried out within the last 3 calendar months of the 12 month period, the new period shall be counted from the moment of completion of the last training.

The advanced training procedures may be associated with significant interruptions in specialist's work or the transition to the autumn–winter or spring–summer periods of the aerodrome operation.

Recurrent training should be determined, as appropriate, based on a comparison of the required training program and the training program already completed by the relevant personnel, taking into account the previous training of the personnel.

Recurrent training programs may differ depending on the categories of specialists to which they apply and are determined based on a comparison of the training completed by the relevant person and the training program to be completed by the person to hold a particular position.

Methods used to conduct testing for students may include: assessment of theoretical knowledge using a computer; practical part; oral or written tests, or a combination of such methods. In this regard, the study analyzed the existing systems for checking the level of knowledge, which are based on the principle of an interactive system.

Currently available and free e-learning systems include: Atutor, Ilias, and Diskurs. The common disadvantages of these systems are as follows: lack of a ready-made solution, lack of technical support, which entails the need for additional costs for the development and adaptation of systems to perform specific tasks. Therefore, many users prefer paid versions that meet the following criteria:

- availability and completeness of information about the system;
- ease of use and possibility to promptly change platform;
- support of materials in the format of educational content;
- availability of tools for the development of educational content (courses, tests);
- reporting principles;
- formation of the structure of users (by groups, years of study).

The system adapted for application in the format of the issue under consideration is the Moodle system, which is known to be a learning management system and has become widespread in recent years since it provides an Internet space for students and teachers to work together. With its flexible interface, the system provides the possibility of configuring layouts and design of individual pages, including tools for communication, collaborative work, and control of training and testing stages (Fig. 3).

Files in the form of texts, images, video, and audio are uploaded into the Moodle repository, where they are used to build up lecture course, as well as tasks for intermediate testing. If a student fails intermediate tests on key course topics and does not gain the required number of points, the system limits its access to subsequent topics.

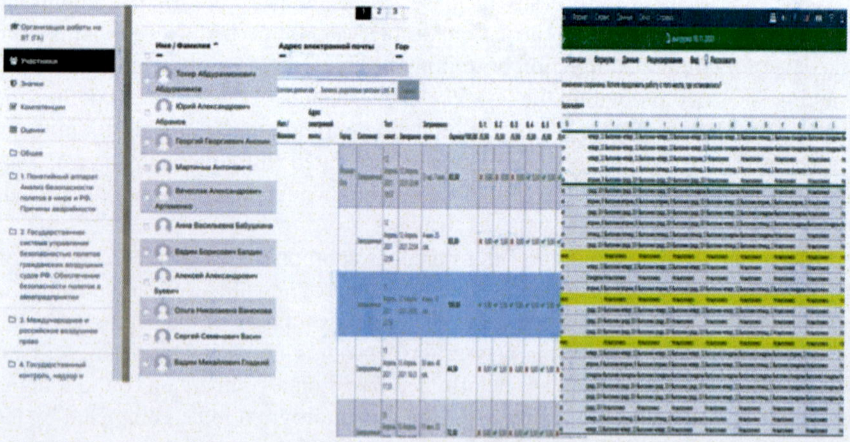

Fig. 3 Work control in the learning management system

Competency is assessed through viewing learning outcomes and built-in analytics system. Therefore, the procedures that are included in every professional competency test should include the following steps.

Step 1. Designation of trainers and examiners who will be involved in the training and professional aptitude testing programs.

Step 2. Providing theoretical briefing by qualified instructors whose theoretical knowledge and work experience correspond to the field of training.

Step 3. Checking and assessing the competency and practical skills of personnel by examiners who are able to assess test results in a qualified manner.

4 Conclusion

As a result of the study, the set goal was achieved, and one of the urgent and poorly studied issues aimed at the development, and adaptation for application of methodological tools for assessing the quality of expert systems was solved.

The following key outcomes were obtained:

1. The scope of primary issues for the formation of basic competencies and their development in an industrial environment, the main of which are as follows: Development of a list of competencies, indicating the stages of their formation; development of modules for assessing professional competencies; development of models and methods for assessing the quality level of expert systems, was justified and formulated.
2. An indicative list of competencies for the training of flight safety supervisors has been developed, which, in contrast to the traditional approach, is presented in the format of generalized groups and subgroups.
3. Conclusions about the advantages and disadvantages of traditional and alternative methods for assessing the quality of education formed the basis of the proposed model for assessing the quality level of expert systems.
4. The results obtained from objective prerequisites for expanding the field of study, concentrating it in the format of practical testing of the proposed model, and the feasibility of its application in practice.

References

1. Bakhvalov SV, Arshinsky LV (2014) Osobennosti primeneniya ekspertnykh sistem dlya otsenki kachestva podgotovki specialistov (Features of the use of expert systems to assess the quality of training specialists). Transportnaya infrastruktura Sibirskogo regiona 1:458–462 (In Russian)
2. Leal MS, Gomez MLO, Toma RB (2021) Conceptual construction of global competence in education. Teoria de la Education. 34(1):83–103
3. Chan CKY, Luk LYY (2021) Academics' beliefs towards holistic competency development and assessment: A case study in engineering education. Studies Educ Eval 72:101102

4. Sellberg C, Wiig AC, Säljö R (2021) Mastering the artful practice of navigation: The situated endorsement of professional competence in post-simulation evaluations. Stud Educ Eval 72:101111
5. Lafiosca P, Fan IS (2021) Review of non-contact methods for automated aircraft inspections. Insight: Non-Destr Test Condition Monit 62(12):692–701
6. Wu H, Peek L, Mathews MC, Mattson N (2021) Cultural competence for hazards and disaster researchers: framework and training module. Nat Hazard Rev 23(1):06021005
7. Boshtyanchich E, Ismagilova FS, Mirolyubova GS, Yansha N (2016) Subjektivnye kriterii kontrolya sobstvennoj deyatel'nosti rossijskikh i slovenskikh rukovoditelej: sravnitel'ny analiz professional'noy kompetentsii (Subjective criteria for monitoring the own activities of Russian and Slovenian managers: a comparative analysis of professional competence). Obrazovanie i nauka. 8(137):66–85 (In Russian)
8. Grab, V.P. (2007). Osobennosti primeneniya ekspertnykh metodov pri otsenivanii kachestva prikladnykh programmnykh sredstv i sistem avtomatizatsii informatsionno-metodicheskogo obespecheniya obrazovatel'nogo processa i upravleniya obrazovatel'nymi uchrezhdeniyami. Trudy mezhdunarodnogo simpoziuma "Nadezhnost' i kachestvo" 2: 223–224. (In Russian)
9. Zaitsev SA, Izmalkov AA (2014) Integral'ny metod otsenki kachestva izdelij (Integral method of product quality assessmen). Agrotekhnika i energoobespechenie. 1(1):174–177 (In Russian)
10. Lisova EV, Koltakova TV, Lipinsky AV (2017) Primenenie ekspertnykh sistem v prakticheskoj deyatel'nosti. Vestnik Voronezhskogo instituta vysokikh tekhnologij. 2(21):90–92 (In Russian)
11. Magomedova KT (2016) Otsenka kachestva elektronnogo obucheniya v obrazovatel'nom uchrezhdenii. Nauka, tekhnika i obrazovanie. 2(20):175–182 (In Russian)
12. Stebenyaeva TV, Lazareva LY (2015) Osnovnye etapy i metody provedeniya ekspertnoj integral'noj otsenki kachestva innovatsionnykh programmnykh produktov. Problemy sovremennoj ekonomiki (Novosibirsk) 23:145–151 (In Russian)
13. Yazidi A, Ivanovska M, Zennaro FM, Lind PG, Viedma EH (2021) A new decision making model based on rank centrality for GDM with fuzzy preference relations. European J Oper Res 297 (3): 1030-1041
14. Paraschi EP, Georgopoulos A, Papanikou M (2021) Safety and security implications of crisis-driven austerity HRM practices in commercial aviation: a structural equation modelling approach. Safety Sci 147:105570
15. Nooralishahi P, Ibarra-Castanedo C, Deane S, Avdelidis NP, Maldague XPV (2021) Drone-based non-destructive inspection of industrial sites: a review and case studies. Drones 5(4):106

Developing a Decision Support Tool for Air Route Planning System

Nhut Ngo, Evgeny Neretin, and Phuong Nguyen

Abstract Following the rapid growth of the aviation industry in the present, air traffic congestion has become a major issue in Asia, especially in the ASEAN region. This issue makes flight delays and cancellations rise rapidly throughout the region. It makes Air Traffic Flow Management (ATFM) become important and necessary to ensure a smooth flow of flights and a balance between traffic demand and capacity while increasing the safety, efficiency, cost-effectiveness, and environmental sustainability of an Air Traffic Management (ATM) System. This project focuses on developing a decision support tool for Air Route Network System at the Strategic ATFM phase. The Flight Route System uses a heuristic algorithm to optimize the domestic flight routes of Vietnam based on the demand of Air Traffic Management Center Vietnam (ATFMC-VN).

Keywords ATFM · ATM · Flight route · Optimization

1 Introduction

In recent years, Vietnam's Civil Aviation activities have been growing at a high rate. According to forecasts of the International Air Transport Association (IATA), the number of flights in Vietnam is predicted to grow by 178% for the next 20 years [1]. With the rapid and strong development of domestic and regional flight activities, the density, as well as the complexity of flight control regulations, has gradually revealed many limitations. According to statistics on flight operations of Vietnamese Airlines in 2019 (Fig. 1), the number of delayed flights has increased compared to 2018 [2]. The flight management industry is supposed to provide better monitoring and evaluation so that problems can be identified and resolved on a timely basis.

N. Ngo (✉) · P. Nguyen
Vietnam Aviation Academy, Ho Chi Minh, Vietnam
e-mail: nhutnm@vaa.edu.vn

E. Neretin
Moscow Aviation Institute (National Research University), Moscow, Russia

O. A. Gorbachev et al. (eds.), *Proceedings of 10th International Conference on Recent Advances in Civil Aviation*, Lecture Notes in Mechanical Engineering, https://doi.org/10.1007/978-981-19-3788-0_10

In addition, the International Civil Aviation Organization (ICAO) is implementing a series of programs, plans, and roadmaps for upgrading and overall development of aviation systems, flight operational assurance services. As an ICAO member state, Vietnam also has to implement plans and actions to perfect the system to meet the ICAO programs and plans set out, especially the air traffic homogenization plan.

This project is a collaboration between the integration Center branch of IRKUT Corporation and VATM. It focuses on developing a decision support tool for Air Route Network System in order to enhance the performance of Flight Operation Activity in Vietnam. The Flight Route Planning System is used to create and optimize the domestic flight routes of Vietnam.

There are three main objectives in this project:

- Create and optimize the flight route to save operation costs for Airlines.
- Develop a tool to support the VATM in making Flight Route Planning.
- Ensure the safety of flight routes without crossing the forbidden zones.

The project has three stages in development (Fig. 2).

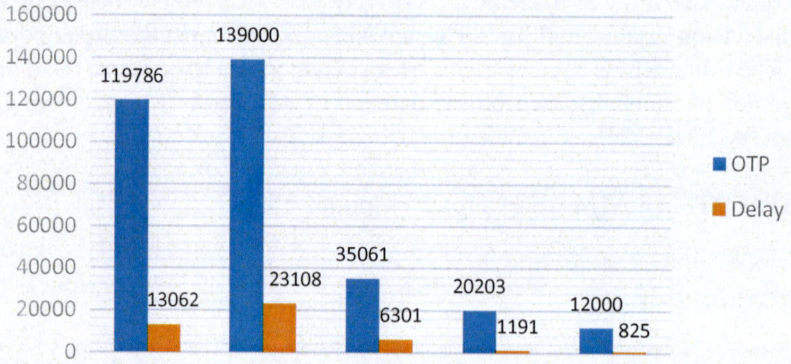

Fig. 1 Flight operations data in 2019 [2]

Fig. 2 Three stages of the project

In the first stage of project, the system focuses on the domestic flights inside Vietnam among three main international airports (HAN, DAD, SGN). It concentrates on the en-route phase. Also, the dangerous weather conditions and terrains factors are considered as the forbidden zones.

2 System Design

Figure 3 presents the architecture of the Flight Route Planning System (FRP). The core of this system is the pathfinding algorithm unit. It computes the best route based on both parameters input by the user and the information provided by Data Processing unit to enhance the efficiency. The raw data is handled by the Data Processing unit into useful information.

2.1 Actors in the System

There are three main actors taking responsibility for the specific functions of the system (Fig. 4).

- Users from Airlines who use the system to calculate the best route.
- VATM—the client who provides data, regulations to the system.
- Geo Map Sub-system provides the map data to the system.

2.2 Database Structure

The system has four tables:

- User: contains all user accounts.
- Fl_route: contains all flight routes created by users.
- Path: association class between fl_route and waypoint.

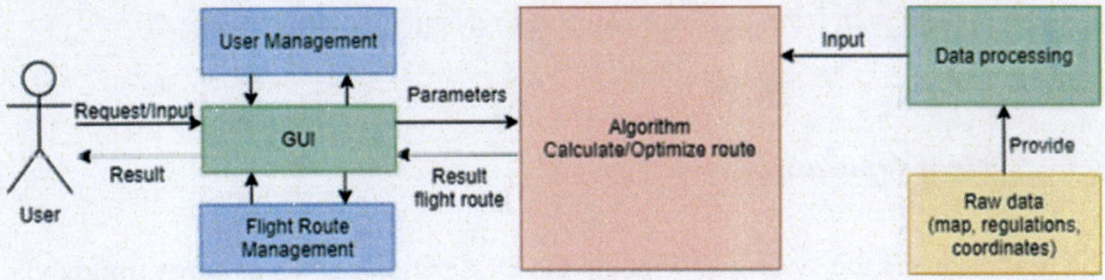

Fig. 3 High-level system architecture

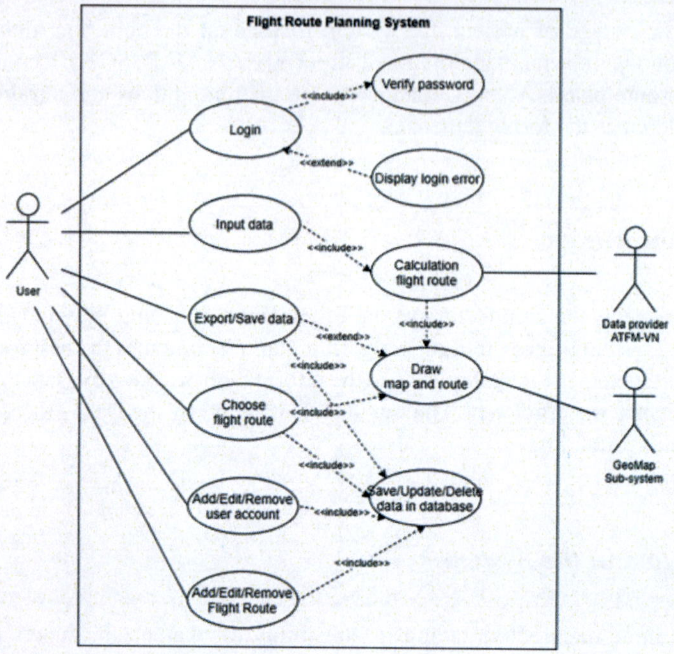

Fig. 4 Interaction of actors in the system

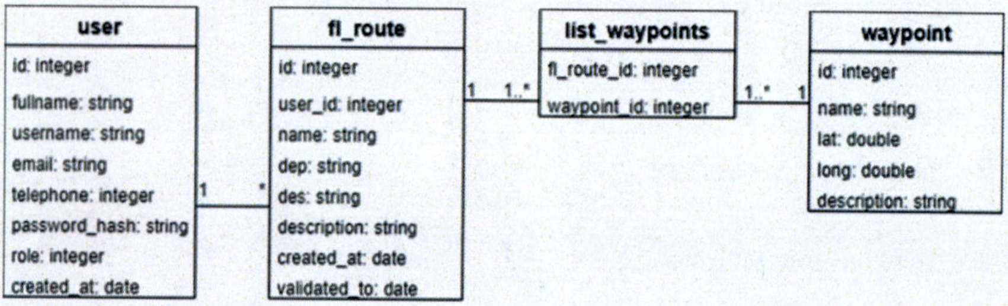

Fig. 5 Relationship of tables

A user can create many flight routes. A flight route can have a list of waypoints inside to make a path from departure airport to destination airport (Fig. 5).

2.3 System Operation

The user accesses the system and inputs the Destination and Departure airports via the interface, then presses request to make the program calculate the route. The GUI will transfer the request to the Flight Route Calculation. The calculation will use

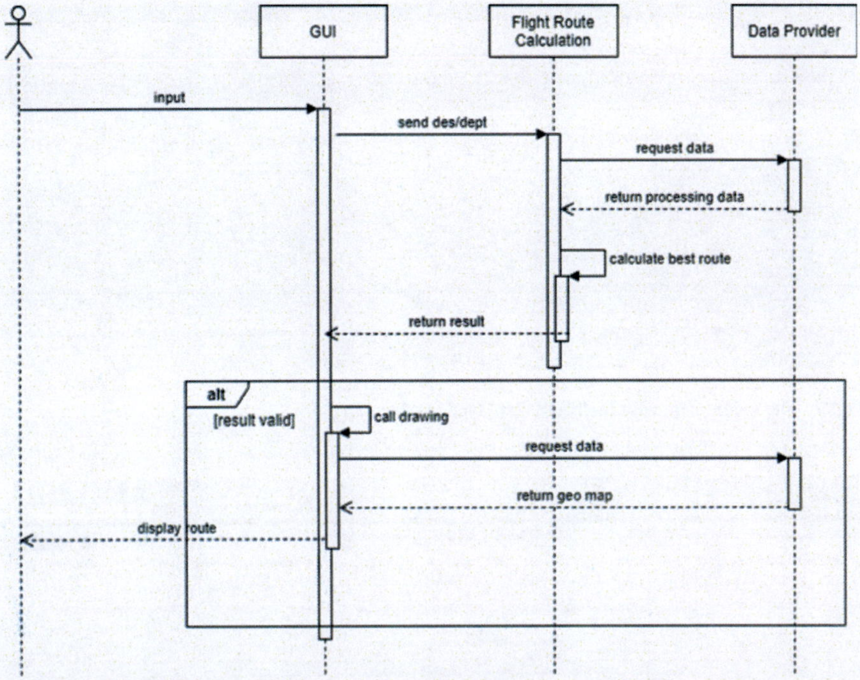

Fig. 6 User finds route and receives the result on the interface

the data provided by Data Provider and the input data from the user to calculate the route. The result will be returned to the interface. The interface will call the drawing function to draw the route and also get the data from Data Provider to display the map (Fig. 6).

2.4 *Software Implementation*

In order to satisfy the criteria of success that the system can be connected by any means of communication and to different OSs (Operating Systems), the system will be built as a web service using Flask which is one of the top 10 popular web frameworks in 2021 [3]. Flask is a Python framework that focuses on building robust web apps. It is used for developing complex web applications with the most flexibility. This framework not only aims to keep the core simple but also extensible. It offers more freedom to developers by including many hooks to customize its behavior.

The application has the index page (Fig. 7) where users can input departure airport and destination airport to find the flight route. The flight route is demonstrated on the map with the information of waypoints along on its path.

Fig. 7 The index page with the route displayed on the map

Fig. 8 The flight route management page

The flight route can be saved into the database and is listed in the flight route management page (Fig. 8). The user can modify or remove the route on this page. Furthermore, the user can use the search box to find the flight route and get it displayed on the map.

3 Pathfinding Algorithm

3.1 Three Phases of Optimization

The process of creating an algorithm to create optimal ARN (Air Route Network) consists of three layers: Navigation, Guidance, and Stabilization. At the time of writing this paper, the project is in Phase 1—Navigation (Fig. 9).

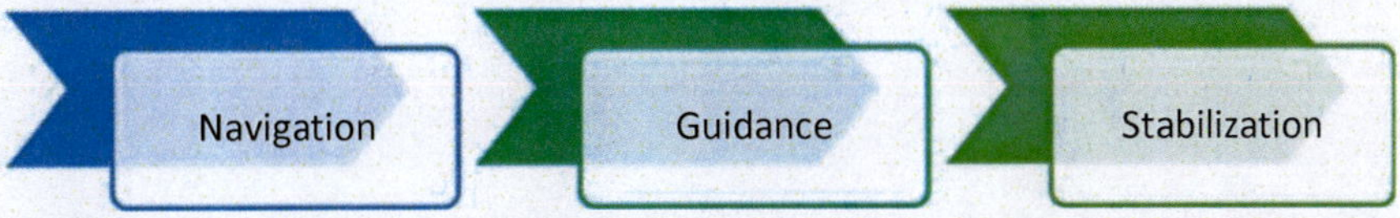

Fig. 9 Three phases of creating a complete optimization algorithm

- Navigation: the trajectory is created by a successive set of points connected to each other. The problem is solved geometrically. At this stage, the conditions of aircraft such as fuel consumption, engine power, mass weight, operation rules, etc. are not included. The trajectory will be the straight lines connecting these points, the route will not be "smooth".
- Guidance: this level will use optimal control methods to optimize the created flight trajectory based on other navigation trajectory methods. It takes into account the aircraft's mass weight, engine power, the moment of inertia, air resistance, hysteresis problems of control surfaces, operation rules, etc. Hence, the trajectory in this layer will be "smoother", since the aircraft conditions are considered as a part of the algorithm, the flight route cannot bend too much.
- Stabilization: in the guidance phase, it is not possible to calculate all the disturbances caused during the flight, as well as the errors of the navigation and measurement systems. Therefore, the fact that the aircraft can deviate from trajectory has been found in the guidance. To get the aircraft back to its designed track, an autopilot system must be used to stabilize.

3.2 Choosing Pathfinding Algorithm

At the navigation phase, the trajectory is needed to be created based on the pathfinding algorithm. The algorithm is chosen based on its cost after completing the search. According to the comparison of algorithms between A*, Dijkstra and Greedy Best-First Search; among three of them, Dijkstra calculates the distance from the start point. Greedy Best-First Search estimates the distance to the goal point. A* is using the sum of those two distances giving the low cost and being faster than Dijkstra and Greedy Best-First Search. Hence, A* algorithm is the best choice in pathfinding that could save cost and time. [2, 4, 5] (Fig. 10).

3.3 A* Algorithm

A* computes using function: $f(n) = g(n) + h(n)$ [4]

- $g(n)$ is the cost of the path from the start to goal.
- $h(n)$ is a heuristic function estimating cost of the cheapest path from n to goal.

Fig. 10 Comparing the cost of three algorithms in finding path [4]

All waypoints in Vietnam are connected and formed into a traversal graph to be used in A* algorithm. Each waypoint is considered as a node in the traversal graph and has total cost of g(n) and h(n). The cost of g(n) is determined by the distance (in kilometers or miles) from one node to another node, using data provided by VATM (Fig. 11). While the cost of h(n) is the increased cost caused by prohibited and dangerous zones (weather conditions, terrains) near the node (Fig. 12).

To let A* find the path, it is necessary to change the traversal graph to the matrix with nodes. Each node has its own estimated cost based on several factors such as

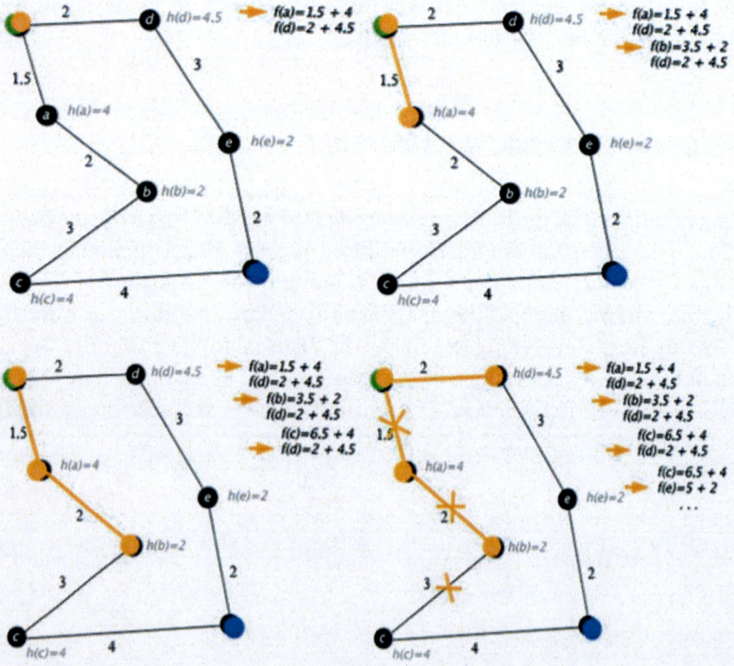

Fig. 11 Explanation how A* algorithm works [6]

Route Designator (RNP Type)	[Route Usage Notes]								
Significant Point Name	Significant Point Coordinates								Remarks
(RNP Type)	Track TRUE ↓↑	Dist (KM)	(COP)	Upper limit / Lower limit	Minimum flight altitude	Lateral limits (KM)	FL series ↓	FL series ↑	Controlling unit (Airspace class) Remarks
W1	Route availability (1) H24								
▲ NOI BAI DVOR/DME (NOB)	211247N 1055006E								
	162 / 342	55		FL285 / FL040	1 200 M	20	Even(1)	Odd(1)	Sector 1 Ha Noi ACC FREQ: 132.3 MHz
▲ LOVBI	204434N 1055948E								
	162 / 342	41		FL285 / FL040	1 200 M	20	Even(1)	Odd(1)	-LOVBI is the transfer of control point between Sector 1 and Sector 2 - Ha Noi ACC - Sector 2 Ha Noi ACC FREQ: 133.65 MHz
▲ NAM HA DVOR/DME (NAH)	202314N 1060704E								
	155 / 335	142		FL285 / FL040	1 200 M	20	Even(1)	Odd(1)	
△ MAREL	191341N 1064137E								
	156 / 336	46		FL285 / FL040	1 200 M	20	Even(1)	Odd(1)	
▲ VIDAD	185052N 1065218E								
	156 / 336	46		FL285 / FL040	1 200 M	20	Even(1)	Odd(1)	-VIDAD is the transfer of control point between Sector 2 and Sector 3 - Ha Noi ACC - Sector 3 Ha Noi ACC FREQ: 125.9 MHz

Fig. 12 Node data in A* traversal graph [7]

matching demand to capacity, aircraft/flight demanding, aircraft/flight total delay, actors/stakeholders involved, etc. Obstacles (prohibited and dangerous zones) also are added on the map [8, 9] (Fig. 13).

The traversal graph used in A* algorithm had been created from the list coordinates of 216 waypoints, 38 forbidden zones, and the border of Vietnam. The trajectory must comply with the following conditions [10–12]:

- It shall not go into the forbidden zones (military zones, dangerous zones, etc.).

CÁC KHU VỰC HẠN CHẾ *RESTRICTED AREAS*		
VVR6		
Một vòng tròn bán kính 15 KM có tọa độ tâm 214318B 1054230Đ	Không hạn chế / Mặt đất	Không vực 10
A circle, 15 KM radius centred at 214318N 1054230E	*UNL / GND*	*Air force training*
VVR7		
Một vòng tròn bán kính 15 KM có tọa độ tâm 212400B 1052700D	6 000 M / Mặt đất	Không vực 11 Giới hạn thấp 3 000 M khi có hoạt động của tàu bay dân dụng
A circle, 15 KM radius centred at 212400N 1052700E	*6 000 M / GND*	*Air force training* *Lower limit 3 000 M when civil aircraft in operation*

Fig. 13 Prohibited zone data in A* traversal graph [7]

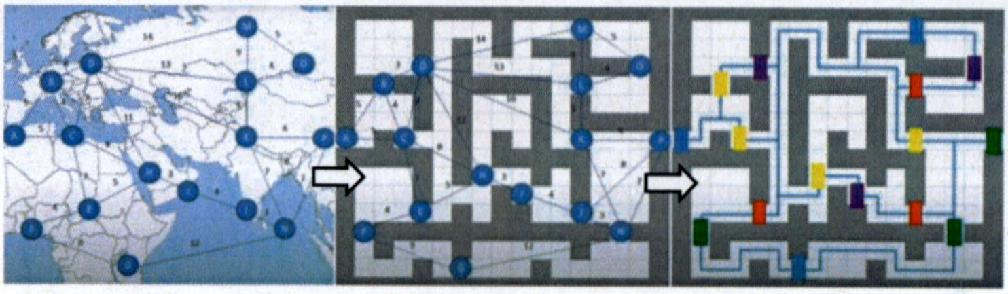

Fig. 14 Graph structure transfer into matrix using A* algorithm [13]

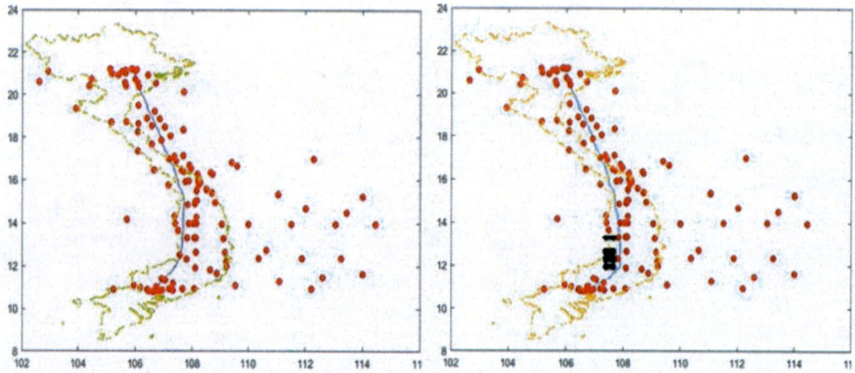

Fig. 15 Flight route from HAN to SGN

- It shall not go out of Vietnam.
- It shall not bend too much causing difficulties for aircraft.

4 Results

The A* algorithm has created the flight route successfully between three main airports in Vietnam in Fig. 14 and Fig. 15. The created flight routes connect the waypoints (red nodes) along their path. In addition, they could avoid the forbidden zones (black nodes). The flight routes archived requirements from Phase 1—Navigation (Fig. 16).

5 Conclusion

The results have achieved the goal of Phase 1—Navigation. But the routes are not really "smooth" in some parts of their path. It will be optimized in Phase 2—Guidance to make it become smoother and more practical as the heuristic function includes aircraft operation conditions, real-time weather conditions, real-time air

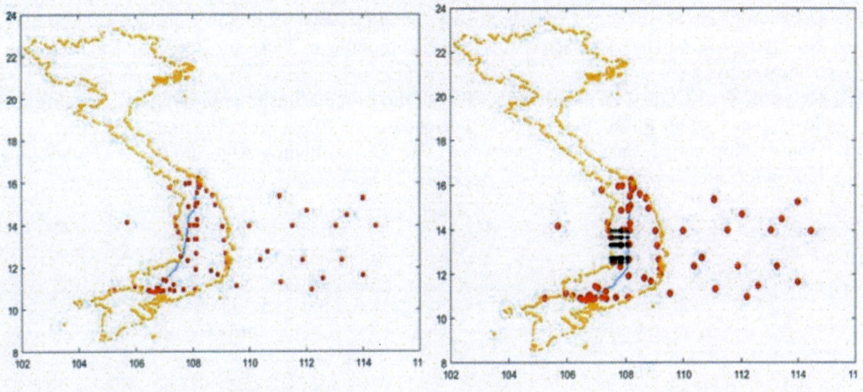

Fig. 16 Flight route from DAD to SGN

traffic congestion, etc. In addition, to cope with the algorithm in Phase 2, the application has to be upgraded thus allowing the user to input more specific conditions to support the algorithm for finding the suitable route with lowest cost.

References

1. IATA (2019) The importance of air transport to Vietnam, pp 4. https://www.iata.org/en/iata-repository/publications/economic-reports/vietnam--value-of-aviation. Accessed 10 May 2021
2. CAA VN (2019) Summary of data on on-time, delayed and canceled flights of Vietnamese airlines in 2019, https://www.caa.gov.vn/bao-cao-thong-ke/tong-hop-so-lieu-khai-thac-cac-chuyen-bay-dung-gio-cham-%20huychuyen-cua-cac-hang-hang-khong-viet-nam-12-thang-nam-2019-20200106090750469.htm. Accessed 10 May 2021
3. Vailshery LS (2021) Most popular web frameworks among developers worldwide 2021. https://www.statista.com/statistics/1124699/worldwide-developer-survey-most-used-frameworks-web/. Accessed 12 May 2021
4. Amit Patel (2021) Introduction to A*, http://theory.stanford.edu/~amitp/GameProgramming/AStarComparison.html#the-a-star-algorithm. Accessed 12 May 2021
5. Idrissi AK, Malapert A, Jolin R (2019) 8th international conferance on operations research and entreprise systems. Feb 2019, Prague, Czech Republic. ffhal-02069634
6. Wikipedia page, A* search algorithm, https://en.wikipedia.org/wiki/A*_search_algorithm. Accessed 12 May 2021
7. VATM, AIP Viet Nam completed version (after AMDT01.20) (2020) PART II—ENR, pp 191–493, Vietnam
8. Janić M (2021) Analysing and modeling performances of a long-haul air route network. Transp Syst Technol 7(1):5–36. https://doi.org/10.17816/transsyst2021715-36
9. Félix Patrón RS, Botez RM (2014) Flight trajectory optimization through genetic algorithms coupling vertical and lateral profiles. https://doi.org/10.1115/IMECE2014-36510
10. Decree No. 125/2015/ND-cp, Date December 04th 2015, Detailing the management of flight operations

11. Decision of the Minister of Defense No. 144/2004/QD-BQP of October 30, 2004 Regulating No-Flying Areas in the Airspace of the Socialist Republic of Vietnam. Правильное ли здесь оформление?
12. Decision No. 21/QD-TTg dated 08/01/2009 of the Prime Minister approving the planning for development of air traffic up to 2020 and orientation to 2030, Air traffic must be a safe
13. Graph Computer Science, Data structure 6. The A* Pathfinding Algorithm, https://www.youtube.com/watch?v=eSOJ3ARN5FM. Accessed 21 May 2021

Using Data-Driven Approach in 4D Trajectory Prediction: A Comparison of Common AI-Based Models

Evgeny Neretin, Man Nguyen, and Phuong Nguyen

Abstract Artificial intelligence (AI) is developing strongly and widely applied in many fields, including in the aviation. At the 13th Air Navigation Conference (AN-Conf/13-WP/232) organized by ICAO in Montreal from 9 to October 19 2018, participants discussed AI benefits and preparation for AI-enabled air traffic management (ATM). In this paper, a comparison was carried out to evaluate common AI-based models: Linear regression (LR), Random forest (RF) regression, Extremely gradient boosting (XGBoost) regression and deep neural network (DNN) models for predicting four-dimensional (4D) flight trajectory under weather uncertainties. The datasets used in this paper contain actual ADS-B historical trajectory data of flights from Ho Chi Minh to Ha Noi for 14 days (November 12–26 2021) and time-synchronized weather data along the waypoints of each flight. After comparing the training performance of LR, RF, XGBoost, DNN models, the best-fit DNN models were chosen for further improvement. By tuning their main hyperparameters, the training results are significantly improved in terms of training time and mean absolute errors.

Keywords Artificial intelligence · AI-based methods · 4D-trajectory prediction · DNN

E. Neretin · M. Nguyen · P. Nguyen (✉)
Moscow Aviation Institute (National Research University), Moscow, Russia
e-mail: phuongntlp@vaa.edu.vn

E. Neretin
e-mail: neretines@mai.ru

M. Nguyen
e-mail: nguyennm@mai.ru

P. Nguyen
Vietnam Aviation Academy, Ho Chi Minh City, Vietnam

O. A. Gorbachev et al. (eds.), *Proceedings of 10th International Conference on Recent Advances in Civil Aviation*, Lecture Notes in Mechanical Engineering, https://doi.org/10.1007/978-981-19-3788-0_11

1 Introduction

ATM modernization and research programs are being developed in many regions of the world such as NextGen [1] in the United States, Single European Sky ATM Research (SESAR) [2] in Europe and Collaborative Actions for Renovation of Air Traffic Systems (CARATS) [3] in Japan, Seamless ASEAN Sky [4] in Southeast Asian Countries. These programs are aligned with a single global vision as agreed in the Global Air Traffic Management Operations Concept of International Civil Aviation Organization (ICAO) [5] and are uniquely tailored to practical situations in each region. One key issue is the transition from Airspace to Trajectory-Based Operations (TBO), whereby each flight was represented by a shared trajectory, used as a common flight plan. Through the System Wide Information Management (SWIM), this common trajectory information is synchronized for aircraft operators, air navigation service providers and related participants to use for strategic planning and tactical decision-making (Fig. 1).

Generally, different trajectory prediction methods have been proposed:

- The physical-based methods: use kinematics and dynamic equations to predict the flight trajectory for different phases (climb, cruise and descent) within the flight dynamics and aircraft performance constraints.
- The filter-based methods: consider the problem of aircraft trajectory prediction as a state estimation problem and use estimation methods such as Kalman filter and hidden Markov model to predict the future trajectory of an aircraft.
- The AI-based methods: these non-parametric methods use large amounts of historical data of aircraft trajectory related to training an aircraft predictor, which does not use any aircraft aerodynamic parameters.

The first two methods need some parameters that are usually unavailable to the ground control system, such as the aircraft takeoff weight, thrust, drag, lift and flight parameters given to the Flight Management Systems. The AI-based methods have recently been the most popular ones and obtained promising performance for

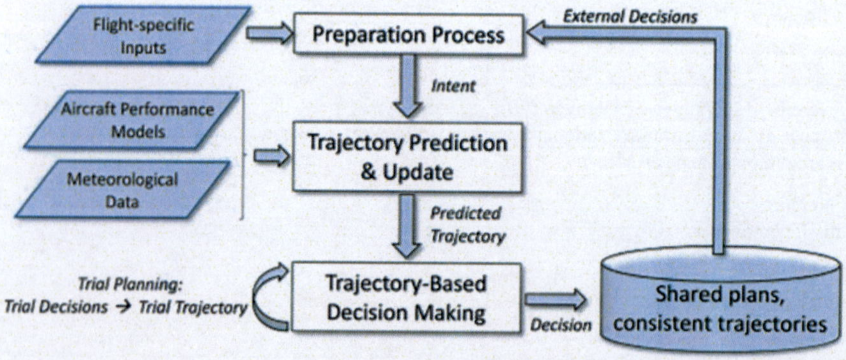

Fig. 1 Using trajectory for decision-making in TBO [6]

dealing with the stochastic flight environment using historical data. This paper aims to evaluate and compare the training performance of Linear regression, Random forest, XGBoost and deep neural network models in 4D-trajectory prediction with weather uncertainties and aircraft intention.

2 Related Works on AI-Based 4D-Trajectory Prediction

In the last decade, the AI-based approach was investigated in many works, projects on 4D-trajectory prediction and proved significant accuracy in performance. Different AI-based models from conventional machine learning to multi-layer deep neural network learning were proposed, such as: Generative Adversarial Imitation learning (GAIL) model in [7]; A hidden Markov model (HMM), Bayesian framework and several Gaussian mixture models (GMMs) are used in [8]; The Deep Q-Network (DQN) model in [9]; Deep-Gaussian process (DGP) model in [10]; Local linear functional regression model in [11]; The hybrid estimation model consists of a hybrid estimation block, an intent inference block and an Interacting Multiple Model trajectory prediction block in [12]; Clustering-based preprocessing and Multi-cells neural network (MCNN)-based machine learning in [13]; A hybrid convolutional neural network and Gated Recurrent Unit (CNN-GRU) along with a 3D-CNN model in [14]; A hybrid convolutional neural network (CNN) and long short-term memory (LSTM) model in [15]; Long short-term memory (LSTM) neural network model in [16]; A multi-stage hybrid approach is employed for a new variation of the core TP task, the so-called *Future Semantic Trajectory Prediction*, including hidden Markov model (HMM), linear regressors, regression trees and feed-forward neural networks in [17]; Convolutional recurrent neural network that consists of a long short-term memory (LSTM) encoder network and a mixture density LSTM decoder network in [18]; Logistic regression, support vector regression, deep neural network, conventional neural network, recurrent neural network and long short-term memory algorithms in [19]; Long short-term memory network in [20]; A Bayesian deep learning-based CNN + RNN + FCNN probabilistic architecture in [21]; The *ensemble-meta estimators* algorithms in [22]; A hybrid clustering/HMM two-phase algorithm in [23]; A novel Conditional Generative Adversarial Network (CGAN) approach in [24]; A combined deep feed-forward neural networks (DNN) and deep long short-term memory (LSTM) neural networks based on Bayesian approach in [25]; The LSTM + MDN network in [26] and, etc.

Seyed M. H. et al. find out the modern algorithms show the higher regression accuracy but have the lowest resiliency against crafted adversarial attacks [19]. So that, the future works on AI-based regression methods are necessary to be developed to avoid adversarial attacks, beside of improvement in the accuracy and training speed.

3 Data Preparation

3.1 Raw Datasets

ADS-B dataset. Historical flight data of actual aircraft trajectory derived from https://flightaware.com/live/findflight give the information about aircraft identification (id), aircraft type (at), origin–destination pair (od), latitude (φ), longitude (φ), altitude (δ), timestamp (t), velocity (v), course (ϑ). Flight 4D-trajectory can be described by the tuple:

$$TD_i^k = \left(\varphi_i^k, \theta_i^k, \delta_i^k, t_i^k\right) \tag{1}$$

$$i = \left[0, T^k - 1\right]; k = [1, N]$$

where T^k is the number of states in kth trajectory, N is the number of trajectories in the dataset. The time interval of states is not equidistant with increments from 10 to 30 s.

Aircraft intention dataset. Aircraft identification (id), origin–destination pair (od), timestamp (t), aircraft type (at) can be described by the tuple

$$I_j^k = \left(id_j^k, at_l^k, od_m^k, t_i^k\right); i = \left[0, T^k - 1\right] \tag{2}$$

$$k = [1, N]; j = [1, k]; l = [1, k]; m = [1, k]$$

where j is the index of aircraft identification in *kth* flight; l is the index of aircraft type in *kth* flight; m is the index of origin–destination pair in *kth* flight (in this study, $m = 1$).

Weather dataset. The real weather data derived from https://api.meteomatics.com/ and time synchronized with each state in every trajectory give information on wind speed, wind direction, temperature, pressure and weather symbol (describe an overall impression of the weather state such as clear sky, rain, fog, drizzle, thunderstorm, sandstorm, snow…) along waypoints of the route.

$$w_i^k = \left(sp_i^k, dr_i^k, tp_i^k, pr_i^k, sb_i^k\right) \tag{3}$$

$$i = \left[0, T^k - 1\right]; k = [1, N]$$

3.2 Enriched Dataset

The data processing was implemented to filter out the duplication and missing data in ADS-B flight dataset. Adding weather variables and aircraft intention in a trajectory state results in a trajectory with *enriched points* or *enriched states*, thus to an *enriched trajectory*:

$$TDE_{ij}^{k} = \left(\varphi_i^k, \theta_i^k, \delta_i^k, t_i^k, sp_i^k, dr_i^k, tp_i^k, pr_i^k, sb_i^k, id_j^k, at_j^k, od_j^k\right) \quad (4)$$

$$i = \left[0, T^k - 1\right]; k = [1, N]; j = [1, k]$$

According to the air traffic control rules in flight information regions (FIRs) of Vietnam, the enriched flight trajectory dataset is classified into 2 classes for terminal airspace (where the altitude is lower 10,000 feet) and en-route airspace (where the altitude is higher included 10,000 feet).

4 Methodology

LR, RF, XGBoost models are implemented based on scikit-learn framework version 1.0 and DNN model is developed based on a very popular deep learning framework TensorFlow, version 2.1.0. The experimental models are built in Python and all trained with GTX 3090 GPU, RAM 48 GB, CPU i7.

4.1 Algorithm

The proposed methodology consists of two stages. During the 1st stage, a baseline comparison of trainging accuracy of following AI-based models takes place: the linear regression model, random forests, XGBoost and artificial neural network models. During the 2nd stage, the best-fit model is chosen and different hyperparameters are tuned to increase the performance in both terms of accuracy and training speed. All AI-based models in this study are trained and evaluated according to the following algorithm 1 described in Table 1.

4.2 Models and Experiments

The 1st stage: The baseline comparison in training accuracy of common AI-based models.

Table 1 General algorithm in 4D-trajectory prediction using AI-based models

Algortihm 1
Read the dataset file by the syntax: *pd.read*
Split the dataset into train:test subset by the syntax: *train_test_split()*
Import necessary modules and functions: *scipy, sklearn, pandas, sklearn.ensemble, numpy, stats, xgboost, sklearn.metrics, tensorflow. keras.models, Dens, Adam, tensorflow.keras.layers, tensorflow.keras.optimizer, tensorflow.keras.callbacks, load_model, Model, Activation, Flatten, Conv1D, Maxpooling1D, Dropout, Input, RMSprop, ModelCheckpoint, mean_square_error, XGBRegressor, LinearRegressor, accuracy_score_mean_absolute_error, RandomForestRegressor*
Input the required parameters:
LR model*: default*
RF model*: random_state = 2508, max_depth = 2*
XGBoost model*: n_estimators = 12, max_depth = 2, eta = 0.1, subsample = 0.7, colsample_bytree = 0.8*
DNN model*: learning rate = 0.001,epochs = 500,batch_size = 32,validation_split = 0.2, optimizer = Adam*
DNN* model*: learning rate = 0.07, epochs = 500, batch_size = 1024, validation_split = 0.2, optimizer = Adam*
DNN** model*: learning rate = 0.001,epochs = 600, batch_size = 3200, validation_split = 0.3, optimizer = Adam*
Fit the model to get the predicted targets by the syntax: *model.fit(train,target)*
Test and evaluate model performance by the syntaxes:
test = np.asarray(test).astype('float16')
pred = predictions[:, 0].tolist()
MAE = mean_absolute_error(gt, pred)
MSE = mean_squared_error(gt, pred, squared = False)

Linear regression model. LR models from scikit-learn framework version 1.0 are used to predict latitude, longitude, altitude, velocity and course of the aircraft by training the actual expert trajectory data in terms of the linear approximation.

Random forest model. RF models from scikit-learn framework version 1.0 are used as a meta estimator with the number of trees = 100, max depth of the tree = 2, random rate = 2508 that fits decision trees on various sub-samples of the dataset and uses averaging to improve the predictive accuracy and control over-fitting.

XGBoost model. XGBoost models form scikit-learn framework version 1.0 provide a parallel tree boosting to solve the data-driven 4D-trajectory prediction with high accuracy and training speed. These XGBoost models use the hyperparameters: n_estimators = 12; max_depth = 2; eta = 0.1; subsample = 0.7; colsample_bytree = 0.8.

Deep neural network model. The 1st DNN model consists of six hidden fully connected layers, fits to both terminal and en-route datasets using Adam optimization algorithm and Rectified Linear Unit (ReLU) activation function in hidden layers, Linear activation function at output layer. The hyperparameters of these 1st DNN models are set as learning rate = 0.0001, batch_size = 32, epoch = 500 and validation rate = 0.2.

The second stage: Tuning different hyperparameters of chosen best fit models.

The 2nd DNN model (DNN*) consists of three hidden fully connected layers that fits to the terminal airspace dataset using Adam optimization algorithm and Rectified Linear Unit (ReLU) activation function in hidden layers, Linear activation function at output layer. We tuned the hyperparameters of DNN* model into learning rate = 0.07, batch_size = 1024, epoch = 500 and validation_split = 0.2, optimizer = Adam.

The 3rd DNN model (DNN**) consists of six hidden fully connected layers that fits to the en-route airspace dataset using Adam optimization algorithm and Rectified Linear Unit (ReLU) activation function in hidden layers, Linear activation function at output layer. We tuned the hyperparameters of DNN** models into learning rate = 0.001, batch_size = 3200, epoch = 600 and validation_split = 0.3, optimizer = Adam.

5 Results and Conclusion

Experimental results of the 1st stage of LR, RF, XGBoost and DNN models prove the much higher accuracy of DNN models in comparison with LR, RF and XGBoost ones. And at the 2nd research stage, we tune different hyperparameters in the chosen DNN models to improve their training accuracy and speed.

After choosing the best-fit DNN models, we conducted many experiements to increase their performace by tuning the following hyperparameters several times: learning rate, batch_size, epoch, validation_split.

In final results, these techniques help to gain the training speed of DNN* and DNN** models in 20.8 times and 5.8 times faster than DNN models respectively without decreasing their accuracy. However, the mean absolute errors of the predicted targets (latitude, longitude, altitude, velocity, course) shown in Table 2 are not as good as required in practical operations. Our future work will aim to develop DNN models with more effective hyperparameters, optimized architecture for solving 4D-trajectory prediction problems.

Table 2 Mean absolute errors of DNN* and DNN** models

Dependent variables (Targets)	MAE	
	DNN* models	**DNN** models**
Lat. (deg)	3.03	2.45
Lon. (deg)	0.09	0.58
Alt. (ft)	2188.61	6021.76
Course (deg)	120.75	117.33
Speed (kts)	31.00	23.34

References

1. FAA, "NextGen Mid-term Concept of Operations for the National Airspace System", Wasington, https://www.faa.gov/nextgen/; Accessed 10 May 2021
2. SESAR Joint Undertaking, "SESAR 2020 Concept of Operations", Brussels, 2017, https://ec.europa.eu/research/participants/documents/downloadPublic?documentIds=080166e5b6d2b912&appId=PPGMS; Accessed 13 May 2021
3. Japan Ministry of Land, Industry and Tourism, "CARATS - Long Term Vision for the Future Air Traffic Systems - Changes to Intelligent Air Traffic Systems", Japan, 2010, https://www.icao.int/Meetings/anconf13/Documents/WP/wp_252_en.pdf; Accessed 14 May 2021
4. ASEAN, "ASEAN Air Navigation Service Master Plan - Seamless ASEAN Sky: One Sky for "One Vision, One Identity, One Community", 2020, https://www.caat.or.th/wp-content/uploads/2018/09/Number-2-Revised-Presentation-1-ASEAN-ATM-Master-Plan-for-ARISE.pdf; Accessed 14 May 2021
5. ICAO, "ICAO, Global TBO Concept - draft material in development by an ICAO expert", 2021, https://www.icao.int/airnavigation/tbo/Pages/Why-Global-TBO-Concept.aspx.; Accessed 24 May 2021
6. Mondoloni, S., Rozen, N. (2020) Aircraft trajectory prediction and synchronization for air traffic management applications. Aerospace Sciences, vol. 119, doi:https://doi.org/10.1016/j.paerosci.2020.100640
7. Bastas, A., Kravaris, T. and Vouros, G. A. (2020) Data Driven Aircraft Trajectory Prediction with Deep Imitation Learning. Computer Science – mathematics, https://arxiv.org/abs/2005.07960.
8. Lin, Y., Zhang, Jw. & Liu, H. (2018) An algorithm for trajectory prediction of flight plan based on relative motion between positions. Frontiers Inf Technol Electronic Eng, vol. 19, doi:https://doi.org/10.1631/FITEE.1700224, pp. 905-916.
9. Spatharis, C., Blekas, K., & Vouros. (2020) Apprenticeship learning of flight trajectories prediction with inverse reinforcement learning. 11th Hellenic Conference on Artificial Intelligence. doi:https://doi.org/10.1145/3411408.3411427.
10. Chen, Z.; Guo, D.; Lin, Y. (2020) A Deep Gaussian Process-Based Flight Trajectory Prediction Approach and Its Application on Conflict Detection. Algorithms, vol. 13, doi:https://doi.org/10.3390/a13110293, p. 293
11. Tastambekov, K., Puechmorel, S., Delahaye, D., Rabut, C. (2014) Aircraft trajectory forecasting using local functional regression in Sobolev space. Transportation research. Part C, Emerging technologies, Elsevier. 39: 1–22. ff10.1016/j.trc.2013.11.013ff. ffhal-00924360f.
12. Song Y, Cheng P, Mu C (2012) An improved trajectory prediction algorithm based on trajectory data mining for air traffic management. IEEE International Conference on Information and Automation. https://doi.org/10.1109/icinfa.2012.6246959
13. Wang, Z., Liang, M., Delahaye, D. (2017) Short-term 4D Trajectory Prediction Using Machine Learning Methods, 7th SESAR Innovation Days, Nov 2017, Belgrade, Serbia. ffhal-01652041
14. Hesam Sahfienya, Amelia C. Regan. (2021)"4D flight trajectory prediction using a hybrid Deep Learning prediction method based on ADS-B technology: a case study of Hartsfield-Jackson Atlanta International Airport(ATL)", https://arxiv.org/ftp/arxiv/papers/2110/2110.07774.pdf; Accessed 25 May 2021
15. Ma, L., & Tian, S. (2020)"A Hybrid CNN-LSTM Model for Aircraft 4D Trajectory Prediction", IEEE Access, vol. 8, doi:https://doi.org/10.1109/access.2020.3010963, p. 134668–134680.
16. Han, P., Yue, J., Fang, C., Shi, Q. and Yang, J. (2020) Short-term 4D trajectory prediction based on LSTM neural network, Proc. SPIE 11427, Second Target Recognition and Artificial IntelligenceSummit Forum, vol. 11427, doi:https://doi.org/10.1117/12.2550425.
17. Georgiou, H., Pelekis, N., Sideridis, S., Scarlatti, D., Theodoridis, Y. (2020) Semantic-aware aircraft trajectory prediction using flight plans, International Journal of Data Science and Analytics, vol. 9, doi:https://doi.org/10.1007/s41060-019-00182-4, pp. 215-228

18. Pang, Y., Liu, Y. (2020) "Probabilistic Aircraft Trajectory Prediction Considering Weather Uncertainties Using Dropout As Bayesian Approximate Variational Inference". AIAA Scitech 2020 Forum. doi:https://doi.org/10.2514/6.2020-1413.
19. Hashemi SM, Botez RM, Grigorie TL (2020) New Reliability Studies of Data-Driven Aircraft Trajectory Prediction. Aerospace 7:145. https://doi.org/10.3390/aerospace7100145
20. Shi Z, Xu M, Pan Q, Yan B, Zhang H (2018) LSTM-based Flight Trajectory Prediction. International Joint Conference on Neural Networks (IJCNN). https://doi.org/10.1109/IJCNN.2018.8489734.pp.1-8
21. Pang, Y., Zhao, X., Yan, H., Liu, Y. (2021) "Data-driven trajectory prediction with weather uncertainties: A Bayesian deep learning approach", Transportation Research Part C, vol. 130, doi:https://doi.org/10.1016/j.trc.2021.103326, p. 103326
22. Hernández, A. M., Casado Magaña, E. J. and Berna, A. G. (2018) "Data-driven Aircraft Trajectory Predictions using Ensemble Meta-Estimators", IEEE/AIAA 37th Digital Avionics Systems Conference, doi:https://doi.org/10.1109/DASC.2018.8569535, pp. 1–10.
23. Fernández, E.C., Cordero, J., Vouros, G., Pelekis, N., Kravaris, T., Georgiou, H., Fuchs, G., Casado, E., Costas, P., Ayhan, S. (2017) "DART: A Machine-Learning Approach to Trajectory Prediction and Demand-Capacity Balancing", 7th SESAR Innovation Days, Belgrade, 2017, https://www.sesarju.eu/sites/default/files/documents/sid/2017/SIDs_2017_paper_65.pdf; Accessed 25 May 2021
24. Pang, Y. & Liu, Y. (2020) "Conditional Generative Adversarial Networks (CGAN) for Aircraft Trajectory Prediction considering weather effects", AIAA Scitech 2020 Forum, DOI: https://doi.org/10.2514/6.2020-1853.
25. Zhang X, Mahadevan S (2020) Bayesian Neural Networks for Flight Trajectory Prediction and Safety Assessment. Decis Support Syst. https://doi.org/10.1016/j.dss.2020.113246,pp.113246
26. Hou, L.-H., Liu, H.-J. (2019) "An End-to-End LSTM-MDN Network for Projectile Trajectory Prediction", in: Advanced Data Mining and Applications. Advanced Data Mining and Applications, pp. 114–125, doi:https://doi.org/10.1007/978-3-030-36204-1_9.

Development of the Perspective Aircraft Cockpit Indication System Simulator

Nikita Silin and Andrey Ivanov

Abstract This article proposes the architecture of the simulation complex of the indication system for the validation of ergonomic solutions in the design of IMF cockpits at an early stage of development of advanced civil aircraft. The use of the proposed solution will significantly reduce material and time costs due to the fact that in the process of developing the software for the indication system, solutions already validated with the flight crew will be used as requirements for the indication, as well as models used as part of the complex that meet the requirements of the R-331 standard /DO-331, which means they can be used in the development of onboard software. Additionally, since the process of validating the indication system using the proposed complex is inextricably linked with the process of developing onboard software, the software and hardware of the prototyping bench can be used as a technical tool for training the crew to work with the IMF of the aircraft being developed, which further reduces aircraft operating costs.

Keywords Civil aircraft · Aircraft cockpit · Ergonomics · Avionics · Logic integration bench

1 Introduction

Flight safety is a fundamental factor considered in the development of civil aircraft. Despite the fact that the number of aviation accidents in the world is constantly decreasing, it still remains at an unsatisfactory level. At the same time, the human factor remains one of the main reasons for their occurrence. Mostly, crew errors are caused by interaction with the aircraft problems [8]. Often, the reason they arise is incorrect interpretation of the displayed information, when the pilot tries to respond with control actions to events in the environment or a change in the characteristics of the aircraft, due to the unpredictability, suddenness and urgency of the development of an unfavorable situation [9].

N. Silin (✉) · A. Ivanov
Moscow Aviation Institute (National Research University), Moscow, Russia
e-mail: SilinND@mai.ru

O. A. Gorbachev et al. (eds.), *Proceedings of 10th International Conference on Recent Advances in Civil Aviation*, Lecture Notes in Mechanical Engineering, https://doi.org/10.1007/978-981-19-3788-0_12

The information and control field (IMF) of the cockpit is one of the most functionally loaded parts of the aircraft. In the busiest parts of a flight, a cockpit human–machine interface (HMI) that is not optimal in terms of crew workload can lead to catastrophic consequences. In this regard, during its development, engineers spend a significant amount of effort on creating an intuitive logic for displaying information, its appearance, colors, font styles, relative position and other aspects that affect the cognitive load on the crew. Another important area is the development of ways to control the information field for reconfiguring the displayed information or interacting with various aircraft systems through interactive indication frames.

To reduce the development time for aircraft (AC), it is necessary to validate the solutions used in the IMF cockpits design at an early stage of the life cycle. For these purposes, various test benches are used, one of which is prototyping benches, which allow you to work out system logic on simulation models. Such benches for testing the IMF are an imitation of the front dashboard and the central cockpit pedestal using personal computer monitors. They can display relevant equipment, such as indicators or control panels, whose logic and appearance are to be evaluated.

This article proposes the architecture of the simulation complex of the indication system of a perspective civil aircraft as part of a prototyping bench. Thanks to the proposed approach, it becomes possible at an early stage of development to validate the logic and appearance of indications, to carry out an ergonomic assessment of the cockpit information field, which, together with the technologies used, leads to a significant reduction in the material and time costs for civil aircraft.

2 Analysis of Existing Solutions

This section presents the results of the analysis of cockpits in terms of display systems for modern aircraft of the transport category. In the standard configuration, the external indicators display the primary flight display (PFD) and the horizontal situation indicator (HSI). The remaining indicators are multifunctional; that is, the information displayed on them can be configured in various ways. All information necessary for the crew to perform flights is contained in the following frames:

- Synoptics (SYN), which displays information about the state of various aircraft systems. As a rule, these are the control system, the power supply system, the hydraulic system, the air conditioning system, the fuel system, the system of doors, hatches and emergency ladders and the wheel braking system;
- complex navigation stop indicator (navigation display (ND)) in various display modes;
- virtual flight control system control panel (flight management system virtual control panel (FMS VCP));
- virtual radio control panel (radio management system virtual control panel (RMS VCP));
- frame with summary information about the state of aircraft systems (STATUS);

- frame of control checks (checklists (CHKL));
- a frame for displaying engine parameters and warning messages (engine and warning display (EWD)), which presents the main parameters of engines, data on mechanization, temperature overboard and in the passenger compartment and other flight data, as well as an area with text messages about failures and recommendations for the actions of the crew.

On the Airbus A350 aircraft, the information field is controlled using the following consoles, shown in Fig. 1 [11]:

- control panel for electronic centralized monitoring function (electronic centralized aircraft monitoring control panel (ECAM CP)), designed to access synoptic display frames, navigate through the configuration menu, manage messages, etc., and located on the central pedestal;
- cursor and keyboard control unit (keyboard and cursor control unit (KCCU)), designed to interact with interactive display elements and located on the central pedestal;
- electronic flight instruments system control panel (EFIS CP) control panel, designed to control PFD and ND and located at the top of the main instrument panel.

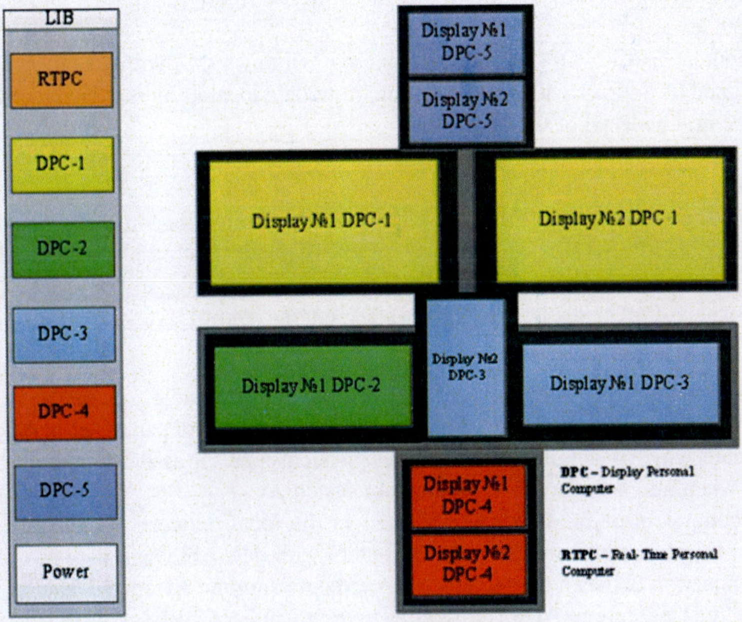

Fig. 1 Block diagram of the prototyping stand

3 Prototyping Bench

The aircraft development process is very time consuming. If an error or an unsuccessful solution is found at a late stage of development, then significant time and material costs are required to eliminate it. Therefore, for the purpose of early validation of the work logic of the component, search modeling benches or prototyping benches are used, designed to work out the logic on simulation models.

Such benches are widely used for the development of the cockpit IMF, as they allow, using test sessions with the flight crew, to evaluate the convenience of working with the cockpit controls in terms of the system's reaction to the operations performed, the logic and appearance of the indication in the conditions of performing flight scenarios at an early stage of life cycle. In this case, pre-validated solutions are taken for development, which significantly reduces the material and time costs for the development of human–machine interfaces of various aircraft systems [11].

Figure 1 shows the structure of the MC-21 cockpit prototyping bench. It is intended for testing HMI systems such as the indication system, crew warning system, aircraft navigation system and others.

The bench includes:

- computer DPC-1 running the operating system (OS) Windows, with monitors for displaying the outside environment using the freely distributed application FlightGear;
- computers DPC-2, DPC-3, DPC-4 running Windows OS, with touch monitors designed to interact with HMI equipment models located on the dashboard and the central pedestal of the cockpit;
- computer DPC-5 running Windows OS, with touch monitors for interaction with the integrated ceiling panel;
- RTPC computer running CentOS OS for real-time simulation, on which non-HMI models, such as navigation sensors, aircraft dynamic motion model and others, work;
- physical aircraft controls: side control stick, engine control lever and pedal control post.

Either an additional monitor can be inserted into the middle part of the central pedestal to display the models of the indication system control panels, or a plywood board plate, which makes it possible to place physical panels to assess their ergonomic quality when interacting with the indication system.

All computers of the bench are located in the local network, which makes it possible to run models on different computers with data exchange between them. Communication between the models is carried out using the Avionics Development System-2nd Generation (ADS2) software environment from TechSat. ADS2 is a real-time testing and simulation environment for the development, verification, integration and validation of avionics in the aviation, space and automotive industries. The ADS2 environment allows you to test the logic of systems with real-time simulation, which makes it possible to work out typical flight scenarios or abnormal situations that occur onboard [13].

4 The Architecture of the Simulation Complex of the Indication System

The proposed architecture of the simulation complex of the indication system is shown in Fig. 2.

It includes:

- models of indicators, in an amount equal to the number of indicators onboard;
- models of control panels;
- simulation environment, which includes a dynamic aircraft motion model for flight simulation and a set of sensor models that generate signals displayed on indicator models (inertial navigation system, radio altimeter, global navigation satellite system, air signal system, etc.);
- physical control panels of the "Rotary Knob" type of the "BMW iDrive" model;
- module for processing data from physical controls;
- ADS2 environment for providing information interaction between the parts of the complex.

The indicators of modern civil aircraft are being developed in accordance with the concept of integrated modular avionics. With this approach, the functions of aircraft systems are allocated into logical sections—functional software [1, 7]. For example, on the MC-21 aircraft, the indicators include three independent functional

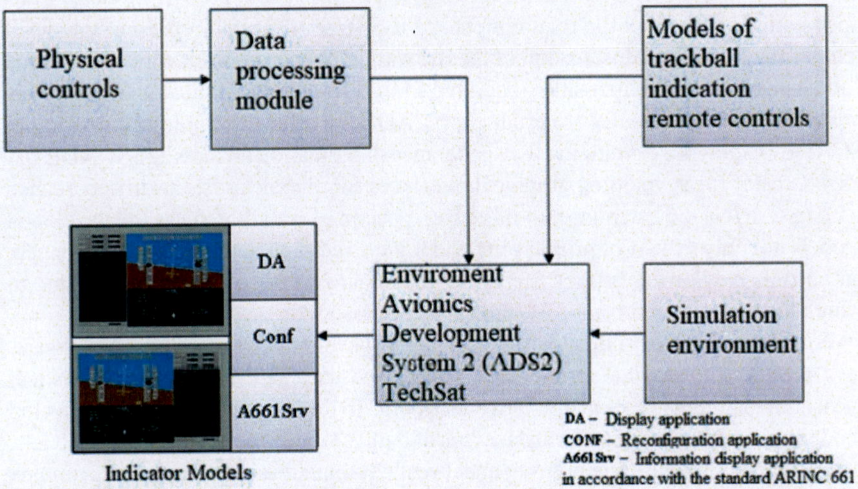

Fig. 2 Block diagram of the simulation complex

software applications that perform the function of image forming, the function of data displaying in accordance with the ARINC 661 standard and the function of reconfiguring indication formats. This separation is also proposed to be used as part of the simulation complex of the indication system, which will make it possible to apply it to aircraft with different architecture of indicators and use the developed logic in the further life cycle after validation with the flight crew. If only the traditional approach is used without applying the ARINC 661 standard, then the corresponding part of the indicator model is not developed, and vice versa.

The ARINC 661 standard defines the rules for interaction between the information display system and user applications, meaning aircraft systems, the data from which must be displayed to the crew. It also defines a set of primary elements called widgets that are used to create indications formats [2, 5].

Currently, the ARINC 661 standard is widely used to develop interactive pages to help pilots configure various systems, such as an aircraft navigation system or a radio control system. One of the perspective directions for the development of civil aviation indication systems is its application to all displayed formats.

The civil aviation software development process is regulated by the KT-178C/DO-178C standard, which defines the activities and objectives that must be met for aviation authorities' approval and admission to the aircraft certification process. Achieving the required goals is a very time-consuming task, and therefore, various approaches are used to automate and simplify this process. One of these is the model-based approach to development, according to which the various software aspects of systems that are used to support the development process or software verification are replaced by their abstract representation, that is, the model. The model-based approach is regulated by the R-331/DO-331 guideline, which is an addition to KT-178C/DO-178C [10].

In the aviation industry, one of the most commonly used tools for model-based software development is the ANSYS SCADE software package. The models developed with its help meet the requirements of R-331/DO-331 and represent low-level requirements and/or a description of the software architecture from which the source code can then be automatically generated [3, 6]. To develop indicator models, it is proposed to use three software products: SCADE Suite for developing control logic, SCADE Display for developing the appearance of indication formats and SCADE UA Page Creator for developing graphical interfaces for avionics systems in accordance with the ARINC 661 standard. In this case, it becomes possible to use the developed models not only as part of prototyping and logic validation at the prototyping bench, but also as part of the further life cycle of onboard software development, due to which the material and time costs are further reduced.

For further use of indicator models in the development of onboard software, it is necessary to provide an identical set of used parameters, their data types and measurement units with real onboard equipment. To do this, it is necessary to develop models of aircraft systems that are sources of data for indicator models. This also lets, using the ADS2 environment, to connect real equipment to the simulation complex instead of models to explore its characteristics.

Depending on the purpose of the simulation complex and the objectives of the test sessions, the set of the systems required for developing models of systems may be different. For example, to validate a flight and navigation frame, models of the following systems are required:

- warning of external threats—a system for preventing collision of aircraft in the air, a system for preventing collision of aircraft with the ground and a weather radar;
- flight and navigation equipment—inertial navigation system, air signal system, satellite navigation system, radio altimeter, etc.;
- radio navigation and landing systems—omnidirectional azimuth radio beacon, ranging system, radio compass, course-glide path system, etc.;
- systems of short-wave and ultra-short-wave communication;
- automatic flight control system.

To simulate a real flight, it is necessary to include in the simulation complex a dynamic aircraft motion model that calculates the main parameters characterizing the flight by continuously solving a system of equations describing the spatial position, models of the control system, and the basic propulsion system. To control the flight, side control sticks, pedal control posts and a block of engine control levers are used.

As already noted, each indicator model consists of three parts:

- function of display formation on the DA indicator;
- function of reconfiguration of indication frames CONF;
- data display function according to the ARINC 661 standard.

Due to the nature of the ADS2 environment, each of these functions is a separate performing application.

Figure 3 shows the DA function model. It consists of two parts—a graphic part responsible for the appearance of the frame and a logical part responsible for converting input parameter values from environment models into parameters for controlling graphic primitives. The interaction between ADS2 and the indicator model is carried out via the TCP protocol [12].

The CONF function model does not have a graphical part; therefore, it is completely developed in SCADE Suite.

Figure 4 shows an interaction scheme between the indication system (cockpit display system (CDS)) and user applications of aircraft systems (user application (UA)) in accordance with ARINC 661. Aircraft system developers define the logic for displaying information in the UA and the appearance in the definition file (definition file (DF)), in which frames are drawn using a set of widgets contained in a widget library. Based on data from other aircraft systems and notifications from the CDS about the actions of the crew, when interacting with frames, the UA generates the necessary commands to update the state of the widgets. The function of drawing frames on the indicator is performed by the ARINC 661 application core, based on the data contained in the widget library, in the definition file and commands from the UA.

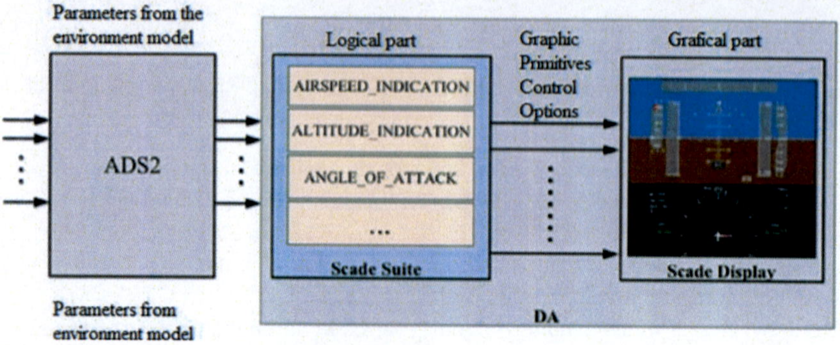

Fig. 3 Block diagram of the display application function

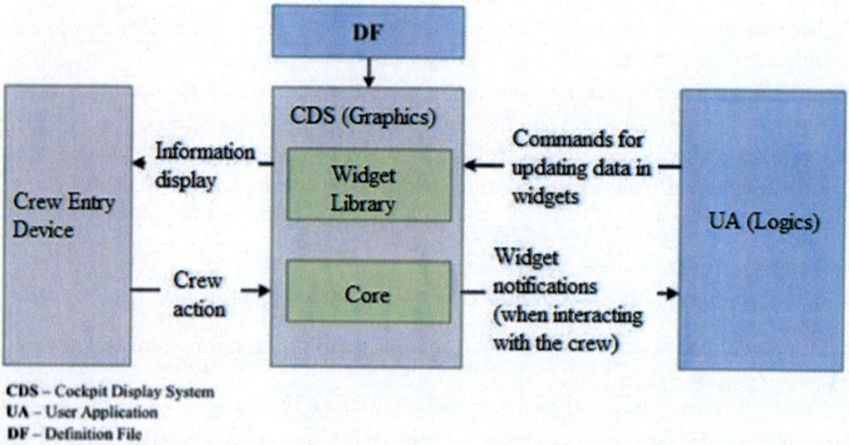

Fig. 4 Interaction in accordance with ARINC 661

As part of the simulation complex of the indication system, the CDS function is performed by the ARINC 661 application simulator. DF is developed in SCADE UA Page Creator, and the UA model in SCADE Suite. Interaction between crew input devices, UA and simulator is also carried out using the ADS2 environment via the TCP protocol.

In addition to evaluating the logic and appearance of the indication, another task of the simulation complex can be an ergonomic evaluation of the information field control logic. For this task, models of control panels are developed. As an example, Fig. 5 shows the models of the indication control panel and the models of the automatic control system panel for the MC-21 aircraft [4].

Remote control models are intended solely for the purpose of simulating the IEP of the aircraft cockpit for testing the indication control logic, and therefore, any tools can be used to develop them. The models shown in Fig. 5 were developed

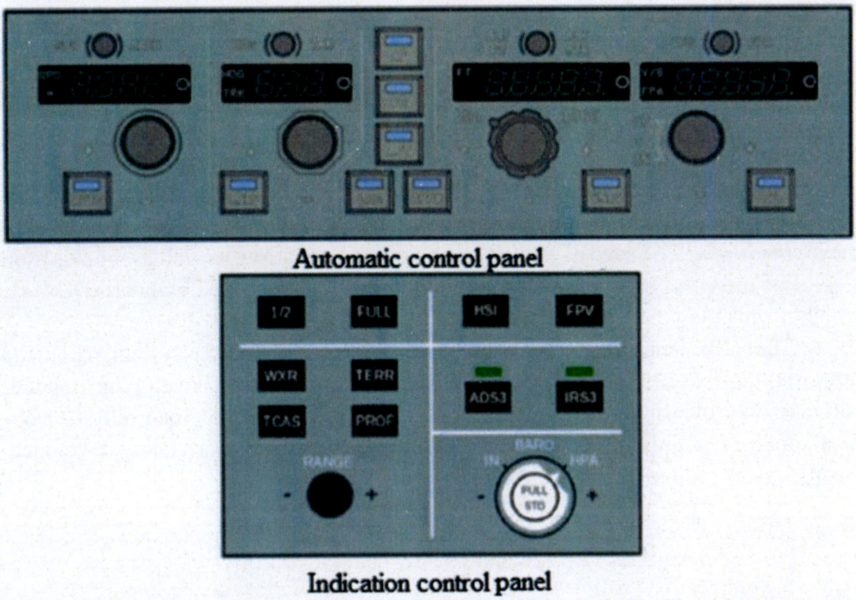

Fig. 5 Models of the indication control panel and the automatic control system panel

using the SCADE Rapid Prototyper tool, designed to develop graphical panels for debugging control logic. This tool was chosen due to the fact that it is built on the basis of SCADE Display, due to which the process of integrating models on the prototyping stand developed in it is identical to the process of integrating SCADE Display models. It also has a set of built-in widgets, such as buttons, racks, and joysticks, which reduces the time for developing models of control panels.

During testing at the prototyping bench, interaction with the controls is carried out by directly pressing the corresponding areas of the touch monitors. In the same way, an assessment can be made of the direct method of controlling the information field through touch screens, the introduction of which is a perspective direction in the development of the IMF of the cockpit. However, in this implementation of the bench, there is no possibility to carry out an ergonomic assessment of some control panels, with which the crew interacts not with single clicks, but performs long manipulations. An example of such a remote control is a trackball remote control, which controls the cursor to interact with interactive pages. In this regard, it is necessary to provide mechanisms for including such physical controls in the simulation complex. This can be implemented by connecting such consoles to the test bench computers using special drivers and introducing additional data processing modules that convert keystrokes into parameters for transmission to the ADS2 environment via the TCP protocol. Also, if the exchange protocol is known, direct connection of consoles to the ADS2 environment through special boards can be implemented.

5 Conclusion

This article proposes the architecture of the simulation complex of the indication system for the validation of ergonomic solutions in the design of IMF cockpits at an early stage of development of advanced civil aircraft. The use of the proposed solution will significantly reduce material and time costs due to the fact that in the process of developing the software for the indication system, solutions already validated with the flight crew will be used as requirements for the indication, as well as models used as part of the complex that meet the requirements of the R-331 standard/DO-331, which means they can be used in the development of onboard software.

Additionally, since the process of validating the indication system using the proposed complex is inextricably linked with the process of developing onboard software, the software and hardware of the prototyping bench can be used as a technical tool for training the crew to work with the IMF of the aircraft being developed, which further reduces aircraft operating costs.

References

1. Aeronautical Radio Inc (2013) ARINC Specification 661-5. Cockpit display system interfaces to user systems, The USA: Annapolis, pp 563
2. AR MAC (2016) Manual R-331. Development and verification based on the model. Supplement to documents KT-178C and KT-278A, pp 119
3. Burkov AY, Nasyrov M (2015) Model-based software development for reliability-critical embedded systems using the SCADE software package from Esterel technologies. Reshetnev Readings 2:209–211
4. Fedosov EA, Kosyanchuk VV, Selvesyuk NI (2015) Integrated modular avionics. Radio-Electron Technol. Moscow: KRET 1: 66–71
5. Kilic B (2019) Stress among ab-initio pilots: A model of contributing factors by AHP/C. Ucler, Özyeğin University, Faculty of aviation and aeronautical sciences, department of professional flight, Istanbul, Turkey
6. Levin DN (2019) Methods for assessing the crew workplace in the process of designing the cabin of a promising aviation complex. Comput Nanotechnol 2:95–100
7. LLC (2017) UAC—Integration Center EC.202.MC-21.14.000568–04 Procedures for working with avionics consoles and controls, pp 156
8. Polyakov VB, Neretin ES, Ivanov AS, Budkov AS, Dyachenko SA, Dudkin SO (2018) Elektron magazine proceedings of the MAI. 100, pp 21
9. Rouwhorst W, Verhoeven R, Suijkerbuijk M (2017) Use of touch screen display applications for aircraft flight control. Bos T. In: 2017 IEEE/AIAA 36th digital avionics systems conference (DASC), St. Petersburg, FL, pp 1–10
10. RTCA Inc (2011) DO-331C Model-based development and verification. Supplement to DO-178C and DO-278A, The USA: Washington, pp 136
11. Sergeev AN (2016) Fundamentals of local computer networks: Textbook, St. Petersburg: Publishing house "Lan", pp 184
12. TechSAT GmbH, ADS2—Avionics development system 2G, Release 4 User Manual, May 2018, pp 37–40
13. Zheng Y, Lei XY (2014) Research and implementation of virtual cockpit panel development platform based on ARINC 661. In: Proceedings of 2014 IEEE Chinese guidance, navigation and control conference. Yantai, China, pp 1–5

Development of Guidelines on the Use of Color Solution for Electronic Indication Systems When Designing Human–Machine Interface of Civil Aircraft Objects

Nikita Silin and Andrey Ivanov

Abstract The article is devoted to improve the flight safety of civil aviation objects by creating and applying recommendations for choosing a color solution for the main graphic information of a human–machine interface (HMI) based on a color design criterion. The ergonomic combination of color pairs plays a major role in HMI design. Considering that up to 7 different text colors can be used in HMI development, it is necessary to choose the main background in such a way that all applicable elements for the task are visible on it. When choosing a text color, it should be considered that the text should be clearly visible on the interface, should not cause irritating feelings for the user, and have a positive effect on the psychological pressure of the crew. The use of the described method when developing the visual design of the civil aircraft HMI will reduce the time spent to perform the required tasks, reduce the workload on the crew, thereby increase flight safety, and provide a competitive advantage of the developed system.

Keywords Civil aircraft · Ergonomics · Avionics · Human–machine interface · Graphic information

1 Introduction

The problems of the human factor in the aviation industry have been considered repeatedly [1–3]. One of the significant factors affecting flight safety is the level of ergonomics of avionics.

Due to the growth in the use of human–machine interfaces (HMIs), avionics design must be guided by techniques that allow to design the interface as efficient and effective as possible. The guidelines available today are mostly recommendations and show what HMI should not be, that is, they do not provide specific rules for designing interfaces. While creating new methods and rules, developers are moving

N. Silin (✉) · A. Ivanov
Moscow Aviation Institute (National Research University), Moscow, Russia
e-mail: Ndsilin@mail.ru; SilinND@mai.ru

O. A. Gorbachev et al. (eds.), *Proceedings of 10th International Conference on Recent Advances in Civil Aviation*, Lecture Notes in Mechanical Engineering, https://doi.org/10.1007/978-981-19-3788-0_13

toward flexibility and improvement of many standards related to HMI. This indicates a lack of clear criteria for evaluating the effectiveness of interfaces.

In many critical areas of industry, difficult working conditions associated with an inefficient display system can lead to catastrophic consequences with loss of life. In this regard, the issue of developing criteria for evaluating the effectiveness of HMI critical systems is currently acute.

The article proposes an assessment of the effectiveness of the visual design parameters of the HMI of aircraft objects according to the criterion of color unification. Using this solution will reduce the time spent on the required tasks and the number of human errors by minimizing the psychological burden on crew members.

2 Parameters Affecting the Quality of Human–Machine Interfaces

HMI is a complex of technical and information software tools, through which an interactive mode of interaction between a human operator and computing facilities is carried out [4].

An analysis of works in the field of requirements for the visual design of the HMI showed that the evaluation of the effectiveness of the HMI depends on the following parameters [5]:

1. Task execution time.
2. The number of involuntary user errors.
3. HMI standardization.
4. Input information volume.

All of the above options can be combined into two groups:

1. Visual design options.
2. Parameters of system functionality.

As the rule, HMI developers consider the functionality of the system separately from its visual user interface, and little consideration is given to the interaction between the user and the system. But, software users, as a rule, do not separate the functionality and user interface, and satisfaction with interaction with a software product is formed directly from working with the HMI. Therefore, visual design parameters are the key characteristics of the interface, and their effectiveness directly affects the quality of work.

3 Evaluation of the Effectiveness of the Visual Design of the HMI According to the Criterion of Color Unification

When developing the visual design of the HMI for civil aircraft in accordance with Guideline 25-11A, which defines the requirements to the electronic indication systems for transport category aircraft, it is recommended to use 8 color shades: black, green, red, blue, amber, white, magenta, and brown [6]. The incorrect combination of these colors can irritate crew members, leading to increased cognitive load, and therefore potentially catastrophic consequences.

Based on the axiom about the potential danger of human activity "A person's reaction to external stimuli may be erroneous and be accompanied by anthropogenic and technogenic hazards," in order to prevent an erroneous reaction of the user, his attention should be limited from sources of irritation [2]. In this regard, in order to select an ergonomic color pair, it is necessary to take into account that:

1. The background color should not interfere with the perception of the imposed information elements;
2. Labels based on the image of controls, menus, symbols, and graphics must remain identifiable and distinguishable.

The proposed assessment of the effectiveness of the HMI visual design is based on the choice of a color scheme for the background and text, taking into account the visibility of the color on the background. Each color combination was evaluated according to the criteria presented in Table 1.

Table 2 shows an analysis of the combination of color pairs. Applying the criteria presented in Table 1, each background color obtained a score. Assuming 100% visibility of all colors, the background was assigned a maximum score of 7 (according to Guideline 25-11A, 7 colors were used for each background). The presence of color with the criterion "Poorly visible" and "Not visible" to the background was added points of 0.5 and 0, respectively. The score column shows the reason for the reduction in the total score for each color.

To select the most ergonomic color solution, Table 3 suggests a percentage that allows you to determine the correct use of background and text colors.

The use of color combinations with 50 and 0% visibility when designing the visual interface of the aviation indicator on the bench test resulted in an increase in task

Table 1 Evaluation criteria

Evaluation criteria	
Visible	1 (100%)
Poorly visible	0, 5 (50%)
Not visible	0 (0%)

Table 2 Combination of color pairs

Backg-round color	Text color							Score, reason	
Black	Green	Red	Blue	Amber	White	Magenta	Brown	6	Brown is not visible
Green	Black	Red	Blue	Amber	White	Magenta	Brown	3,5	Blue and red are not visible; amber, magenta, brown are poorly visible
Red	Black	Green	Blue	Amber	White	Magenta	Brown	3	Green, magenta, brown are not visible, amber and blue colors are poorly visible
Blue	Black	Red	Green	Amber	White	Magenta	Brown	3,5	Green, brown are not visible, red, amber, magenta colors are poorly visible
Amber	Black	Red	Blue	Green	White	Magenta	Brown	3,5	Blue is not visible, green, red, brown, white and magenta colors are poorly visible
White	Black	Red	Blue	Amber	Green	Magenta	Brown	6	Blue and amber colors are poorly visible
Magenta	Black	Red	Blue	Amber	White	Green	Brown	3	Red, amber, brown are not visible, blue, green colors are poorly visible
Brown	Black	Red	Blue	Amber	White	Magenta	Green	3	Red, magenta, black are not visible, blue, green colors are poorly visible

Table 3 Evaluation of background and text colors

-	Black text (%)	Green text (%)	Red text (%)	Blue text (%)	Amber text (%)	White text (%)	Magenta text (%)	Brown text (%)
Black background	–	100	100	100	100	100	100	0
Green background	100	–	0	0	50	100	50	50
Red background	100	0	–	50	50	100	0	0
Blue background	100	0	50	–	50	100	50	0
Amber background	100	50	50	0	–	50	50	50
White background	100	100	100	50	50	–	100	100
Magenta background	100	50	0	50	0	100	–	0
Brown background	0	50	0	50	100	100	0	–

completion time and human error. In this regard, it is allowed to use colors that have only 100% visibility. These color pairs are described in Table 4 and are offered as an addition to the document [6].

4 Analysis of Time Characteristics When Working with the User Interface

To analyze the time characteristics of the proposed solution, it is proposed to use a variation of the GOMS–KLM method. This method allows you to simulate the performance of the necessary task by the user, and based on the model, evaluate the quality of the interface according to the criterion of time parameters [7].

An observable GOMS method is present, such as analysis using the GOMS critical path method (GOMS critical path method, CPM-GOMS) or the most common version that uses only non-target operators, methods, or keystroke-level model (KLM) selection rules. The KLM method is the most efficient and fastest method for assessing the time characteristics required by an expert and is suitable for comparing alternative solutions in terms of tasks for a wide range of users.

To model the process, a list of key time characteristics was created, shown in Table 5, which will be used when performing a certain action.

Table 4 Additional colors solution

In accordance with paragraph 4.3.5.8 of manual 25-11A [5], the following pairs of colors should be avoided:	Additional colors solutions are suggested:
1. Intense red and blue 2. Intense red and green 3. Intense blue and green 4. Purple on green 5. Purple on black 6. Green on white 7. Blue on black 8. Red on black	*Background use*: 1. Black background for green, red, blue, amber, white, and magenta text 2. Green background for black and white text 3. Red background for black and white text 4. Blue background for black and white text 5. Amber background for black text 6. White background for black, green, red, magenta and brown text 7. Magenta background for black and white text 8. Brown background for amber and white text *Text use*: 1. Black text for all background colors 2. Green text for black and white backgrounds 3. Red text for black and white backgrounds 4. Blue text for black background 5. Amber text for black and brown backgrounds 6. White text for black, green, red, blue, magenta and brown backgrounds 7. Magenta text for black and white backgrounds 8. Brown text for white background

Table 5 Time characteristics

Operator	Action description
A	The time it takes to press the control panel
B(n)−n * A (s)	The time it takes to press a set of keys in sequence on the control panel
C	The time it takes for an operator to point to a position on the user interface
D	The time it takes for an operator to press a trackball button
F	The time it takes for a user to move their hand from the keyboard to the trackball or from the trackball to the keyboard
G	The time required for the user to prepare for the next step
H(t)	System response time

The proposed color solution was used to design the HMI for the aircraft navigation system and tested on the test benches of avionics together with test pilots, line pilots, developers, and people who do not have experience with this interface.

For testing, 4 groups of experts with different backgrounds, 5 people each, were assembled. Table 6 describes the characteristics of the expert group.

Table 6 Expert group

Expert number	Sex	Age	Experience in this position
Group 1—test pilots			
Expert 1	M	47	12
Expert 2	M	51	20
Expert 3	M	44	12
Expert 4	M	53	18
Expert 5	M	49	19
Group 2—line pilots			
Expert 6	M	44	22
Expert 7	M	42	15
Expert 8	M	51	20
Expert 9	M	50	19
Expert 10	M	42	11
Group 3—developers			
Expert 11	M	38	12
Expert 12	F	34	5
Expert 13	M	27	8
Expert 14	F	24	2
Expert 15	F	33	8
Group 4—people who do not have experience with this interface			
Expert 16	F	24	–
Expert 17	F	27	–
Expert 18	M	32	–
Expert 19	F	34	–
Expert 20	M	22	–

During testing, 10 series of 50 experiments each were carried out. Each experiment included a set of actions on the HMI of the aircraft navigation system. Test results revealed that, compared with the assumption according to the P25-11A document, the proposed color scheme allowed to reduce the time spent on the required task.

Table 7 shows the results of testing against time criteria for one of the action sets "selection of departure and destination aerodrome." After conducting all the series, the average time was obtained, which was spent on performing one operation.

From the amount of the spent time, it can be seen that using the proposed color scheme, the total time to complete the action was reduced by 2.5 s.

Table 7 Results of approbation of time characteristics

Action: selection of departure and destination aerodrome on the civil aircraft navigation system				
No.	Type of transaction	Average operation time according to P25-11A	Average operation time according to the proposed method	Description
1	F	0, 4	0, 4	Move your hand to the trackball
2	G	1, 2	0, 9	Thinking
3	C	1, 1	0, 9	Move the trackball pointer to the flight plan field in the menu section
4	D	0, 2	0, 2	Pressing the trackball button
5	G	1, 2	1, 0	Thinking
6	C	1, 1	1, 0	Move the trackball pointer to the departure airfield
7	D	0, 2	0, 2	Pressing the trackball button
8	G	2, 2	1, 5	Selection of the departure aerodrome from the proposed list
9	C	1, 2	1, 0	Move the trackball pointer to the destination airfield
10	G	2, 1	1, 4	Selecting the destination aerodrome from the proposed list
11	D	0, 2	0, 2	Pressing the trackball button
12	C	1, 1	1, 0	Move the mouse pointer to the "EXEC" button to confirm changes to the flight plan
13	D	0, 2	0, 2	Pressing the trackball button
Sum		12, 4	9, 9	

5 Analysis of the Number of Errors Made When Working with the User Interface

Table 8 shows an analysis of the number of human errors when performing the same action "selection of departure and destination aerodrome." Based on the results of all series of tests, the average value of the number of errors was obtained.

Table 8 Number of committed human errors

Action: selection of departure and destination aerodrome on the civil aircraft navigation system				
No.	Type of transaction	Average operation time according to P25-11A	Average operation time according to the proposed method	Description
1	F	0	0	Move your hand to the trackball
2	G	0	0	Thinking
3	C	2	1	Move the trackball pointer to the flight plan field in the menu section
4	D	0	0	Pressing the trackball button
5	G	0	0	Thinking
6	C	2	1	Move the trackball pointer to the departure airfield
7	D	0	0	Pressing the trackball button
8	G	3	2	Selection of the departure aerodrome from the proposed list
9	C	2	1	Move the trackball pointer to the destination airfield
10	G	3	2	Selecting the destination aerodrome from the proposed list
11	D	0	0	Pressing the trackball button

(continued)

Table 8 (continued)

Action: selection of departure and destination aerodrome on the civil aircraft navigation system				
No.	Type of transaction	Average operation time according to P25-11A	Average operation time according to the proposed method	Description
12	C	1	1	Move the mouse pointer to the "EXEC" button to confirm changes to the flight plan
13	D	0	0	Pressing the trackball button
Sum		13	8	

From the sum of the mistakes made in one action, it can be seen that the use of the proposed color solution made it possible to reduce the number of human errors by 39%.

6 Conclusion

The ergonomic combination of color pairs plays a major role in HMI design. Considering that up to 7 different text colors can be used in HMI development, it is necessary to choose the main background in such a way that all applicable elements for the task are visible on it.

When choosing a text color, it should be considered that the text should be clearly visible on the interface, should not cause irritating feelings for the user, and have a positive effect on the psychological pressure of the crew.

The use of the described method when developing the visual design of the civil aircraft HMI will reduce the time spent to perform the required tasks, reduce the workload on the crew, thereby increase flight safety, and provide a competitive advantage of the developed system.

References

1. Artemov AD, Lysakov ND, Lysakova EN (2018) The human factor in the operation of aviation equipment. Monograph
2. Belov SV, Belova SV (2014) Noxology: textbook and workshop for bachelors, 3rd edn. Revised and additional. Moscow: Yurayt Publishing House, pp 451. Series: bachelor, basic course
3. Papadimitrio E, Schneider C, Tello AJ, Damen W, Vrouenraets LM, Broeke A (2020) Transport safety and human factors in the era of automation: What can transport maodes learn from each other? Delft University of Technology. Faculty of Technology, Policy and Management, Jaffalaan 5, 2628 BX, Delft, the Netherlands

4. Sergeev SF, Paderno PI, Nazarenko NA (2011) Introduction to the design of intelligent interfaces: Textbook. St. Petersburg: SPbGU ITMO, pp 108
5. Paderno PI, Burkov EA, Nazarenko NA (2015) The quality of information systems: a textbook for students of higher educational institutions studying in the direction of undergraduate training "Information systems and technologies." Academy, Moscow, p 218
6. AR IAC (2009) Guidelines 25–11A for the certification of electronic indication systems for transport category aircraft, pp 70
7. Kieras DA (2002) Guide to GOMS model usability evaluation using GOMSL and GLEAN3. University of Michigan (ftp.eecs.umich.edu/people/kieras)
8. Lysakov ND (2014) Psychological aspects of the human factor in aviation. Bulletin of the University 2:250–253

Radio Navigation Systems

The chapter focuses on urgent issues and prospects for developing the radio electronic equipment: aircraft radio navigation, radar and communication systems and complexes, implementing modern technologies in air traffic management, supporting the operation of air transport radio equipment.

Keywords Radio navigation · Radio detection and ranging · Air traffic management · Flight safety · Communication systems · Radio-technical support of flights · International Civil Aviation Organization

Peculiarities of Applying Pseudolites for Increase in GNSS Positioning Accuracy

Oleg Skrypnik, Natalya Arefyeva, Roman Arefyev, and Tatyana Portnova

Abstract The authors consider main elements of the global navigation satellite system (GNSS) and peculiarities of its operation in the multi-system mode and with a pseudolite as an augmentation. A simulation modeling complex for orbital groupings of systems forming the GNSS was presented. By comparing the results of natural experiments and mathematical modeling the performance of the complex with regard to the BeiDou system was estimated. The vertical dilution of precision was estimated for all possible modes of GNSS application as well as with a pseudolite used. The technique of finding an optimal pseudolite position was proposed and efficiency of its use in all possible modes of GNSS application was estimated.

Keywords Global navigation satellite system · Multi-system mode · Augmentation · Pseudolite · Vertical dilution of precision · Optimization

1 Introduction

One of the key factors for the efficient operation of the air transport system is creation of a safe, highly accurate, reliable, and environmentally sound air navigation system at the global, regional, and national levels. It is the global navigation satellite system (GNSS) standardized by ICAO which is most appropriate to modern requirements for air navigation service. The GNSS is the most important tool in implementing the performance-based navigation (PBN) concept and contributes to safer and more efficient flying all over the world [1].

Traditional GNSS applications in civil aviation (GA) mainly use a single frequency of one GNSS satellite constellation, as the rule, the L1 frequency of the GPS system (USA) which serves as the basis for global implementation of the PBN

O. Skrypnik (✉)
Belarusian State Academy of Aviation, Minsk, Republic of Belarus
e-mail: skripnikon@yandex.ru

N. Arefyeva · R. Arefyev (✉) · T. Portnova
Irkutsk Branch of Moscow State Technical University of Civil Aviation, Irkutsk, Russia
e-mail: aqua160905@mail.ru

O. A. Gorbachev et al. (eds.), *Proceedings of 10th International Conference on Recent Advances in Civil Aviation*, Lecture Notes in Mechanical Engineering, https://doi.org/10.1007/978-981-19-3788-0_14

and Automatic Dependent Surveillance (ADS) [2–4]. For use in civil aviation, the ICAO also recommends fully deployed GLONASS (Russia) and Beidou (China) systems which are GNSS elements, and in the future—the Galileo system (The European Union) as well.

Apart from the PBN and ADS, the GNSS is also used in many other areas of aviation that require information on the location of mobile objects and time (e.g., GPWS, ACAS, MLAT, A-SMGCS, etc.).

As new technologies of air navigation service and airspace use are developed and implemented, the dependence on the location and time information provided by the GNSS significantly increases and this trend will continue as the traditional navigation infrastructure is rationalized.

The experience in operating separate GNSS components identified a number of restrictions on their application for separate phases of the flight, first of all, landing. It is possible to remove some restrictions by using satellite- and ground-based augmentation systems. The augmented GNSS is an essential technical means which provides efficiency and reliability in implementing the CNS/ATM concept and all its applications.

In performing the task of position-fixing the ICAO specialists consider transition to multi-frequency (dual-frequency) receivers and use of several satellite constellations simultaneously (a multi-system mode) as tools for achieving even greater operational benefits from GNSS application [5–8].

Another tool for improving the GNSS efficiency could be the use of augmentations in the form of pseudolites (PLs)—stationary or mobile transmitters with known coordinates transmitting signals in the format of GNSS elements and compatible with existing equipment [9–11].

These technologies create opportunities for further improvement of GNSS navigation performance and gain of additional operational benefits such as improved accuracy, integrity, availability and continuity of service in use, reduced GNSS vulnerability and sensitivity to intended and unintended interference.

Of particular interest in this regard are studies aimed at assessing the advantages in positioning accuracy that can be obtained when operating the GNSS in a multi-system mode and using pseudolites as an augmentation.

2 Research Tools and Methods

For conducting research in the LabView environment, a complex was developed for simulating the orbital constellations of the systems that form the GNSS [12], adding to it the module for simulating the BeiDou orbital constellation. The appearance of the complex interface panel is shown in Fig. 1.

A special BeiDou feature is a hybrid orbital constellation composed of 3 orbit types (MEO, GEO, and GSO) located at different heights. This system has a much greater number of navigation satellites (NSs) than GLONASS or the GPS.

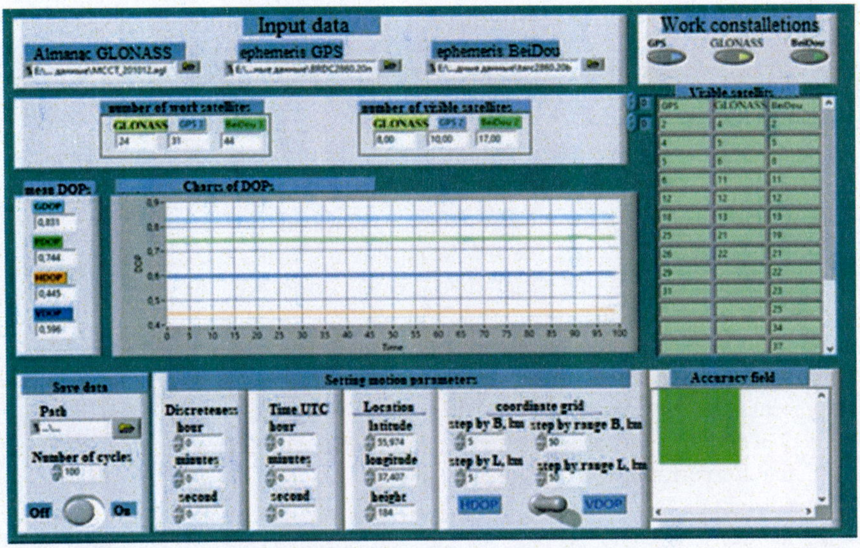

Fig. 1 Interface panel of the simulation complex

The adequacy assessment of the BeiDou orbital constellation module was carried out by comparing the results of a natural experiment with the ATGM336H 5 N receiver and the results of simulation. For comparison, the dilution of precision (DOP) values at the observation point (52.09 N, 104.34 E) were used. During the natural experiment, the positioning accuracy provided by the receiver was also evaluated. The experiment was carried out for an 8 h time interval with a measurement resolution of 1 s.

The peculiarity of the receiver used is the ability to work with a limited NS number of the BeiDou system (only up to the 20th one). Moreover, during the experiment, no more than 11 NSs from this number could be seen. The website of the Chinese Office for Satellite Systems contains information about the ephemerides of the complete BeiDou grouping composed of 49 NSs. For mathematical modeling, the ephemerides of only those satellites were taken that were visible by the receiver during the natural experiment.

When assessing the PL effectiveness, the final approach to an airdrome located at high latitudes was considered. Two PL locations were considered: the PL location at a fixed point in the terminal area and its optimal location which provided a minimal mean DOP for the entire flexible approach trajectory. The optimization was carried out with help of Hooke—Jeeves method [13].

3 The Investigation Results and Their Discussion

The results of the experiments on research of the BeiDou positioning characteristics are shown in Figs. 2 and 3.

Figure 2 shows that the trend of the VDOP variation which was obtained when simulating (curve 1) corresponds to the results of the natural experiment (curve 2) which means adequacy of the program module operation.

The presented diagram of scattering the positioning errors in the horizontal plane (Fig. 3a) and the graph of the height measurement error (Fig. 3b) show that, when observing 11 NSs, the BeiDou system provides sufficiently high (only a few meters) positioning accuracy compatible to GLONASS and GPS accuracy.

The results obtained for the BeiDou system as well as the results obtained earlier and presented in [14] for GLONASS, GPS, and multi-system modes, show that the accuracy of determining the height for all the considered GNSS applications remains insufficient for performing a categorized landing.

Since height measurement accuracy is related to VDOP, effectiveness of different GNSS applications could be compared by the mean VDOP over a fairly long time interval. The summarized results for a day's observation interval obtained by simulating are shown in Table 1.

The analysis of the results obtained shows that the BeiDou system has potentially the best height measurement accuracy. Operation in the multi-system mode, especially when BeiDou is included, can significantly improve positioning and height measurement accuracy. The presence of satellite shadings, stimulated by an increase in the mask angle from 5 (typical value) to 10°, degrades the mean VDOP by 25–30%.

One of the possible ways for further improvement of GNSS accuracy in the vertical channel is PL use. In this case, the problem of finding an optimal PL position in the terminal area arises since time change in the configuration of GNSS orbital constellations leads to a change in the DOP and, accordingly, in the positioning

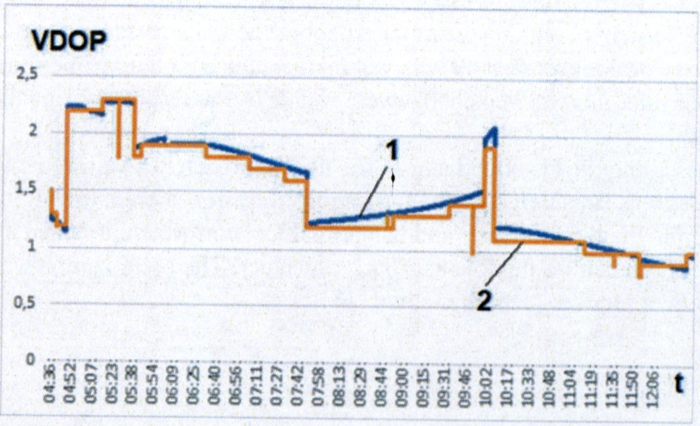

Fig. 2 VDOP change

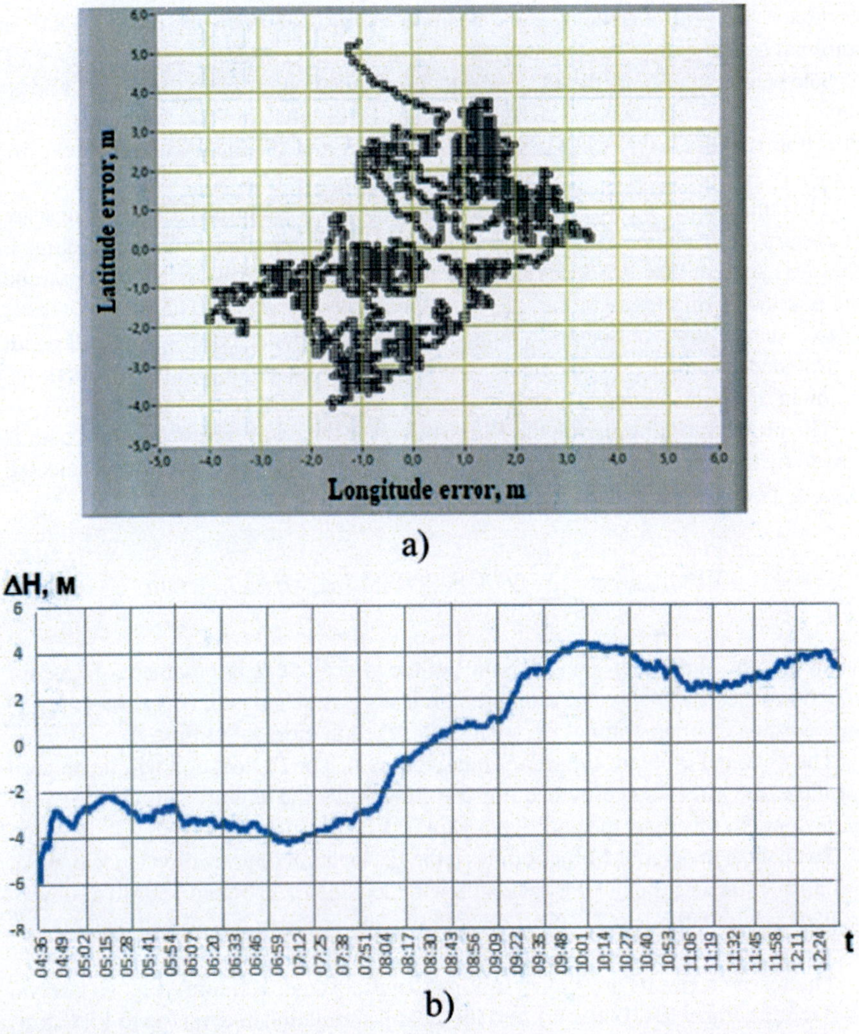

Fig. 3 BeiDou position accuracy: **a**—in the horizontal plane; **b**—the height measurement error

Table 1 The mean VDOP values with no PL used

Satellites system	GLONASS	GPS	BeiDou	GLONASS/BeiDou
Mean VDOP	1.51	1.72	0.93	0.74
Satellites system	GLONASS/GPS	GPS/BeiDou	GLONASS/GPS/BeiDou (mask angle of 5°)	GLONASS/GPS/BeiDou (mask angle of 10°)
Mean *VDOP*	1.02	0.76	0.65	0.81

accuracy. We will consider the PL position as optimal when the mean VDOP is minimal over the observation interval.

The results of solving the PL placement optimization problem for a fixed glideslope point with coordinates 72.015 N, 114.051 E (distance of 7 km from the runway threshold), with GLONASS/GPS/BeiDou operating in the multi-system mode, are shown in Fig. 4.

The analysis of the results obtained shows that the optimal PL locations are concentrated below the glideslope point relative to which the problem of optimization was solved. The RMS deviation of the position scattering is about 16 m and the maximum difference in the VDOP value at these points is 0.85%. This allows us to conclude that, for operation in the multi-system mode, a near-to-optimal result is provided when a PL is installed below the glideslope point relative to which the problem of optimization is solved.

The problem of optimizing the PL position for the entire approach flight path is solved by finding a point that provides the minimal VDOP averaged for all selected discrete points of the glideslope over the specified time interval.

$$\mathrm{VDOP}_{\mathrm{mn}_k} = \frac{1}{n}\sum_{1}^{n} \mathrm{VDOP}_{k,n}(B_{\mathrm{PL}_k}, L_{\mathrm{PL}_k}, H_{PL_k}) \rightarrow \min$$

where n is the number of selected points of the glideslope; k is a moment of time for which the mean VDOP value along the glideslope is calculated; $(B_{\mathrm{PLk}}, L_{\mathrm{PLk}}, H_{\mathrm{PLk}})$ are coordinates of an optimal PL position for the kth moment of time.

The obtained problem solution is quasi-optimal. The PL location will be optimal for the entire glideslope provided it is traveling while the aircraft is flying along the glideslope. In this case, the additional gain in VDOP value is 8% [15].

To analyze in detail the functioning of the PL location optimization algorithm, let us consider the cases of a GPS approach with heading of 270° and 180° to a runway located at high latitudes (72° N). The optimization problem was solved over 10 min intervals.

The research results are shown in Fig. 5 (initial stage, 100 min), Fig. 6 (interval 0–200 min), Fig. 7 (0–1000 min), and for the full simulation interval (up to 1500 min) Fig. 8.

In the figures you can see (a)—the graph of VDOP changes, (b)—PL positions when approaching with a heading of 270°, (c)—PL positions when approaching with a heading of 180°.

As can be seen from the presented results the VDOP degradation caused by a decrease in the number of visible NSs to 5 leads to a significant shift in the optimal PL position. The majority of the optimal PL positions are located near the glideslope and the optimal positions themselves depend on the approach heading.

The generalized results for the mean VDOP, when a GNSS is augmented with a PL installed at an optimal point, are shown in Table 2.

The analysis of the results obtained shows that the efficiency of applying a PL in the multi-system GNSS operation mode is decreased compared to using one of the

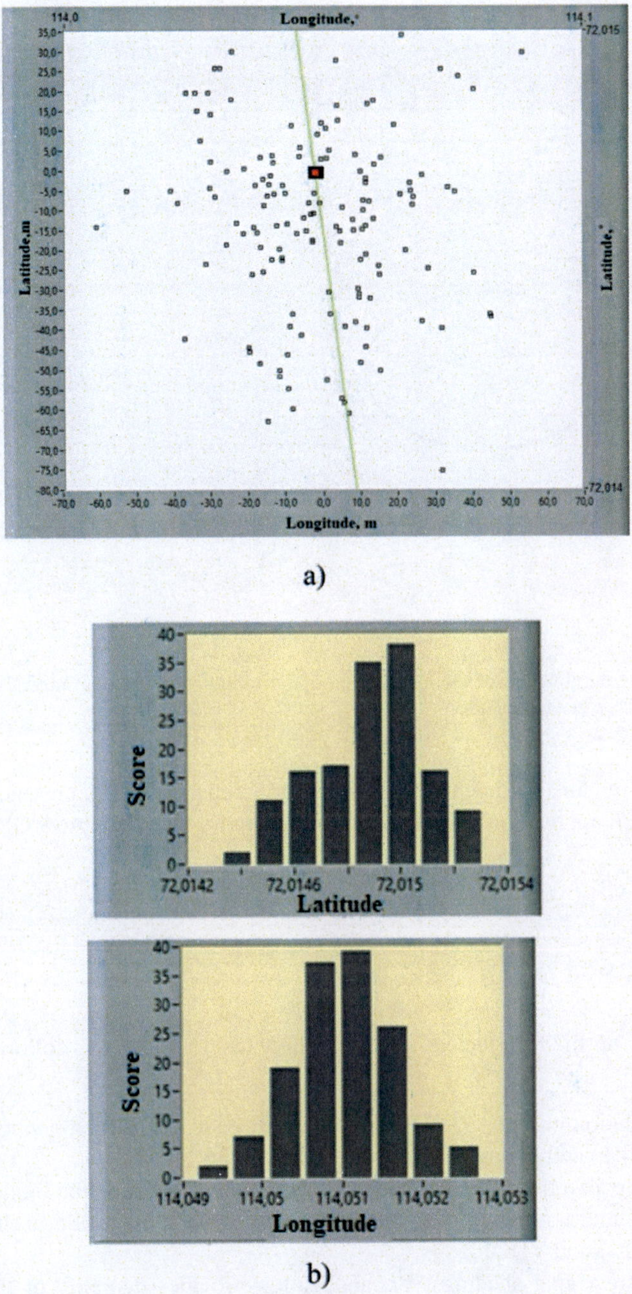

Fig. 4 The results of solving the optimization problem for a fixed point: **a** diagram of scattering the PL positions relative to the point; **b** histograms of latitude and longitude distribution of the PL positions

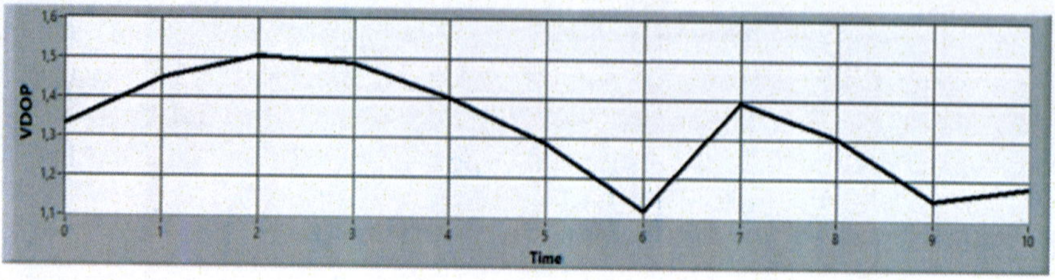

a)

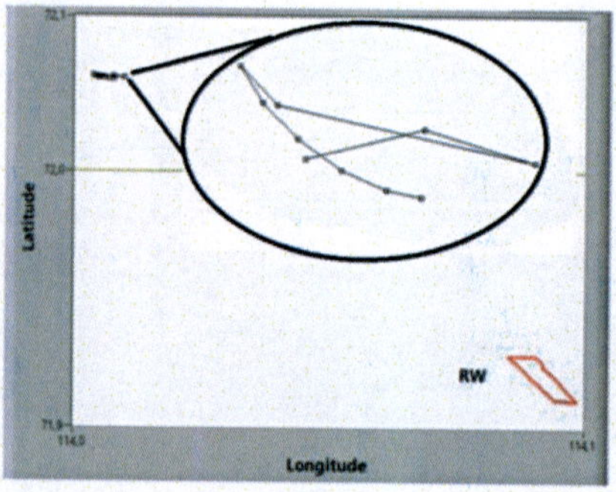

b)

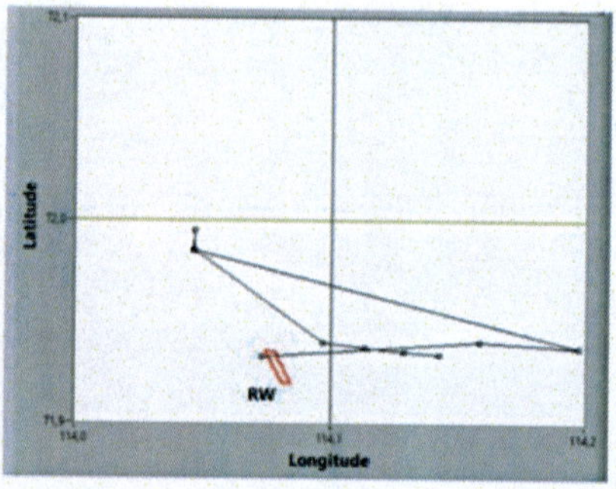

c)

Fig. 5 Solving the problem of the PL position optimization on a 100 min interval **a** VDOP, **b** heading of 270°, **c** heading of 180°

systems but remains sufficiently significant for practical use. The presence of shadings makes PL application more effective including for the GLONASS/GPS/BeiDou mode.

4 Conclusion

The results of the conducted research allow us to draw the following main conclusions:

- PL application provides VDOP improvement both when using a single system (by 32–54%) and in multi-system mode (by 22–36%);
- the relative gain from PL application is decreased when using the Beidou system and in the multi-system mode, i.e., in those cases when the GNSS itself provides sufficiently good VDOP;
- if there are signal shadings, PL application provides damping of the VDOP degradation including when using the multi-system mode;
- for a fixed point of the glideslope the optimal PL position is below this point within the zone not exceeding 30 m;

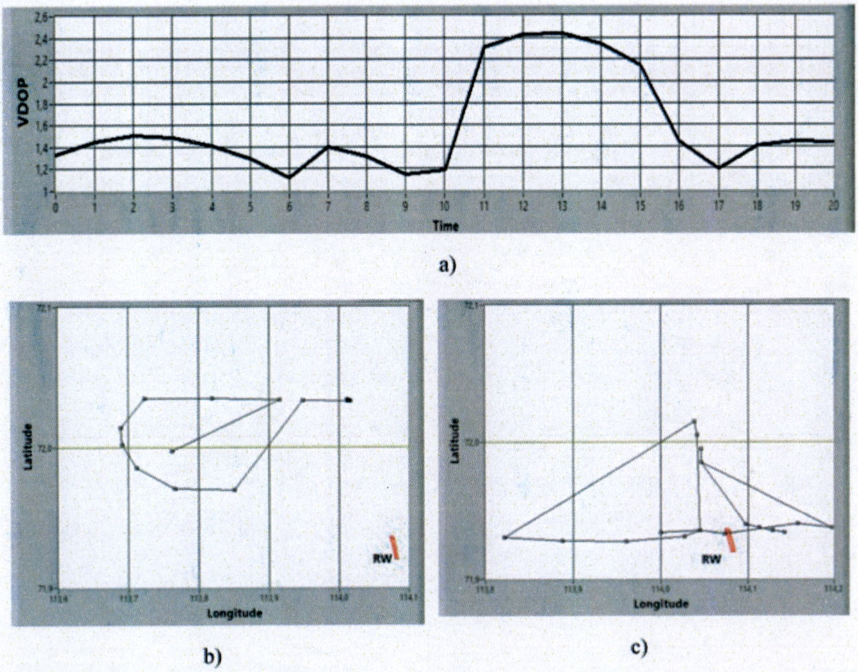

Fig. 6 Solving the problem of the PL position optimization on a 200 min interval **a** VDOP, **b** heading of 270°, **c** heading of 180°

- this allows the PL position to be determined without solving the optimization problem for each particular airdrome;
- in high latitudes, the zone of scattering the PL optimal positions has the largest size when using the GPS and is decreased significantly when using the multi-system mode.

The efficiency, and hence, usefulness of applying the procedure for optimizing the PL location is determined by the variation range of DOP values provided by the GNSS in the considered space area. If the DOP values vary insignificantly, the optimization procedure becomes ineffective and in this case, the PL should be installed below the glideslope at any point suitable for the location conditions.

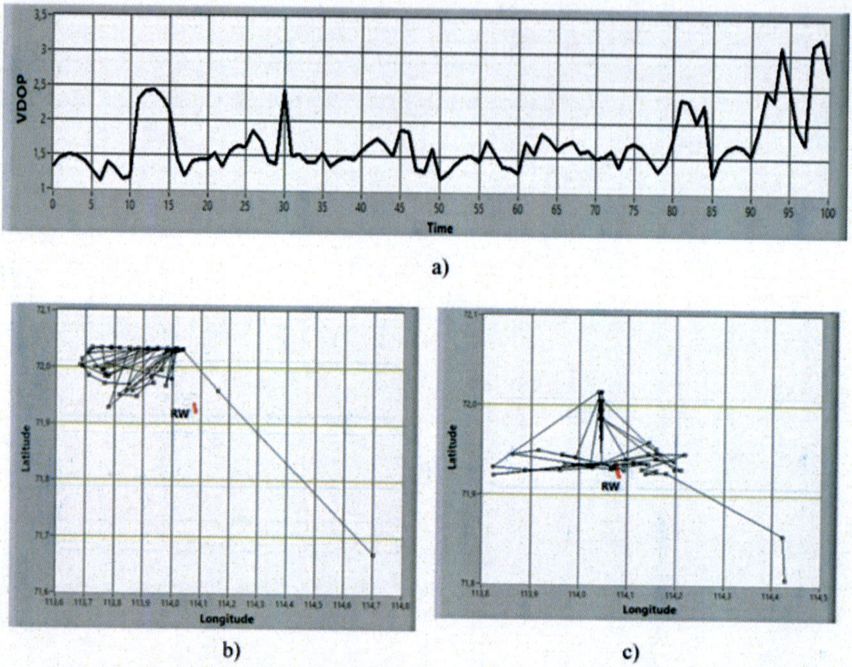

Fig. 7 Solving the problem of the PL position optimization on a 1000 min interval **a** VDOP, **b** heading of 270°, **c** heading of 180°

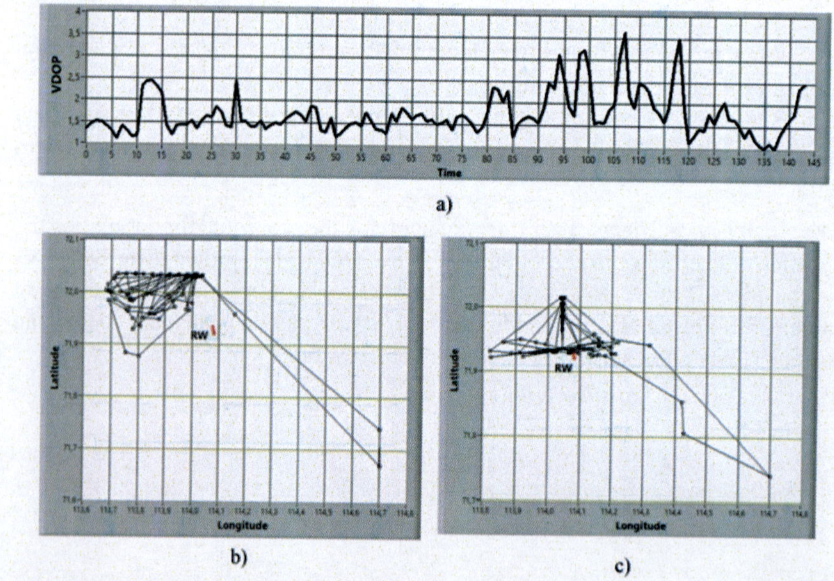

Fig. 8 Solving the problem of the PL position optimization on a 1500 min interval **a** VDOP, **b** heading of 270°, **c** heading of 180°

Table 2 Mean VDOP values with a PL used

Satellite system	GLONASS	GPS	BeiDou	GLONASS/BeiDou
Mean *VDOP*	0.74	0.8	0.63	0.55
Gain, %, compared to Table 1	51	53.5	32.2	25.7
Satellite system	GLONASS/GPS	GPS/BeiDou	GLONASS/GPS/BeiDou (mask angle of 5°)	GLONASS/GPS/BeiDou (mask angle of 10°)
Mean *VDOP*	0.65	0.56	0.51	0.57
gain, %, compared to Table 1	36.3	26.4	21.5	29.6

References

1. Rukovodstvo po navigacii, osnovannoj na harakteristikah (PBN) (Performance-based Navigation (PBN)) Doc. 9613. 4-e izd, pp 444 (In Russian)
2. Global Navigation Satellite System (GNSS) Manual (2013). Doc. 9849. 2nd edn. ICAO, pp 90
3. Sauta OI, Shatrakov AY, Shatrakov YG, Zavalishin OI (2019) Satellite radio navigation systems. In: Principles of radio navigation for ground and ship-based aircrafts. Springer Aerospace Technology. Springer, Singapore. https://doi.org/10.1007/978-981-13-8293-2_6
4. Akmaykin DA, Bolelov EA, Kozlov AI, Lezhankin BV, Svistunov AE, Shatrakov YG (2021) Pseudo-ranging radio navigation systems. In: Theoretical foundations of radar location and radio navigation. Springer Aerospace Technology. Springer, Singapore. https://doi.org/10.1007/978-981-33-6514-8_14
5. Concept of operations (CONOPS) for dual-frequency multi-constellation (DFMC) global navigation satellite system (GNSS) (2018). ICAO navigation systems panel. CONOPS V6.4: 49
6. Hofmann-Wellenhof B, Lichtenegger H, Wasle E (2008) GNSS–global navigation satellite systems: GPS, GLONASS, Galileo, and more. Springer Science and Business Media. https://doi.org/10.1007/978-3-211-73017-1
7. Kaplan E, Hegarty C. (2005) Differential GPS. Understanding GPS: Principles and applications, 2nd edn. Artech House: Norwood, MA, USA.—C, pp 379–454
8. Zheng X, Liu Y, Fan G, Zhao J, Xu C (2018) Analyses of the sensitivity of multi-constellation advanced receiver autonomous integrity monitoring vertical protection level availability to error parameters and a failure model over China. Adv Mech Eng. https://doi.org/10.1177/1687814018776191
9. Cobb HS (1997) GPS pseudolites: theory, design, and applications—Ph.D. Dissertation, Stanford University, September
10. Raquet J et al (1996) Development and testing of a mobile pseudolite concept for precise positioning. Navigation 43(2):149–165
11. Lee HK et al (2002) Analysis of pseudolite augmentation for GPS airborne applications. In: Proceedings 15th international technology of the meeting of the satellite division of the US Institute of Navigation, Portland, Oregon, pp 24–27
12. Skrypnik ON, Aref'ev RO, Astrakhanceva NG (2015) Eksperimental'nye issledovaniya harakteristik sovmeshchennogo GNSS priemnika v vysokih shirotah (Experimental researches of the combined GNSS receiver characteristics in high latitudes). Nauchnyj Vestnik Moskovskogo

Gosudarstvennogo Tehnicheskogo Universiteta Grazhdanskoj Aviacii. Moskva: MGTU GA. 221(11):35–42. (In Russian)
13. Skrypnik ON, Arefev RO, Arefeva NG (2017) Sovershenstvovanie aeronavigacionnogo obespecheniya etapa posadki putem optimizacii razmeshcheniya psevdosputnikov GLONASS (Improvement of air navigation support of the landing phase by optimizing the allocation of pseudolites GLONASS). Trudy MAI 92:28 (In Russian)
14. Skrypnik ON, Arefyev RO, Arefyeva (Astrakhanceva) NG (2020) The assessment of positioning error characteristics of combined GLONASS/GPS receivers. Crede Experto: Transp Soc Educ Lang 1:44–58
15. Skrypnik ON, Arefev RO (2020) Optimization of a mobile pseudolite track for increasing accuracy of the GLONASS integrated navigation-and-time field. Sovremennye Naukoemkie Tehnologii 2:51–58

On the Laws of the Scatter Matrix Elements Probabilities Distribution

Eduard Bolelov, Anatoly Kozlov, Nikolay Voskresensky, and Vyacheslav Erokhin

Abstract An extensive application of the radio polarimetry methods to solve the most varied tasks of radar location, radar navigation and air traffic control has occurred recently which is inevitably faced with the relevance of describing the reflectance profile for surveillance objects of such ambience as (terrain, vegetation, atmosphere), above-ground facilities, air targets (aircraft, helicopters, unmanned aerial vehicles), etc. On the whole, in all the events formulated solutions are based on the evaluation and dynamics of altering the principal radio polarimetry characteristic of surveillance objects, i.e., their scatter matrix (SM). The present paper is devoted to the laws of the SM elements probabilities distribution.

Keywords Radio polarimetry · Scatter matrix (SM) · Law of distribution · Polarization characteristic

1 Introduction

In a number of practical studies, for example, air traffic control [1, 2] in the vast majority of cases the radar stations and targets (aircraft) have relatively long distance, which allows us to consider the incident wave in the vicinity of the target and the scattered wave of antenna as plane waves, which can be described using only one E-field vector **E**. If we narrow it to the case of monostatic radio location, when

E. Bolelov (✉) · A. Kozlov · N. Voskresensky
Moscow State Technical University of Civil Aviation, Moscow, Russia
e-mail: e.bolelov@mstuca.aero

A. Kozlov
e-mail: a.kozlov@mstuca.aero

N. Voskresensky
e-mail: n.voskresensky@mstuca.aero

V. Erokhin (✉)
Irkutsk Branch of Moscow State Technical University of Civil Aviation, Irkutsk, Russia
e-mail: ww_erohin@mail.ru

O. A. Gorbachev et al. (eds.), *Proceedings of 10th International Conference on Recent Advances in Civil Aviation*, Lecture Notes in Mechanical Engineering, https://doi.org/10.1007/978-981-19-3788-0_15

the receiver and transmitter antennae are superimposed in space, which is typical of radar surveillance systems, the problem of obtaining information about radar targets, finally is driven to the comparison of the two E-field vectors of the emitted wave $\mathbf{E}_{emit}$ and received $\mathbf{E}_{rec}$ electromagnetic waves of the antenna.

The distance to the radar is critical for such a comparison, and it can be easily considered, as the radiation field decreases as 1/R, thus, in order not to overcharge the obtained ratios, the field scattered by the target in the direction of the target (the incident wave), may be interpreted as $\mathbf{E}_{rec.}$

For radar signals the following representation is applicable [3, 4]:

$$S(t) = A(t)\cos(\omega t + \varphi(t) + \varphi_0) \tag{1}$$

where $A(t)$ and $\varphi(t)$—some slowly shifting functions within a period of high frequency $T = 2\pi/\omega$, φ_0—the initial phase, which is not essential for this particular case, thus considered as $\varphi_0 = 0$.

In a number of sources [5, 6] the representation (1) is often replaced by its complex form

$$S(t) = A(t)e^{j\omega t}e^{j\varphi(t)} \tag{2}$$

based on

$$A\cos(\omega t + \beta) = \mathrm{Re}\left(Ae^{j\omega t}e^{j\beta}\right) \tag{3}$$

Such an approach allows us to dwell on both the amplitude and the phase of the signals subject to research. As generally the antenna emits the elliptically polarized radiation (in particular, linear, circular) and the reflected wave is also the case, it is reasonable to use two orthogonal components—the horizontal—E^{hr} and the vertical—E^{vert} instead of vector $\mathbf{E}$. It is clear, that every component of the emitted wave $\mathbf{E}_{\mathrm{emit}}$ "breeds" the both components of the scattered wave $E^{\mathrm{hr}}_{\mathrm{rec}}$ and $E^{\mathrm{vert}}_{\mathrm{rec}}$, at the same time each of them will be different from the radiation pattern component that has bred them both in phase and amplitude.

As the electromagnetic wave scattering is a linear process, there is a linear dependence between the independent components of the scattered and emitted waves. It allows us, considering the being said, to associate the components of the E-field vector of the emitted wave in the vicinity of the target with the corresponding components of the scattered wave of the antenna using the following equations [6–9]:

$$\begin{cases} E^{\mathrm{hr}}_{\mathrm{rec}} = S_{\Gamma\Gamma}e^{j\Psi_{\Gamma\Gamma}}E^{\mathrm{hr}}_{\mathrm{rec}} + S_{\Gamma\mathrm{B}}e^{j\Psi_{\Gamma\mathrm{B}}}E^{\mathrm{hr}}_{\mathrm{emit}}, \\ E^{\mathrm{vert}}_{\mathrm{rec}} = S_{\mathrm{B}\Gamma}e^{j\Psi_{\mathrm{B}\Gamma}}E^{\mathrm{hr}}_{\mathrm{emit}} + S_{\mathrm{BB}}e^{j\Psi_{\mathrm{BB}}}E^{\mathrm{vert}}_{\mathrm{emit}}, \end{cases} \tag{4}$$

where $E^{\mathrm{hr}}_{\mathrm{emit}}$, $E^{\mathrm{vert}}_{\mathrm{emit}}$ —the horizontal and vertical components of the emitted wave, a $\mathrm{E}^{\mathrm{hr}}_{\mathrm{rec}}$, $\mathrm{E}^{\mathrm{hr}}_{\mathrm{rec}}$ —of the received wave, $S_{m,n}$—some coefficients ($m,n = \Gamma,\mathrm{B}$), showing the amplitude of the incident wave, and $\Psi_{m,n}$—the change of phase which takes place during the reflection.

Naturally comes the introduction of complex coefficients $\dot{S}_{\text{mn}} = S_{\text{mn}} e^{j\psi_{\text{mn}}}$ then:

$$\begin{cases} E_{\text{rec}}^{\text{hr}} = \dot{S}_{\Gamma\Gamma} E_{\text{emit}}^{\text{hr}} + \dot{S}_{\Gamma\text{B}} E_{\text{emit}}^{\text{vert}}, \\ E_{\text{rec}}^{\text{vert}} = \dot{S}_{\text{B}\Gamma} E_{\text{emit}}^{\text{hr}} + \dot{S}_{\text{BB}} E_{\text{emit}}^{\text{vert}}. \end{cases} \tag{5}$$

The relation (5) may be represented as the vector–matrix equation:

$$\boldsymbol{E}_{\text{rec}} = \dot{\boldsymbol{S}} \boldsymbol{E}_{\text{emit}} \tag{6}$$

where $E_{\text{rec}} = \begin{bmatrix} E_{1\text{rec}} \\ E_{2\text{rec}} \end{bmatrix}$, $E_{emit} = \begin{bmatrix} E_{1\text{emit}} \\ E_{2\text{emit}} \end{bmatrix}$—column matrices of the received and emitted waves (index 1 corresponds to x-component, index 2 corresponds to y-component):

$$\dot{\text{S}} = \begin{bmatrix} \dot{S}_{11} & \dot{S}_{12} \\ \dot{S}_{21} & \dot{S}_{22} \end{bmatrix} = \begin{bmatrix} S_{11} \text{e}^{\text{j}\psi_{11}} & S_{12} \text{e}^{\text{j}\psi_{21}} \\ S_{21} \text{e}^{\text{j}\psi_{21}} & S_{22} \text{e}^{\text{j}\psi_{21}} \end{bmatrix} \tag{7}$$

The scatter matrix.

The scatter matrix $\dot{S}$ is the basic characteristic of the radar target.

As it is clear from the relation (7), at each moment in time the radar target is defined by four complex and eight real numbers. However, as the laws of electrodynamics suggest, the off-diagonal elements of a scatter matrix are equal to each other, i.e.$\dot{S}_{12} = \dot{S}_{21}$ or $S_{12} = S_{21}$ and $\psi_{12} = \psi_{21}$.Thus, the quantity of values that define the target is diminished to two complex and six real numbers.

Let us discuss the physical aspect of these values. Let the polarized wave be emitted horizontally, i.e. $E_{\text{emit}}^{\text{vert}} = 0$, then we shall have [6–10]:

$$\begin{cases} E_{\text{rec}}^{\text{vert}} = S_{\Gamma\Gamma} e^{j\Psi_{\Gamma\Gamma}} \text{E}_{\text{emit}}^{\text{hr}}, \\ E_{\text{rec}}^{\text{vert}} = S_{\text{B}\Gamma} e^{j\Psi_{\text{B}\Gamma}} \text{E}_{\text{emit}}^{\text{hr}}, \end{cases} \tag{8}$$

from where we obtain for $S_{\text{гг}}$ and $S_{\text{вг}}$:

$$\begin{cases} S_{\Gamma\Gamma} = \left| E_{\text{rec}}^{\text{hr}} \right| \cdot \left| E_{\text{rec}}^{\text{hr}} \right|^{-1}, \\ S_{\text{B}\Gamma} = \left| E_{\text{rec}}^{\text{hr}} \right| \cdot \left| E_{\text{emit}}^{\text{hr}} \right|^{-1}, \end{cases} \tag{9}$$

and for phases $\Psi_{\text{гг}}$ и $\Psi_{\text{вг}}$:

$$\begin{cases} \Psi_{\Gamma\Gamma} = \arg \dfrac{E_{\text{rec}}^{\text{hr}}}{E_{\text{emit}}^{\text{hr}}} \\ \Psi_{\text{B}\Gamma} = \arg \dfrac{E_{\text{rec}}^{\text{vert}}}{E_{\text{emit}}^{\text{hr}}} \end{cases} \tag{10}$$

The argument of the relation of the received wave horizontal component to the similar value of the emitted wave at the target equals to the phase $\Psi_{\Gamma\Gamma}$, and the relation module equals to $S_{\Gamma\Gamma}$. The argument of the received wave vertical component to the horizontal component of the emitted wave ($E_{\text{emit}}^{\text{vert}} = 0$) equals to $\Psi_{\text{B}\Gamma}$, and the relation module equals to $S_{\text{B}\Gamma}$. S_{BB} and $S_{\Gamma\text{B}}$ make similar sense, supposing the vertically polarized wave is being emitted ($E_{\text{emit}}^{\text{hr}} = 0$).

The scatter matrix entries S_{11} and S_{22} are considered as main, and S_{12}—is considered as cross entry.

The scatter matrix (7) may be represented as:

$$\text{S} = \begin{bmatrix} \text{S}_{11} & \text{S}_{12} \\ \text{S}_{12} & \text{S}_{22} \end{bmatrix} \tag{11}$$

where $S_{ij} = \left|S_{ij}\right| e^{i\phi_{ij}};\ \ i, j = 1, 2$—the scatter matrix entries.

However, for practical purpose it is more expedient to be concerned with not the absolute values of the SM elements but with the magnitudes that are appropriately normalized with respect to the amplitude and phase which reduces the number of independent characteristics by two units.

Consequently, we can say that in the radio polarimetry any object under observation (target) is described by means of four numbers which principally differs it from the classical radio location where a description of a visual target is conducted using merely one number, to be precise, by a scattering cross-section [5].

It is simply evident that depending on a target observation angle, the stated set of four numbers as well as a scattering cross-section vary. Moreover, this variation bears crucially random nature.

2 The Laws of Scatter Matrix Elements Probability Distribution

The most comprehensive characteristics, describing the quadruple variation, is the appropriate four-dimensional SM elements probability density function (PDF), that is:

$$W\left(\frac{S_{12}}{S_{11}}, \frac{S_{22}}{S_{11}}, \varphi_{12} - \varphi_{11}, \varphi_{22} - \varphi_{11}\right) \tag{12}$$

For the solution of the problem to set up the desired distribution density (DD), let us use the relations associating the SM eigenvalues λ_1 and λ_2 and characteristics α, γ (angles on Poincare sphere determining the position of the basis within which the study is executed), relating the polarization basis within which the dimension with its own polarization basis is executed [7]:

$$\begin{aligned} S_{11} &= e^{2i\eta} e^{2i\alpha} \left(\lambda_1 e^{-2i\alpha} \cos^2 \gamma + \lambda_2 e^{2i\alpha} \sin^2 \gamma \right) \\ S_{12} &= \left(-\lambda_1 e^{-2i\alpha} + \lambda_2 e^{2i\alpha} \right) \sin \gamma \cos \gamma; \\ S_{22} &= e^{-2i\eta} e^{-2i\alpha} \left(\lambda_1 e^{-2i\alpha} \sin^2 \gamma + \lambda_2 e^{2i\alpha} \cos^2 \gamma \right). \end{aligned} \tag{13}$$

Apart from that, let us add here so-called fundamental equation for the SM elements:

$$\left(S_{11} e^{-2i\beta} - S_{22} e^{2i\beta} \right) \sin 2\gamma - 2S_{12} \cos 2\gamma = 0 \tag{14}$$

where $\beta = \alpha + \eta$.

Relying upon the deduced equations, we can be engaged in defining the desired densities of probabilities for the SM elements that are directly the measured parameters.

The assumption concerning a proportional distribution of angles α and γ and the normal law for the distribution of the SM eigenvalues—λ_1 and λ_2 seems quite reasonable [11–15].

The desired values are DD (distribution densities) of such random variables as S_{ij}, their real and imaginary parts $\mathrm{Re}S_{ij}$ and ImS_{ij}, the generalized phase.

$$\psi_{ij} ImS_{ij} = \tan^{-1} \frac{ImS_{ij}}{\mathrm{Re}S_{ij}} \tag{15}$$

the SM elements modules.

$$\left| S_{ij} \right| = \sqrt{\left(\mathrm{Re}S_{ij} \right)^2 + \left(ImS_{ij} \right)^2} \tag{16}$$

the complete scattering cross-section.

$$\sigma_{\Sigma}^2 = \left| \lambda_1^2 \right| + \left| \lambda_2^2 \right| \tag{17}$$

Polarization anisotropy coefficient q.

Search for the laws of distribution was conducted within the framework of a set of Pearson curves. As a result, the PDFs of all the stated above characteristics were deduced [10].

For illustration the Figs. 1, 2 and 3 show PDFs $W(\psi_{11})$, $W(\mathrm{Re}S_{11})$ and $W(\mathrm{Re}S_{12})$ at the various SM eigenvalues λ_1 and λ_2.

Fig. 1 PDF (ψ_{11}), $\overline{\lambda_1} = 10, 25$. $\overline{\lambda_2} = 0, 1$

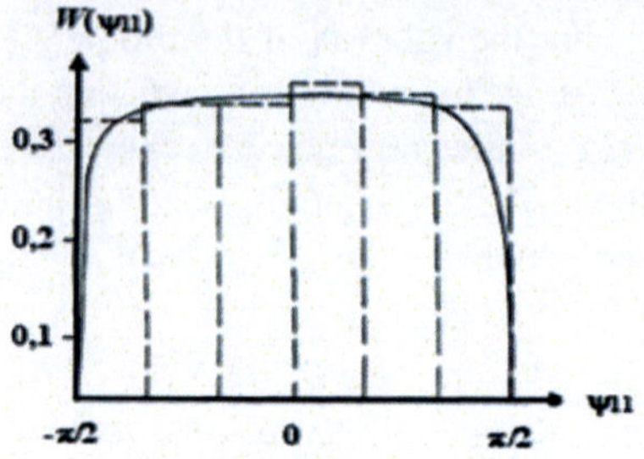

Fig. 2 PDF $W(\mathrm{Re}\, s_{12})$

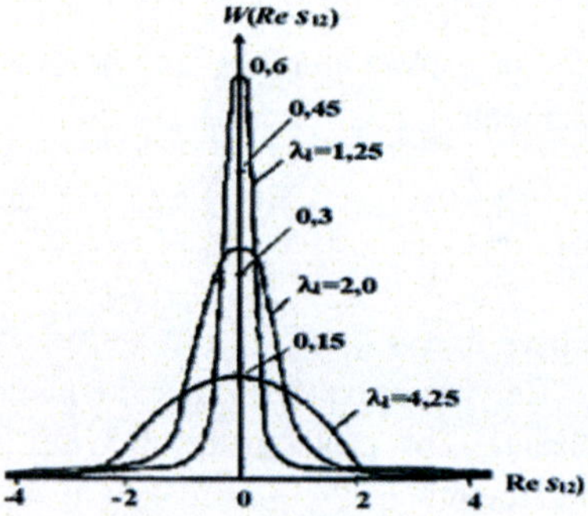

Fig. 3 PDF $W(\mathrm{Re}\, s_{11})$

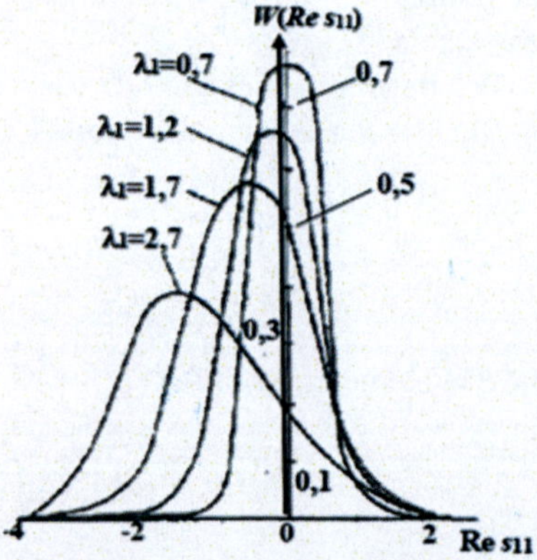

Let us also focus on the fact that PDF $|S_{11}|$ substantially depends on the values of the initial parameters and can vary significantly transforming from one type of Pearson curves to another one.

For example, if $\overline{\lambda_1} = 2, 25$ and $\overline{\lambda_2} = 0$ (see Fig. 4), then the first type of Pearson curves takes place, i.e., the dependence is present as $W(S_{11}) = A(a + S_{11})^m (b - S_{11})^n$.

If $\overline{\lambda_1} = 0, 7$—and $\overline{\lambda_2} = 0, 5$, then already five types of Pearson curves are present as $W(S_{11}) = AS_{11}^{-m} \exp\left\{-\frac{\gamma}{S_{11}}\right\}$ (see Fig. 5).

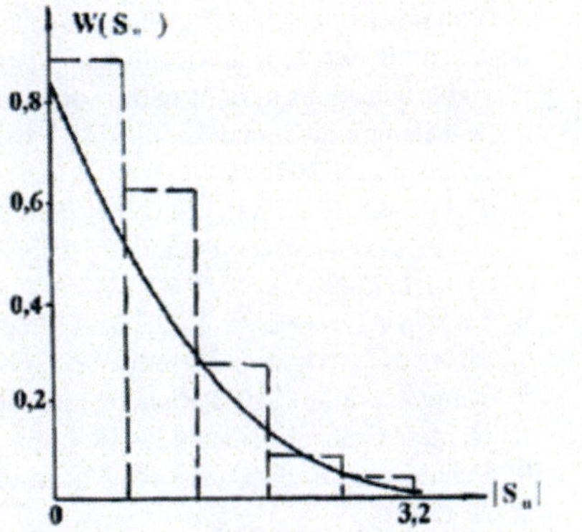

Fig. 4 PDF $W(|S_{11}|)$

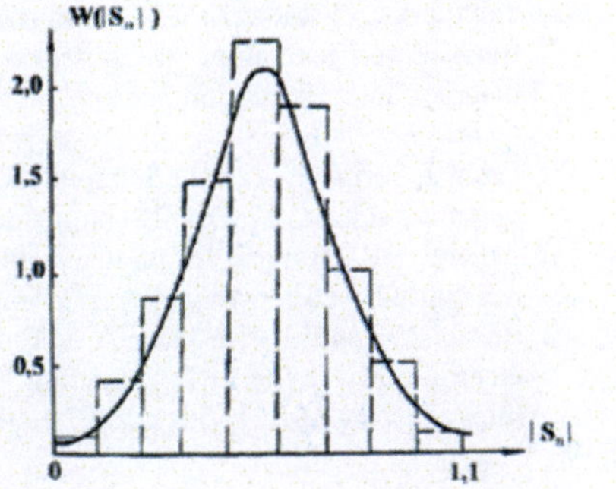

Fig. 5 PDF $W(|S_{11}|)$

Thus, use of relations (1) and (2) allows us to define PDFs of many polarization characteristics corresponding with the SM elements of a radio location object.

References

1. Akmaykin DA, Bolelov EA, Kozlov AI, Lezhankin BV, Svistunov AE, Shatrakov YG (2021) Principal physics of radar location and radio-navigation. In: Theoretical foundations of radar location and radio navigation. Springer Aerospace Technology. Springer, Singapore. https://doi.org/10.1007/978-981-33-6514-8_2
2. Ivic IR (2014) Assessment of censoring using coherency-based detectors on dual-polarized weather radar. J Atmos Oceanic Tech 31:1694–1703. https://doi.org/10.1175/jtech-d-13-00074.1
3. Sen AK, Bhattacharya AB (2019) Radar systems and radio aids to navigation. Dulles, VA: Mercury Learning & Information
4. Akmaykin DA, Bolelov EA, Kozlov AI, Lezhankin BV, Svistunov AE, Shatrakov YG (2021) Radar systems of air transport. In: Theoretical foundations of radar location and radio navigation. Springer Aerospace Technology. Springer, Singapore. https://doi.org/10.1007/978-981-33-6514-8_11
5. Kozlov AI, Logvin AI, Sarychev VA (2005) Polyarizaciya radiovoln: polyarizac. struktura radiolokac. signalov (Radio-wave polarization). Moskva: Radiotekhnika, 2005. 704 s.—.

Polyarizaciya radiovoln, Nauchnaya seriya (In Russian)
6. Chen* B, Wu Y, Han M, Zhang Q (2018) A novel architecture of millimeter-wave full-duplex radio-over-fiber system with source-free bs based on polarization division multiplexing and wavelength division multiplexing. Prog Electromagnet Res C 80: 103–110. https://doi.org/10.2528/pierc17102201
7. Bai X, Jin R, Liu L, Geng J, Liang X (2015) Generation of OAM radio waves with three polarizations using circular horn antenna array. Int J Antennas Propag. https://doi.org/10.1155/2015/132549
8. Logvin AI, Sarychev VA, others (2020) Theoretical introduction into the theory of radiopolarimetric navigation systems. Springer Aerospace Technology.
9. Kutoyants YA (2020) Parameter estimation for continuous time hiden markov processes. Autom Remote Control 81(3):445–468. https://doi.org/10.1134/S0005117920030054
10. Panangaden P (2009) Labelled Markov processes. Imperial College Press. https://doi.org/10.1142/P595
11. Bruce P, Gibbs A (2011) Advanced Kalman filtering, least-squares and modeling: a practical handbook. John Wiley & Sons Inc., Hoboken, New Jersey
12. Tawfiq AA, el-Raouf M, El-Gawad AA, Farahat MA (2021) Realiability assessment for electrical power generation system based on advanced Markov process combined with blocks diagram. Int J Electr Comput Eng 11(5):3647–3659. https://doi.org/10.11591/ijece.v11i5.pp3647-3659
13. Pachl J, Terraf PA (2021) Semipullbacks of labelled Markov processes. Logical methods in computer science 17(2):1–12. https://doi.org/10.23638/lmcs-17(2:3)2021
14. Amaral RR, Borges JA, Gomes HM (2021) Proportional topology optimization under reliability-based constraints. Appl Comput Mech https://doi.org/10.22055/jacm.2021.38440.3.
15. Ehsani A, Ranjbar AM, Jafari A, M.F.-Firuzabad, (2008) Reliability evaluation of deregulated electric power systems for planning applications. Reliab Eng Syst Saf 93(10):1473–1484. https://doi.org/10.1016/j.ress.2007.10.005

The Technique of Determining the Operability Scope of an Airborne Flight Navigation Complex on a Set of Invariant Control Ratios

Eduard Bolelov, Stepan Shalupin, and Nikolay Malisov

Abstract In the process of using airborne flight navigation complexes (AFNCs) for their intended purpose, a change in the operating conditions or the failure of its elements leads to various consequences, depending on the structure of the complex or the class of tasks being solved. Firstly, a change in the operating conditions (failure of individual elements of the complex) leads to a sharp decrease in the efficiency of solving flight and navigation tasks. At the same time, the AFNC as a whole can exist in one of two states: operable or failure. However, this model is typical only for particular problems. Secondly, the AFNC is reconfigured when the operating conditions change. This allows to temporary solve flight and navigation tasks with acceptable, albeit lower quality. This is the most typical situation for navigation tasks. Both in the first and in the second cases, the technical condition monitoring algorithms of the AFNC should inform the user about the violation of the operating conditions within a specified time with required detection and false alarm probabilities. The paper presents a technique for determining the operability scope of an airborne flight navigation complex on a set of invariant control ratios, which allows development of optimal control algorithms.

Keywords Technical condition · Control algorithm · Airborne flight navigation system · Technical condition diagnostics · Invariant ratio

1 Introduction

Modern airborne flight navigation complexes (AFNCs) have a complex hierarchical structure [1–4]. Such a structure is a combination of two types of structures—centralized and decentralized with several levels of hierarchy of management, information

E. Bolelov (✉) · S. Shalupin
Moscow State Technical University of Civil Aviation, Moscow, Russia
e-mail: e.bolelov@mstuca.aero

N. Malisov (✉)
Irkutsk Branch of Moscow State Technical University of Civil Aviation, Irkutsk, Russia
e-mail: malisovnik@mail.ru

O. A. Gorbachev et al. (eds.), *Proceedings of 10th International Conference on Recent Advances in Civil Aviation*, Lecture Notes in Mechanical Engineering, https://doi.org/10.1007/978-981-19-3788-0_16

processing, and information exchange. The structure of the AFNC is determined on the foundation of an analysis of the tasks solved by modern civil aviation aircraft and the corresponding onboard systems, the principles of their construction and functioning, composition, peculiarity of provided information, functional and structural execution, as well as their software.

As part of a modern AFNC, one can distinguish:

- airborne aerobatic complex (AAC);
- airborne navigation complex (ANC);
- communication and information transmission complex (CITC).

Functionally, the AFNC is interconnected with other interfaced onboard aircraft systems (power unit, crew and passenger life support system, electrical system, autopilot, functional aircraft systems, etc.). By radio channels, individual onboard systems and AFNC as a whole are interconnected with ground stations, radio beacons, and automated air traffic control systems in the process of solving flight and navigation tasks. The basis of the integrated type bank is an airborne computer system (ACS), which includes onboard computer systems AAC, ANC, CITC, and specialized processors.

The following features are characteristic of modern AFNC:

1. A high level of integration of onboard systems and computing systems based on the widespread use of the principles of information and functional integration, as well as the modularity of hardware and software, the utilization of a sufficiently large number of complex systems for various purposes as part of the AFNC.
2. The presence of common computing resources and information exchange channels in AFNC, as well as common antennas in a number of radio electronic onboard systems.
3. Effective human–machine interaction by the advanced "crew-to-AFNC" interface.
4. The use in display systems, management, and control multifunctional indicators, color graphic displays with keyboard-programmable, tactile and voice switch, tactile and voice control systems; implementation of the information output about targets coming from various onboard systems to a common indicator in a uniform coordinate system; implementation of robotic target surveillance systems.
5. Implementation of hardware based on a prospective element base.
6. Widespread use of embedded monitoring systems and applied diagnostic programs in ASC software as part of the AFNC.

The given information allows us to form the image of the AFNC model as an object of operation (see Fig. 1).

The following designations are shown in the figure:

- $T_i, i = \overline{1, I}$ is a list of solvable problems describing the multifunctionality of the complex;
- $R_j, j = \overline{1, J}$ is a list of AFNC operating modes that provide a solution to the entire set of problems;

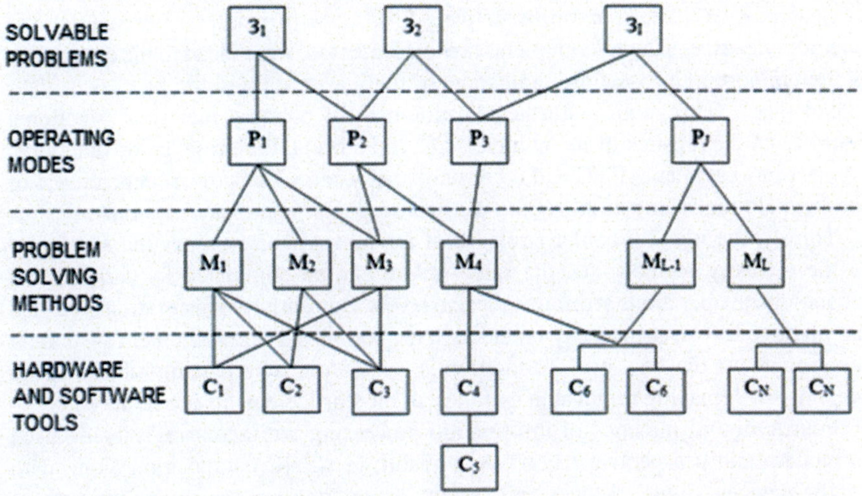

Fig. 1 AFNC models as objects of operation

- $M_l, l = \overline{1, L}$ is list of methods used for the abovementioned modes;
- $C_n, n = \overline{1, N}$ is a list of hardware and software elements.

The number of modes J and methods L determines the functional redundancy of AFNC.

The algorithms of the AFNC functioning determine the nature of the relationship between problems, modes, methods, and hardware and software. At the same time, it is possible that the same problem can be solved in several AFNC functioning modes, and separate modes provide the solution of several problems at once. That is, there is not only functional but also information redundancy. So, in particular, in relation to Fig. 1, the first problem can be solved in the first and second modes and the second in the second and third modes. Therefore, the second mode provides a solution for two problems. The same can be said about the relationships between both modes and methods, as well as methods and hardware and software. An example of structural redundancy (see Fig. 1) is the presence of two C_N elements connected in parallel in reliability, which provide the implementation of the M_L method.

Thus, the analysis of AFNC as an object of operation revealed the following features.

Firstly, modern and prospective AFNCs are multifunctional complexes with information, functional, and structural redundancy, which allows solving a wide range of diverse tasks in several modes of operation, applying various methods of solving the problem in one mode or another, using structural redundancy to increase the reliability of individual devices and systems. All of this, in the end, provides the possibility of operating AFNCs with several failures and damages, especially in conditions of limited resources.

Secondly, in modern and prospective AFNC, new principles of organizing the interaction between their components are implemented, which lead to the emergence of new important connections, and some of them arise only in the process of their intended use. This requires further development of onboard means of functional control and diagnostics of the complex TC. In addition, the presence of functional connections between AFNC and other onboard systems leads to the dependence of the normal functioning of AFNC on the functioning of conjugated systems.

Thirdly, the block-modular structure of modern AFNC assumes the ambiguity of the recovery method, i.e., the same failure can be eliminated by carrying out technological operations at different design levels. Currently, the restoration is carried out on three levels, structurally included in the complex TC management system on the ground. At the first level, restoration is carried out by replacing blocks, at the second—by replacing removable modules, at the third, elements are replaced.

Fourth, digital methods of information processing are increasingly being used in modern and prospective AFNC. As a result, in AFNC composition along with analog devices, digital devices and systems have appeared which, by the type of their implementation, can be divided into hardware and software parts.

It is obvious that for modern AFNC, the most urgent is the development of new methods and means of monitoring and diagnosing the TC. At the same time, it is especially important to determine the scope of operability of the AFNC. Let us consider the methodology for determining the scope of operability of an airborne flight and navigation complex on a set of invariant control ratios.

2 The Technique of Determining the Operability Scope of an Airborne Flight Navigation Complex

There are two main approaches to solve the problems of technical condition monitoring of airborne flight navigation systems during their intended usage.

1. The statistical dynamics of failure is known (for example, failure occurs due to a voltage surge in a controlled process). Control of the AFNC technical condition in this case is reduced to establishing the fact of the presence or absence of an abrupt signal in the controlled process
2. Due to the variety of causes and factors causing abnormal AFNC operating modes, the construction of reliable a priori models of measurement errors in such operating modes is usually impossible. Therefore, in order to monitor the AFNC technical condition, it is necessary to create a procedure for detecting abnormal operating modes based only on models of observation vectors and conditions in nominal operating conditions.

The experience of AFNC technical operation for aircrafts, existing methods, tools for monitoring and diagnosing of the AFNC technical condition has shown [5, 6]. However, the monitoring and diagnostic tools currently used do not contribute to improve the AFNC efficiency and lead to unreasonable resource consumption during on-ground inspections and restorations of failed devices included in the AFNC. Thus, the problem of creating the AFNC control system capable of assessing its technical condition (TC) with high reliability both in flight and on the ground is currently urgent. This task requires, at first, to form a methodology for determining the operability scope of the complex. Let us define the concepts of the operability condition and the operability scope.

Operability conditions are certain ratios (usually inequalities) between the measured parameters (or characteristics), under which the system belongs to the class of operable. Thus, the operability criterion shows which systems should be recognized as operable, while the operability conditions show which systems are actually recognized as operable during monitoring. One of the tasks of the synthesis of the AFNC control system is to find the conditions of operability that most fully and accurately correspond to a given operability criterion. In most cases, it is advisable to calculate the operability conditions (according to the measured parameters) based on the construction of the operability scope.

The operability scope is a set of points in the space of internal parameters of the control object corresponding to operable systems. The internal parameters of the control object must fully describe it. The set of these parameters for the AFNC should be sufficiently representative. Then, the operability scope will exactly match the operability criterion. Unfortunately, the set of internal parameters of the controlled system is never complete, and therefore, it is possible to build only an approximate operability scope. It should be noted that only for simple control objects specified in the form of transfer functions, it is possible to find an analytical dependence of the operability criterion with the parameters of the controlled system, thereby solving the problem of finding the operability scope.

Currently, the direction of functional control based on mathematical methods of invariance theory is considered promising [7–9]. Idea is that due to the structural, functional, and informational redundancy of the AFNC, control ratios are determined. Those ratios are invariant to input influences, which uniquely characterize both a set of technical parameters $\{K\}$ and a set of tactical and technical characteristics $\{\Phi\}$, and do not depend on the characteristics of unknown input signals $X(t)$ and interfering disturbances $\theta(t)$ of arbitrary type. It can significantly increase the methodological reliability of monitoring and diagnosis. The peculiarity of the application of invariance theory in relation to functional control (FC) tasks is that in order to determine TC, it is necessary to exclude from the measurement results not only interfering disturbances $\theta(t)$, but also unknown input effects (information signals) $X(t)$.

In order to reduce the sources of methodological error, the determination of the TC of AFNC is performed on a set $\{\Phi\}$. Then, the synthesis of invariant control ratios is based on the use of functional and structural redundancy, when the solution of one problem is provided in different ways, i.e., the same state variables are evaluated by different components of AFNC.

Control ratios or redundancy ratios can be written mathematically as follows [3, 4, 9]:

$$r_i(t) = \hat{\psi}_i\big(X^*(t-pT), X^*(t-(p-1)T), ..., X*(t)\big), \quad (1)$$

where r_i is the control ratio, the statistical characteristics of which are determined by the set $\{\Phi\}$ and by the type of functional transformation $\hat{\psi}_i(X(t-pT), X(t-(p-1)T), ..., X(t)) = 0$, $i = \overline{1, s}$, p is the order of the control ratio, T is the period of updating information.

When using control ratios $r_i, i = \overline{1, s}$ a rigid functional relationship is formed between the performance indicators of the AFNC functioning and the tolerances for them [6]. In essence, the value of the invariant control ratio $r_i, i = \overline{1, s}$ at a fixed moment t is a known function of the coordinate values $\Phi(t)$ at that moment, i.e., from the coordinates of the TC vector on the set $\{\Phi\}$. In contrast to the control and diagnosis on the set {K}, in which the coordinates and dimension of the vector $K(t)$ are selected in accordance with the AFNC structural components, the vector $\Phi(t)$ is set and the failed AFNC structural parts are in accordance with invalid estimates in coordinates $X^*(t)$.

The advantages of functional control on a set of control ratios (1) are as follows:

- the structure of the TC management system of the AFNC does not depend on the AFNC functioning model, which will be used only for diagnosis, i.e., determining the component part of the AFNC that forms an unacceptable assessment of the state variable;
- validation of the control ratios fulfillment for state variables estimates, in essence, means solving the control problem on a set $\{\Phi\}$ since unacceptable deviations of control ratios are determined by the statistical characteristics of the estimates and the tolerances on them, which are indicators of the efficiency of the AFNC operating in each specific mode;
- if the set of control ratios contains all the estimated coordinates $X(t)$ and their derivatives, then the control completeness is equal to one.

When distinguishing the types of TC of the AFNC, it is necessary to determine the tolerances for the vector of control ratios R so that they minimize (according to one of the criteria) the probability of an erroneous decision about the type of TC based on the results of control and diagnosis.

Assignment of tolerances $\left[r_i^1, r_i^2\right]$, $i = \overline{1, n-1}$, defines the boundaries of the operability scope on the set of control ratios {R}. Thus, the type of functional transformation $\widehat{\psi}_i, i = \overline{1, n-1}$, tolerances $\left[\phi_i^1, \phi_i^2\right]$, $i = \overline{1, n}$, the criterion for making a decision based on the value $r_i, i = \overline{1, n-1}$ about the type of the TC allows to find a set $\hat{Q}_R$ corresponding to a operational type of TC radio electronic complexes (REC) $\hat{Q}_R = \{r_i | r_i \in \left[r_i^1, r_i^2\right]\}, i = \overline{1, n-1}$.

The proposed technique for determining tolerances is based on the varied behavior of the distribution density R in the operable and inoperable states of the AFNC.

The AFNC technical condition is determined by the values of the coordinates of the evaluation vector $X^*(t)$. Therefore, we take the variances of measurement errors of the vector $X(t)$ as an indicator of the quality of AFNC functioning $\Phi(t)$. To obtain concrete results, it is necessary to determine the type of functional transformation

$\hat{\psi}_i, i = \overline{1, n-1}$ over the vector of estimates $X^*(t)$. Failure detection is carried out by the magnitude of the difference signal:

$$\hat{\psi}(\cdot) = f_i\left(X^*, t\right) - \varphi_i\left(X^*, t\right) = r_i(t), \tag{2}$$

where $f_i(\cdot), \varphi_i(\cdot)$ are known functional dependencies that determine the type of transformation over estimates obtained through various information channels $i = \overline{1, n-1}$.

The operability scope Q_{R} on the set $\{\mathrm{R}\}$ is determined using the restrictions the Φ_0 on the indicator $\Phi\left(x_1^*, x_2^*, \ldots, x_n^*, t\right)$ of the functioning quality of the AFNC:

$$Q_R = \left\{r_i(t), i = \overline{1, n-1} | \Phi\left(x_1^*, x_2^*, \ldots, x_n^*, t\right) \leq \Phi_0\right\}. \tag{3}$$

When using the tolerance control technique of the TC REC, the entire operability scope Q_R is approximated by an $(n-1)$-dimensional parallelepiped $\hat{Q}_{\mathrm{R}}$, the sides of which correspond to independent tolerances $\left[r_i^1, r_i^2\right]$ on $r_i(t), i = \overline{1, s}$.

The process of assigning tolerances $\left[r_i^1, r_i^2\right]$, in essence, determines the criteria for making a decision on the AFNC technical condition.

In most technical applications, a simple loss function is used, so the tolerances $\left[r_i^1, r_i^2\right]$ for control ratios are determined based on the minimum probability of an erroneous decision:

$$P_{err} = P_{\text{ло}} + P_{\text{но}} \rightarrow \min_{[r_i^1, r_i^2]} \tag{4}$$

When using the measurement error vector $\Phi(t)$ as coordinates, the control ratio can be represented as

$$r(t) = \Delta_1(t) - \Delta_2(t), \tag{5}$$

where $\Delta_{1,2}$ are the measurement errors of the components of the control ratio.

Setting deterministic tolerances $\left[\Delta_1^1, \Delta_1^2\right], \left[\Delta_2^1, \Delta_2^2\right]$ allows to select a operable type of TC on a set of values $\Delta_1(t), \Delta_2(t)$, which are a subset of

$$Q_0 = \left\{\Delta_1, \Delta_2 | \Delta_1 \in \left[\Delta_1^1, \Delta_1^2\right], \Delta_2 \in \left[\Delta_2^1, \Delta_2^2\right]\right\}, \tag{6}$$

and to determine the probability of the AFNC to stay in this state:

$$P(Q_0) = \iint_{Q_0} W_1(\Delta_1) W_2(\Delta_2) \mathrm{d}\,\Delta_1 \mathrm{d}\,\Delta_2 \tag{7}$$

In the operation of intricate technical complexes, such as AFNC, the most relevant is to determine the operability scope under conditions of a priori uncertainty, i.e., at

$P(Q_0)P\left(\overline{Q}_0\right)0,5$. This is a typical situation when using the AFNC in a constantly changing operating environment, both external and internal; when the interference situation changes; the appearance of sudden failures, etc. [10–15]. The presence of a methodological error P_{err} is a disadvantage of the developed FC method although monitoring and diagnosis is performed on a set $\{\Phi\}$. However, the magnitude of this error will be significantly less than the methodological error of the existing FC system due to the following factors:

- the usage of control completeness equal to one, independent of the type of model, and the functioning of the components of the AFNC;
- decrease in TC types on a set of control ratios.

Determination of the operability scope on a set of control ratios with deterministic tolerances assumes that its boundaries will remain unchanged when the operating conditions change. Therefore, in the process of functioning AFNC, it is more reasonable to set the a priori probability of its operable state, which will allow determining the operability scope based on the actual requirements for the tactical and technical characteristics of the AFNC.

3 Conclusion

As the AFNC operational experience shows, one of the main directions for improving the efficiency of the existing system of technical operation is to resolve the existing discrepancy between the level of development of modern AFNC and means of monitoring their technical condition. The creation of functional control systems based on the principles of invariance will not only eliminate this discrepancy, but also ensure the fullest possible use of the potential capabilities of modern and prospective AFNC.

References

1. Setoguchi T, Ogino Y, Ouchida R, Takahashi A, Ayukawa Y, Koyano K (2021) Retraction: Setoguchi et al. accuracy assessment of implant placement in a newly developed dynamic navigation system: a pilot study. https://doi.org/10.3390/app112210756
2. Kuznetsov SV (2021) Sistema ekspluatacionnogo kontrolya bortovogo oborudovaniya vozdushnyh sudov grazhdanskoj aviacii i nauchnye osnovy ee formirovaniya (operational control system of civil aircraft airborne equipment and scientific basis of its formation). Nauchnyj vestnik MGTU GA 24(3):31–41. https://doi.org/10.26467/2079-0619-2021-24-3-31-41 (In Russian)
3. Bolelov EA, Tsykarev AV, Sbitnev AV (2015) Algoritm kontrolya tekhnicheskogo sostoyaniya bortovogo pilotazhno-navigacionnogo kompleksa, uchityvayushchij informacionnuyu izbytochnost' kompleksa (The algorithm of control of technical condition the onboard flight-navigation complex, taking into account the information redundancy of the complex). Nauchnyj vestnik MGTU GA 222:175–181 (In Russian)

4. Rajamani R (2019) Condition-based maintenance in aviation: the history. The Business and the Technology, Warrendale, Pennsylvania: SAE International
5. Valdivia de Matos HL (2018) Model-based specification of integrated modular avionics systems using object-process methodology. In: 2018 IEEE/AIAA 37th digital avionics systems conference (DASC). https://doi.org/10.1109/DASC.2018.8569855
6. Li X, Chen X (2018) Civil aviation operation information processing research. Journal of Physics: Conference Series, vol 1069. In: 3rd annual international conference on information system and artificial intelligence (ISAI2018) 22–24 June 2018, Suzhou. https://doi.org/10.1088/1742-6596/1069/1/012065
7. Wu Y, Xiao G, Wang G, He F, Dai Z, Wang Y (2018) Research on safety analysis method of functional integrated avionics systems. In: 2018 IEEE/AIAA 37th digital avionics systems conference (DASC). https://doi.org/10.1109/DASC.2018.8569355
8. Bolelov EA, Sbitnev AV, Chaliapin SV (2014) Matematicheskaya model' signalov na vyhode bortovyh radionavigacionnyh sistem, uchityvayushchaya ih vnezapnye otkazy (mathematical model of output signals of airborne radio navigation systems with respect of sudden failures). Nauchnyj vestnik MGTU GA 210:160–162 (In Russian)
9. Wang M, Xiao G, Liu X, Wang G (2019) Integrated modular avionics system design based on formal dynamic organization. In: 2019 IEEE/AIAA 38th digital avionics systems conference (DASC). https://doi.org/10.1109/DASC43569.2019.9081755
10. Bogachev AS, Kudinov AT et al (2019) Markovskie modeli tekhnicheskogo sostoyaniya perspektivnyh bortovyh radiolokacionnyh sistem vozdushnyh sudov (Markov models of the technical condition of prospective airborne radar systems). Nauchnyj vestnik GosNII GA 26(337):113–126 (In Russian)
11. Prisacaru A, Gromala PJ, Jeronimo MB, Han B (2017) Prognostics and health monitoring of electronic system: a review. In: 18th international conference on thermal, mechanical and multi-physics simulation and experiments in microelectronics and microsystems (EuroSimE). https://doi.org/10.1109/EuroSimE.2017.7926248
12. Zeitler A (2019) Challenges of certification and integration of new hardware into legacy avionics architectures. In: IEEE/AIAA 38th digital avionics systems conference (DASC). https://doi.org/10.1109/DASC43569.2019.9081621
13. Cevher S (2018) A fault tolerant software defined networking architecture for integrated modular avionics. In: IEEE/AIAA 38th digital avionics systems conference (DASC). https://doi.org/10.1109/DASC.2018.8569681
14. Luis P, Gaëlle L, Yue M, Chantal R (2018) Knowledge discovery for avionics maintenance support. In: IEEE/AIAA 37th digital avionics systems conference (DASC). https://doi.org/10.1109/DASC.2018.8569856
15. Erokhin V, Lezhankin B, Portnova T (2021) Bi-criteria aircraft trajectory optimization in implementing the area navigation concept. Int J Aeronaut Space Sci 22(4):948–962. https://doi.org/10.1007/s42405-021-00353-3

Applying LoRa Technology in Unmanned Aircraft Systems

Muslim Mezhetov, Anna Tikhova, Uliana Vakhrusheva, and Andrey Fedorov

Abstract Evolution in the methods of forming and receiving signals is continuously related to development of the information transmission media by using air. Information transmission radio channels have certain requirements for the hardware providing reliable data delivery from the information source to the end-user. One may attribute the presence of artificial and natural interferences that considerably deteriorate the quality of the received information to radio channel features. Especially, this is true for the systems transmitting control instructions. One of such systems is the ground—unmanned aerial vehicle (UAV) data-transmission channel, whose performance quality affects accomplishment of the mission set to the UAV and the resilience of latter. In this paper, we address application of the LoRa modulation in the UAV control channel and present a model for a signal shaper and a receiver in the Labview2010 software.

Keywords Unmanned aircraft system · Unmanned aerial vehicle · LFM · LoRa

1 Introduction

Rapid advance of unmanned aircraft systems (UAS) is, first of all, related to a wide range of their missions. Such missions involve locality mapping, reconnaissance of various objects, environmental monitoring, traffic police surveillance, as well as their use to deliver various products [1–3].

A wide proliferation of UAVs is related to the fact that solving the above missions is done at minimal costs. The latter basically involves acquisition and maintenance of the equipment and upkeep of the operator controlling the UAV.

M. Mezhetov (✉) · A. Tikhova · U. Vakhrusheva
Irkutsk Branch of Moscow State Technical University of Civil Aviation, Irkutsk, Russia
e-mail: milsumka@mail.ru

A. Fedorov
Saint Petersburg State University of Civil Aviation, Saint Petersburg, Russia

O. A. Gorbachev et al. (eds.), *Proceedings of 10th International Conference on Recent Advances in Civil Aviation*, Lecture Notes in Mechanical Engineering, https://doi.org/10.1007/978-981-19-3788-0_17

Present-day UAVs are hi-tech devices, whose cost makes a considerable amount depending on their class and mission. Therefore, a UAV loss while in service leads to, first, losses of helpful information, and, second, contingencies when solving the mission. The more expensive the UAV, the more difficult it is for the operator to fulfill the mission because the fear of the UAV loss keeps him/her in constant stress, which negatively affects his/her moral and psychological state. But, in some situations, the UAV loss is not related to the operator's actions at all. Such situations involve ice covering, power-unit unreliability, power failure; collision with birds and other objects, navigation system malfunction.

Any UAS, when accomplishing its mission, uses, first of all, a restricted airspace to acquire, store, and transmit information. The information is transmitted through air. Herewith, the UAV telemetry signals may be transmitted in the common flow or use a separate communication channel. In any case, the useful information and the telemetry signals are transmitted on a channel that is subject to the impact from intentional and unintentional interferences (Fig. 1) considerably reducing the UAV efficiency, up to its loss [4, 5].

Fig. 1 Communications within a UAS

2 Method

Therefore, not to lose a UAS, one should use the most anti-interference modulation methods that are capable of resisting a wide range of radio interferences. Herewith, such methods are to ensure a minimal power-consuming from the UAV supply, to be autonomous in case of the battery fails, which will enable to find the lost vehicle and to have the signal transmitting range equal to or longer than the UAV mission track. It is broadband signals that meet such requirements. In radar-tracking systems, a linear frequency modulated (LFM) signal (or chirp) has been long used to improve the resolution due to the spectral range extension. In data-transmission systems, a chirp was first used in the LoRa technology [6, 7].

LoRa is a technology and a modulation method as well. It was patented by the *Semtech* company. The method is based on a spectrum extension technique, in which data are encoded by a frequency change inside a pulse. Herewith, the frequency may increase or decrease over a time interval Fig. 2.

The information packet comprises a preamble and a data series.

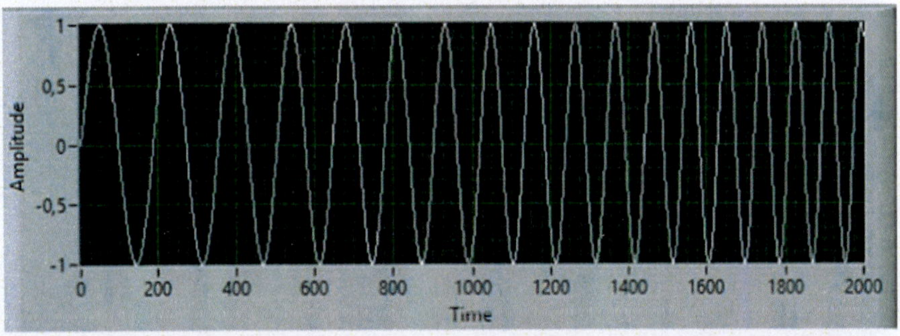

a)

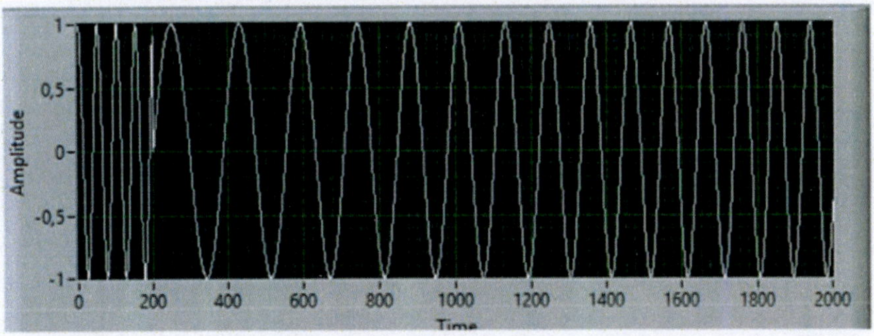

b)

Fig. 2 LFM reference signal (**a**), intelligence signal (**b**)

The preamble detector operation is based on using a matched filter (MF), whose pulse response is complexly interfaced with a chirp in the frequency domain and has its mirror reflection in time [8]:

$$h(t) = A_1 * \cos\left(\omega_H * (T_{\text{sym}} - t) - \frac{\mu}{2} * (T_{\text{sym}} - t)^2\right), 0 \le t < T_{\text{sym}}$$

The principle of transmitting information symbols of the physical layer (PHY DATA UNIT) data packet by a LoRa RF broadband signal is in a frequency shift $e^{j*\Delta\omega*k*t}$ relative to the reference chirp $e^{j*(\omega_H*t+\mu*t^2)}$, where $k = 0, 1, 2, \ldots, 2SF$ is an SF-bit information symbol [9–11]:

$$x(t) = \begin{cases} A_0 * \cos\left(\omega_{\text{H}} * t + \Delta\omega * k * t + \frac{\mu}{2} * t^2\right), 0 \le t < T_0 \\ A_0 * \cos\left(\omega_{\text{H}} * t + \Delta\omega * k * t - \text{BW} * t + \frac{\mu}{2} * t^2\right), 0 \le t < T_0 \end{cases}$$

Figure 3 presents a possible block diagram for a LoRa signal receiver transferring a physical-layer data packet.

Here:

$c(t) = A_1 * \cos\left(\omega_{\text{H}} * t + \frac{\mu}{2} * t^2\right), 0 \le t < T_{\text{sym}}$ is a benchmark chirp,

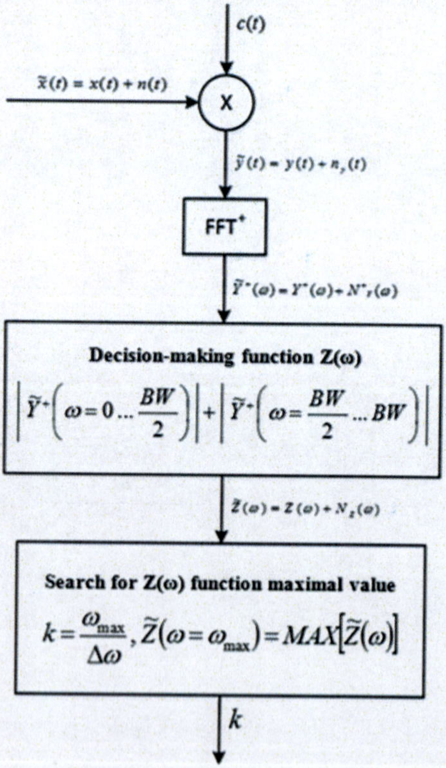

Fig. 3 Possible block diagram for a LoRa signal receiver

$n(t), 0 \leq t < T_{\text{sym}}$ is additive white Gaussian noise.

De-chirped signal looks like:

$$y(t) = x(t) * c(t)$$
$$= \frac{A_0 * A_1}{2} * \begin{cases} \cos(\Delta\omega * k * t) + \cos\left(2 * \omega_H * t + \Delta\omega * k * t + \mu * t^2\right), \\ 0 \leq t < T_0 \\ \cos(BW - \Delta\omega * k * t) + \cos\left(2 * \omega_H * t + \Delta\omega * k * t - Bw * t + \mu * t^2\right), \\ 0 \leq t < T_{\text{sym}} \end{cases}$$

3 Results

When working on this study, we developed a LoRa virtual demodulator, whose algorithm is described in [12]. Figure 4 presents its appearance.

Figure 5 presents the information chirps for different codes.

By removing the addends in braces (as high-frequency components) in the expression for $y(t)$,

$$y(t) = \frac{A_0 * A_1}{2} * \begin{cases} \cos(\Delta\omega * k * t), 0 \leq t < T_0 \\ \cos([Bw - \Delta\omega * k] * t), T_0 \leq t < T_{\text{sym}} \end{cases}$$

We obtain the following complex signal at the Fourier transform (FFT+) unit output:

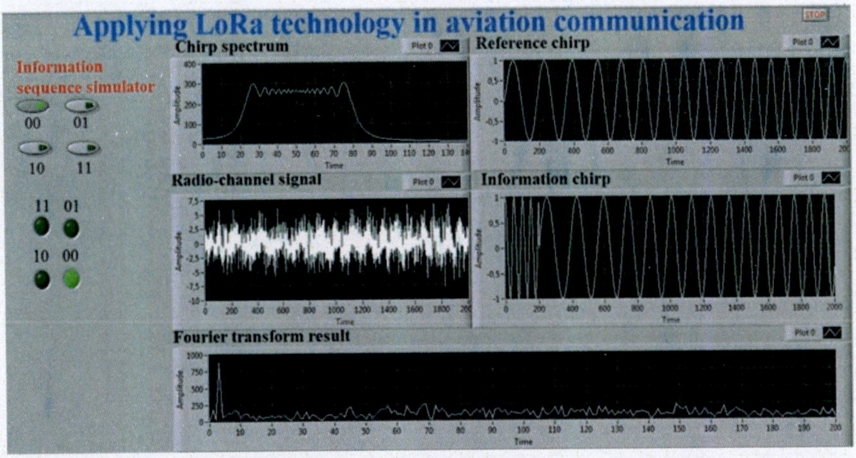

Fig. 4 Virtual LoRa demodulator

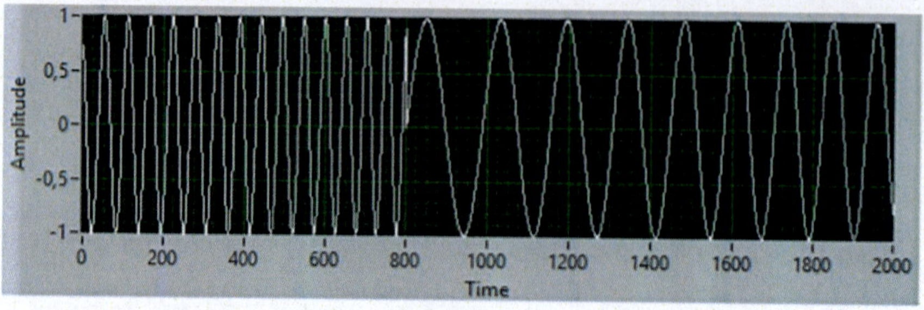

a)

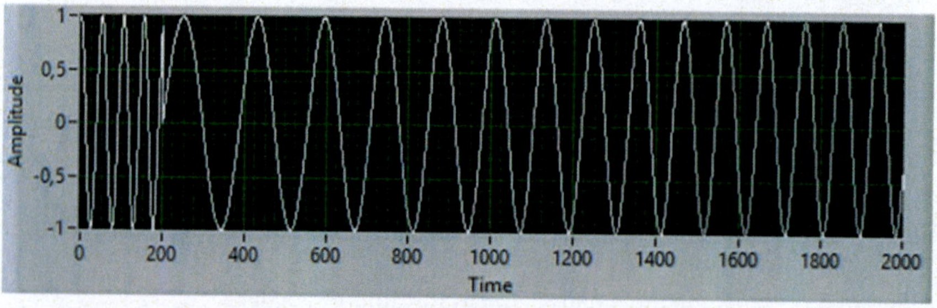

b)

Fig. 5 Information chirp (**a**—for the «00» code, **b**—for the «11» code)

$$
\begin{aligned}
Y(\omega) = \int_{-\infty}^{\infty} y(t) * e^{-i\omega t}\mathrm{d}t = \frac{A_0 * A_1}{4} * e^{-j(\omega-\Delta\omega k)\frac{T_0}{2}} * T_0 \\
* \frac{\sin\left[(\omega - \Delta\omega k) * \frac{T_0}{2}\right]}{(\omega - \Delta\omega k) * \frac{T_0}{2}} + \frac{A_0 * A_1}{4} \\
* e^{-j(\omega-(BW-\Delta\omega k))\frac{T_0+T_{\text{sym}}}{2}} * (T_{\text{sym}} - T_0) \\
* \frac{\sin\left[(\omega - (BW - \Delta\omega k)) * (T_{\text{sym}} - T_0)/2\right]}{(\omega - (BW - \Delta\omega k)) * (T_{\text{sym}} - T_0)/2} \\
+ \frac{A_0 * A_1}{4} * e^{-j(\omega+\Delta\omega k)\frac{T_0}{2}} * T_0 * \frac{\sin\left[(\omega + \Delta\omega k) * \frac{T_0}{2}\right]}{(\omega + \Delta\omega k) * \frac{T_0}{2}} \\
+ \frac{A_0 * A_1}{4} * e^{-j(\omega+(BW-\Delta\omega k))\frac{T_0+T_{\text{sym}}}{2}} \\
* (T_{\text{sym}} - T_0) * \frac{\sin\left[(\omega + (BW - \Delta\omega k)) * (T_{\text{sym}} - T_0)/2\right]}{(\omega + (BW - \Delta\omega k)) * (T_{\text{sym}} - T_0)/2}
\end{aligned}
$$

Further, we discard two last addends having a major effect within negative frequencies and minor within the positive:

$$Y^{+}(\omega) = Y_1(\omega) + Y_2(\omega), \text{ where}$$

$$Y_1(\omega) = \frac{A_0 * A_1}{4} * e^{-j(\omega-\Delta\omega k)\frac{T_0}{2}} * T_0 * \frac{\sin\left[(\omega-\Delta\omega k) * \frac{T_0}{2}\right]}{(\omega-\Delta\omega k) * \frac{T_0}{2}} +$$

$$Y_2(\omega) = \frac{A_0 * A_1}{4} * e^{-j(\omega-(\mathrm{BW}-\Delta\omega k))\frac{T_0+T_{\mathrm{sym}}}{2}} * (T_{\mathrm{sym}} - T_0)$$
$$* \frac{\sin\left[(\omega-(\mathrm{BW}-\Delta\omega k)) * (T_{\mathrm{sym}} - T_0)/2\right]}{(\omega-(\mathrm{BW}-\Delta\omega k)) * (T_{\mathrm{sym}} - T_0)/2}$$

To avoid overlapping of two addends $Y^{+}(\omega)$ at different k, the inequality $\Delta\omega < \frac{\mathrm{BW}}{2*k}$ should be fulfilled.

Consequently,

$$\Delta\omega = \frac{\mathrm{BW}}{2^{\mathrm{SF}+1}}$$

$$T_0 = \frac{\left(2^{\mathrm{SF}+1} - \mathrm{k}\right)}{2^{\mathrm{SF}+1}} * T_{\mathrm{sym}}$$

At the next stage, we calculate the decision-making function $Z(\omega)$, representing the total of the modules for the function $Y_1(\omega)$ and for the function $Y_2(\omega)$ reflected relative to the point $\omega = \mathrm{BW}$:

$$Z(\omega) = Y_1(\omega) + Y_2(BW - \omega) \approx Y^{+}(\omega) + Y^{+}(BW - \omega),$$

where $\omega = 0\ldots\frac{\mathrm{BW}}{2}$.

A key feature for the LoRa (as mentioned above) is its high anti-interference. Figure 6 demonstrate the LoRa signal detector operation under the conditions of additive white Gaussian noise [13, 14].

We determine the value for the k information character decoded by the receiver. For this purpose, we find the frequency ω, at which decision-making function $Z(\omega)$ accepts the maximal value ($\omega_{\max}$):

$$Z(\omega = \omega_{\max}) = \mathrm{MAX}[\mathrm{Z}(\omega)]$$
$$k = \frac{\omega_{\max}}{\Delta\omega}$$

4 Discussion and Conclusions

Thus, the communication system between a drone and the control point is one of the crucial UAV systems [15]. It transmits the control commands to the flight operator, data on the current state of main systems, as well as the flight vehicle location

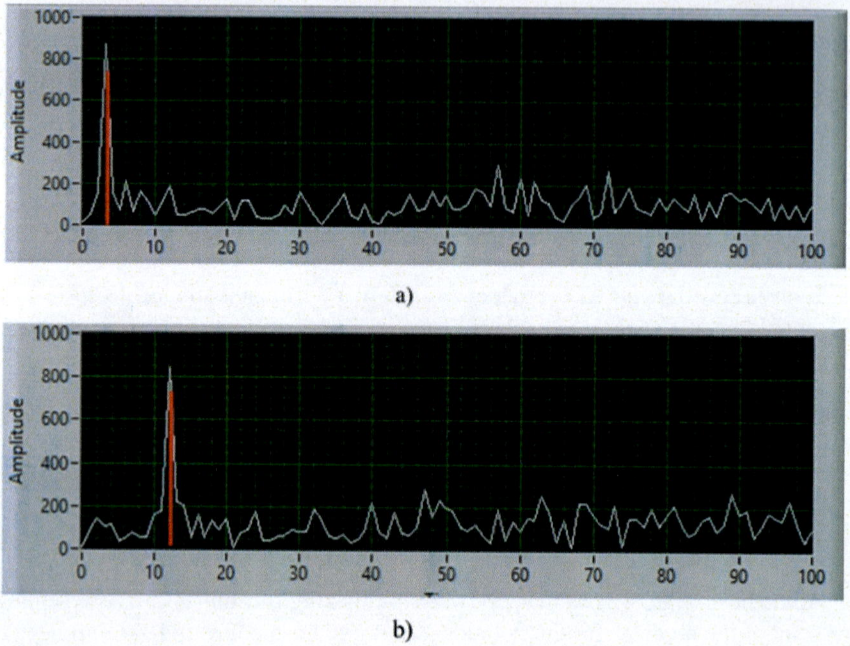

Fig. 6 LFM information spectrum as a result of the Fourier transform (**a**—for the «00» code, **b**—for the «11» code), where $k = 0, 1, 2, \ldots, 2^{SF}$ is an SF-bit information symbol

information. All the above kinds of messages, interchanging with which occurs during any flight, may be integrated into one group: the telemetry data.

The LoRa technology that applies a broadband LFM provides the anti-interference necessary for the UAV, consuming, at the same time, the least energy from the supply.

In this study, we used a developed virtual instrument, enabling to visualize reception and demodulation of a LoRa signal. The instrument will allow us to study the anti-interference of this technology.

References

1. Finn RL, Wright D (2012) Unmanned aircraft systems: surveillance, ethics and privacy in civil applications. Comp Law Secur Rev 28:184–194
2. Linchant J, Lejeune P, Bouché P, Vermeulen C (2013) Aerial surveys using an unmanned aerial system (UAS): comparison of different methods for estimating the surface area of sampling strips. Trop Conserv Sci 6(4):506–520. https://doi.org/10.1177/194008291300600405. https://www.researchgate.net/publication/281343893_Are_unmanned_aircraft_systems_UAS_the_future_of_wildlife_monitoring_A_review_of_accomplishments_and_challengesLisein
3. Akmaykin DA, Bolelov EA, Kozlov AI, Lezhankin BV, Svistunov AE, Shatrakov YG (2021) Principal physics of radar location and radio-navigation. In: Theoretical foundations of radar

location and radio navigation. Springer Aerospace Technology. Springer, Singapore. https://doi.org/10.1007/978-981-33-6514-8_2

4. Skrypnik O, Arefyev R, Arefyeva N, Portnova T (2021) Peculiarities of using a mobile pseudolite for increase of positioning accuracy on aircraft landing. Crede Experto. https://doi.org/10.51955/2312-1327_2021_3_14
5. Ermakov AK, Portnova TY, Lezhankin BV, Erokhin VV (2021) Trajectory control algorithms for unmanned aircraft complexes flying in formation. Wave Electron Appl Inform Telecommun Syst. https://doi.org/10.1109/WECONF51603.2021.9470714
6. Augustin A, Yi J, Clausen T, Townsley WM (2016) A study of LoRa: long range and low power networks for the internet of things. Sensors 16(9):1466. https://doi.org/10.3390/s16091466
7. Pham C, Bounceur A, Noreen U, Ehsan M (2020) Radio channel access challenges in LoRa low-power wide-area networks. LPWAN Technol IoT M2M Appl. https://doi.org/10.1016/B978-0-12-818880-4.00004-1. https://www.sciencedirect.com/science/article/pii/B9780128188804000041
8. Gardner FM (1986) A BPSK/QPSK timing-error detector for sampled receivers. IEEE Trans Commun 34:423–429
9. Petajajarvi J, Mikhaylov K, Pettissalo M, Janhunenl J, Iinatti J (2017) Performance of a low-power wide-area network based on LoRa technology: doppler robustness scalability and coverage. Int J Distrib Sens Netw. https://doi.org/10.1177/1550147717699412
10. Alyafawi I, Dimitrova DC, Braun T (2014) Real-time passive capturing of the GSM radio. In: IEEE international conference on communications (ICC). https://doi.org/10.1109/ICC.2014.6884013
11. Hamida EB, Chelius G, Gorce JM (2009) Impact of the physical layer modeling on the accuracy and scalability of wireless network simulation. Simulation 85(9):574–588. https://doi.org/10.1177/0037549709106633
12. Staniec K, Kowal M (2018) LoRa performance under variable interference and heavy-multipath conditions. Hindawi Wireless Commun Mobile Comput. https://doi.org/10.1155/2018/6931083
13. Raychowdhury A, Pramanik A (2020) Survey on LoRa technology: solution for internet of things. In: Thampi S, et al (eds) Intelligent systems, technologies and applications. Advances in intelligent systems and computing, vol 1148. Springer, Singapore. https://doi.org/10.1007/978-981-15-3914-5_20
14. Sornin N, Luis M, Eirich T, Kramp T, Hersent O (2016) LoraWan specification version 1.0. LoraWan Alliance Specification Document. https://lora-alliance.org/wp-content/uploads/2020/11/lorawan1_0_2-20161012_1398_1.pdf. Accessed 12 Nov 2021
15. Erokhin V, Lezhankin B, Portnova T (2021) Bi-criteria aircraft trajectory optimization in implementing the area navigation concept. Int J Aeronaut Space Sci 22(4):948–962. https://doi.org/10.1007/s42405-021-00353-3

Method of Determining Statistical Characteristics of Temperature Profile Measurement Errors for the Aircraft Take-off and Landing Phases

Nikolay Voskresensky and Vitaliy Karachentsev

Abstract The air temperature and its profile is the most essential parameter of the atmospherse subject to be measured and predicted. It has a significant impact on engine thrust and fuel consumption; aircraft's aerodynamic characteristics, takeoff and landing; the result of weather and dangerous meteorological phenomena forecasts, etc. Currently, civil aviation airfields are beginning being equipped with the systems of remote temperature profile measuring. These systems include the MTR-5 microwave temperature profiler. Experimental comparative studies of the MTR-5 profiler and radio sounding data allow us to make a conclusion about one significant disadvantage of the MTR-5, which is associated with the necessity of its calibration. The lack of calibration or insufficient frequency of its execution may leads to substantial deviations of a measured temperature profile from a real one. In order to increase the efficiency of the MTR-5 application, it is required to develop a mathematical model of the MTR-5 temperature profile measurement errors. This paper deals with the temperature profile measuring statistical errors determining the methodology and the temperature profile measuring errors model.

Keywords Temperature profile · Measurement error · Remote temperature measuring · Radio sounding

1 Introduction

Various facilities are utilized to conduct the weather observation and measurements in the vicinity of an aerodrome. Each aid possesses particular disadvantages concerned with the principle of their operation, operational and other limitations [1]. In order

N. Voskresensky (✉)
Moscow State Technical University of Civil Aviation, Moscow, Russia
e-mail: n.voskresensky@mstuca.aero

V. Karachentsev
Irkutsk Branch of Moscow State Technical University of Civil Aviation, Irkutsk, Russia

O. A. Gorbachev et al. (eds.), *Proceedings of 10th International Conference on Recent Advances in Civil Aviation*, Lecture Notes in Mechanical Engineering, https://doi.org/10.1007/978-981-19-3788-0_18

to measure the air temperature, probes, integrated into the composition of aerodrome no-maintenance meteorological stations, are commonly used. Notably, the air temperature measurement is executed on the ground level $T(h = 0)$, but if to speak more precisely, on the level of 2 m from the ground surface. The temperature profile in the vicinity of an aerodrome is computed, using empirical formulae [2]. Nowadays, aerodromes of civil aviation are beginning to be equipped with systems of temperature profile measurement. The MTR-5 microwave temperature profiler is referred to such gauges [3, 4]. The use of temperature profilers allows us to increase the forecast reliability for an aerodrome, enhance flight safety on takeoff-landing phases. Let us provide several examples, confirming this fact.

Temperature inversions in the pre-ground layer pose the biggest hazard for aircraft takeoff or landing. These phenomena, provided their hidden nature, cause the hazard factors for aircraft takeoff, landing. When an aircraft enters the layers of warmer air, the thrust decay occurs; therefore, the vertical speed decreases (see Fig. 1). It was established [1] that when the temperature rises by 10°, the vertical speed decreases by 10–15%. The layer of temperature inversion on a glide slope can result in thrust rollback, unpredictable for a flight crew, followed by the aircraft altitude loss, which, in terms of low altitudes, can result in a disaster (see Fig. 2).

The data on the temperature profile coupled with the data on other atmosphere parameters make it possible to forecast the conditions of weather hazards occurrences. At this rate, for example, the conducted investigations [1, 4] allow us to state that such a hazardous phenomenon, as aircraft icing, can be observed within a wide range of sub-zero temperatures; however, its probability is ultimate within a relatively narrow range of temperature from −5 °C up to −10 °C and RH > 85%. Beyond these intervals, the probability of icing reduces quite fast. A qualitative forecast of an icing area requires the data about $T(h)$ temperature and RH (h) humidity profile. On the basis of the data, a range of altitudes, at which the icing criteria are implemented, is defined. The practice established that while icing the air temperature, surrounding an aircraft, based on its values, is usually below 0° and equal or below T_{ice} saturation temperature over ice. A cloud layer, where $T < 0$ and simultaneously $T < T_{ice}$, is a layer of potential icing.

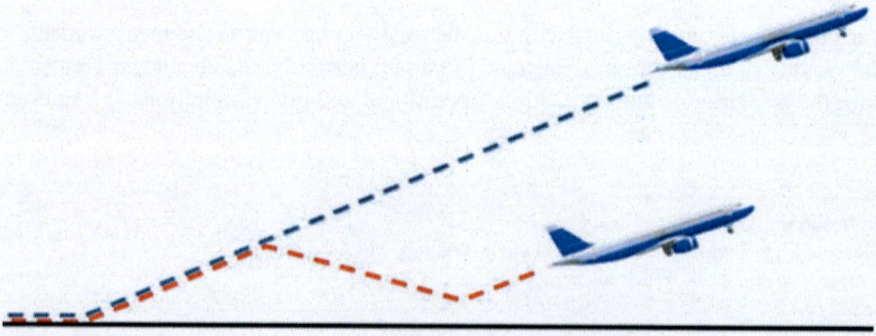

Fig. 1 Influence of the temperature inversion on the aircraft takeoff trajectory

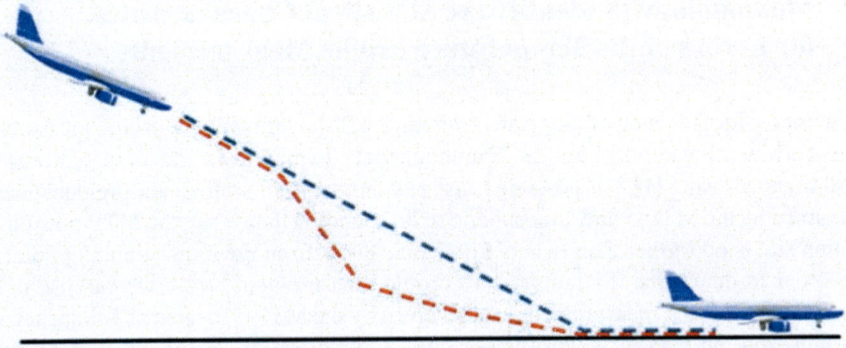

Fig. 2 Influence of the temperature inversion on the aircraft landing trajectory

A task of forecasting fogs also requires reliable information about the temperature profile. For departing, approaching and landing aircraft, radiative fogs, formed over the aerodrome surface, present a major hazard. Such fogs emerge most frequently over the land and can be observed as an ice field. Radiative fogs occur under clear skies and slight wind (up to 2 m/s) [1]. A fog forecast is carried out, using the data about $T(h)$ temperature profile and its $\mathrm{d}T(h)$ gradient, RH $(h = 0)$ humidity, and $U(h = 0)$ wind velocity in the surface layer.

Reliable data about the temperature profile in the vicinity of an airport are of paramount importance to define the zero isotherm height [5, 6]. The position of zero isotherm is the necessary initial data, used by aerodrome meteorological radars to identify the structure of cloud formations, and identification of an extent of their hazard.

The experimental comparative research of MTP-5 profiler and data of radio sounding [4–7] enables us to make a conclusion about a main disadvantage of MTP-5 related with the necessity of its calibration. The lack of calibration or insufficient frequency of its execution can lead to the situation when the measured temperature profile will have essential deviations from the real values of temperature at heights. It can appear to be impracticable to forecast weather hazards. In [2], in order to ensure the calibration, it is proposed to deliver measured data about the environment temperature from a temperature probe of the aerodrome meteorological station T_1 $(h = 0)$ to the temperature profiler. Based on the data, the temperature profiler performs the temperature comparison that is obtained, using sounding T_2 $(h = 0)$ and the temperature measurements by a probe of an aerodrome meteorological station $T_1(h = 0)$. The calculated difference $\Delta = T_1(h = 0) - T_2(h = 0)$ is recommended to use for calibration of a temperature profiler. The given method of calibration has an obvious disadvantage, as it does not ensure calibration about all the measured temperature profile.

In order to enhance the efficiency of MTP-5 application, a mathematical model for errors of MTP-5 temperature profile measurement is necessary to develop, using the data of temperature-wind radio sounding of atmosphere [8].

2 Methodology to Identify the Statistical Characteristics for Errors of the Temperature Profile Measurement

Let us consider the methodology determining a MTP-5 temperature profile measurement errors using atmospheric radio sounding data. In practical terms of meteorological flight ensuring [1], it is possible to use real temperature profile measurement data obtained by the MTP-5 and atmospheric radio sounding data when the MTP-5 installation site is no farther than twenty kilometers away from the radio sounding point. In order to determine the temperature profile measurement errors we can use the temperature profile measurement data received by the MTP-5 located at Tolmachevo airfield and the data of Novosibirsk radio sounding station. Radio sound data are used as reference measurements. Figure 3 shows the location of Novosibirsk radio sounding station and the MTP-5 location site at Tolmachevo airfield.

As a result of preliminary processing of the obtained information on temperature profiles, arrays of temperature profile difference values measured by the MTR-5 and temperature-wind probing data of the radio sounding station [6, 8] are formed [5]. The obtained differences of the temperature profile $\Delta T(k)$ and its variation gradient $\Delta DT(k)$ are defined by the following ratios:

$$\Delta T(k) = T_{\mathrm{MTP}}(k) - T_{\mathrm{p}}(k), \quad (1)$$

$$\Delta \mathrm{d}T(k) = \mathrm{d}T_{\mathrm{MTP}}(k) - \mathrm{d}T_{\mathrm{p}}(k), \quad (2)$$

where $T_{\mathrm{MTP}}(k)$, $\mathrm{d}T_{\mathrm{MTP}}(k)$ are the temperature profile and its rate of change obtained from the MTR-5 data; $T_{\mathrm{p}}(k)$, $\mathrm{d}T_{\mathrm{p}}(k)$ are the temperature profile and its rate of change obtained from the data of temperature-wind sensing.

Fig. 3 Novosibirsk radio sounding station location and the MTR-5 installation site at Tolmachevo airfield (Novosibirsk)

Since the arrays of $\Delta T(k)$ and $\Delta dT(k)$ values are the implementations of random processes, their further processing is carried out using mathematical statistics [9, 10]. To determine the statistical characteristics of $\Delta T(k)$ and $\Delta dT(k)$ values, the received data records are processed [3].

The method of determining the statistical characteristics of temperature profile measurement errors for aircraft takeoff and landing phases is based on finding the parameters optimal estimates of the auto-regression model of Gaussian processes according to experimental data.

Let us assume that as a result of the conducted experimental studies, an array of sample values of a random vector stationary ergodic process is obtained:

$$\boldsymbol{Y}_1^N = \{\boldsymbol{Y}(k), k = 1, 2, \ldots, N\}. \tag{3}$$

The sample volume is considered to be sufficient so that applying the usual methods of mathematical statistics makes it possible to determine reliable estimates of the distribution law, mathematical expectation, variance, excess, and asymmetry coefficients, as well as the correlation function of each component of $\boldsymbol{Y}(k)$ process.

Suppose that, according to these estimates Y(k), vector process is Gaussian and has the dimension of $(n \times 1)$. The statistical dynamics of the stationary Gaussian sequence can be generally described by the auto-regression (AR) equation as:

$$\boldsymbol{Y}(k) = \sum_{l=1}^{N} \boldsymbol{B}_l \boldsymbol{Y}(k-l) + \Gamma \boldsymbol{N}(k) \tag{4}$$

where $\boldsymbol{B}_l$ are the matrices with $(n \times n)$ dimension of constant unknown coefficients;$\boldsymbol{N}(k)$ is the vector of independent random Gaussian variables with zero mathematical expectation and unit variance; $\boldsymbol{\Gamma}$ is an unknown matrix $(n \times n)$, which, without the limitation of generality, is supposed to be lower triangular and non-degenerate.

The procedures for testing hypotheses about the order of the auto-regression p-process and accounting for transient trends in model (4), solved by the weighted sum of the known time functions, are similar to the scalar case and, therefore, are not considered further.

Taking into account (4), it can be shown that the final joint a posteriori distribution of the estimated parameters, provided that the entire implementation (3) is observed, has the form:

$$p\left(B_1, \ldots, B_l, G|Y_{p+1}^N\right) = \frac{C}{(\det G)^{(N-p)/2}} \times$$
$$\exp\left\{-\frac{1}{2}\sum_{k=p+1}^{N}\left[Y(k) - \sum_{l=1}^{p} B_l Y(k-l)\right]^T G^{-1}\left[Y(k) - \sum_{l=1}^{p} B_l Y(k-l)\right]\right\}$$

where **C** is a constant that does not depend on the estimated parameters, and **G** matrix is defined by the relation:

$$\boldsymbol{G} = \boldsymbol{\Gamma}\boldsymbol{\Gamma}^T. \tag{5}$$

The optimal estimates of $\boldsymbol{B}_l$ matrix coefficients are based on the necessary condition of the quadratic form extremum:

$$F = \left\{ -\frac{1}{2} \sum_{k=p+1}^{N} \left[\boldsymbol{Y}(k) - \sum_{l=1}^{p} \boldsymbol{B}_l \boldsymbol{Y}(k-l) \right]^T G^{-1} \left[\boldsymbol{Y}(k) - \sum_{l=1}^{p} \boldsymbol{B}_l \boldsymbol{Y}(k-l) \right] \right\}, \tag{6}$$

which is presented as:

$$\frac{\partial F}{\partial \left(b_{ij}\right)_m} = 0, m = 1, \ldots, p, i, j = 1, \ldots, n. \tag{7}$$

The matrix derivative for any of its elements:

$$\frac{\partial \boldsymbol{B}_m}{\partial \left(b_{ij}\right)_m} = E_{i,j}, m = 1, \ldots, p. \tag{8}$$

is a matrix in which all the elements are zero except the element with (i, j) indices equal to one.

Taking into account (8), the equality (7) can be represented as:

$$\sum_{k=p+1}^{N} \left[\boldsymbol{E}_{(ij)} \boldsymbol{Y}(k-m) \right]^T G^{-1} \left[\boldsymbol{Y}(k) - \sum_{l=1}^{p} \boldsymbol{B}_l \boldsymbol{Y}(k-l) \right] = 0, \tag{9}$$

or

$$\sum_{k=p+1}^{N} \left[\boldsymbol{E}_{(ij)} \boldsymbol{Y}(k-m) \right]^T G^{-1} \boldsymbol{Y}(k) = \sum_{k=p+1}^{N} \left[\boldsymbol{E}_{(ij)} \boldsymbol{Y}(k-m) \right]^T G^{-1} \left[\sum_{l=1}^{p} \boldsymbol{B}_l \boldsymbol{Y}(k-l) \right], \tag{10}$$

Having performed multiplication in square brackets of the left part of equality (10), taking into account (8) and the transposition operation, we obtain:

$$\sum_{k=p+1}^{N} \left[\boldsymbol{E}_{(ij)}\boldsymbol{Y}(k-m)\right]^{T}\boldsymbol{G}^{-1}\boldsymbol{Y}(k) = \sum_{k=p+1}^{N} \boldsymbol{V}_{1}^{T}\boldsymbol{Y}(k-m)\boldsymbol{Y}^{T}(k)\boldsymbol{G}^{-1}\boldsymbol{V}_{1},$$

where: V_1 is a column vector with all the elements equal to one.

Performing the same with the right part (10), to determine the optimal estimates of $\boldsymbol{B}_l^*$ matrix, we obtain the following system of linear matrix equations:

$$\sum_{l=1}^{p} \boldsymbol{B}_l^* \sum_{k=p+1}^{N} \left[\boldsymbol{Y}(k-l)\boldsymbol{Y}^{T}(k-m)\right] = \sum_{k=p+1}^{N} \boldsymbol{Y}(N)\boldsymbol{Y}^{T}(k-m), \tag{11}$$

To obtain the optimal estimates of $\boldsymbol{\Gamma}$ matrix coefficients, let us represent the right-hand part of the final a posteriori distribution formula as follows:

$$R = (\det G)^{-\frac{N-p}{2}} \exp\left\{-\frac{1}{2}\sum_{k=p+1}^{N}\left[Z^{T}(k)G^{-1}Z(k)\right]\right\}, \tag{12}$$

where the following symbol is introduced to shorten the entry:

$$\boldsymbol{Z}(k) = Y(k) - \sum_{l=1}^{p} \boldsymbol{B}_l^* \boldsymbol{Y}(k-l). \tag{13}$$

Let us find the logarithm of the equality both parts (12). Since the logarithm is a monotonically increasing function, from the necessary extreme condition on **G** matrix coefficients, we have

$$(N-p)\boldsymbol{G}^{-1} + \left[\frac{\partial}{\partial g_{ij}} \sum_{k=p+1}^{N} \boldsymbol{Z}^{T}(k)\boldsymbol{G}^{-1}\boldsymbol{Z}(k)\right] = 0. \tag{14}$$

When writing (14), we take into account that the derivative of f scalar function from the matrix is the matrix:

$$\frac{\mathrm{d}f(G)}{\mathrm{d}G} = \left[\frac{\mathrm{d}f(G)}{\mathrm{d}g_{ij}}\right] \text{ and } \frac{\mathrm{d}}{\mathrm{d}G}[\ln(\det G)] = G^{-1}.$$

Besides, on the basis of the matrix identity of $\boldsymbol{G}\boldsymbol{G}^{-1} = \boldsymbol{I}$, where $\boldsymbol{I}$ is a single matrix that can be deduced as:

$$\frac{\mathrm{d}\boldsymbol{G}^{-1}}{\mathrm{d}g_{ij}} = -\boldsymbol{G}^{-1}\boldsymbol{E}_{ij}\boldsymbol{G}^{-1}. \tag{15}$$

Therefore, for optimal estimates of the inverse matrix elements, we acquire:

$$\left(g^{-1}\right)^{*}_{ij} = (N-p)\left[\sum_{k=p+1}^{N}\left(\mathbf{Z}^{T}(k)\right)_{i}\left(\mathbf{Z}(k)\right)_{j}\right]^{-1}, i, j = 1, \ldots, n. \quad (16)$$

In conformity with the known rules, we find the lower triangular matrix Γ^*optimal estimate, having determined all the elements of the inverse matrix on the basis of the formulae (16) and performed the operation of its inversion.

In conformity with the known rules, we find the lower triangular matrix Γ^*optimal estimate, having determined all the elements of the inverse matrix on the basis of the formulae (16) and performed the operation of its inversion:

- recorded arrays of sample values are presented in accordance with (3);
- generated arrays are used as initial data in (11);
- in conformity with (11), the values of ***B*** optimal estimates for AR equations of the given order are defined;
- taking into account the estimates obtained in accordance with (11), for the AR equations of the given order, using (13) with the help of (16), the values of Γ^* optimal estimates are determined;
- relying on the expression (4), using the obtained matrix ***B*** and Γ^* optimal estimates for the AR equations of the given order, the models of the vector Gaussian random process are formed;
- based on the obtained models, implementations of the vector Gaussian random process are generated;
- statistical manipulation of the vector Gaussian random process generated implementations is performed to check validity of the developed models.

3 Conclusion

The considered methodology allows us to enhance the efficiency of MTP-5 temperature profiler application. Maximum efficiency of the temperature profiler can be obtained by data fusion about the temperature profile from its output with the temperature profile data from other probes, the operation of which is based on other physical principles. For this purpose, the implementation of developing such complex systems to measure the temperature profile stipulates the successive accomplishment of the following stages:

- identification of a list of complexing systems and gauges;
- synthesis of algorithms for the complex data processing about the temperature profile in the vicinity of an aerodrome;
- assessment of synthesized algorithms quality for the complex data processing of the temperature profile in the vicinity of an aerodrome.

The implementation of the first stage is associated with defining an extent of influence of measurement accuracy of the temperature profile on flight safety and the composition of elements for the information part of the complex data processing system. At this stage, a risk assessment for a forthcoming flight by means of factors related with the temperature profile data is vital [11].

The second stage is concerned with selecting and justifying a theoretical base for the synthesis of the algorithms for complex data processing about the temperature profile stage in the vicinity of an aerodrome, correct mathematical statement of the problem of algorithm synthesis of complex data processing about the temperature profile and the problem solution of synthesis. This stage is principal.

On the third stage the development of the methodology for an assessment of the algorithm, quality of the complex data processing about the temperature profile in the vicinity of an aerodrome is implemented. Moreover, the mathematical or simulation modeling, as well as the experimental assessment of quality for the algorithms of complex data processing about the temperature profile in the vicinity of an aerodrome for aerodromes located in different geographical areas and under various conditions of the atmosphere [12–15], is carried out.

References

1. Shun CLI, McLeod C, Johnston LK (2009) Meteorological services to aviation. WMO Bulletin 58(2):94–103
2. Korablev YN (2018) Operativnoe informirovanie ekipazhej vozdushnyh sudov ob opasnyh meteoyavleniyah v rajonah arkticheskih posadochnyh ploshchadok (operational notifying aircraft crew about dangerous meteorological phenomena in the zones of arctic landing sites). Nauchnyj vestnik MGTU GA 21(5):137–149. https://doi.org/10.26467/2079-0619-2018-21-5-137-149 (In Russian)
3. Kadygrov EN, Miller EA, Troitsky AV (2013) Study of atmospheric boundary layer thermodynamics during total solar eclipses. IEEE Trans Geosci Remote Sens 51(9):4672–4677. https://doi.org/10.1109/TGRS.2013.2248014
4. Zuev VV, Nakhtigalova DP, Shelekhov AP et al (2016) Application of MTP-5PE meteorological temperature profiler in an airport for determining spatial zones of possible aircraft Icing. Atmos Ocean Opt. https://doi.org/10.1134/S1024856016020159
5. Bolelov EA, Vasiliev OV, Galaeva KI, Ziabkin SA (2020) Analiz raznosti vysot nulevoj izotermy po dannym dvuh temperaturnyh profilemerov (analysis of the height difference of the zero isotherm according to two temperature profilers). Nauchnyj vestnik MGTU GA 23(1):19–27. https://doi.org/10.26467/2079-0619-2020-23-1-19-27 (In Russian)
6. Bolelov EA, Ermoshenko YM, Fridzon MB, Korablev YN (2017) Dinamicheskie pogreshnosti datchikov temperatury pri radiozondirovanii atmosfery (dynamic error of the temperature sensors with the sounding of the atmosphere). Nauchnyj vestnik MGTU GA 20(5):88–97. https://doi.org/10.26467/2079-0619-2017-20-5-88-97 (In Russian)
7. Bolelov EA (2019) Povyshenie opravdyvaemosti meteoprognozov po aerodromu putem kompleksirovaniya izmeritelej meteoparametrov atmosfery (increasing of the forecast success rate for an aerodrome by complexing the meteorologistical parameters profilers of the atmosphere). Nauchnyj vestnik MGTU GA 22(5):43–53. https://doi.org/10.26467/2079-0619-2019-22-5-43-53 (In Russian)
8. Perov AI (2016) Synthesis of optimal incoherent subcarrier tracking of BOC (1,1) navigation signals. Am J Appl Sci 13(9):984–995. https://doi.org/10.3844/ajassp.2016.984.995
9. Kailath T, Poor HV (1998) Detection of stochastic processes. IEEE Trans Inf Theory 44(6):2230–2231. https://doi.org/10.1109/18.720538

10. Rybalkina AL, Trusova EI, Sharov VD (2018) Metodika ocenki riska predstoyashchego poleta dlya vertoletov s uchetom neblagopriyatnyh meteouslovij (risk assessment methodology for a forthcoming flight of helicopters taking into account unfavorable meteorological conditions). Nauchnyj vestnik MGTU GA 21(6):124–140. https://doi.org/10.26467/2079-0619-2018-21-6-124-140 (In Russian)
11. Caballero R (2014) Physics of the atmosphere. IOP Publishing, Bristol, UK. https://doi.org/10.1088/978-0-7503-1052-9
12. Akmaykin DA, Bolelov EA, Kozlov AI, Lezhankin BV, Svistunov AE, Shatrakov YG (2021) Principal physics of radar location and radio-navigation. In: Theoretical foundations of radar location and radio navigation. Springer Aerospace Technology. Springer, Singapore. https://doi.org/10.1007/978-981-33-6514-8_2
13. Jun-i Y, Mizuhata Y, Tokitoh N (2021) Template synthesis of novel norcorrole complexes with a phenyl-substituted phosphorus center. Eur J Org Chem. https://doi.org/10.1002/ejoc.202101312
14. Gorbachev OA, Lezhankin BV, Erokhin VV, Povarenkin NV (2021) Algorithm of optimal control over the output power of the transmitter of an automatic dependent surveillance broadcast transponder. In: Wave electronics and its application in information and telecommunication systems, WECONF 2021—conference proceedings, St. Petersburg, p 9470720. https://doi.org/10.1109/WECONF51603.2021.9470720
15. Ermakov AK, Portnova TY, Lezhankin BV, Erokhin VV (2021) Trajectory control algorithms for unmanned aircraft complexes flying in formation. In: Wave electronics and its application in information and telecommunication systems, WECONF 2021—conference proceedings, St. Petersburg, 31 May–04 June 2021. St. Petersburg, 2021. p 9470714. https://doi.org/10.1109/WECONF51603.2021.9470714

Creation of Weather Datasets for Simulation of the Near Airfield Zone Weather Radars

Sergei Zyabkin, Oleg Vasiliev, and Boris Lezhankin

Abstract The problems of modeling the signal-interference situation for solving the problems of certification of airborne meteorological radars are analyzed. The possibilities of using similar tools for the development and validation of ground-based weather radar complexes of the near airfield zone are considered. The method of forming the initial set of weather phenomena situation data for creating simulation scenarios of dangerous meteorological phenomena is shown. Created weather data includes set of radar reflectivity measurements, located on a regular three-dimensional grid. The essence of the proposed technique consists in direct interpolation of radar reflectivity data using the method of barycentric coordinates followed by spatial median filtering. Initial reflectivity factor values are obtained by using the weather radar of the near airfield zone. The simulation result is demonstrated in accordance with the proposed algorithm.

Keywords Weather radar · Near airfield zone · Flight safety · Modeling of weather phenomena situation · Radar reflectivity

1 Introduction

A unique source of highly-discrete meteorological data is a ground-based weather radar [1–3]. However, modern weather radars are quite complex and bulky systems, even in a portable version. The most promising in the conditions of remote airfields and helicopter pads, as well as airfields with seasonal operation in the distant areas of the country is the use of a ground-based near airfield zone weather radar (WR NAF)

S. Zyabkin (✉)
Moscow State Technical University of Civil Aviation, Moscow, Russia
e-mail: s.zyabkin@ians.aero

O. Vasiliev
"Experimental Workshop NaukaSoft" LLC, Moscow, Russia

B. Lezhankin
Irkutsk Branch of Moscow State Technical University of Civil Aviation, Irkutsk, Russia

O. A. Gorbachev et al. (eds.), *Proceedings of 10th International Conference on Recent Advances in Civil Aviation*, Lecture Notes in Mechanical Engineering, https://doi.org/10.1007/978-981-19-3788-0_19

[4]. At the same time, to date, Russian WR NAF are practically absent in operation [5].

In order to correct this situation, it is necessary to create additional tools for both development and testing, as well as certification of such systems. This paper describes an approach to the generation of a set of meteorological data based on in situ observations made by a ground-based single-position weather radar complex of the near airfield zone to provide a methodological basis for the development and testing of new aerodrome weather radars.

2 Weather Radar Simulation

The task of adequate modeling of the signal and interference situation is crucial in the certification of airborne meteorological navigation (MN) radars according to the RTCA DO-220 standard [6, 7]. Among other things, this standard requires to assess the effectiveness of detection of wind shear and turbulence dangerous to aircraft using a set of models of hazardous weather events dangerous to aviation that caused accidents in the past. The simulation is performed using the Airborne Doppler Weather Radar Simulation (ADWRS) software package developed by the Langley Research Center.

Because airborne weather radars and ground-based WR NAF solve similar tasks of detecting weather phenomena dangerous to aviation in the sectors of take-off and landing of aircraft, it is advisable to use the ADWRS software package also for analyzing, monitoring and validating the operation of WR NAF. However, the simulation scenarios proposed in the RTCA DO-220 standard for airborne weather radars must be adapted to ground-based systems.

In accordance with the ADWRS user manual [8], each simulation scenario corresponds to a set of weather phenomena situation data, including a set of radar reflectivity factor and three components of the wind velocity vector measurements, located on a regular three-dimensional grid with a 50 m step. Figure 1 shows a set of reflectivity data No. 349 at a height of 150 m from the RTCA DO-220 standard [6].

Other input data define a radar scanning method, an empirical radiation pattern in two orthogonal planes, simulation parameters, radar parameters, parameters to define clutter characteristics and parameters to define the signal processing algorithms to be used.

3 Procedure Description

Let us examine the procedure for creating a similar set of radar reflectivity data. The radar reflectivity obtained by the WR NAF "Monocle" with single complete survey

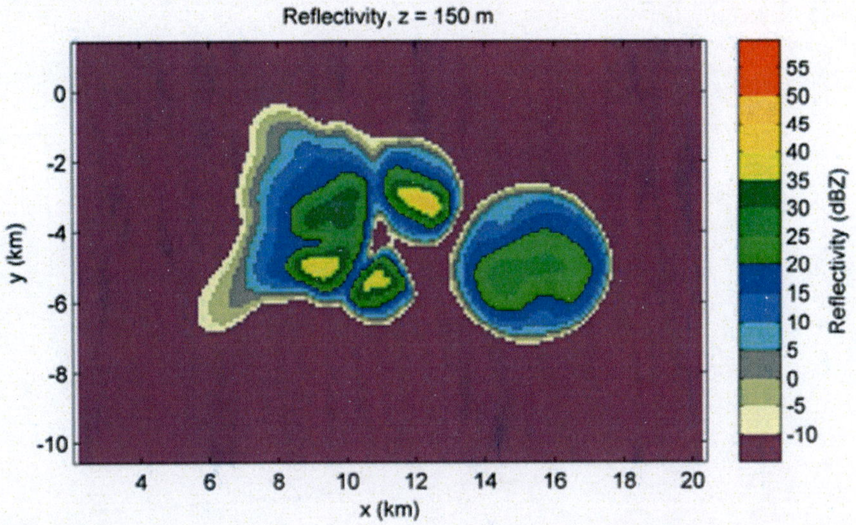

Fig. 1 Reflectivity data set No. 349 at 150 m elevation. Figure is taken from RTCA DO-220, page 136

in the 100 km range mode, which takes 10 min, will be used for modeling. The parameters of the source data used are presented in Table 1.

The procedure of multidimensional interpolation for spatial data on a regular grid is described in detail [9, 10]. However, the measurements of radar reflectivity factor dBZ, obtained by ground-based Doppler weather radar, are not located on a regular grid, but diverge concentrically from the installation site, and the coordinates of the measurement points are described in the coordinates of azimuth φ, elevation angle ψ and range R, as shown in Fig. 2.

Table 1 Parameter of the original reflectivity data set

Parameter	Value
Signal range resolution, m	2700
Range bin size, m	202.5
Number of range bins	512
Azimuth scan increment for data, °	1.6
Main lobe horizontal width, °	3.2
Elevation angle scan increment for data, °	from 0.7 to 3.5
Main lobe vertical width, °	3.5
Azimuth scanning sector, °	360
Elevation angle scanning sector, °	90

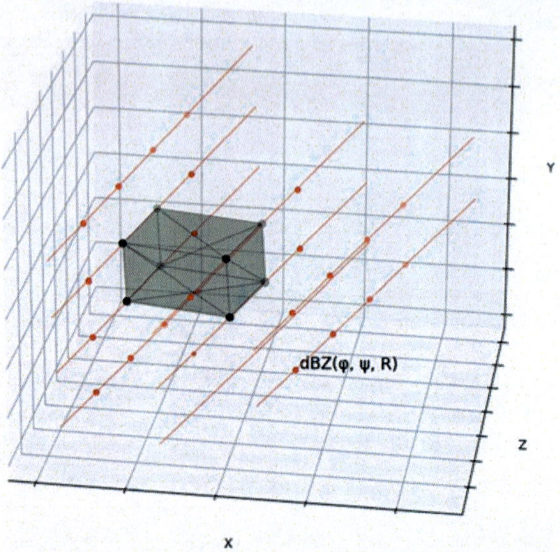

Fig. 2 Reflectivity data measurement coordinates

Several data beams emanating from the weather radar site are shown in red on Fig. 2. An elementary prismatic region is marked in black. Inside it, a multidimensional interpolation procedure takes place to search for those values of the desired regular grid that fall inside a given volume. Let us consider the general solution of this problem, schematically presented in Fig. 3.

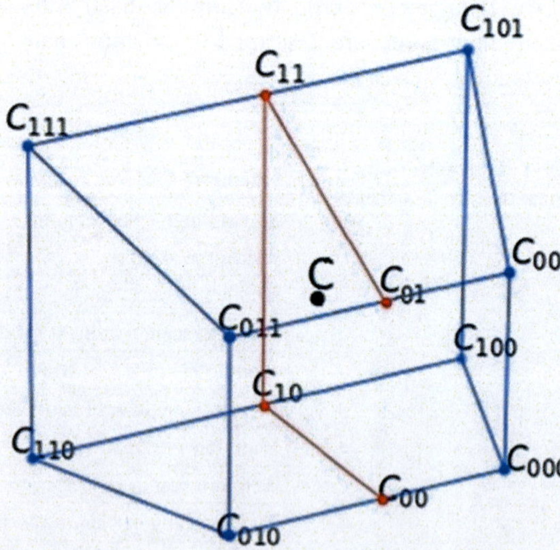

Fig. 3 Spatial interpolation of the value at point *C* by eight measurements at the peaks of the of prismatic volume

The technique is as follows—at the beginning, the prismatic volume is intersected by a vertical plane passing through both the weather radar installation location and the point at which the interpolation is required. In Fig. 3, the points of intersection of this plane with the edges of the prismatic volume are designated as C_{xx}. Then the radar reflectivity factor values at intermediate points C_{11}, C_{10}, C_{01} and C_{00} is linearly interpolated from the data at points C_{111} and C_{101}, C_{110} and C_{100}, C_{011} and C_{001}, C_{010} and C_{000} respectively. The expression for calculating one of the values is as follows:

$$Z(C_{11}) = \frac{C_{111}C_{11}}{C_{111}C_{101}} Z(C_{111}) + \frac{C_{101}C_{11}}{C_{111}C_{101}} Z(C_{101}), \tag{1}$$

where $Z(C_{xx})$ is the radar reflectivity factor value at the point and $C_y C_{y'}$ is the distance between the peaks C_y and $C_{y'}$. The remaining values are calculated similarly.

These intermediate points form two triangles, and such that any desired interpolation location lies inside at least one of them. Thus, the problem is reduced to the trivial problem of linear interpolation on the plane by three points or, simply put, linear interpolation of the function of two variables [11, 12].

This problem can be solved using method of barycentric coordinates. First, the coordinates of the intermediate points that make up the interpolation triangle and the interpolation point itself are translated into two-dimensional local coordinate system (x, y). Then the geometric meaning of barycentric coordinates (α, β, γ) is that they are equal to the ratio of the areas S_{xx} of the triangles shown below (see Fig. 4). The barycentric coordinates can be calculated as follows.

$$S = S_{12} + S_{13} + S_{23}; \alpha = \frac{S_{12}}{S}; \beta = \frac{S_{13}}{S}; \gamma = \frac{S_{23}}{S}. \tag{2}$$

The solution of this system of equalities is the following expressions:

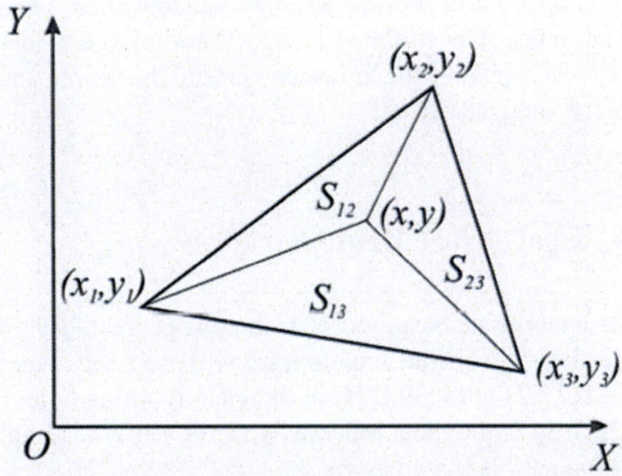

Fig. 4 The geometric meaning of barycentric coordinates as triangles area ratios

$$\alpha = \frac{(x_2y_3 - x_3y_2) + x(y_2 - y_3) + y(x_3 - x_2)}{(x_2 - x_1)(y_3 - y_1) - (x_3 - x_1)(y_2 - y_1)}; \tag{3}$$

$$\beta = \frac{(x_3y_1 - x_1y_3) + x(y_3 - y_1) + y(x_1 - x_3)}{(x_2 - x_1)(y_3 - y_1) - (x_3 - x_1)(y_2 - y_1)}; \tag{4}$$

$$\gamma = 1 - \alpha - \beta. \tag{5}$$

After the barycentric coordinates of the location (x, y) are obtained, the value of the function in it, i.e. the desired radar reflectivity factor value $Z(C)$ in single node of the regular grid, is calculated using the following equation:

$$Z(C) = \alpha Z(C_{00}) + \beta Z(C_{10}) + \gamma Z(C_{11}). \tag{6}$$

Then the same procedure is repeated for all points of a regular three-dimensional grid to obtain a complete data set for reflectivity Z.

However, the resulting data set does not take into account neither the influence of spatial resolution in range due to the parameters of the selected signal, nor the influence of spatial angular resolution due to the radiation pattern characteristics. There is also a possibility of obtaining erroneous measurements of radar reflectivity factor due to interference. Traditionally, meteorological radar solves these problems by using spatial median filtration [13, 14]. As a median filtration window for a data set obtained using WR NAF "Monocle" with the characteristics from Table 1, it is advisable to use a window with dimensions of no more than 1050 m in the horizontal plane and 250 m in the vertical:

$$Z_{i,j,k} = \text{median}\left(Z_{i-N,j-N,k-M}, \ldots, Z_{i,j,k}, Z_{i+1,j,k}, \ldots, Z_{i+N,j+N,k+M}\right) \tag{7}$$

where $Z_{i,j,k}$ is the radar reflectivity factor for a node with indices (i, j, k); N, M are the maximum values of horizontal and vertical indices depending on the median filtering window. For the selected window, these values are 10 and 2, respectively. The choice of these dimensions for the window provides correction of possible surges of measurements due to inaccuracies in the source data while preserving the nature of the source data [15].

4 Usage Case Demonstration

To demonstrate the proposed technique of creating a reflectivity model dataset of the weather phenomena situation, we will use a set of meteorological data obtained on 09/16/2021 at 09:39 UTC in the vicinity of the Orlovka airfield in the Tver region. The map of the radar reflectivity factor at 1 km. altitude measured by the WR NAF "Monocle" is shown in Fig. 5.

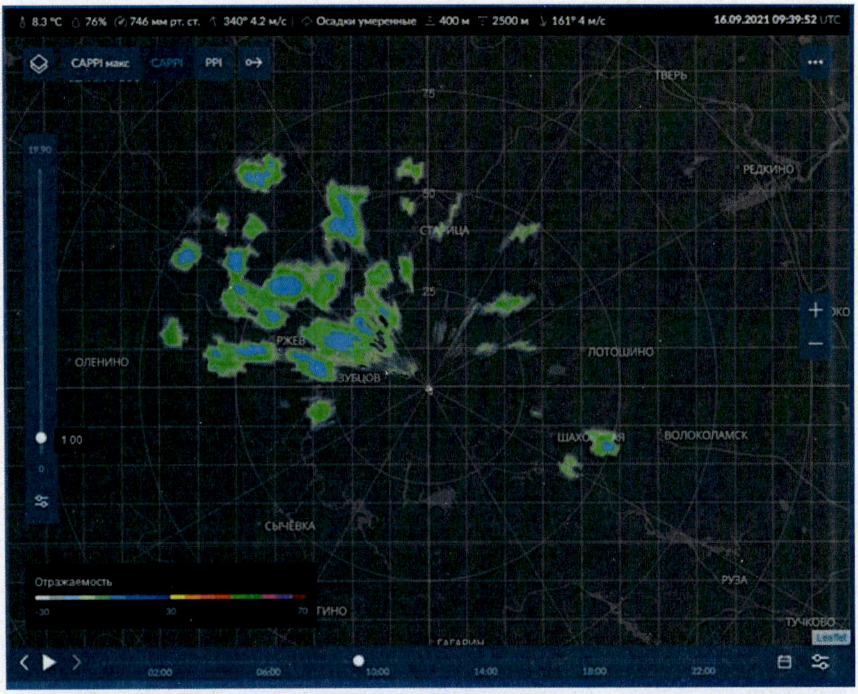

Fig. 5 Map of radar reflectivity factor at 1 km. altitude measured and displayed by the WR NAF "Monocle". 09/16/2021 09:39 UTC. Orlovka airfield. Tver region

The result of generating a set of weather phenomena data for the simulation scenario based on the selected source and the algorithm described in this paper is shown in Fig. 6.

As one can be seen, the resulting data set from Fig. 5 correctly repeats the less detailed image provided by the WR NAF "Monocle", which has a spatial image resolution of 1 km. The result of the formation in the future can be further processed and adjusted manually by the user to obtain the desired simulation model of the weather phenomena situation.

5 Conclusion

The article proposes to use the experience of certification of airborne meteorological radars to solve the problems of developing and testing ground-based weather radar systems of the near airfield zone. One of the challenges that must be overcome is the creation of a database of simulation scenarios that are specific to remote regions of the Russian Federation, where the use of new systems is most needed.

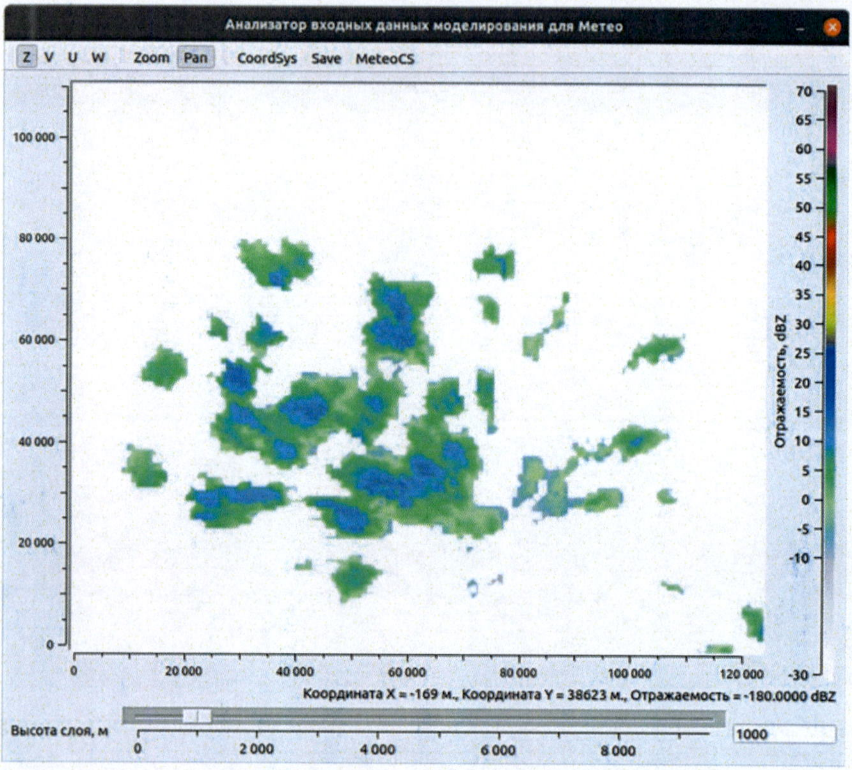

Fig. 6 A set of generated reflectivity data for the ADWRS based on the model data from Fig. 4. The step of the regular grid is 50 m. After generation, the data is shifted by 80 km. on the X-axis and 20 km on the Y-axis

The article discusses the creation of a dangerous reflectivity scenario based on direct measurements using a WR NAF. The obtained radar reflectivity factor data on an irregular grid are subjected to interpolation processing based on the method of barycentric coordinates. The final data further undergoes spatial median filtering. The use of the proposed method is demonstrated on data obtained in the vicinity of the Orlovka airfield in the Tver region.

In the future, further work is required both for the direct creation of a database of scenarios, and the development of a methodology for creating dangerous weather scenarios based on the wind characteristics of the atmosphere.

Acknowledgements The reported study was funded by RFBR, project number 20-38-90139.

References

1. Montopoli M, Marzano FS (2011) Meteorological radar systems. Integrated ground-based observing systems. Springer, Berlin. https://doi.org/10.1007/978-3-642-12968-1_2
2. Matrosov SY, Cifelli R, Kennedy PC, Nesbitt SW, Rutledge SA, Bringi VN, Martner BE (2006) A comparative study of rainfall retrievals based on specific differential phase shifts at X- and S-band radar frequencies. J Atmos Oceanic Tech 23(7):952
3. Doviak RJ, Zrnic DS (1993) Doppler radar weather observations, 2nd edn. Academic Press, London/San Diego
4. Galaeva KI (2020) Analiz rezul'tatov ispytanij i sertifikacii meteorologicheskogo radiolokacionnogo kompleksa blizhnej aerodromnoj zony (Results analysis of the tests and certification of near-airfield meteorological radar complex). Nauchnyj vestnik MGTU GA 23(1):28–40. https://doi.org/10.26467/2079-0619-2020-23-1-28-40 (In Russian)
5. Zhukov VY, Shchukin GG (2014) The state and prospects of the network of Doppler weather radars. Russ Meteorol Hydrol 39:126–131. https://doi.org/10.3103/S1068373914020095
6. Minimum Operational Performance Standard (MOPS) For Airborne Weather Radar Systems, RTCA. https://standards.globalspec.com/std/10003550/rtca. Accessed 20 Feb 2021
7. Shchukin GG, Bulkin VV (2011) Meteorological passive-active radar observations. J Commun Technol Electron 56(5):509–530
8. Lupidi1 A, Moscardini C, Garzelli A, Berizzi F, Cuccoli F (2011) Polarimetry applied to avionic weather radar: improvement on meteorological phenomena detection and classification. In: Proceedings of ESAV'11, Capri, Italy, 12–14 Sept 2011
9. Vrcej B, Vaidyanathan PP (2001) Efficient implementation of all-digital interpolation. IEEE Trans Image Process 10(11):1639–1646. https://doi.org/10.1109/83.967392
10. Getreuer P (2011) Linear methods for image interpolation. Image Process on Line. https://doi.org/10.5201/ipol.2011.g_lmii
11. Shafarevich IR, Remizov AO (2012) Linear transformations of a vector space to itself. In: Linear algebra and geometry. Springer, Berlin. https://doi.org/10.1007/978-3-642-30994-6_4
12. Abraham Albert Ungar (2010) Barycentric calculus in euclidean and hyperbolic geometry: a comparative introduction. World Scientific. https://doi.org/10.1142/7740
13. Arunachalam S, Khairnar SM, Desale BS (2013) The fast fourier transform algorithm and its application in digital image processing. Math Theor Model 3:267–273
14. Aleshechkin AM, Erokhin VV (2017) Trajectory optimization of dynamically controlled objects in INS/GNSS integrated navigation system. Gyrosc Navig 8(1):15–23. https://doi.org/10.1134/S2075108716040027
15. Bolelov EA, Vasiliev OV, Galaeva KI, Ziabkin SA (2020) Analiz raznosti vysot nulevoj izotermy po dannym dvuh temperaturnyh profilemerov (Analysis of the height difference of the zero isotherm according to two temperature profilers). Nauchnyj vestnik MGTU GA 23(1):19–27. https://doi.org/10.26467/2079-0619-2020-23-1-19-27 (In Russian)

Correlator of the Preamble of an Automated Dependent Surveillance Signal

Vyacheslav Erokhin, Muslim Mezhetov, Tatyana Portnova, and Sergey Turintsev

Abstract A problem of implementing the methods of optimal signal reception, particularly the algorithms of optimal filtration and smoothing, is in selecting the parameters of message and observation models corresponding to real data. An urgent scientific and practical problem, therefore, is research of peculiarities of the methods for optimal information processing when applied to real signals of modern radio-technical facilities. To solve the problem, a scientific and experimental apparatus composed of an antenna system, an attenuator, a digital receiver and a PC was constructed. The authors formulated the essential problems of the optimal radio reception theory regarding SSR and ADS-B signals. The paper presents the research results of correlation processing when applied to real RTF signals. The paper demonstrates the effectiveness of applying a correlator for detection a desired signal in the accepted implementation and determination of delay time.

Keywords Automated dependent surveillance · Correlator · Signal preamble · A posteriori probability · Signal delay · Estimation error · Correlation function

1 Introduction

Radio-technical facilities (RTF) of different purpose are, first of all, information systems designed to transmit and extract useful information. In real conditions of RTF operation, the problem of receiving and extracting information is solved in the context of signal distortion because of noise and disturbances. These distortions reduce the credibility of extracted information; therefore, it is necessary to take measures to eliminate the influence of the factors, i.e., to solve a problem of RTF optimization [1]. In the statistical RTF theory, the result of extracting information from a received signal (observation) is treated as making a decision of message parameters either at each time instant or at specified ones. One of the essential issues is selection of the best (to some extent) system. A methodological basis for this selection is statistical

V. Erokhin (✉) · M. Mezhetov · T. Portnova · S. Turintsev
Irkutsk Branch of Moscow State Technical University of Civil Aviation, Irkutsk, Russia
e-mail: ww_erohin@mail.ru

O. A. Gorbachev et al. (eds.), *Proceedings of 10th International Conference on Recent Advances in Civil Aviation*, Lecture Notes in Mechanical Engineering, https://doi.org/10.1007/978-981-19-3788-0_20

decision theory developed by A. Wald. Formulation of the principal points of the theory is not generally related to specific practical tasks and contains the general approach to solution of problems of statistical analysis and synthesis of particular RTF [1].

The paper deals with issues of applying the methods of statistical analysis and RTF synthesis to the modern systems based on the secondary surveillance radar (SSR) as well as on the automated dependent surveillance broadcast (ADS-B)—a system onboard a dynamic object (aircraft) which supplies periodically information of the state vector (location coordinates, speed constituents) and other information from onboard systems [2–4].

The aim of the paper is to solve an urgent applied scientific problem of implementing the methods of optimal signal reception under the influence of random disturbances when applied to modern radio-technical facilities.

2 Experimental Research

The methods of optimal signal reception, in particular, algorithms for optimal filtration and smoothing are difficult to be put into practice because of need for selection of the parameters of message and observation models which have to correspond to real data. For instance, the parameters of a desired signal and measurement noise should, to the maximum extent possible, correspond to real processes. But the entire statistical description of such processes is, as a rule, impossible [5, 6]. That is why one of the urgent scientific and practical problems is research of peculiarities of applying the algorithms for optimal information processing to real signals of modern radio-technical facilities.

To solve the problem, the authors have developed a scientific experimental apparatus (its block diagram you can see in Fig. 1) composed of an antenna system, an attenuator, a digital receiver and a PC.

Figure 2 demonstrates a generalized block diagram of a digital receiver which received ES-1090 signals transmitted by an ADS-B ground station. The band-pass filter (BPF) of the receiver is designed to pass the signals of a desired frequency band

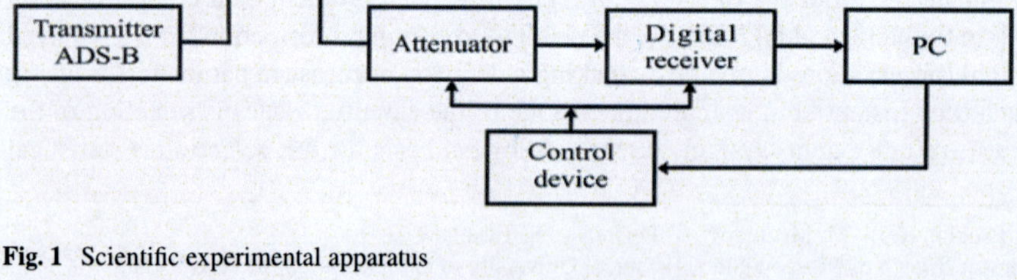

Fig. 1 Scientific experimental apparatus

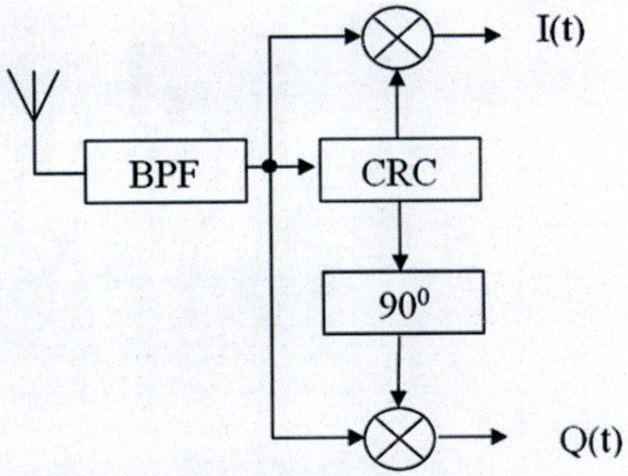

Fig. 2 Generalized block diagram of a digital receiver

and to block the signals of all other frequencies. The carrier recovery circuit (CRC) tunes the internal receiver generator to the carrier of the input signal.

At the receiver output, we obtain an imaginary and a real constituents $I(t)$ and $Q(t)$ (Fig. 3a, b, correspondingly) of the input signal. Having performed operation [7, 8].

$$U(t) = \sqrt{I(t)^2 + Q(t)^2},$$

We obtain a signal envelope $U(t)$ which is shown in Fig. 4.

3 Problem Statement

Let us formulate the main problems of the theory of optimal radio reception methods with regard to observable SSR and ADS-B signals whose solution will be given in the paper. They are: (1) detection of a desired signal in the accepted particular implementation; (2) estimation of signal delay time. When solving the detection problem, we have to find out in an optimal way if there is a signal $S(t, \lambda)$ on the interval $[0, T]$ in the accepted signal implementation $y(t)$. In addition to the desired signal, there is also a disturbance $n(t)$ in the accepted implementation. Numerous problems of the optimal reception theory have found their complete solution for a disturbance presented as white Gaussian noise (WGN) with characteristics [1]

$$M\{n(t)\} = 0, \qquad M\{n(t_1)n(t_2)\} = \frac{N_0}{2}\delta(t_2 - t_1)$$

where N_0 is physically measurable spectral noise density, and $\delta(x)$ is a delta function.

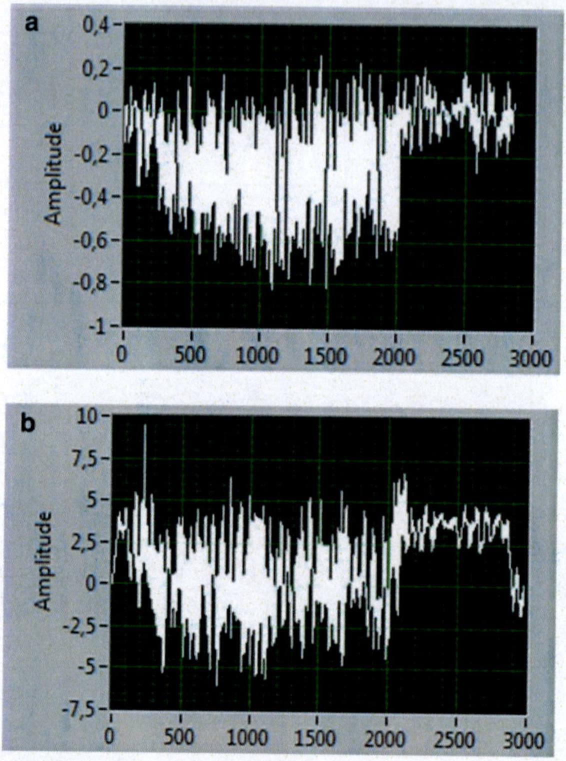

Fig. 3 Oscillograms of the constituents **a** $I(t)$, **b** $Q(t)$

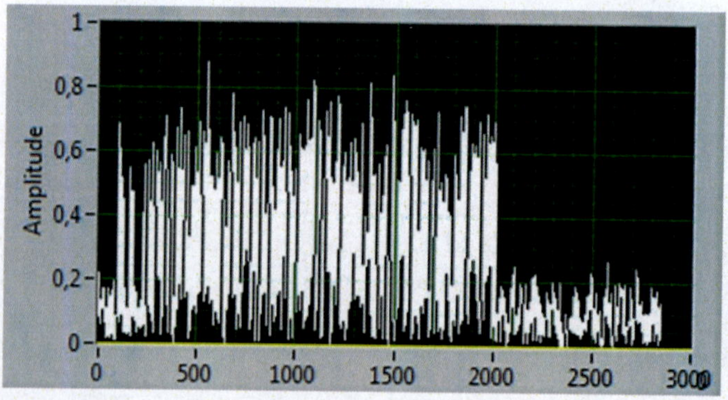

Fig. 4 Signal envelope

These are the disturbances which are considered in the paper. Then, the received additive signal–noise mixture can be written as

$$\xi(t) = S(t) + n(t), t_0 \leq t \leq t_0 + T$$

where T is a final time interval on which observation $\xi(t)$ takes place.

In the theory of optimal reception, the considered problem statement is classified as a problem of detecting a deterministic signal, i.e., a signal with fully known parameters, against the background of additive WGN $n(t)$ with two-sided spectral density $N_0/2$. The analysis of results of synthesizing optimal detectors (Bayesian and non-Bayesian ones) shows that they all calculate the likelihood ratio and compare it with the threshold. The likelihood ratio $p\left(Y_0^T\right)$ can be found through conditional probability densities (PD) of the observed implementation in the context of presence or absence of a desired signal [1]

$$p\left(Y_0^T\right) = \exp\left\{\frac{2}{N_0}\int\limits_0^T S(t,\lambda)\left(y(t) - \frac{1}{2}S(t,\lambda)\right)\mathrm{d}t\right\}.$$

The likelihood ratio determines the structure of the optimal detector. The decisive rule can be written as:

$$\frac{2}{N_0}\int\limits_0^T y(t)S(t)\mathrm{d}t \geq \frac{E}{N_0} + \ln(h_0) = h, \tag{1}$$

where $E = \frac{2}{N_0}\int_0^T S^2(t)\mathrm{d}t$ is energy of the received signal.

The expression in the left part of formula (1) determines an optimal algorithm for processing the observed implementation $y(t)$ which is known as an optimal receiver (OR). For the OR output process, we obtain

$$u_{\mathrm{OR}}(t) = \frac{2}{N_0}\int\limits_0^T y(\tau)S(\tau)\mathrm{d}\tau, \quad t \in [0, T].$$

Practical implementation of the optimal receiver is possible in two forms: either as a correlator or as a matched filter.

4 Detection and Estimation of ADS-B Signal Delay Time

By method of exchanging information, the ADS-B system of VDL Mode 4 is a synchronous system where each object emits signals at times known to all objects in advance [2], thus allowing the pseudorange to be measured provided that the time of arriving the transmitted message is determined. The measured pseudorange d is determined by a delay of time τ_i of receiving a signal by the ith object relative to the known time τ_j of emitting this signal by the jth object [9]

$$d = c(\tau_i - \tau_j) = c\tau,$$

where c is propagation speed of a radio signal along the radio path, and τ is delay time of signal propagation.

With regard to the case under consideration, the preamble of an ADS-B signal has the form shown in Fig. 5.

Let us denote the reference signal of the preamble by $S(t)$ and the signal received from an object by $S(t - \tau)$. Possible values of the parameter τ are within the interval $[0, T]$ with known a priori probability density $p_{\text{pr}}(\tau)$. The received oscillation is a random process in the form of additive mixture of a desired signal and noise [10]:

$$y(t) = S(t) + n(t),$$

$n(t)$ is a disturbance in the form of white Gaussian noise (WGN),

$$S(t) = \begin{cases} A, t_1 \le t \le t_1 + \tau_и, t_2 \le t \le t_2 + \tau_и, t_3 \le t \le t_3 + \tau_и, t_4 \le t \le t_4 + \tau_и \\ 0, \text{for other } t \end{cases}$$

is the preamble of a desired signal.

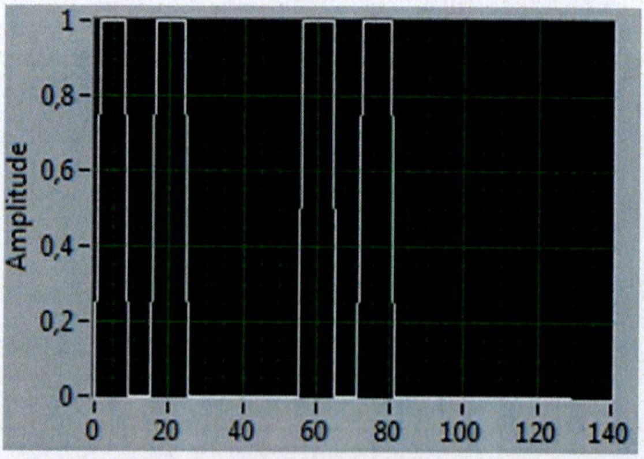

Fig. 5 Preamble standard

In our case, the receiver is designed to determine the value τ with a minimal error by analyzing the received oscillation $y(t)$. The information of τ is included in the a posteriori probability density (APD) which, with regard to the case under consideration, has the following form:

$$p_{\mathrm{ps}}(\tau) = kp_{\mathrm{pr}}(\tau)\exp\left\{-\frac{1}{N_0}\int_0^T [y(t) - S(t-\tau)]^2 \mathrm{d}t\right\}$$

where $k = p^{-1}(x)$ is a coefficient dependent on the sample results but independent on the parameter τ, and $p_{\mathrm{pr}}(\tau)$ is a priory probability density.

To simplify the expression, we write the exponent as

$$-\frac{1}{N_0}\int_0^T [y(t) - S(t-\tau)]^2 \mathrm{d}t = -\frac{1}{N_0}\int_0^T y^2(t)\mathrm{d}t$$
$$+\frac{2}{N_0}\int_0^T y(t)S(t-\tau)\mathrm{d}t - \frac{E}{N_0}. \tag{2}$$

In the right part of (2) a summand equal to the ratio of energy of the received oscillation to the spectral noise density is not dependent on τ so, in the APD formula, multiplier

$$\exp\left(-\frac{1}{N_0}\int_0^T y^2(t)\mathrm{d}t\right)$$

is included in the constant k, and then, it is possible to write

$$p_{\mathrm{ps}}(\tau) = kp_{\mathrm{pr}}(\tau)\exp\left(-\frac{E}{N_0}\right)\exp[q(\tau)],$$

where

$$q(\tau) = \frac{2}{N_0}\int_0^T y(t)S(t-\tau)\mathrm{d}t. \tag{3}$$

The multiplier $\exp(-E/N_0)$ should also be included in the constant k which is reasonable in our case as the signal energy does not depend on its delay τ. Hence, we obtain

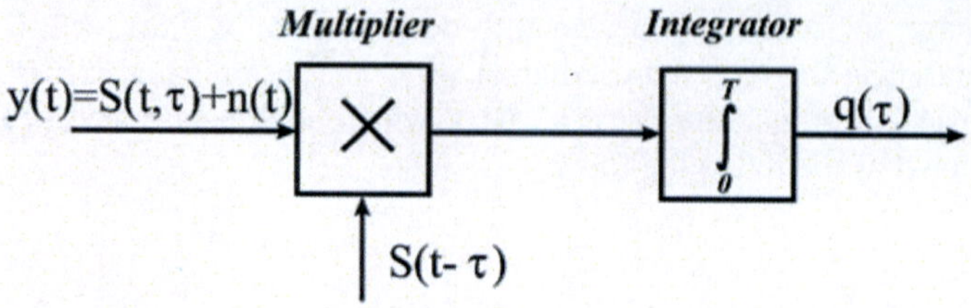

Fig. 6 Simplified block diagram of a correlator

$$p_{ps}(\tau) = kp_{pr}(\tau)\exp[q(\tau)].$$

The main procedure of processing the received oscillation $y(t)$ with a view to obtaining the APD parameter τ consists in producing a function $q(\tau)$ and that is what defines the operation to be performed in order to extract information of the parameter τ. To determine $q(\tau)$, it is necessary to form a copy of the signal $S(t)$ on the receiving side. Figure 6 demonstrates a simplified block diagram of a receiver which produces, for different τ, a function $q(\tau)$ composed of a multiplier and an integrator.

The function $q(\tau)$ characterizes the ratio of cross-correlation between the received oscillation $y(t)$ and the desired signal $S(t-\tau)$ as formula (3) coincides with the expression for cross-correlation function (CCF) between $y(t)$ and $S(t-\tau)$ (accurate to the multiplying constant), and therefore, the device shown in the figure is known as a cross-correlator or just correlator [11–13]. Integral (3) is referred to as a correlation integral.

In practice, the cross-correlation function (CCF) $q(\tau)$ [14] for fixed values of τ can be produced by the following device (Fig. 7). Here, ν is a number of delay line taps, and Δ is a difference in delays between adjacent delay line taps and $\nu\Delta = T$. A comparison circuit determines the channel with the biggest correlation integral. It is the delay time corresponding to this channel which will be considered estimative.

Thus, based on the accepted implementation $y(t)$ and the known form of a desired signal, it is necessary to find out in an optimal way whether there is a signal $S(t)$ and unknown time signal location τ_0. For this purpose, the function $q(\tau)$ should be produced by integrating the result of multiplying $y(t)$ by signal $S(t-\tau)$.

5 Modeling and Research Results

Implementation peculiarities of correlation processing were researched with regard to real RTF signals which were received by the installation shown in Fig. 1. To determine the beginning of an information message (the beginning of the data unit, Fig. 8) [15], it is necessary to find correlation between the input signal (Fig. 4) and the preamble standard (Fig. 5).

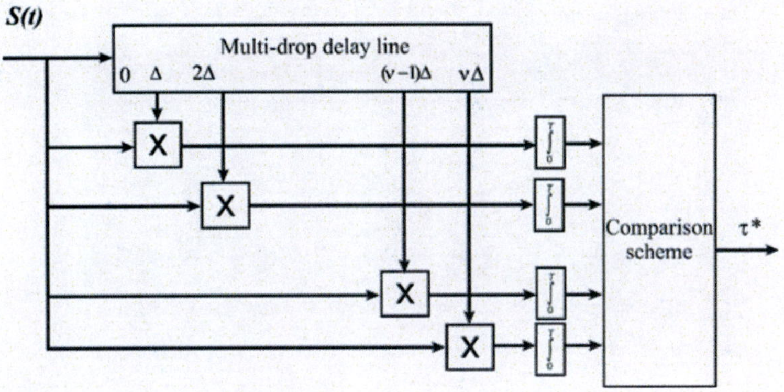

Fig. 7 Correlator for determining delay time

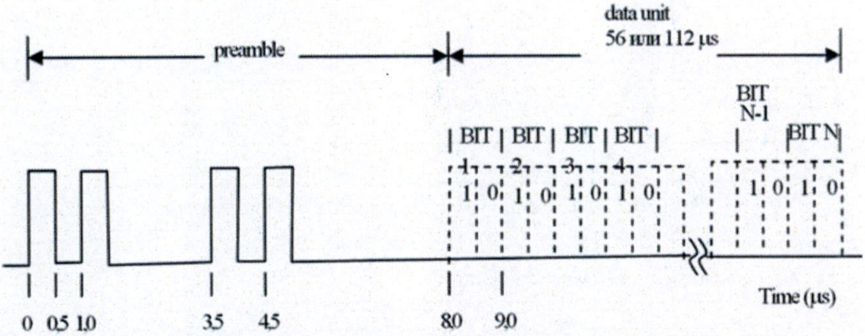

Fig. 8 Structure of a reply signal in *S* mode

The signal implementation at the correlator output is presented in Fig. 9.

The analysis of the received signal (Fig. 9) shows that the maximal function value corresponds to the 225th sample, i.e., the first symbol of the data unit begins from

Fig. 9 Output signal of the correlator

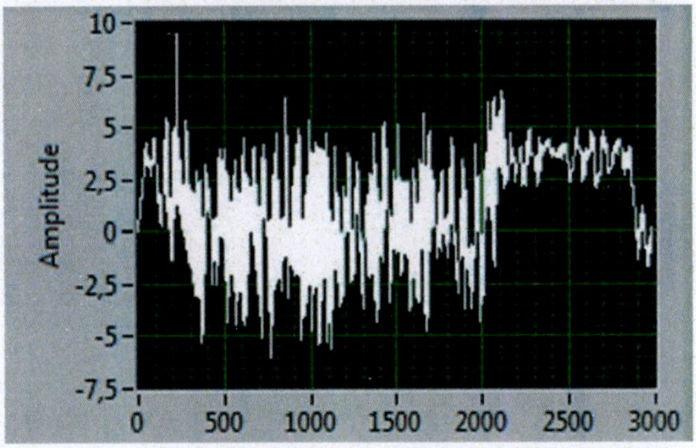

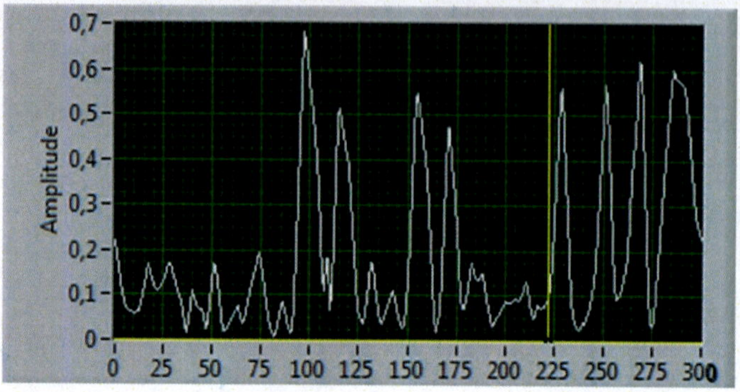

Fig. 10 Indication of the beginning of the data unit

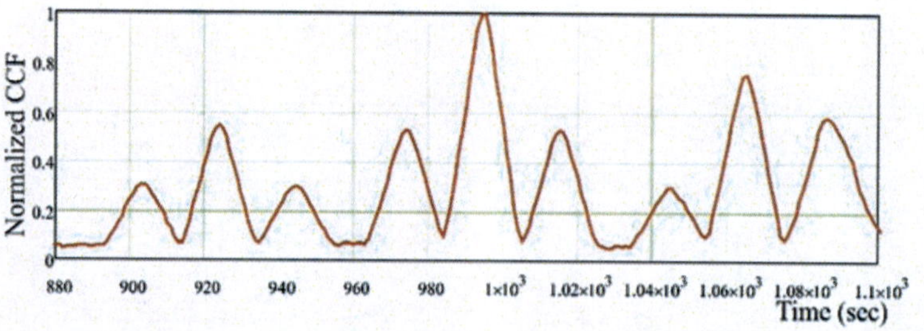

Fig. 11 Normalized cross-correlation function

the 225th sample (Fig. 10—it is indicated by a yellow vertical line). The number of samples for a preamble is equal to 128, discretization interval $\Delta t = 6.25 * 10^{-8}$.

Figure 11 presents a graph of a normalized CCF between the received signal and the reference preamble signal.

By the CCF peak and its position on the time axis, it is possible to determine the presence of a desired signal in the accepted implementation and to determine the time location.

Based on the data of the parameters and preamble of an information signal, the authors developed a signal model in the form of a train of video pulses with known time location. The developed model allowed us to simulate the received oscillation which was a random process in the form of additive mixture of a desired signal and noise (Fig. 12):

$$\xi(t) = S(t) + n(t),$$

$$S(t) = \begin{cases} A, t_1 \le t \le t_1 + \tau_и, t_2 \le t \le t_2 + \tau_и, t_3 \le t \le t_3 + \tau_и, t_4 \le t \le t_4 + \tau_и \\ 0, \text{for other } t \end{cases}$$

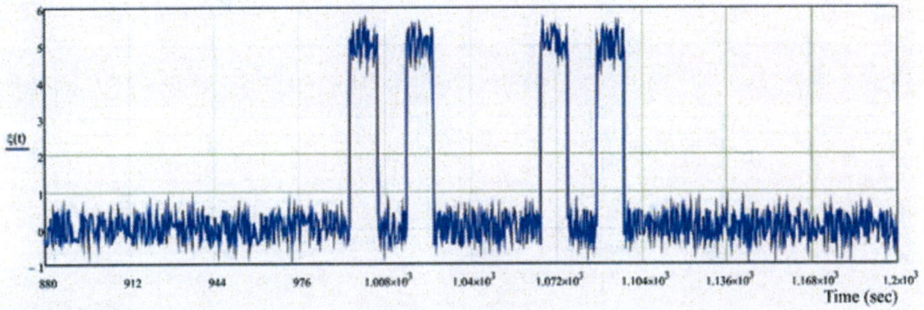

Fig. 12 Implementation of the additive mixture of a video pulse train and noise

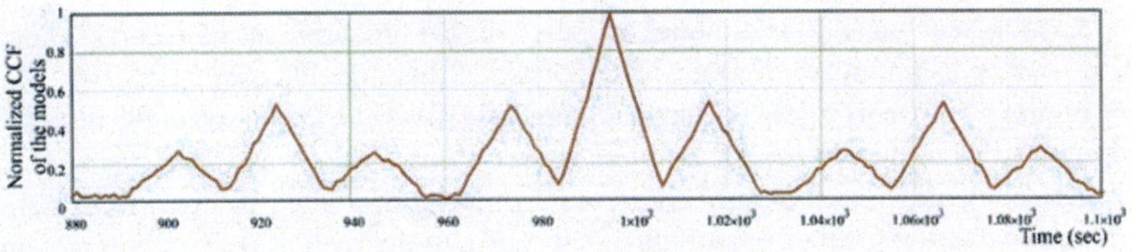

Fig. 13 Normalized CCF of the models of the accepted implementation and preamble signal

is the preamble of a desired signal, $n(t)$ is a disturbance in the form of WGN.

Figure 13 presents a graph of a normalized CCF between the model of the received additive mixture of a desired signal with noise and the model of the preamble reference signal.

Based on the data of the parameters and preamble of an information signal, the authors developed a signal model in the form of a radio pulse with account of high-frequency (HF) filling. The developed model allowed us to simulate the received oscillation which was a random process in the form of additive mixture of a desired signal and noise (Fig. 14):

$$\xi(t) = S_{\text{M}}(t) + n(t),$$

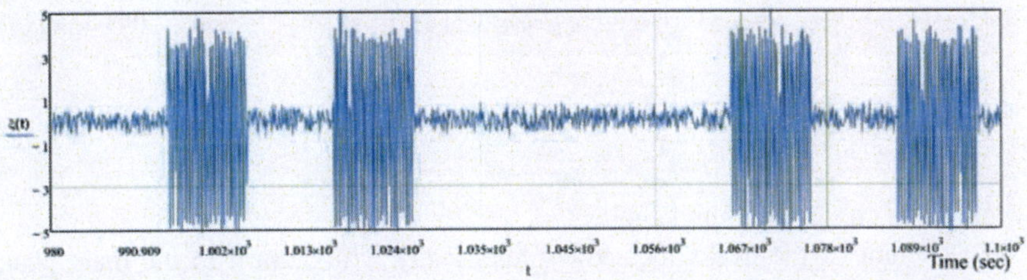

Fig. 14 Implementation of the model of additive signal–noise mixture

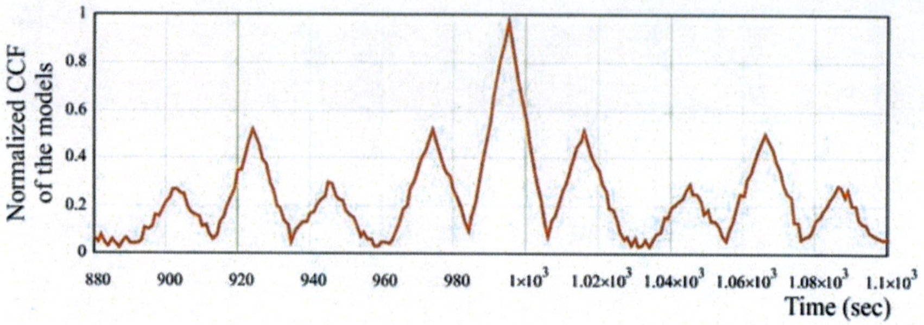

Fig. 15 Normalized CCF of the models of accepted implementation and preamble reference signal

$S_{\text{M}}(t) = Acos(\omega t + \varphi)$ is a model of an HF signal preamble at the receiver input, and A, ω, φ is amplitude, frequency and phase of the modeled oscillation.

Figure 15 presents a graph of a normalized CCF between the model of the received signal and the model of the preamble reference signal.

The comparative analysis of the research results presented in Figs. 11, 13 and 15 shows that the CCF of modelled signals has a more explicit peak in comparison with the CCF of real observed signals owing to diffused leading and trailing edges of preamble pulses.

The function $q(\tau)$ can generally be written as a sum of two summands (signal and noise functions):

$$q(\tau) = q_{\text{s}}(\tau) + q_{\text{n}}(\tau),$$

where

$$q_{\text{s}}(\tau) = \frac{2}{N_0}\int_0^{nT} S(t - \tau_0)S(t - \tau)dt,$$

$$q_{\text{n}}(\tau) = \frac{2}{N_0}\int_0^{nT} n(t)S(t - \tau)dt.$$

The signal function $q_{\text{s}}(\tau)$ reaches its highest peak when $\tau = \tau_0$ and has a height

$$q_{s\max}(\tau_0) = \frac{n2E_1}{N_0},$$

where E_1 is energy of a single pulse.

The height of the peaks located symmetrically with respect to the main peak gradually reduces. The noise function $q_{\text{n}}(\tau)$ has a zero mean value, and its dispersion is equal to

$$\sigma_{\text{n}}^2 = \frac{n2E_1}{N_0}$$

The ratio of the maximal peak of the signal function to the mean square value of the noise function

$$\frac{q_{s\,\max}(\tau_0)}{\sigma_{\text{n}}} = \sqrt{\frac{n2E_1}{N_0}}.$$

Thus, if the signal/noise voltage ratio at the correlator output for a single pulse is $\sqrt{\frac{2E_1}{N_0}}$, then, when n such pulses are received, this ratio is multiplied by $\sqrt{n}$. The result is valid not only for video pulses but also for a train of n coherent radio pulses. For a similar train of non-coherent radio pulses, the signal/noise ratio will be lower. Therefore, in the neighborhood of τ_0, both the useful peak of the function $q(\tau)$ and that of APD stand out against the noise more clearly in a pulse train than in a single pulse.

Increase in number of pulses can help achieve the required signal/noise voltage ratio at the correlator output. In our case, an n-fold increase in observation (integration) time leads to $\sqrt{n}$-fold improvement of the signal/noise voltage ratio. This result can be used in practice both for detecting periodic pulses against the background of intense disturbances and for estimating unknown parameters. Unlike a single pulse, here the decision is made with the time delay n times longer.

The results of mathematical modeling of a correlator and its performance were evaluated on the criterion

$$E_1 = \text{M}[|\varepsilon(t_1)|], 0 \le t_1 \le T_{\text{н}},$$

where M is a symbol of mathematical expectation of a correlator error; $\varepsilon(t_1) = \tau(t_1) - \hat{\tau}(t_1)$ is an error of signal delay time at the moment $t = t_1$ caused by disturbance $n(t)$, and $T_{\text{н}}$ is time of process realization.

To obtain the desired estimate by mathematical modeling, you can use the following algorithm [16]:

$$E_2 = \frac{1}{N}\sum_{j=1}^{N} \left|\varepsilon_j(t_1)\right|,$$

Thus, for estimating the efficiency E_2, it is necessary to have a mathematical model of a correlator to obtain implementations of instantaneous error values ($\tau_j(t)$ is assumed to be known):

$$\varepsilon_j(t) = \tau_j(t) - \widehat{\tau_j}(t).$$

The analysis of the results of calculating the estimation error of the signal arrival time $\varepsilon_j(t)$ for different correlator operation conditions, in particular for different values of the signal/noise ratio, has shown high efficiency of applying the proposed method ($\varepsilon_j(t) \cong 0$) to optimal reception of signals in real RTF.

The obtained results of estimating the signal arrival time prove that the operation of creating the cross-correlation function provides the best filtration of a desired signal against the background of noise and gives, in comparison with other possible operations, the maximal peak signal/noise ratio.

In general, there are difficulties encountered in applying correlators practically. As follows from formula (3), for producing the CCF, it is necessary to know the form of the desired signal $S(t)$ on the receiving side. In practice, however, this is not always possible, and this fact limits application of correlation reception. In our case, the characteristics of an ADS-B signal preamble are known which allows us to apply the correlation reception methods for solving the problems of detecting a signal and determining the delay time. Difficulties of practical implementation are related to formation of a reference signal as the transmitting and receiving devices are spaced-apart; moreover, ADS-B transponders are installed on aircraft and other movable vehicles. Making the assumption that phase fluctuations of a radio signal are slow because of propagation of radio waves through a turbulent medium, a special synchronization channel based on different variants of phase synchronization is applied for forming a reference signal from a received oscillation.

6 Conclusion

Thus, in solving an urgent applied scientific problem on researching the methods of optimal signal reception under the action of random disturbances with respect to modern radio-technical facilities, the efficiency of applying a correlator is shown for detection of a desired signal in the accepted implementation and delay time determination.

References

1. Akmaikin DA et al (2021) Theoretical foundations of radar location and radio navigation. Springer Aeros Technol
2. Abdulaziz A, Yaro AS, Adam AA, Kabir MT, Salau HB (2015) Optimum receiver for decoding automatic dependent surveillance broadcast (ADS-B) signals. Am J Signal Process 5(2):23–31
3. Chen YH, Lo S, Akos DM, Wong G, Enge PA (2013) Testbed for studying automatic dependent surveillance broadcast (ADS-B) based range and positioning performance to support alternative position navigation and timing (APNT) In: Proceedings of the 26th international technical meeting of the satellite division of the institute of navigation (ION GNSS+ 2013), Nashville, TN: 09-2013, pp 263–273
4. Strohmeier M, Lenders V, Martinovic I (2014) On the security of the automatic dependent surveillance-broadcast protocol. IEEE Commun Surv Tutor 17(2):1066–1087

5. Sage AP, Melsa JL, Steinway WJ (1972) Estimation theory with applications to communication and control. IEEE Trans Syst Man Cybern. https://doi.org/10.1109/TSMC.1971.4308330
6. Bruce P, Gibbs A (2011) Advanced Kalman filtering, least-squares and modeling: a practical handbook. John Wiley & Sons Inc., Hoboken, New Jersey
7. Liu F, Yan Y, Wang W, Tan B, Chen L, Qu C, Li J (2019) A digital correlation receiver for the mingantu spectral radioheliograph. Publ Astron Soc Austral 36:E043. https://doi.org/10.1017/pasa.2019.35
8. Bondarenko VN et al (2013) Noise immunity of a quasi optimal correlation receiver of noiselike signals with minimum frequency-shift keying. J Commun Technol Electron 58(12):1194–1199. https://doi.org/10.1134/S1064226913070048
9. Rakesh Y, Sri Rama Krishna K (2020) Short time fourier transform with coefficient optimization for detecting salient regions in stereoscopic 3D images: GSDU. Multimedia Tools Appl Springer Science+Business Media (79):8801–8824. https://doi.org/10.1007/s11042-018-6686-x
10. Taufik Abrro et al (2019) Low-complexity massive MIMO detectors under spatial correlation and channel error estimates. Wireless Pers Commun Springer Science+Business Media (106):2335–2358. https://doi.org/10.1007/s11277-019-06320-2
11. Fadhil MJ (2014) Digital communication receiver based on FPGA. Des Int J Sci Eng Res 5(2):154–160
12. Vestenicky P, Vestenicky M (2019) Optimization of receiving window width of the correlation receiver for radiofrequency identification marker localization. Int J Distrib Sens Netw 15(9):1–12
13. Zhang Q, Qin J (2014) Random correlation-based receiver for impulse radio communications in UWB channels. Wireless Pers Commun, Springer Science+Business Media New York (79):21–30. https://doi.org/10.1007/s11277-014-1838-4
14. Hassan ES, Mustafa AI, Awadalla KH et al (2018) A new modified short-multipath-insensitive code loop discriminator. Wireless Pers Commun 103:1391–1407. https://doi.org/10.1007/s11277-018-5844-9
15. International Civil Aviation Organization (2007) Aeronautical telecommunications, vol III. University Street, Montréal, Quebec, Canada, Communication Systems
16. Linty N, Dovis F (2019) An open-loop receiver architecture for monitoring of ionospheric scintillations by means of GNSS signals. Appl Sci 9(12):2482. https://doi.org/10.3390/app9122482

Aircraft Position-Fixing in a Multilateration System

Nikolay Povarenkin, Boris Lezhankin, Tatyana Portnova, and Vyacheslav Erokhin

Abstract The problem of increasing the level of civil aircraft flight safety in zones with no radar control for economic and geographic reasons is solved by applying a multi-position surveillance system. A promising means of data exchange between an aircraft and an air traffic controller (ATC) is automatic dependent surveillance-broadcast (ADS-B) which provides transmission of aircraft navigation information. The problem of aircraft position-fixing in a multi-position surveillance system on the basis of multilateration technology in a stochastic statement has been formulated and its solution with the help of modern methods of optimal signal filtration has been proposed. To implement range tracking and pseudorange measurements in a multilateration (MLAT) system, the authors propose applying a second-order filter. An a priori model of signal delay dynamics has been proposed in order to synthesize a filter of the tracking system. Using the selected model for signal delay time, a Kalman-filter-based algorithm of optimal estimation and its characteristics were examined. The paper describes theoretical and experimental studies of a multilateration surveillance system and factors influencing its operation. The accuracy characteristics within the system coverage in accordance with the presumed location were analyzed. The research on characteristics of an MLAT system with a variant of station location in the Bodaybo region of Irkutsk regional center of air traffic management (ATM) was carried out.

Keywords Multi-position system · Multilateration · A priori MODEL · Two-stage processing · Signal delay · Kalman filter · Radio navigation · Transponder

---

N. Povarenkin (✉)
Saint Petersburg State University of Aerospace Instrumentation, Saint Petersburg, Russia
e-mail: povarenkin.nv@guap.ru

B. Lezhankin · T. Portnova · V. Erokhin
Irkutsk Branch of Moscow State Technical University of Civil Aviation, Irkutsk, Russia
e-mail: taty-port@yandex.ru

O. A. Gorbachev et al. (eds.), *Proceedings of 10th International Conference on Recent Advances in Civil Aviation*, Lecture Notes in Mechanical Engineering, https://doi.org/10.1007/978-981-19-3788-0_21

1 Introduction

Modern air traffic growth causes the aircraft crews to solve navigation problems in limited time periods while ensuring the required level of flight safety which is achieved, among other things, through use of integrated surveillance systems [1]. As recommended by the International Civil Aviation Organization (ICAO), such systems should use primary and secondary surveillance radars (SSR) as a main information source for current situation. Nowadays, the most promising direction is implementation of a multi-position surveillance system based on the ADS-B system [2, 3]. In the scientific and technical literature, these systems are known as multilateration (MLAT) ones [4]. In the MLAT systems, the aircraft position is determined by measuring distances between the aircraft and some navigation reference points where receivers are located capable of receiving signals radiated by the aircraft. The ADS-B and MLAT systems provide additional opportunities for radio-electronic surveillance which should be applied for increasing the efficiency of air traffic service [2, 3]. According to the CNS/ATM (Communications, Navigation, Surveillance/Air Traffic Management) concept of expanded implementation of modern surveillance systems, application of primary means of air navigation will be reduced [5].

The ADS-B elements are designed for surveillance of objects by receiving and processing the information of flight path parameters transmitted by an aircraft through datalink in a broadcast mode: SSR Mode S, extended squitter "1090ES", Mode 4, and others. All information is transmitted to the ATC unit responsible for guiding the aircraft. The MLAT advantages are ability to be deployed in a mountainous area regardless of takeoff angles, joint operation with Mode S, ability to determine the aircraft position regardless of the aircraft navigation equipment. Data exchange in the ADS-B datalink is performed in such a manner as to continuously monitor the aircraft flight path parameters. Development of efficient algorithms for processing information is essential to implementation of such systems. While synthesizing and researching an MLAT system it is necessary to take into account the interference environment in order to provide the efficient operation under conditions of continuous random disturbances. This will allow the performance of the navigation parameter tracking system to be improved in comparison with a system synthesized in the classic deterministic statement. The authors of the work [4] proposed some estimation algorithms at the level of secondary processing of information, but synthesis and research of algorithms for filtering navigational parameters at the level of primary processing were not considered. However, errors at the primary stage lead to errors of estimating the navigational parameters which is why increase in accuracy of time determinations in data exchange channels is urgent.

The purpose of the paper is to solve an urgent scientific and applied problem of increasing the efficiency of aircraft position-fixing under the action of random disturbances using the multilateration technology and to research operation peculiarities of a multilateration system.

2 Multi-position Surveillance System

A multilateration system of air situation surveillance is based on an interrogation method when one of the ground stations generates an interrogation signal on a frequency of 1030 MHz. This signal serves as a synchronization signal (start of time counting) for three ground stations and as an interrogation signal for the onboard equipment to generate a reply signal in accordance with which all ground stations measure the information parameter. The design of a receiving station of "MERA" multi-position surveillance system and the appearance of its antenna system you can see in Fig. 1 [6].

The transmitting station of the multi-position surveillance system (Fig. 2) forms and amplifies HF pulse signals for control commands supplied by the server-concentrator. It performs interrogations in Modes A/C/S for the desired range and with maximum pulse power [6].

The server-concentrator (Fig. 3) is a computing unit that provides aircraft position-fixing on the basis of miltilateration using the time difference of signals fixed by soft and hardware of the receiving stations [6].

To perform the task of position-fixing, it is necessary to have four receiving stations. The server-concentrator also provides control and management of the multi-position surveillance system as well as tracking and information output in the automated ATC system. The system is controlled through a terminal located in close proximity to the server-concentrator which allows the technical staff to perform monitoring and diagnostics of the system on-site. The control and reference responder (CRR) of the multi-position surveillance system consists of a CRR unit and an antenna (Fig. 4) [6].

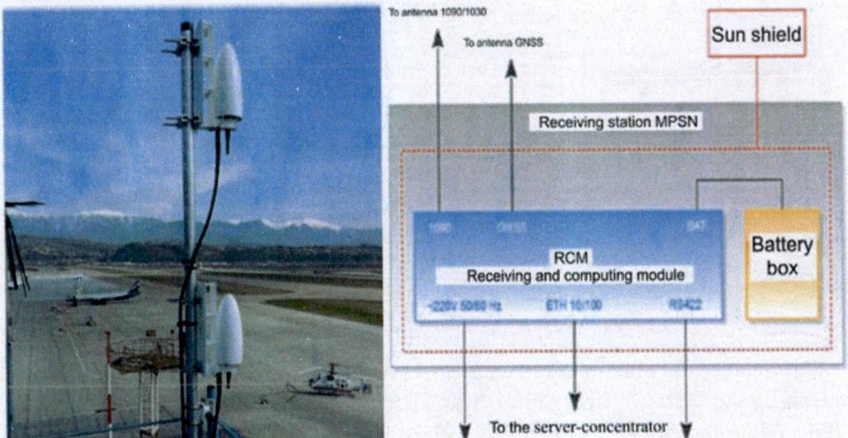

Fig. 1 Receiving station of the multi-position surveillance system: **a** antenna of the receiving station; **b** composition of the receiving station

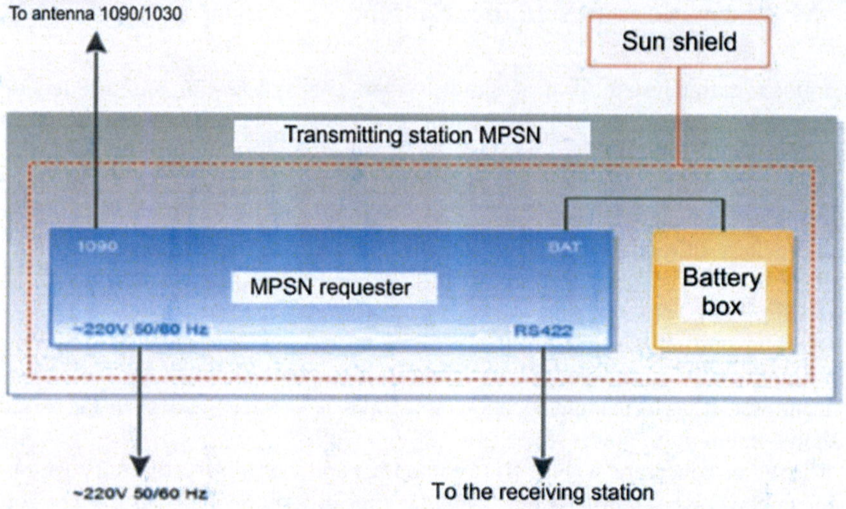

Fig. 2 Composition of the transmitting station of the multi-position surveillance system

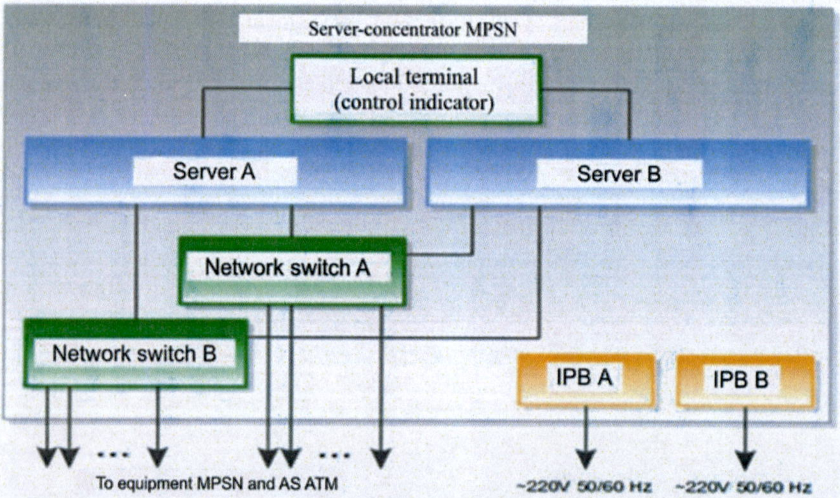

Fig. 3 Composition of the server-concentrator of the multi-position surveillance system

CRR functionality includes procedures of synchronizing the receiving stations and controlling the system's operability. The CRR implements algorithms of receiving and decoding interrogation signals as well as generation, coding, and transmission of reply signals for the receivers of the multi-position surveillance system and is a simulator of an aircraft transponder [6].

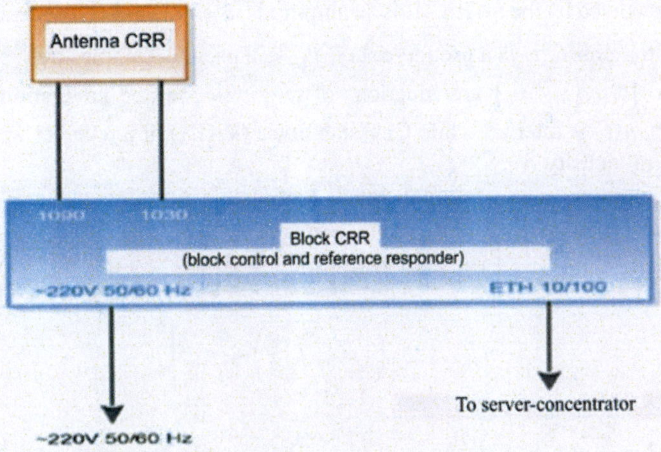

Fig. 4 Control and reference responder of the multi-position surveillance system

3 Research Methods

The problem of position-fixing on the basis of multilateration is reduced to calculating the object range and direction. To solve the problem a differential distance-measuring method is applied which helps to determine the aircraft coordinates by using range differences at the points of location of spaced receiving stations.

Signals emitted by an ADS-D transponder are received by the antennas of peripheral (spaced) receiving stations (RSs) located in visual range. The input signal of an RS transponder can be written as [7]:

$$u_{\text{in}}(t) = \sum_{i=1}^{n} S_i\left(t, \breve{\tau}_i, \breve{f}_{\text{D}i}\right) + n(t), \tag{1}$$

where

$$\begin{aligned} S_i\left(t, \breve{\tau}_i, \breve{f}_{\text{D}i}\right) &= A_i G_{\text{RC}}\left(t - \breve{\tau}_i\right) G_{\text{NS}}\left(t - \breve{\tau}_i\right) \\ &\quad \cos\left(\omega_{0i}\left(t - \breve{\tau}_i\right) + \phi_{0i}\right) \\ &= A_i G_{\text{RC}}\left(t - \breve{\tau}_i\right) G_{\text{NS}}\left(t - \breve{\tau}_i\right) \\ &\quad \cos\left(\omega_{0i} t + 2\pi \int_0^t \breve{f}_{\text{D}i}(\nu) \text{d}\nu + \phi_{0i}\right) \end{aligned}$$

is a signal received by the ith RS, A_i is an amplitude, ω_{0i} is a carrier frequency, ϕ_{0i} is a random initial phase, $\breve{\tau}_i$ is a pseudo-delay, $\breve{f}_{Di}$ is a pseudo-Doppler frequency shift, $G_{RC}\left(t-\breve{\tau}_i\right)G_{NS}\left(t-\breve{\tau}_i\right)$ are functions of ranging-code and navigation-message modulation, $n(t)$ is internal white Gaussian noise (WGN) of a receiver with single-sided spectral density N_0.

At the output, a vector of digital signals n of visible aircraft is created:

$$y(t_v) = \left[S_1\left(t, \breve{\tau}_1, \breve{f}_{D1}\right) + n_{j1}, S_2\left(t, \breve{\tau}_2, \breve{f}_{D2}\right)\right.$$
$$\left. + n_{j2}, \ldots, S_n\left(t, \breve{\tau}_n, \breve{f}_{Dn}\right) + n_{jn}\right]^T,$$

where $n_{jl}, l = \overline{1, n}$ are independent discrete WGNs with zero mathematical expectations and equal dispersions $\sigma_n^2 = N_0/(2T_d)$, $T_d = t_j - t_{j-1}$ is a discretization interval.

Because of variation of signal time delay during the flight and presence of measurement noise, the methods of optimal filtration theory should be used for creating estimates of state variables. Optimal (quasi-optimal) estimation of the information process carried by a radio signal is created by a tracking system composed of a discriminator, a filter, and a reference signal generator. When receiving additive mixture of a desired signal and WGN (1), let us write the discriminator output process as

$$u_{dv}\left(\tilde{\lambda}_v\right) = \left(\frac{\partial \ln f\left(Y_1^M, \lambda\right)}{\partial \lambda}\right)^T \Bigg|_{\lambda = \tilde{\lambda}_v},$$

where $Y_1^{\mathrm{M}} = \left\{y_{v-1,1}, y_{v-1,2}, \ldots, y_{v-1,M}\right\}$ is implementation of observations on the interval $\left[t_{v-1,1}, t_{v-1,M}\right]$, $f\left(Y_1^{\mathrm{M}}, \lambda\right)$ is a function of observed implementation Y_1^{M} and estimated process $\lambda = |\tau f_{\mathrm{D}}|^T$.

In our case $f\left(Y_1^{\mathrm{M}}, \lambda\right)$ is a likelihood function of a posteriori probability (APP) $p\left(Y_1^{\mathrm{M}}|\lambda\right)$ which is defined by expression

$$p\left(Y_1^M|\lambda\right) \cong \exp\left(\sum_{l=1}^{M} S^T\left(t_{k-1,l}, \breve{\tau}, \breve{f}_D, \vartheta_{NS}\right)\right.$$
$$\left. D_{\mathrm{n}}^{-1}\left(y\left(t_{k-1,l}\right) - \frac{1}{2} S\left(t_{k-1,l}, \breve{\tau}, \breve{f}_D, \vartheta_{NS}\right)\right)\right), \tag{2}$$

where D_{n}^{-1} is an observation noise variance matrix, ϑ_{NS} is a navigation message.

Based on (2) it is possible to calculate estimates of the delay time $\hat{\tau}$ and Doppler frequency shift $\hat{f}_{\mathrm{D}}$ corresponding to the APP maximum:

$$\left\{\hat{\tau},\, \hat{f}_{\mathrm{D}}\right\} = \max_{\tau, f_{\mathrm{D}}} p\left(\tau,\, f_{\mathrm{D}} \middle| Y_{t_0}^{t_0+T}\right).$$

In case of measuring pseudorange from the aircraft to the ith receiving station, the measured range $\tilde{D}_i$ differs from the true (geometric) range.

D_i by an unknown value $\tilde{D}_i$ which can be presented as [8, 9]

$$\tilde{D}_i = c\tilde{\tau}_i = D_i + \tilde{D}_i = \sqrt{(x_i - x)^2 + (y_i - y)^2 + (z_i - z)^2} + \tilde{D}_i, \qquad (3)$$

where D_i is the true range from the ith receiving station to the aircraft; x, y, z are coordinates of the aircraft position; $\breve{\tau}_i = t_{\mathrm{NRP(UTS)}} - t_{\mathrm{AC(OBTS)}}$ is time of propagating a signal from the aircraft to the receiving station, $t_{\mathrm{NRP(UTS)}}$ is time of receiving the signal by ADS-B navigation reference point (NRP) specified in the user's timescale (UTS), $t_{\mathrm{AC(OBTS)}}$ is time of radiating the signal by the aircraft specified in the onboard timescale (OBTS); c is speed of radio wave propagation.

With that,

$$\tilde{D}_i = c\tilde{\tau}_i,$$

where $\tilde{\tau}_i$ is discrepancy between the UTS and the OBTS.

The aircraft position is determined based on the totality of available measurements $\left\{\tilde{D}_i\right\}$ which are a set of pseudoranges (3). Solution of this system for unknown x, y, z means performing the position-fixing task. Presence of synchronization errors leads to the errors in estimating navigation parameters which makes it necessary to increase the estimation accuracy of signal arrival time.

To implement the multilateration algorithms, it is essential that all objects of a MLAT system should synchronize their watches according to the signals of GNSS satellites. We think of a GNSS TS as being a single one for which single errors of time synchronization will be used. In implementing distance measurements the second-order filter should be used. To synthesize the filter of a signal delay tracking system, let us apply an a priori model which is determined by equations written as [7]:

$$\tau_\nu = \tau_{\nu-1} + T V_{\tau,\nu-1}, \qquad (4)$$

$$V_{\tau,\nu} = V_{\tau,\nu-1} + n_{\tau,\nu-1}, \qquad (5)$$

where T is an interval of signal observation; $n_{\tau,\nu}$ is discrete WGN with known dispersion D_{n_τ}; τ is delay time of a signal envelope; V_τ is rate of change of delay time of a signal envelope; ν is a time index.

Let us present the observation equation as:

$$\xi_\nu = \tau_\nu + n_{\xi,\nu}, \tag{6}$$

where $n_{\xi,\nu}$ is discrete WGN with dispersion D_{n_ξ}.

Based on selected models of signal delay time dynamics (4), (5) and observation (6) we will obtain an algorithm of optimal filtration as follows:

$$\left.\begin{aligned}
&\hat{\tau}_\nu = \tilde{\tau}_\nu + K_{1,\nu}(\xi_\nu - \tilde{\tau}_\nu),\\
&\tilde{\tau}_\nu = \hat{\tau}_{\nu-1} + T\hat{V}_{\tau,\nu-1},\\
&\hat{V}_{\tau,\nu} = \hat{V}_{\tau,\nu-1} + K_{2,\nu}(\xi_\nu - \tilde{\tau}_\nu),\\
&K_{1,\nu} = \frac{R_{11,\nu}}{D_{n_\xi}},\quad K_{2,\nu} = \frac{R_{12,\nu}}{D_{n_\xi}}
\end{aligned}\right\} \tag{7}$$

where R_{ij} are elements of a filtration error matrix $\mathbf{R}$ of the vector $\mathbf{x} = |\tau\, V_\tau|^T$.

The procedure of calculating the matrix $\mathbf{R}$ is reduced to finding solution of expressions:

$$\begin{aligned}
&\mathbf{R}_\nu^{-1} = \tilde{\mathbf{R}}_\nu^{-1} + \mathbf{H}^T\mathbf{H}/D_{n_\xi},\\
&\tilde{\mathbf{R}}_\nu = \Phi_{\nu,\nu-1}\tilde{\mathbf{R}}_{\nu-1}\Phi_{\nu,\nu-1}^T + \mathbf{G}\mathbf{G}^T D_{n_\tau},
\end{aligned}$$

where $\tilde{\mathbf{R}}_\nu$ is an extrapolated matrix.

In the vector–matrix form, the formalized problem to be solved is written as:

$$\begin{aligned}
&\mathbf{x}_\nu =_{\nu,\nu-1} \mathbf{x}_{\nu-1} + \mathbf{G}_{\nu,\nu-1}\mathbf{n}_{\mathrm{x},\nu-1},\ \boldsymbol{\xi}_\nu = \mathbf{H}_\nu\mathbf{x}_\nu + \mathbf{n}_{\xi,\nu},\\
&\boldsymbol{\Phi} = \begin{vmatrix} 1 & T \\ 0 & 1 \end{vmatrix}, \mathbf{G} = \begin{vmatrix} 0 \\ 1 \end{vmatrix}, \mathbf{H} = |\,1\;0\,|.
\end{aligned}$$

Expressions (7) are Kalman filter equations for optimal estimation by criterion [9]:

$$J_\nu = tr M\left[(\mathbf{x}_\nu - \hat{\mathbf{x}}_\nu)(\mathbf{x}_\nu - \hat{\mathbf{x}}_\nu)^T\right] = \min, \tag{8}$$

where tr is a matrix trace.

Equations (7) allowing $\hat{\mathbf{x}}$ to be determined as the object state information is received are a filter of the tracking system. In view of the foregoing, it is reasonable to use expression (8) as an operating performance indicator.

To determine the aircraft position in an MLAT system, they deal with differential measurements of pseudoranges formed by subtracting the measurements obtained at different stations [10–12]. Then the mathematical model of pseudorange difference can be written as:

$$\Delta D_{ij}(t_k) = D_j(t_k) - D_i(t_k) + n_{ij}(t_k),$$

where $n_{ij}(t_k)$ is an error of forming pseudorange difference approximated with WGN with the known characteristics.

In the mathematical model of measuring pseudorange difference, the aircraft coordinates x, y, z are unknown parameters. By combining the measurements of all receivers of the MLAT ground stations, we will obtain a system of non-linear equations in the following form [13]

$$\begin{cases} \Delta D_{i,1}(t_k) = \sqrt{(x_1 - x)^2 + (y_1 - y)^2 + (z_1 - z)^2} \\ -\sqrt{(x_i - x)^2 + (y_i - y)^2 + (z_i - z)^2} + n_{i1}(t_k) \\ \vdots \\ \Delta D_{i,j}(t_k) = \sqrt{\left(x_j - x\right)^2 + \left(y_j - y\right)^2 + \left(z_j - z\right)^2} \\ -\sqrt{(x_i - x)^2 + (y_i - y)^2 + (z_i - z)^2} + n_{i,j}(t_k) \\ \vdots \\ \Delta D_{i,N-1}(t_k) = \sqrt{(x_{N-1} - x)^2 + (y_{N-1} - y)^2 + (z_{N-1} - z)^2} \\ -\sqrt{(x_i - x)^2 + (y_i - y)^2 + (z_i - z)^2} + n_{i,N-1}(t_k) \end{cases}$$

To determine the vector being estimated, it is reasonable to apply a mathematical apparatus of Kalman filtration. This approach implies using the information of the system state at the previous time moment which provides a more accurate estimate of the vector of unknowns [14].

For convenience, let us present expressions (5), (6) in the matrix form

$$\mathbf{Z}_k = \mathbf{H}_k(\boldsymbol{\lambda}_k) + \mathbf{N}_k, \tag{9}$$

where $\mathbf{Z}_k = \left[\Delta D_{i,1}(t_k) \cdots \Delta D_{i,N-1}(t_k)\right]^T$ is an observation vector (vector of parameters being measured), $\boldsymbol{\lambda} = [xyz]^T$ is a state vector containing the unknown aircraft coordinates, $\mathbf{N}_k = \left[n_{i1}(t_k) \cdots n_{i,N-1}(t_k)\right]^T$ is a vector of additive noise with the known matrix of measurement noise $\mathbf{V}_k$; $\mathbf{H}_k(\boldsymbol{\lambda}_k)$ is a non-linear vector function of relating the vector being estimated $\boldsymbol{\lambda}_k$ to the vector of measurements $\mathbf{Z}_k$

$$\mathbf{H}_k(\boldsymbol{\lambda}_k) = = \begin{bmatrix} \sqrt{(x_1 - x)^2 + (y_1 - y)^2 + (z_1 - z)^2} \\ -\sqrt{(x_i - x)^2 + (y_i - y)^2 + (z_i - z)^2} \\ \vdots \\ \sqrt{(x_j - x)^2 + (y_j - y)^2 + (z_j - z)^2} \\ -\sqrt{(x_i - x)^2 + (y_i - y)^2 + (z_i - z)^2} \\ \vdots \\ \sqrt{(x_{N-1} - x)^2 + (y_{N-1} - y)^2 + (z_{N-1} - z)^2} \\ -\sqrt{(x_i - x)^2 + (y_i - y)^2 + (z_i - z)^2} \end{bmatrix}$$

Change of the state vector is generally described by a differential equation, however, in transition to discrete instants of time, the evolution of the system state vector can be described by a non-linear equation which can most conveniently be written as [15]:

$$\boldsymbol{\lambda}_k = \mathbf{F}(\boldsymbol{\lambda}_k) + \mathbf{W}_k, \tag{10}$$

where $\mathbf{F}(\boldsymbol{\lambda}_k)$ is a non-linear system dynamics function; $\mathbf{W}_k$ is a vector of additive WGN with the known matrix of single-sided noise spectral densities $\mathbf{Q}_k$.

As seen from expression (9), the measured range differences are non-linearly dependent on the object coordinates included in the estimated state vector. Nowadays, for solution of navigation problems concerning Gaussian noises of a system and measurements, they widely apply Kalman algorithms based on Gaussian approximation of a posteriori density with matrices of dynamics $\mathbf{F}(\boldsymbol{\lambda}_k)$ and measurements $\mathbf{H}_k(\boldsymbol{\lambda}_k)$ being expanded by Taylor series [15].

This approach is based on the function linearization by using a linear Kalman filter. The obtained filtration algorithm is known as the extended Kalman filter (EKF) and is reduced to a two-step procedure of prediction correction. The algorithm is sufficiently well-known, so let us consider its application for the aircraft position-fixing in the MLAT system.

The EKF algorithm requires the initial value of the vector of parameters being estimated $\boldsymbol{\lambda}_{k-1}$ (or its value at the previous step of filtration) as well as that of the covariance matrix of the process $\mathbf{P}_{k-1}$ provided the random process $\boldsymbol{\lambda}_{k-1}$ is a Gaussian one with covariance matrices of prediction noise $\mathbf{Q}_k$ and measurement noise $\mathbf{V}_k$. In the initial time moment, the covariance matrix $\mathbf{P}_0$ can have a form of a diagonal matrix with elements equal to assumed dispersion of each component of the vector $\boldsymbol{\lambda}_0$ known a priori. The prediction noise matrix $\mathbf{Q}_k$ can be presented similarly to the matrix $\mathbf{P}_{k-1}$.

The state vector estimates obtained through EKF operation are not optimal as they are a result of expanding the non-linear functions by Taylor series limited by members to the power not higher than one. This approach however allows us to address the problem of non-linear filtration formally and to obtain quasi-optimal estimates with acceptable accuracy.

Having applied the technique to the multidimensional problem of non-linear filtration in discrete-time and having used the equations of linear filtration, we obtain the following EKF-based algorithm for optimal estimation of the state vector which minimizes criterion

$$\mathbf{P}_k = M\left\{\left(\boldsymbol{\lambda}_k - \hat{\lambda}_k\right)\left(\boldsymbol{\lambda}_k - \hat{\lambda}_k\right)^T\right\}, \tag{11}$$

written as [14, 15]:

$$\hat{\lambda}_k = \mathbf{F}_k\hat{\lambda}_{k-1} + \mathbf{K}_k\left(\mathbf{Z}_k - \mathbf{H}_k\mathbf{F}_k\hat{\lambda}_{k-1}\right), \tag{12}$$

$$\mathbf{K}_k = \mathbf{P}_{k/k-1}\mathbf{H}_k^T\left[\mathbf{H}_k\mathbf{P}_{k/k-1}\mathbf{H}_k^T + \mathbf{V}_k\right]^{-1}, \tag{13}$$

$$\mathbf{P}_k = [\mathbf{I}_k - \mathbf{K}_k\mathbf{H}_k]\mathbf{P}_{k/k-1}[\mathbf{I}_k - \mathbf{K}_k\mathbf{H}_k]^T + \mathbf{K}_k\mathbf{V}_k\mathbf{K}_k^T \tag{14}$$

where $\mathbf{I}_k$ is a unity matrix.

In correlations (11), (14) $\hat{\lambda}_k$ and $\mathbf{P}_k$ are a vector of optimal estimates and a covariance matrix of estimation errors.

The expression for a filtration error can be written as

$$\boldsymbol{\varepsilon}_k = \boldsymbol{\lambda} g_k - \hat{\lambda}_k,$$

where $\boldsymbol{\lambda} g_k$ is a vector of true aircraft coordinates.

The filtration error has both deterministic and fluctuating constituents. The deterministic constituent depends on the structure of the information processing system which is determined by the contents of internal linkages, hence:

$$M\|\boldsymbol{\varepsilon}_k\|^2 = \|\overline{\varepsilon}_k\|^2 + tr\{\mathbf{P}_k\},$$

where $\overline{\varepsilon}_k$ is mathematical expectation $\boldsymbol{\varepsilon}_k$; $tr\{\mathbf{P}_k\}$ is a covariance matrix trace.

Errors of aircraft position-fixing in the MLAT system depend on noise and interference intensity as well as on dilution of precision (DOP). For estimation of influence of a flight route on positioning accuracy, DOP K is used and this is a coefficient taking account of mutual position of navigation reference points and an aircraft. The DOP value is determined by the formula [16, 17]:

$$K = \left(tr\left[\left(\mathbf{H}^T\mathbf{H}\right)^{-1}\right]\right)^{1/2}.$$

The elements of the matrix $\mathbf{H} = \left[\frac{\partial \Delta D_{i,1}}{\partial \boldsymbol{\lambda}}\right]$ are calculated as partial derivatives of range difference using the corresponding coordinates.

The radial root-mean-square error σ_{RMSE} (RMSE) and the spherical root-mean-square deviation σ_{RMSD} (RMSD) are determined by the formulas:

$$\sigma_{\mathrm{RMSE}} = \sigma_{\mathrm{PR}} K_{XY}; \sigma_{\mathrm{RMSD}} = \sigma_{\mathrm{PR}} K,$$

where σ_{PR} is a mean square error (MSE) of pseudorange (PR) measurement.

In our case, as a trajectory optimization criterion, we can use the Horizontal Dilution of Precision (HDOP) whose value in equal observations is $\mathrm{HDOP} = [\sigma_x^2 + \sigma_y^2]^{1/2}/\sigma_{\mathrm{PR}}$. The radial root-mean-square (RMS) error σ_{R} of horizontal accuracy is written as

$$\sigma_{\mathrm{R}} = \sigma_{\mathrm{PR}} \mathrm{HDOP}.$$

Since observability of aircraft state variables is essential for their qualitative evaluation, the MLAT system has to be checked for observability before synthesizing a navigational Kalman-filter-based processor. A characteristic of accuracy of estimating navigation parameters is observability which, provided the observation time is infinitely small $\Delta t = t_k - t_0$, is determined by expression:

$$\Gamma = \int_{t_0}^{t_k} \mathbf{H}^T \mathbf{H} dt. \tag{11}$$

It is obvious that the larger the value of det Γ, the higher the estimation accuracy $\hat{x}(t)$. If only a part of the elements $\boldsymbol{\lambda}_k$ is observed, then det $\Gamma = 0$, and hence, vector $x(t)$ cannot be determined.

4 Algorithm Modeling and Analysis

When analyzing the optimal filtration algorithm (7) of a signal delay tracking system, we assume that the equipment of interacting objects is linked via a digital radio channel. Ground stations perform differential distance measurements. Through statistic simulation, the characteristics of the synthesized algorithm of optimal filtration (7) were examined. When modeling the errors the authors used approximations that are given in calculation models (3)–(5), correspondingly.

As initial data the following parameters were selected:

- measurement noise (1 ΔD): $\sigma_{D,\mathrm{ADS-B}\nu} = 3$ m, $\sigma_{\dot{D},\mathrm{ADS-B}\nu} = 0.05$ m/s.
- discreteness of flight counting $dt = 0.001$ s (1000 Hz);
- discreteness of data entry and measurement processing in the Kalman filter $dT_1 = 0.2$ s (5 Hz).

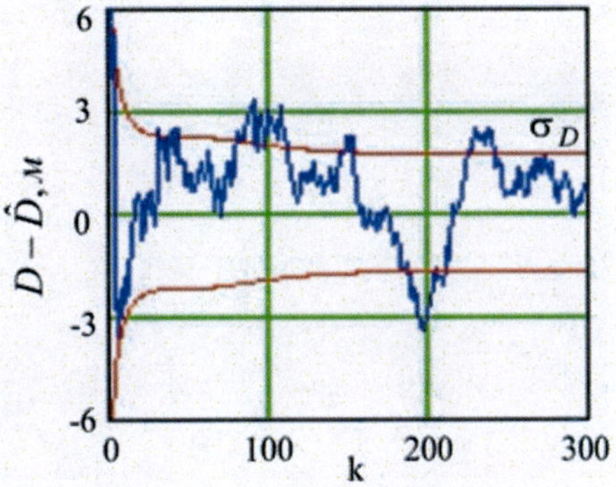

Fig. 5 Range estimation error

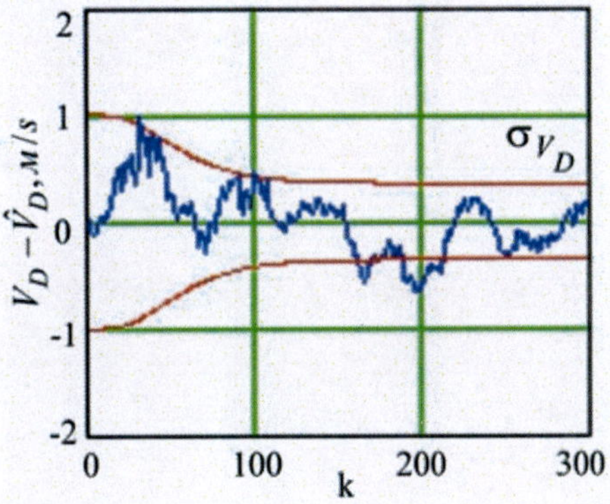

Fig. 6 Error in estimating the rate of range change

The research results for above modeling conditions are shown in Figs. 5 and 6: the estimation error $D - \hat{D}$ and σ_{D} are presented in Fig. 5; the estimation error graphs $V_D - \hat{V}_D$ and the mean square deviation σ_{VD} are shown in Fig. 6

The analysis of the modeling results demonstrates high accuracy of range estimation which will allow the effectiveness of aircraft position-fixing under the action of random disturbances to be increased due to the multilateration technology.

5 A Variant of Multilateration System Location

The ATC automated systems provide aircraft motion control at any time in different weather conditions, among other things, in the context of high-intensity traffic when there are many aircraft in a limited air space simultaneously. On the scheme of overlapping radar fields, you can notice that there exists an air space zone with no radar control because of mountainous terrain between sectors RC−5 and RC−7 (Fig. 7). Radar support and control in this zone is provided due to the information received from the crew while flying over predetermined reference points. This limits

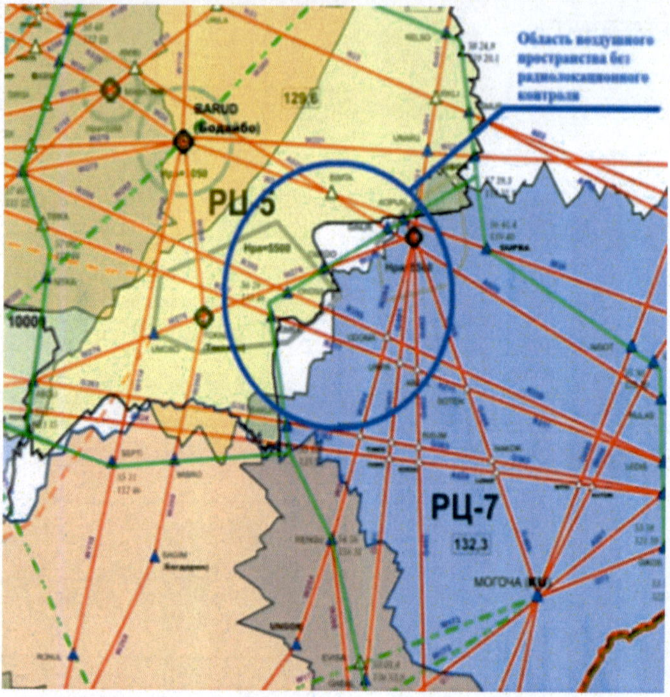

Fig. 7 Area of air space with no radar control

capacity of the air space zone since there is a need for larger separation distances which greatly influences flight safety. As the radar deployment in the region is not economically and geographically sound, application of an MLAT system is proposed.

It is proposed to locate MLAT stations near the settlement of Taksimo (N56 2133; E1145520) and the receiving stations near the settlements of Nerpo (N57 28 13; E115 18 39), Chara (N56 54 27; E118 15 29), Bambuyka (N55 46 40; E115 46 51) and Kuanda (N56 19 39; E116 7 14) as shown in Fig. 8.

6 Modeling and Investigating a Multilateration System

The authors investigated features of an MLAT system with stations located in such a way as to overlap the radar dead zone. Figures 12, 13, 14 and 15 demonstrate the modeling results of the algorithm for aircraft position-fixing in the MLAT system on an example of selected location of a transceiving and receiving stations for specified aircraft trajectories along routes W276 (Figs. 9 and 10) and R211 (Figs. 11 and 12), correspondingly.

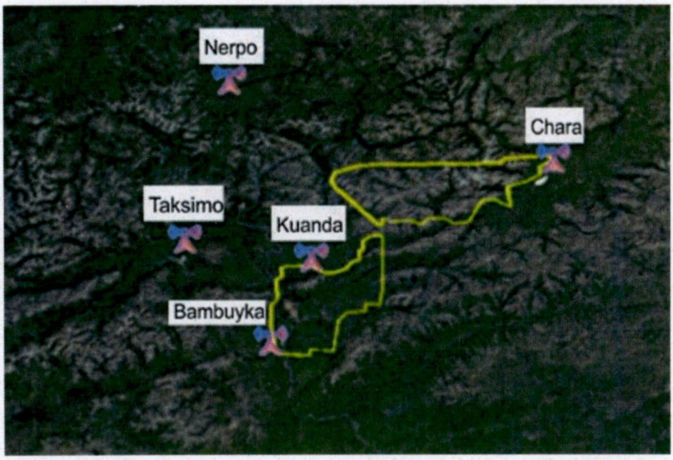

Fig. 8 Positions of MLAT stations on a geographical map

Fig. 9 Error in estimating the coordinate *X* when flying on route W276

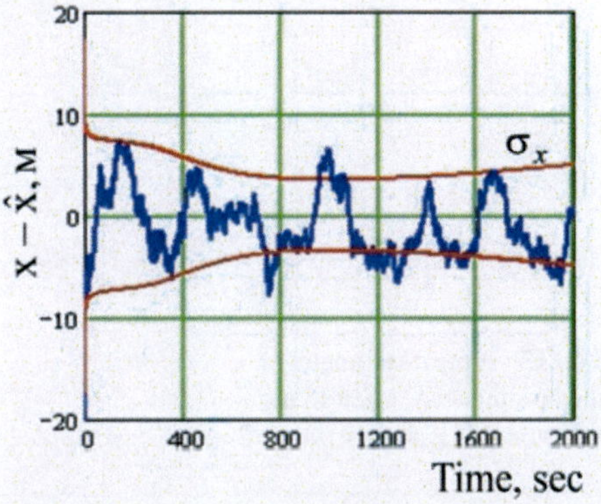

Fig. 10 Error in estimating the coordinate *Y* when flying on route W276

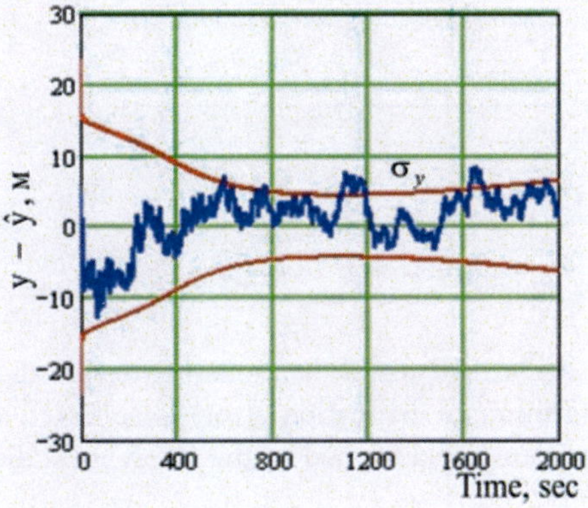

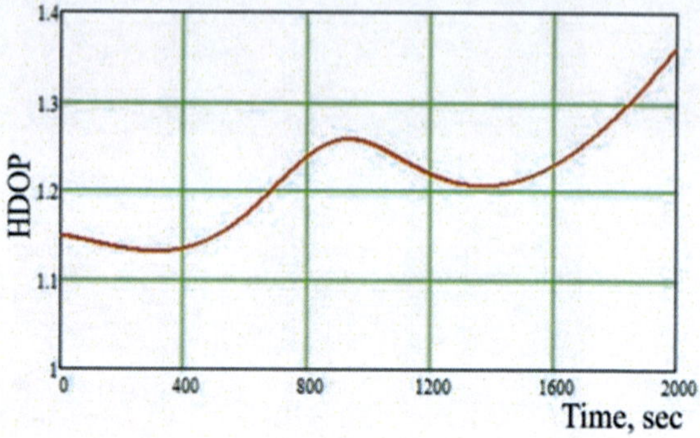

Fig. 11 Horizontal DOP when flying on route W276

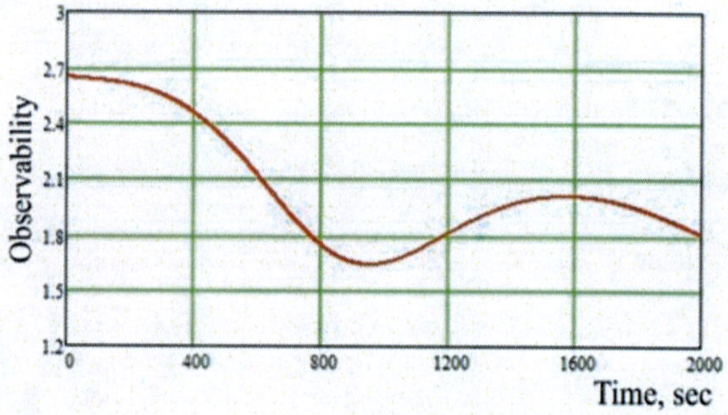

Fig. 12 Observability when flying on route W276

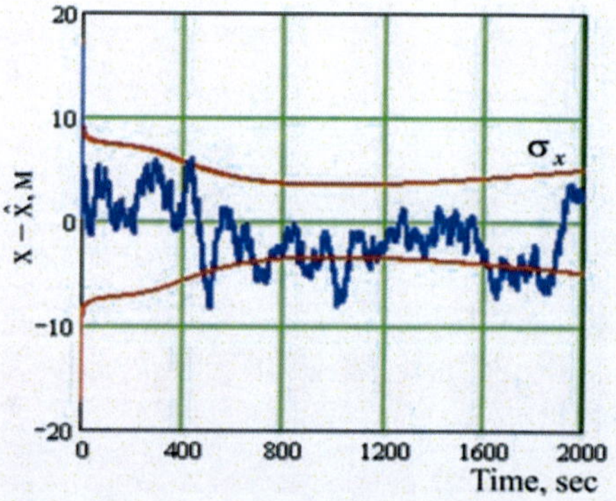

Fig. 13 Error in estimating the coordinate X when flying on route R211

The analysis of the modeling results shows that the highest positioning accuracy (a minimal estimation error) is achieved when an aircraft is in the middle of the working area formed by the receiving stations.

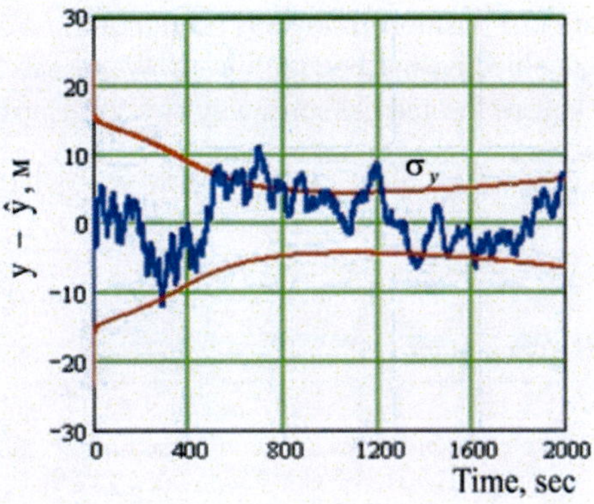

Fig. 14 Error in estimating the coordinate *Y* when flying on route R211

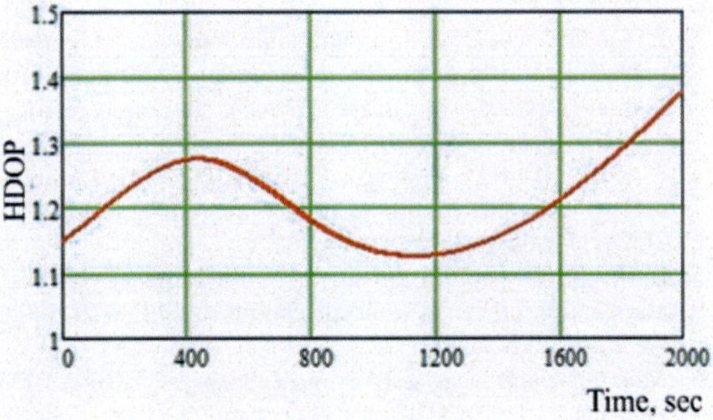

Fig. 15 Horizontal DOP when flying on route R211

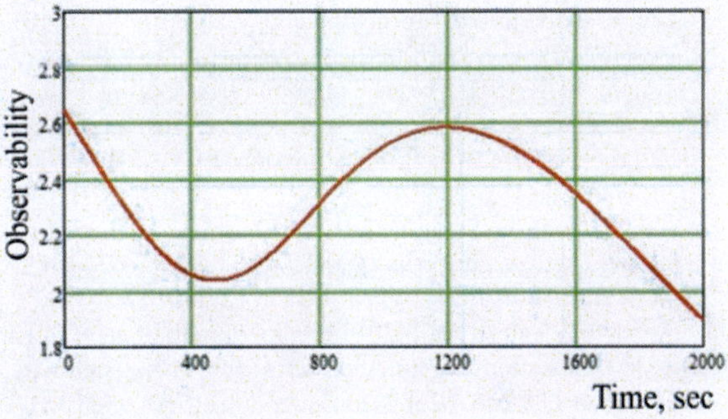

Fig. 16 Observability when flying on route R211

7 Conclusion

Implementation of a multilateration system provides an increase in accuracy of aircraft position-fixing. To improve pseudorange measurements, it is proposed to use a second-order filter in the transponders of receiving stations. The analysis of the obtained results allows us to conclude on high accuracy of pseudorange measurements under the action of random disturbances which indicates that the efficiency of aircraft position-fixing has been increased due to the multilateration technology, i.e., the objective has been achieved. We researched the features of an MLAT system with stations located in such a way as to overlap an airspace zone. The highest positioning accuracy (a minimal estimation error) is achieved when an aircraft is in the middle of the working area formed by the receiving stations.

References

1. International Civil Aviation Organization (2013) 2013–2028 global air navigation plan. Doc 9750-AN/963, 4th edn. Montréal, p 147
2. International Civil Aviation Organization (2013) Assessment of ADS-B and multilateration surveillance to support air traffic services and guidelines for implementation. Cir 326-AN/188, Montréal
3. RTCA DO-260B (2009) Minimum operational performance standards for 1090 MHz extended squitter automatic dependent surveillance—broadcast (ADS-B) and traffic information services—broadcast (TIS-B). https://standards.globalspec.com/std/14356196/rtca-do-260. Accessed 12 Nov 2021
4. Mantilla-Gaviria I, Leonardi M, Galati G (2015) Localization algorithms for multilateration (MLAT) systems in airport surface surveillance. SIViP 9(7):1549–1558. https://doi.org/10.1007/s11760-013-0608-1
5. Eurocontrol Specification for ATM Surveillance System Performance (2015) https://www.eurocontrol.int/sites/default/files/publication/files/201509-eassp-specification-vol1-v1.1.pdf. Accessed 20 Nov 2021
6. Mnogopozicionnaya sistema nablyudeniya “MERA” (MPSN “MERA”) Nomenklaturnyj perechen’ izdelij (2020) http://www.vniira.ru/ru/products/790/811/1179?text=basic-purpose. Accessed 12 Sept 2020 (in Russian)
7. Perov AI, Kharisov VN et al (2010) GLONASS. Principy postroeniya i funkcionirovaniya, 4th edn, Radiotekhnika, Moscow, p 800
8. Frisch D, Hanebeck U (2020) Correction to: TDOA versus ATDOA for wide area multilateration system. EURASIP J Wirel Commun Netw. Springer
9. Hu Y, Zhahg L, Gao L, Ma X, Ding E (2016) Linear system construction of multilateration based on error propagation estimation. EURASIP J Wirel Commun Netw. https://doi.org/10.1186/s13638-016-0657-6
10. Liu W, Dong E, Song Y (2015) Robustness analysis for node multilateration localization in wireless sensor networks. Wireless Netw 21:1473–1483. Springer Science+Business Media New York. https://doi.org/10.1007/s11276-014-0865-0
11. Annepu V, Rajesh A (2020) Implementation of an efficient artificial bee colony algorithm for node localization in unmanned aerial vehicle assisted wireless sensor networks. Wireless Pers Commun 114:2663–2680. Springer. https://doi.org/10.1007/s11277-020-07496-8
12. Chiputa M, Xiangyang L (2018) Real time Wi-Fi indoor positioning system based on RSSI measurements: a distributed load approach with the fusion of three positioning algorithms. Wireless Pers Commun 99:67–83. Springer Science+Business Media. https://doi.org/10.1007/s11277-017-5037-y
13. Monakov AA (2018) Localization algorithm for multilateration systems. J Russian Universities Radioelectron. https://doi.org/10.32603/1993-8985-2018-21-4-38-46
14. Bruce P, Gibbs A (2011) Advanced Kalman filtering, least-squares and modeling: a practical handbook. John Wiley & Sons Inc., Hoboken, New Jersey
15. Optiz F, Kausch T (2004) UKF controlled variable-structure IMM algorithms using coordinated turn model. In: Proceedings of the 7th international conference on information fusion, Stockholm

16. Xu ZL, He DL, Li J, Tang Y, Yang JH (2016) Analysis of location ambiguity, GDOP and station layout of threestation in multilateration. MATEC Web Conf 44:01036. https://doi.org/10.1051/matecconf/20164401036
17. Xu G (2016) GPS—theory. Springer, Algorithms and Applications

Universal Generator for Signal Constellations

Muslim Mezhetov, Anna Tikhova, Vladislav Karelin, and Nikolay Povarenkin

Abstract Present-day digital data transmission systems should provide a high accuracy for information transmission, maintaining, at the same time, the maximal rate. The systems that apply spectral-effective signal modulation methods enabling to reasonably use the selected section of the frequency spectrum. The orthogonal frequency-division multiplexing (OFDM) is among spectral-effective. This kind of modulation uses its own signal constellation at each subcarrier when transmitting information. The system noise immunity generally depends on the signal point arrangement of a signal constellation. At low noise and interference levels, the signal point arrangement increases, which leads to an increase in the information rate. As the noise and interference level increases, the signal point arrangement decreases, which leads to the transmission rate decrease. In the digital information transmission systems, particularly interesting are asymmetric signal constellations, applying which enable generally improve the system operation quality. Therefore, studying signal constellations is a vital problem nowadays. In this paper, we address the structure and the operation principle of a generator for signal constellations, implemented in the NI LabVIEW software. Using such a generator enables to create signal constellations with an extended configuration, which provides a possibility to investigate the system noise immunity by applying symmetric and asymmetric diagrams (maps).

Keywords OFDM · Signal constellation · LabVIEW · QPSK · 16 QAM

M. Mezhetov (✉) · A. Tikhova · V. Karelin
Irkutsk Branch of Moscow State Technical University of Civil Aviation, Irkutsk, Russia
e-mail: milsumka@mail.ru

N. Povarenkin
State University of Aerospace Instrumentation (SUAI), Saint Petersburg, Russia

O. A. Gorbachev et al. (eds.), *Proceedings of 10th International Conference on Recent Advances in Civil Aviation*, Lecture Notes in Mechanical Engineering, https://doi.org/10.1007/978-981-19-3788-0_22

1 Introduction

Evolution in data transmission originates from the systems, in which analog means of forming signal transmission and reception underlay. Low noise immunity, ineffective use of spectral resources, impossibility to regenerate signal at its retransmission, vulnerability of the transmission channel from an unapproved access, difficulty in arranging multiple access at the signal time-division—these are the main disadvantages of such channels [1–3]. The formed spectrum mask (spectrum envelope) in the analog data transmission systems has, as a rule, a shape not corresponding to the rectangular, because the filters applied in such systems have a non-ideal amplitude-frequency response. This results in a necessity to reserve wide guard intervals between the adjacent channels, which essentially reduces the spectrum use efficiency [4–6].

Present-day digital data transmission systems allow using spectral resources of the accessible frequency band and providing relative spectrum purity, due to forming the spectrum envelope shape close to the rectangular. This enables to improve the channel frequency distribution and to reduce the guard interval width. Now, quadrature methods for generating and receiving digital information have proliferated widely, because such methods enable to use the frequency band most effectively. In such systems, one often applies signal constellations of different signal point arrangements to modulate the in-phase (I) and the quadrature (Q) components of a signal (Fig. 1 and 2) [7, 8].

Here, the higher the signal point arrangement (higher M-ary), the more information is channelized, but less the noise immunity of such a system becomes.

When using the orthogonal frequency-division multiplexing (OFDM), the levels of orthogonal quadrature subcarriers are also determined by the signal constellations participating in data transmission. And the signal point arrangement of the signal

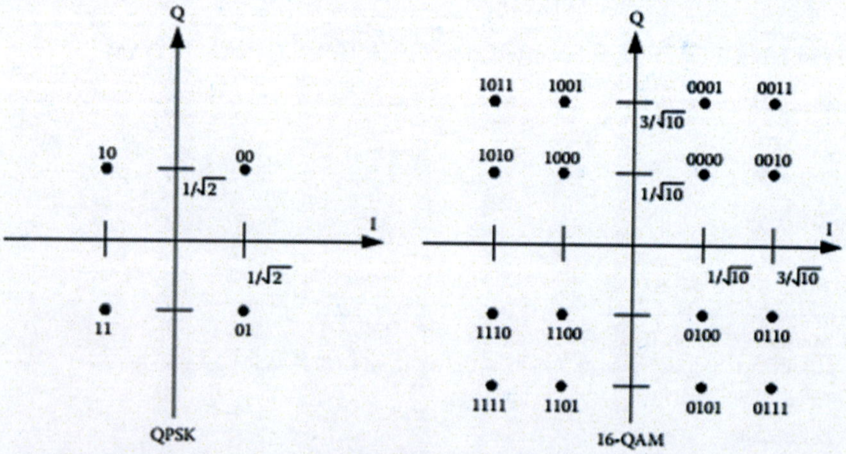

Fig. 1 Example of QPSK signal constellations: QPSK (left) and 16-QAM (right)

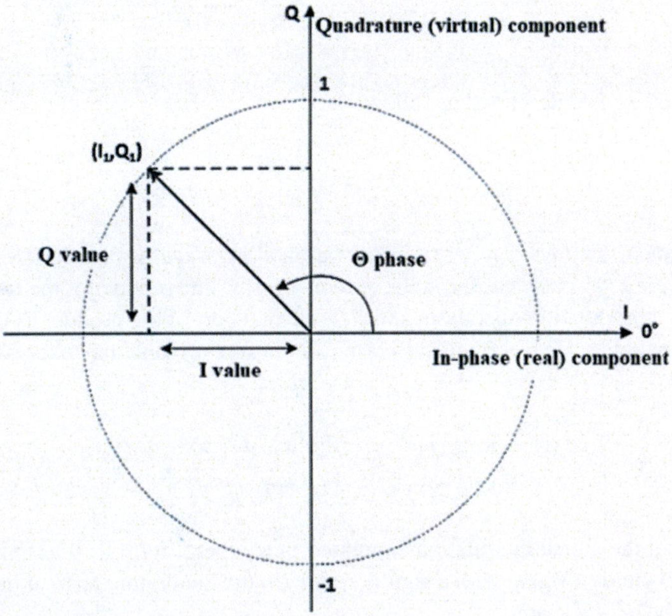

Fig. 2 Presentation for the Q and I components

constellation (SC) may be adapted to the noise and interference level in the channel [9].

In data transmission digital systems, symmetric diagrams of signal constellations (studied sufficiently well) are mainly used [10]. The share of constellation asymmetric diagrams is quite small, but they are interesting to study. For that purpose, we have modeled a virtual instrument—a Universal Generator for Signal Constellations, or UGSC,—in the NI LabVIEW software. The instrument can be used not only to study constellation asymmetric diagrams, but also as an additional software for other virtual instruments, in which ASK, FSK, PSK, BPSK, DPSK, and other modulators and demodulators are implemented. When modeling such modulators and demodulators, one uses, as a rule, a complex envelope for description.

Describing real modulators and demodulators is facilitated by using a complex notation. One may present any band signal $s(t)$ in the complex form like:

$$s(t) = \mathrm{Re}\left\{g(t)e^{i\omega_0 t}\right\} \tag{1}$$

where $g(t)$ is the complex envelope that can be written as follows:

$$g(t) = x(t) + iy(t) = |g(t)|e^{i\theta(t)} = \mathrm{Re}(t)e^{i\theta(t)} \tag{2}$$

The complex envelope amplitude is expressed like:

$$R(t) = |g(t)| = \sqrt{x^2(t) + y^2(t)} \tag{3}$$

while the phase is determined as follows:

$$\theta(t) = \mathrm{arctg}\frac{y(t)}{x(t)} \tag{4}$$

In Formula (1), one may term the $g(t)$ a band message or data in the complex form, and the $e^{i\omega_0 t}$ is the carrier in the complex form. The product of the two values represents a modulation operation, and $s(t)$, the real part of the product, is the transmitted signal [11]. Hence, by using Formulas (1) and (2), one may express the $s(t)$ as follows:

$$\begin{aligned} s(t) &= \mathrm{Re}[x(t) + iy(t)][\cos\omega_0 t + i\sin\omega_0 t] \\ &= x(t)\cos\omega_0 t - y(t)\sin\omega_0 t \end{aligned} \tag{5}$$

Note that the signal modulation expressed in a general form $(a + ib)$ multiplied by $(c + id)$ yields a signal with a sign *variable* (in the quadrature term of the carrier wave) of the *ac–bd* kind.

Let us address a video signal $g(t)$ presented by a sequence of ideal pulses $x(t)$ and $y(t)$ transmitted at discrete instants $k = 1, 2, \ldots$. Thus, one may write $g(t)$, $x(t)$, and $y(t)$ in Eq. (2) as g_k, x_k, and y_k. Let the pulses' amplitude values equal $x_k = y_k =$ 0,707A. Here, one may express the complex envelope in a discrete form as follows:

$$g_k = x_k + iy_k = 0,707A + i0,707A \tag{6}$$

From complex algebra, we know that $i = \sqrt{-1}$, but, in practical terms, i may be addressed as a "mark" reminding that we cannot use ordinary addition when grouping the terms in Formula (6). Further, we will address the I and Q modulations, x_k and y_k, as an ordered couple. Figure 3 shows a modulator implemented by the quadrature principle. One can see here that the pulse x_k is multiplied by $\cos\omega_0 t$ (the carrier I component), and the pulse y_k is multiplied by $\sin\omega_0 t$ (the carrier Q component).

We may briefly describe the modulation process as multiplication of the complex envelope by $e^{i\omega_0 t}$ with the subsequent transmission of the product's real part. So, we write the following:

$$\begin{aligned} s(t) &= \mathrm{Re}\{g_k e^{i\omega_0 t}\} = \mathrm{Re}\{(x_k + iy_k)(\cos\omega_0 t + i\sin\omega_0 t)\} \\ &= x_k\cos\omega_0 t + 0,707A\sin\omega_0 t = A\cos\left(\omega_0 + \frac{\pi}{4}\right) \end{aligned} \tag{7}$$

We remind again that the carrier-wave quadrature term changes its sign during the modulation. If one uses 0,707A $\cos\omega_0 t$ as a reference signal, then the transmitted $s(t)$ in Eq. (7) advances the reference by $\pi/4$. If one uses $-0{,}707A\sin\omega_0 t$ as a reference

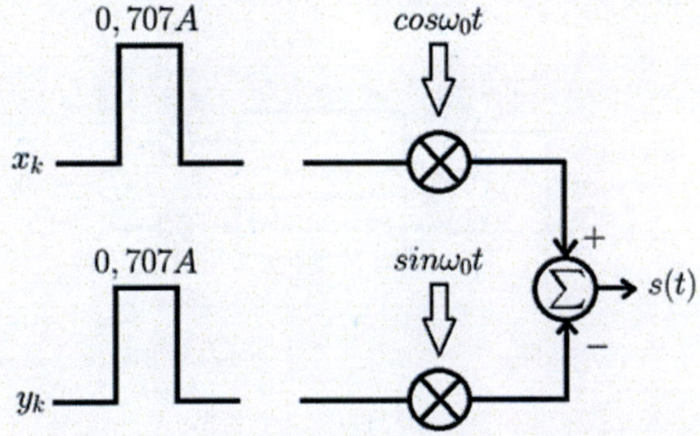

Fig. 3 Modulator operating by the quadrature principle

signal, then the $s(t)$ in Eq. (7) lags the reference by $\pi/4$. Figure 4 provides the graphic illustration of the above.

Let us address an example for a differential *8*-ary phase shift keying (D8PSK) modulator. Figure 5 presents the D8PSK quadrature implementation. Because the modulation is 8-ary, each phase $\Delta\varphi_k$ is assigned a 3-bit message (x_k, y_k, z_k). Because the modulation is differential, for each *k*th transmission time, we obtain the data vector φ_k that can be written as

$$\varphi_k = \Delta\varphi_k + \varphi_{k-1} \tag{8}$$

Addition of the current coded message expressed by a phase difference $\Delta\varphi_k$, with the previous phase $\Delta\varphi_{k-1}$ provides differential message coding. One may note that,

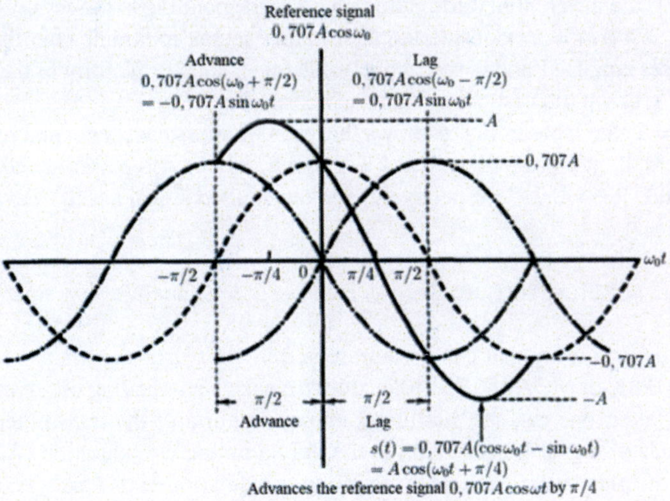

Fig. 4 Advance/lag of sines

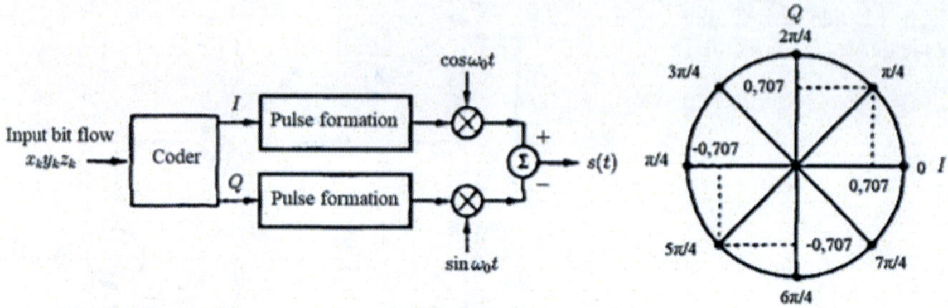

Data coding

x_k	y_k	z_k	Δf_k
0	0	0	0
0	0	1	π/4
0	1	1	2π/4
0	1	0	3π/4
1	1	0	4π/4
1	1	1	5π/4
1	0	1	6π/4
1	0	0	7π/4

Differential information vector

$\phi_k = \phi_{k-1} + \Delta\phi_k$

Assume $\phi_0 = 0$	$k = 1$	$k = 2$	$k = 3$	$k = 4$
$x_k y_k z_k$	110	001	110	010
$\Delta\phi_k$	π	π/4	π	3π/4
ϕ_k	π	5π/4	π/4	π
I	-1	-0,707	0,707	-1
Q	0	-0,707	0,707	0

Fig. 5 D8PSK modulator quadrature implementation

as a result of the $\Delta\varphi_k$ coding by 3-bit sequences, we do not obtain a binary 000–111 sequence, but a special Gray code [12, 13].

Let information sequences 110, 001, 110, 010 arrive at the input of the modulator in Fig. 5 at instants $k = 1, 2, 3, 4$. Further, we use the data encoding table in Fig. 5, Formula (8), and, besides, we assume the initial phase ($k = 0$) is equal to zero: $\varphi_0 = 0$. At $k = 1$, the differential information phase corresponding to the set $x_1 y_1 z_1 = 110$ equals $\varphi_1 = 4\pi/4 = \pi$. Considering the rotating vector amplitude unit, the I and Q video pulses equal -1 and 0, respectively. As Fig. 5 shows, the form of these pulses is usually set by a filter (such, as the raised cosine filter).

For $k = 2$, the table in Fig. 5 shows that the 001 message is encoded by a phase shift $\varphi_2 = \pi + \pi/4 = 5\pi/4$, and, at $k = 2$, the I and Q video pulses are equal: $x_k = -0{,}707$ and $y_k = -0{,}707$, respectively. The transmitted signal has the view given in Formula (5):

$$s(t) = \text{Re}\{(x_k + i y_k)(\cos\omega_0 t + i \sin\omega_0 t)\} = x_k \cos\omega_0 t - y_k \sin\omega_0 t \qquad (9)$$

For the signal array that can be represented in the *phase-amplitude* coordinates (such as MPSK or MQAM), Eq. (9) enables to make a compelling observation [14]. From the latter, one can see that the Q implementation of the transmitter reduces all the kinds of signal transmission to the only amplitude modulation (AM). Each vector on the plane is transmitted by amplitude modulation of its I and Q projections onto the sine and cosine components of its carrier. In each case, the pulse formation is regarded ideal, i.e., the information pulses are assumed to have ideal orthogonal

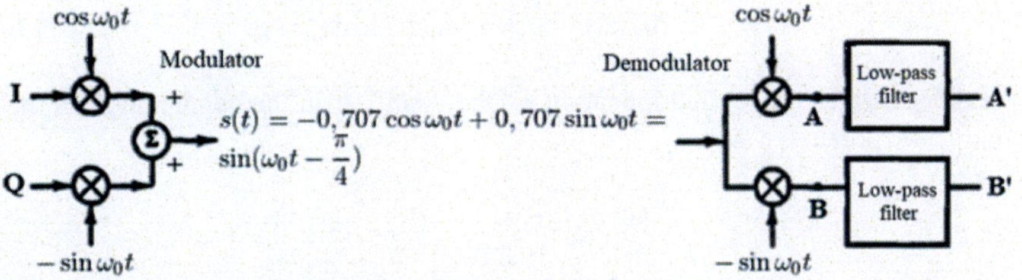

Fig. 6 Example for modulator/demodulator

shapes. Thus, by using Eq. (9) for $k = 2$, at $x_k = -0{,}707$ and $y_k = -0{,}707$, we can write the transmitted signal $s(t)$ as follows:

$$s(t) = -0,707\cos\omega_0 t + 0,707\sin\omega_0 t = \sin\left(\omega_0 t - \frac{\pi}{4}\right) \tag{10}$$

After the signal modulation and transmission, there arises the goal to receive and demodulate the signal. Let us address the D8PSK demodulator example. We start describing the Q implementation of the modulator started with multiplying the complex envelope (of the low-frequency message) by the $e^{i\omega_0 t}$ with a subsequent *transmission* of the real part of the product *s(t)* described in Formula (7). The demodulator of a similar scheme involves the inverse process, i.e., multiplying the received band signal by the $e^{-i\omega_0 t}$ to recover the low-frequency signal. Figure 6 (left part) shows the Fig. 5 modulator in a simplified view, and the signal $s(t) = \sin(\omega_0 t - \pi/4)$ transmitted at $k = 2$ (we continue using the example described earlier). The right part of Fig. 6 presents the Q implementation of the demodulator.

Note a subtle difference between the $-\sin\omega_0 t$ terms in the modulator and in the demodulator. In the modulator, the minus sign appears when determining the real part of the complex signal (product of the complex envelope and the complex carrier). In the demodulator, the $-\sin\omega_0 t$ term appears when multiplying the band signal by the conjugate $e^{-i\omega_0 t}$ of the modulator carrier. The demodulation is coherent if the phase is recovered. To simplify recording the fundamental ratios of the process, we neglect noise. So, after the I multiplication by the $\cos\omega_0 t$ in the demodulator in the point A, we obtain the following signal:

$$\begin{aligned} A &= (-0,707\cos\omega_0 t + 0,707\sin\omega_0 t)\cos\omega_0 t \\ &= -0,707\cos^2\omega_0 t + 0,707\sin\omega_0 t\cos\omega_0 t \end{aligned} \tag{11}$$

By using trigonometric relations, we obtain the following:

$$A = \frac{-0,707}{2}(1 + \cos 2\omega_0 t) + \frac{0,707}{2}\sin 2\omega_0 t \tag{12}$$

After low-pass filtering (LPF) in the point A', the ideal negative pulse is recovered.

$$A\prime = -0,707(\text{accurate within the scale factor}) \tag{13}$$

Likewise, after the Q multiplication by the $-\sin\,\omega_0 t$ in the demodulator in the point B, we obtain the signal:

$$\begin{aligned} B &= (-0,707\cos\omega_0 t + 0,707\sin\omega_0 t)(-\sin\omega_0 t) \\ &= \frac{0,707}{2}\sin 2\omega_0 t - \frac{0,707}{2}(1-\cos 2\omega_0 t) \end{aligned} \tag{14}$$

Thus, in the points A' and B', the (ideal) differential information pulses for the I and Q channels equal –0,707. Because the modulator/demodulator is differential, then, for our example ($k = 2$), we obtain the following:

$$\Delta\varphi_{k=2} = \varphi_{k=2} - \varphi_{k=1} \tag{15}$$

Let us regard that, at the previous instant $k = 1$, the demodulator correctly determined that the signal phase equals π. Then, from Formula (15), we can obtain the following:

$$\Delta\varphi_{k=2} = \frac{5\pi}{4} - \pi = \frac{\pi}{4} \tag{16}$$

Having returned to the modulation table in Fig. 5, one can see that the information sequence $x_2y_2z_2 = 001$ corresponds to the given phase, which agrees with the data transmitted at $k = 2$ [15].

2 Universal Generator for Signal Constellations Structure and Operation Principle

Figure 7 presents the structure of the universal generator for signal constellations, which involves:

- Initial Signal Point Generator (ISPG);
- Signal Point Selective Filter (SPSF);
- Signal Point Sorter (SPS);

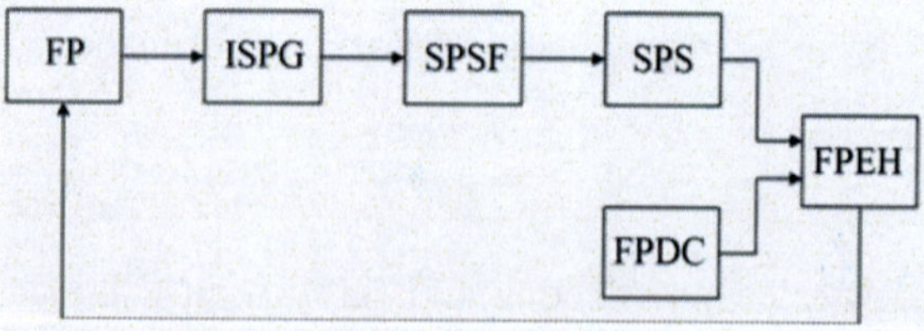

Fig. 7 UGSC block diagram

- Front Panel Display Configurator (FPDC);
- Front Panel Event Handler (FPEH) [16].

The ISPG sets two arrays of all possible values for the signal point coordinates (I and Q, respectively), accounting for the interval between them and the bias of the entire constellation. We calculate the signal point coordinates as follows:

$$i = 0, 1, \ldots S_{\text{map}}$$
$$Q_i = -\frac{\Delta Q}{2} \cdot S_{\text{map}} + \frac{\Delta Q}{2} + Q_{\text{bias}} + i \cdot \Delta Q,$$
$$I_i = -\frac{\Delta I}{2} \cdot S_{\text{map}} + \frac{\Delta I}{2} + I_{\text{bias}} + i \cdot \Delta I, \tag{17}$$

where: ΔQ is the interval on the virtual-value axis; ΔI is the interval on the real-value axis; Q_{bias} is the bias on the virtual-value axis; I_{bias} is the bias on real-value axis; S_{map} is the SC initial size.

Further, (of those dots) we form the array of clusters involving all possible signal points $\text{IQ}_{\text{cluster}}$ with the coordinates $[[S_{\text{map}} * -[\Delta Q] + Q_{\text{bias}} \ldots S_{\text{map}} * [\Delta Q] + Q_{\text{bias}}], [S_{\text{map}} * -[\Delta I] + I_{\text{bias}} \ldots S_{\text{map}} * [\Delta I] + I_{\text{bias}}]]$ of the size $S_{\text{map}}{}^2$. Figure 8 presents the block diagram illustrating the algorithm for forming the given array of clusters.

The ISPG creates a new array of signal points, but not like a 2D array, but like a row vector, and already with the account for the given binary mask (also converted

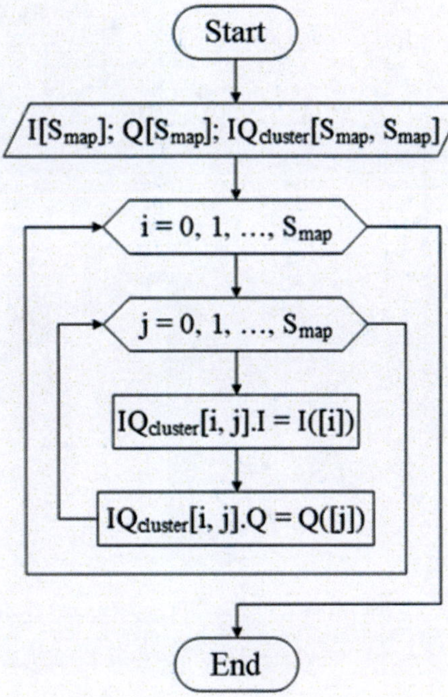

Fig. 8 Block diagram illustrating the algorithm for forming the *IQ* signal point coordinates

to a row vector). This implies that, further, signal points are added to the new array, proceeding from the result of comparing the binary mask elements with the logic unity ("1"), whose indices correspond to those of the elements of all the possible signal points (Fig. 9). The signal point array after the ISPG is a final signal point set for the SC.

Further, the SPS sorts the versions of bit trains (BTs) with the elements of the SC signal point final array. In the implemented SC, there are two sorting versions accessible: ascending sort (left top right down) and descending sort (in the inverse order). Also, there is a manual variation in the BT version relations with repeatability elimination due to checking. Herewith, one may randomly shuffle the order of the BT relations to signal points (Fig. 10).

The FPDC configures the display (the *Graph* indication object) so, that one may see, to which signal point which BT relates.

The FPEH is responsible for handling the event caused by the user's interaction with the front panel controls (saving, SC loading, SC exporting to apply in the ready

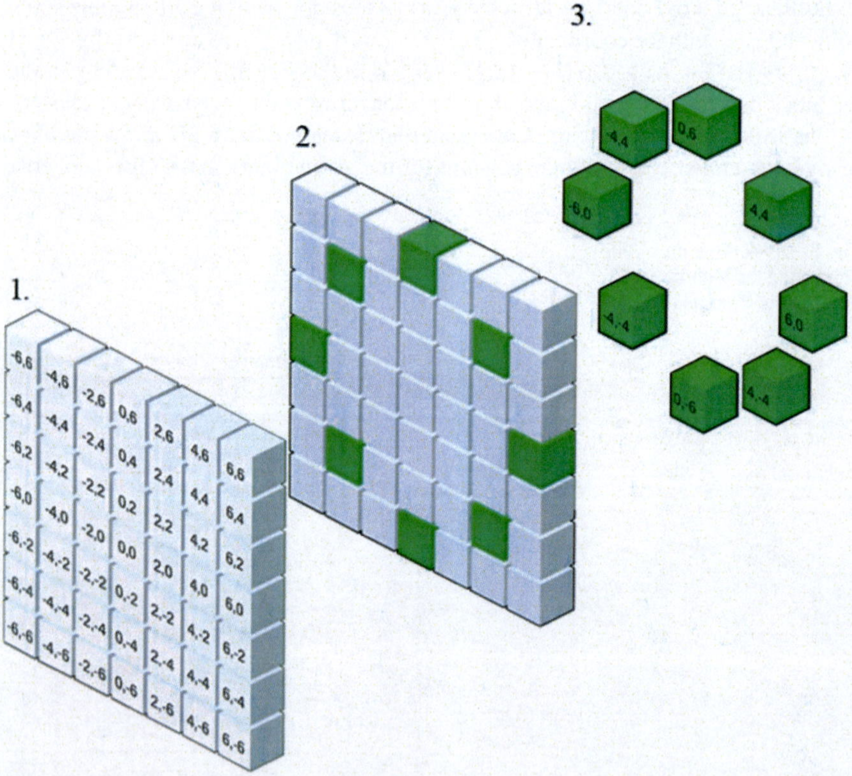

Fig. 9 Visualization for the ISPG operation: (1) initial array of signal points; (2) binary mask; (3) result (the final array of signal points)

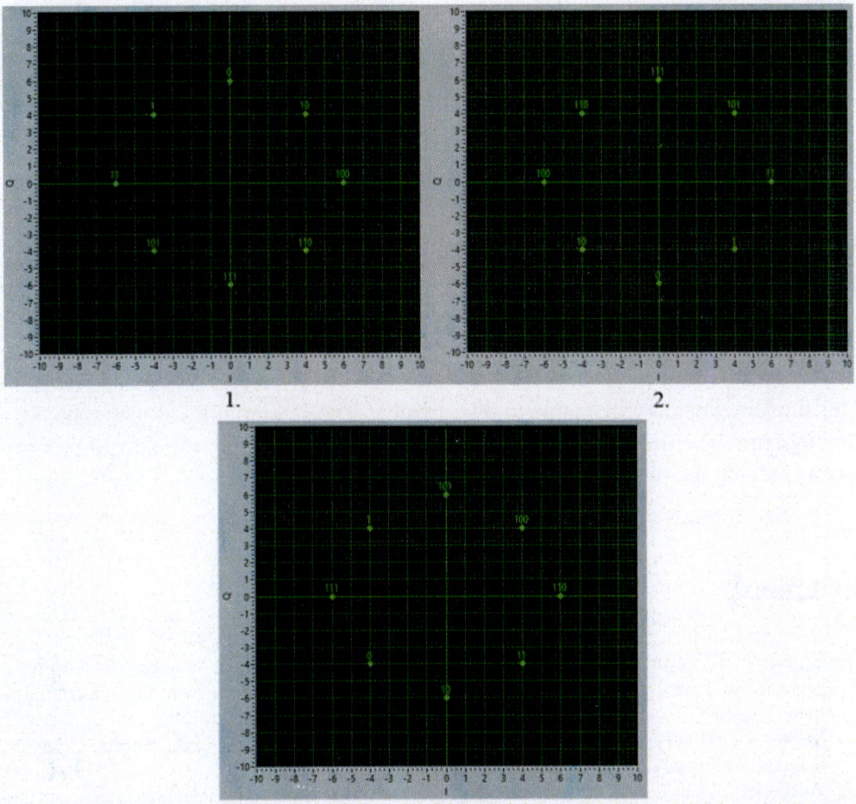

Fig. 10 SC front panel display: (1) ascending sort (2) descending sort (3) setting the BT relations manually

blocks for BT marking, changing the intervals and I/Q biases, changing the SC binary mask, etc.). Figure 11 presents the virtual instrument front panel itself.

In the *BT generator starlets* cluster array, there are the *dots* obtained by generating random BTs and their delivery to the marking block.

3 Conclusion

Thus, we have developed a structure for the universal generator of signal constellations and implemented it like a program model, namely: a virtual instrument allowing one to set almost any configuration of signal constellations. In future, one may further update the structure by rotating the signal constellation at a certain angle (in degrees).

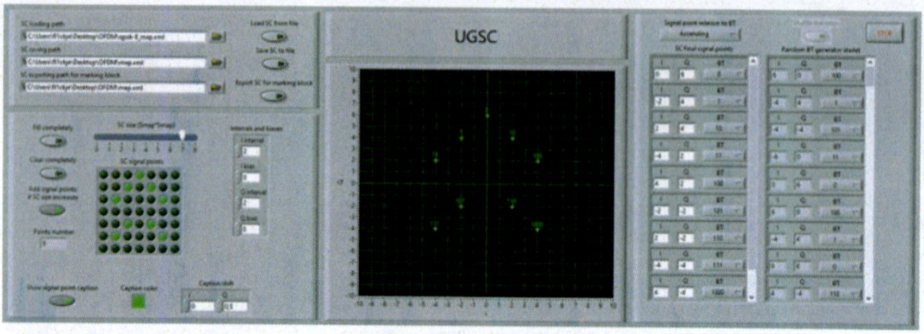

Fig. 11 UGSC front panel

This implementation will enable to investigate the noise immunity of data transmission systems. The developed virtual instrument may be used in educational process when studying the principles of digital data transmission.

References

1. Ermakov AK, Portnova TY, Lezhankin BV, Erokhin VV (2021) Trajectory control algorithms for unmanned aircraft complexes flying in formation. Wave Electron Appl Inf Telecommun Syst (WECONF). https://doi.org/10.1109/WECONF51603.2021.9470714
2. Sheikh A (2012) Principles of transmission and detection of digital signals. Digital Communication, C. Palanisamy, IntechOpen. https://doi.org/10.5772/36640
3. Ansari IR, Uddin S, Naaz S (2016) Use of matched filter in direct sequence spread spectrum techniques. Indian J Sci Technol. https://doi.org/10.17485/ijst/2016/v9i47/100967
4. LaSorte N, Barnes WJ, Refai HH (2008) The history of orthogonal frequency division multiplexing. In: IEEE GLOBECOM 2008—2008 IEEE global telecommunications conference. https://doi.org/10.1109/GLOCOM.2008.ECP.690
5. Akhtar MW, Hassan SA, Ghaffar R et al (2020) The shift to 6G communications: vision and requirements. Hum Cent Comput Inf Sci 10:53. https://doi.org/10.1186/s13673-020-00258-2
6. Mourad A, Yang R, Lehne PH, De La Oliva A (2020) A baseline roadmap for advanced wireless research beyond 5G. Electronics 9(2):351. https://doi.org/10.3390/electronics9020351
7. Makarov SB, Liu M, Ovsyannikova AS, Zavjalov SV, Lavrenyuk I, Xue W, Xu Y (2021) A reduction of peak-to-average power ratio based faster-than-nyquist quadrature signals for satellite communication. Symmetry 13(2):346. https://doi.org/10.3390/sym13020346
8. Xing Z, Liu K, Tang B, Liu Y (2020) Novel PAPR reduction scheme based on piecewise nonlinear companding transform in OFDM systems. IEEE Commun Lett 24(8):1757–1761. https://doi.org/10.1109/LCOMM.2020.2993022
9. Hu Z, Yang J, Guo P, Li Q (2021) Orthogonal frequency division multiplexing with cascade index modulation. IET Commun. https://doi.org/10.1049/cmu2.12313
10. Porath J-E, Aulin T (2003) Design of multidimensional signal constellations. IEE Proc Commun 150(5):317–323. https://doi.org/10.1049/ip-com:20030652
11. Sklar B (2014) Digital communications: fundamentals and applications, 2nd edn. Pearson, New York
12. Yang Y, Cheng L, Li Z (2009) An optical differential 8-PSK modulator using cascaded QPSK Modulators. ECOC 2009, Vienna, Austria. Paper P3.19
13. Aubert DE, Wu YS, Nerurkar VW (1990) Unbalanced quadrature PSK modulator-limiter. US Patent 4,940,954

14. Khormuji MN, Rizvi UH, Janssen GJM, Slimane SB (2006) Rotation optimization for MPSK/MQAM signal constellations over rayleigh fading channels. In: 10th IEEE Singapore international conference on communication systems. https://doi.org/10.1109/ICCS.2006.301497
15. Junsuo Q (2012) A algorithm of fast digital phase modulation signal recognition. TELKOMNIKA Indonesian J Electr Eng 10(8):234–243. https://doi.org/10.11591/telkomnika.v10i8.1666
16. Skrypnik ON, Arefyev RO, Arefyeva NG, Portnova TY (2021) Peculiarities of using a mobile pseudolite for increase of positioning accuracy on aircraft landing. Crede Experto Transp Soc Educ Lang 3:14–26. https://doi.org/10.51955/2312-1327_2021_3_14

Developing a Virtual Device to Identify Signals for Aviation Communication Monitoring System

Muslim Mezhetov, Anna Tikhova, and Alexey Shalayev

Abstract Rapid loading of the used-frequency bands inevitably leads to an increase in the number of unintended interferences related to the system operation features. Also, there exists a danger of unapproved access to the professional communication frequency band. This is most critical for the metric-wave band, because it is here, where the main communication between an aircraft and the ground control occurs at the take-off and the landing. Data losses, in this case, may lead to a terrain accident, because the take-off and the landing are the most complex flight stages demanding prompt information exchange. Due to this, it is important to quickly find and identify the preventing effect impeding the radio communication. The goal of this paper is to develop an algorithm for detecting and identifying the signals in the set band, based on analyzing their in-phase (I) and quadrature (Q) components, which enables to essentially increase air safety.

Keywords Air monitoring · Radio intelligence · Signal analysis · Interference detection

1 Introduction

Over the last two decades, radio-frequency (RF) technology has made a grand leap in its evolution. It has been marked not only by emergence of new standards and information communication methods but also by a wide proliferation of radio-transmitting devices for diverse purposes. Availability of transceivers made them a part of almost all branches of production and household. Cellular phones, pocket and static radio stations, Wi-Fi, various GSM senders, and radio modules—all of these are widely popular with all strata of the global population.

This progress has not avoided specific professional industries. There emerged new methods for information transmission in aviation: air-traffic and flight control systems depend increasingly on the information exchange with both the crew and the aircraft

M. Mezhetov (✉) · A. Tikhova · A. Shalayev
Irkutsk Branch of Moscow State Technical University of Civil Aviation, Irkutsk, Russia
e-mail: milsumka@mail.ru

O. A. Gorbachev et al. (eds.), *Proceedings of 10th International Conference on Recent Advances in Civil Aviation*, Lecture Notes in Mechanical Engineering, https://doi.org/10.1007/978-981-19-3788-0_23

systems. In cellular communication, there emerged a new data-transmission standard discovering new possibilities to use the Internet. The satellite industry develops swiftly and finds its increasing application when solving numerous problems [1].

Along with this, further upgrade of various radio communication systems tends toward a greater automation of establishing communication channels by interfacing various systems across the subscriber-to-subscriber path, and by increasing the message delivery rate. At the same time, the messages are to keep their demanded safety, based on applying multi-parameter adaptation and by using complex anti-interference signal-code designs, as well as optimal methods to receive those messages. Already now, not only static networks of nationwide value possess such possibilities, but also various departmental and commercial, including aviation radio communication systems [2].

Therefore, present-day communication systems are an extremely complex object to study, which, in turn, makes the digital processing of radio signals used in them complicated.

2 Materials and Methods

When put together, the development of information communication and proliferation of radio-transmitting tools leads to an increase in the loading of the used-frequency bands, which inevitably results in increasing the number of unintended interferences. Considering that development and introduction of RF technologies continue to accelerate, the issue of monitoring and controlling the use of air is as acute as it has never been before. This especially refers to the aviation industry, where the safety depends on the correct information exchange [3].

In this paper, we address a version to implement automated monitoring of air [4].

Modulation—variations in a parameter or parameters of the carrier oscillation by information message law—underlies the information transfer through air [5]. There are three main types of modulation: amplitude, phase, and frequency [6]. Each modulation type features its advantages and disadvantages and is applied in different systems as per its features. Hence, by determining the modulation type, one may draw a conclusion of a possible signal belonging to this or that radio system.

A typical feature of present-day systems and communication networks is the use of different adaptation methods. Applying the latter in communication channels may lead to the variation (for different reasons) in the modulation type and in the parameters of the used signals during one communication session. The most indicative example is the variation in the information transmission rate, depending on the quality of communication channels. However, there are even more complex adaptation methods. Besides, under the conditions of incomplete a priori information on the parameters of the received signal, solving some problems of its processing (such as establishing the receiver's main selection bands, tuning the demodulator, the demodulator's making a correct decision on the transmitted value of the modulation parameter) requires maximally exact knowledge of the arriving signal modulation structure. These factors cause a necessity to include some tools in the processing channels to automatically determine the parameters and modulation types for the received signals. Determining the a priori unknown modulation parameters within a wide group of the received signals is, generally, a complicated problem requiring considerable computing resources. This creates difficulties when implementing automated procedures (operating in real-time) to identify the modulation type and modulation parameters. However, in some cases, implementing such procedures is possible by applying the corresponding mathematical methods and present-day computer hardware. Solving the problem of identifying the signal modulation structure implies determining a set of signatures featuring the required modulation types and their parameters, detecting the given set of signatures from the received mixture of the signal and jamming, and processing the signatures to decide on the modulation parameters of the analyzed signal. The block diagram for the subsystem performing such an analysis is shown in Fig. 1.

Existing methods to detect and determine the signal modulation type are based on the spectrum analysis principle [7]. One sets a trigger threshold by the spectral component amplitude, and, at its excess, the program starts analyzing the spectrum shape and its width. For example, AM signals will have a large component at the carrier frequency, relative to which symmetrical information lobes will be located; an FM signal is more broadband, without notable spectrum outbursts, and the signal with a balance modulation contains two symmetrical lobes with no carrier. By analyzing the features of the spectrum shape and its location, the program displays the information on signal presence and on their modulation type in the examined spectrum. Also, the program is able to compare the analyzed spectra with the database (compiled in

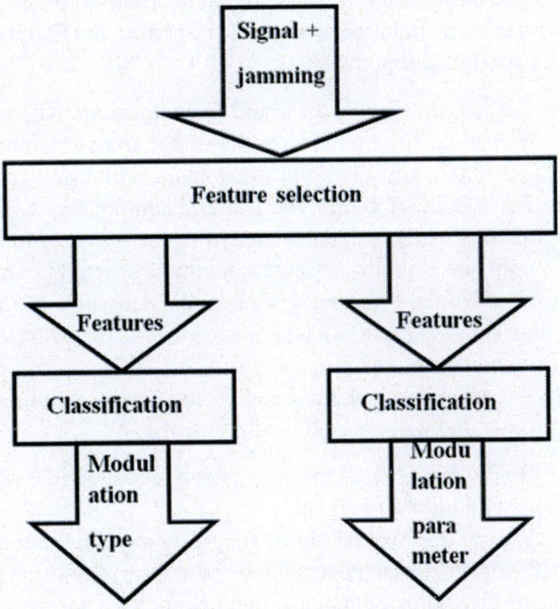

Fig. 1 General block diagram for the analysis subsystem operation

advance) to determine the particular function of the signal and its belonging to the known systems.

Such an approach to the spectrum monitoring has a number of disadvantages:

- Complexity of software implementation. The algorithm for the spectra comparison should consider the presence of noise components (distorting the pattern) in the spectrum.
- Absence of the additional information on the signal. Because the algorithm does not demodulate the signal, it is almost impossible to exactly specify its function (unless it is a pre-known signal recorded in the database).
- Possible false triggerings when analyzing non-conventional signals. This is especially crucial when analyzing the spectrum with single-sideband signals, whose shape may be arbitrary. Also, the program is not able to distinguish the narrow-band AM from an FM signal with a small frequency deviation.

Thus, to analyze the signal modulation structure within radio communication systems, using classical methods for an optimal estimate of parameters and the statistical theory for pattern recognition is ineffective. The requirement for efficiency of tuning the processing equipment causes a necessity to develop special methods to determine the modulation parameters by a non-classified fixed-volume sample.

All the above essentially restrict applying similar algorithms under real radio conditions, due to which the bulk of radio monitoring problems is solved manually at the moment.

But the algorithms based on the analyzing the in-phase and quadrature components of a signal do not possess such disadvantages and can be easily implemented on present-day SDR platforms.

The basic idea, on which signal recognition by its components is based, is that, with each modulation type, the I (in-phase) and Q (quadrature) components vary as per special, unique rules:

- For an AM signal, the *I* and *Q* components will always be elevated relative to 0, and, as long as the phases of the transmission and of the receiving device oscillators are equal, the components will repeat each other's shape.
- For a DSBSC signal, the *I* and *Q* components will have no displacement, and, as long as the oscillator phases of the transmission and of the receiving device coincide, will also repeat each other's shape, like in an AM signal.
- For FM and PM analog signals, the derivative of the *I* and *Q* components' sum of squares (*S* of *S*) will always equal 0, regardless of the oscillators' phase coincidence.
- For an SSB signal, the *I* and *Q* components will be phase-shifted relative to each other and will have no displacement.

Below, we provide the figures demonstrating the above features for the components of all types of signals.

Figures 2, 3, 4 and 5 shows the *I*/*Q* components obtained at the phase coincidence of the oscillators at the receiving and at the transmitting parties. The phase equality of the receiver and of the transmitter heterodynes is an optimal, but not an indispensable

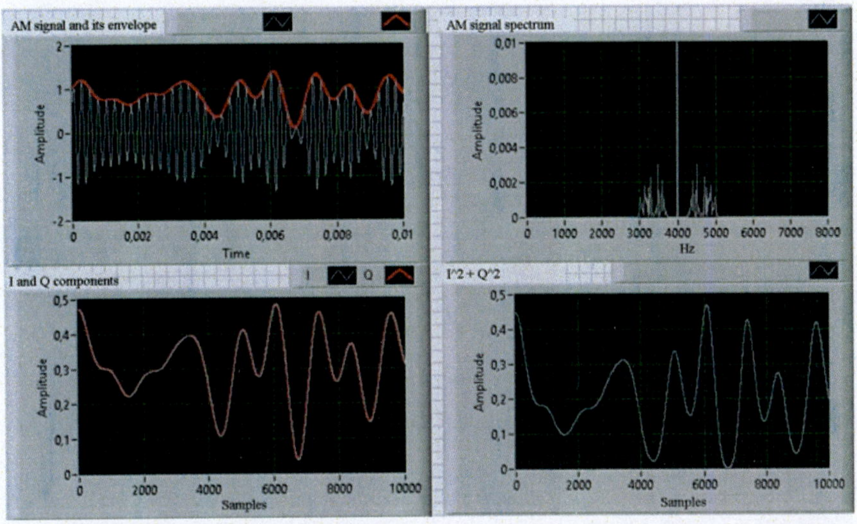

Fig. 2 Features of AM signal in the I/Q representation. Signal, its spectrum, components, and their S of S

condition when analyzing the signal modulation type. As long as the heterodynes are phase-desynchronized, there occurs the I/Q power re-distribution, which manifests itself in an amplitude increase for one component and in proportional amplitude decrease of the other. This is expressed most profoundly with AM and DSBSC signals, in which, at the 45° phase difference between the heterodynes, one of the components becomes equal to 0.

However, this does not prevent the operation of the modulation recognition algorithm that looks like the following Fig. 6.

The first condition explicitly determines an FM analog signal, because their components' S of S will always yield a constant value, regardless of the phase difference between the oscillators of the transmitter and of the receiver. As long as first condition is not fulfilled, the components are then normalized for their further shape comparison. Because for DSBSC and AM the components will always have the same shape, and the heterodyne phase difference leads to re-distribution of the components' power contribution to the signal, the normalization and comparison enable to explicitly distinguish between DSBSC and AM signals from the remaining modulation types. The absence of a component is also a signature of the AM and DSBSC signals and evidences the ±45° mistiming for the heterodynes' initial phases. Then, the algorithm checks the presence of a constant in the components. If the components are always above 0, this evidences an AM signal. If the components have no displacement, this evidences a DSBSC signal. The SSB is special, because the components are not only unequal to each other but also do not repeat each other's shape, as it is the case with FM and PM signals. However, despite this, their S of S is not equal

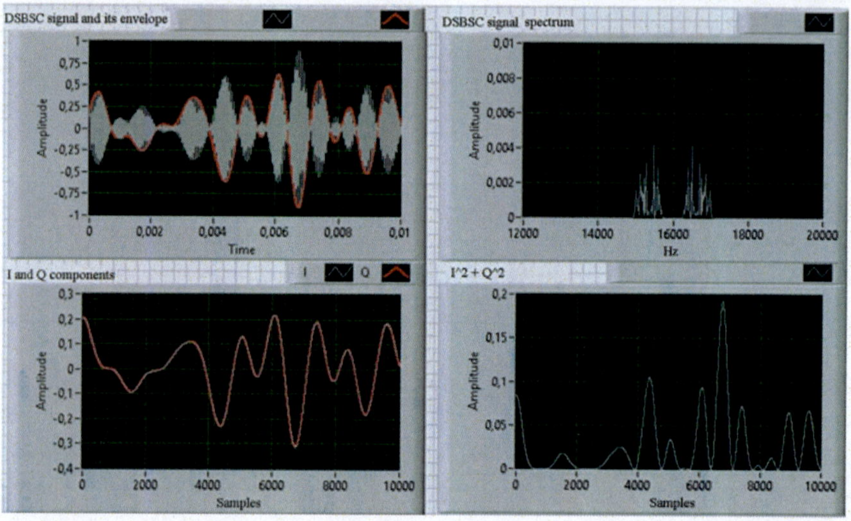

Fig. 3 Features of DSBSC signal in the *I*/*Q* representation. Signal, its spectrum, components, and their *S* of *S*

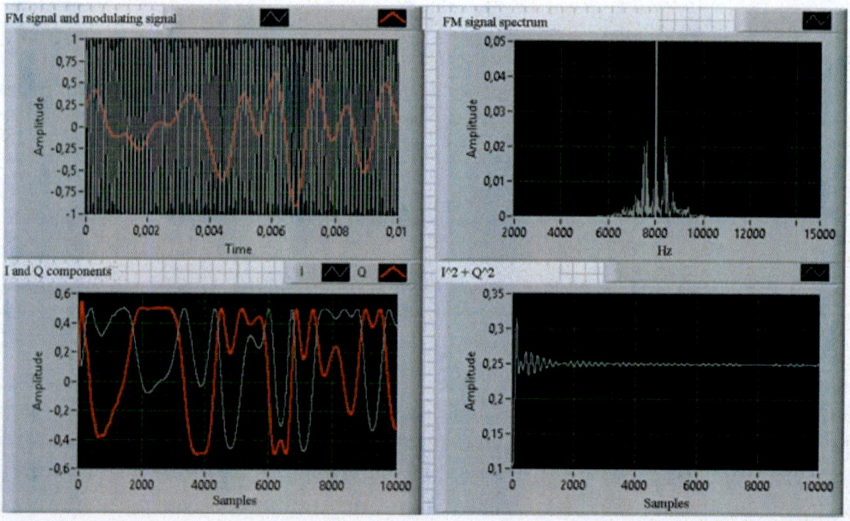

Fig. 4 Features of FM signal in the *I*/*Q* representation. Signal, its spectrum, components, and their *S* of *S*

to the constant, and varies with time, like the AM and DSBSC. Correspondingly, as long as the first two conditions are not fulfilled, one may assert an SSB signal.

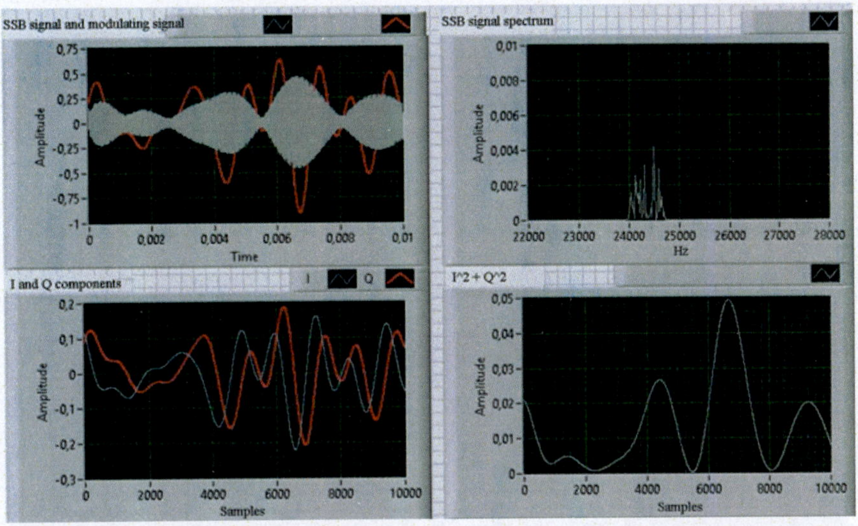

Fig. 5 Features of SSB signal in the *I*/*Q* representation. Signal, its spectrum, components, and their *S* of *S*

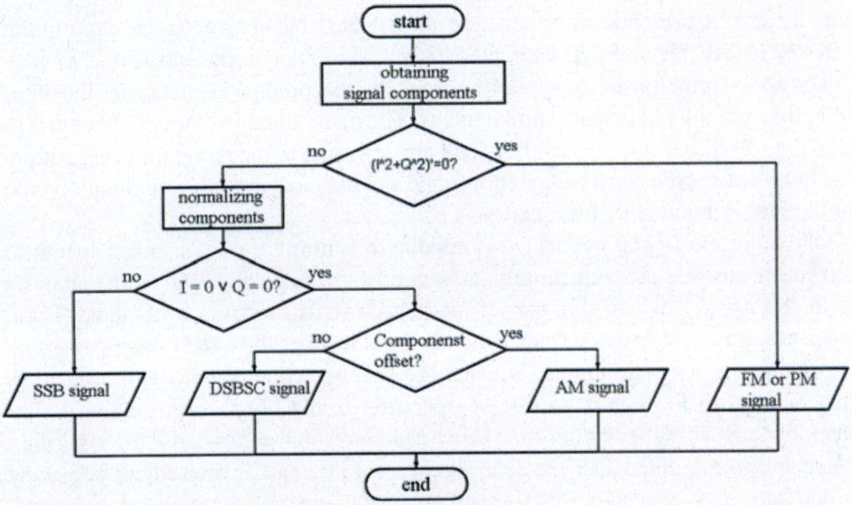

Fig. 6 Algorithm to identify the analog modulation type

An important virtue of such an algorithm is that one can supplement it with the conditions recognizing not only the modulation type but also the data-transmission formats referring to particular systems. For example, for the ACARS data-transmission system, one may detect not only the AM signatures (the *I*/*Q*

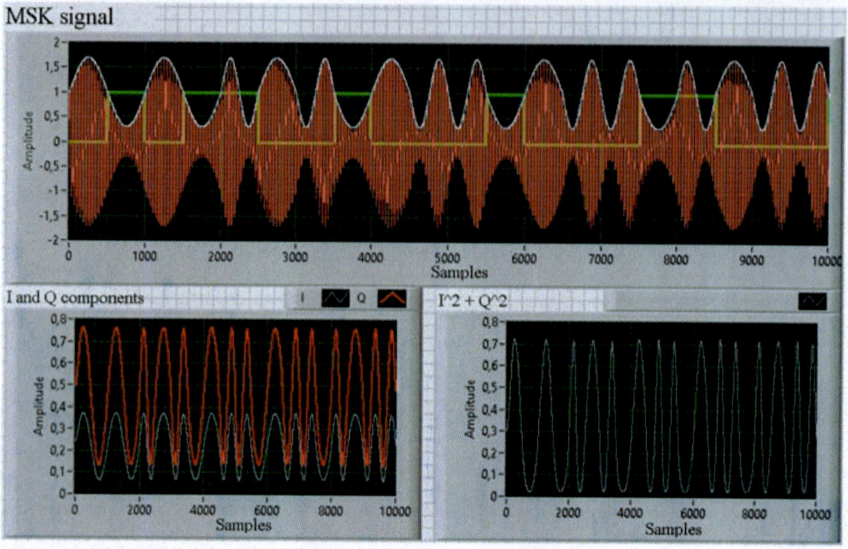

Fig. 7 MSK signal components. Signal, components, and their *S* of *S*

equality and the presence of the constant component), but also recognize a minimum shift key (MSK) signal by its characteristic shape of the components (Fig. 7).

The above principle to recognize the signal by its components underlies the algorithm for studying radio-spectrum segments. The main function of this algorithm is to form the *I*/*Q* signal components that then are delivered to the recognition algorithm. The latter returns the carrier frequency value and the most probable modulation type for the radio station at that frequency.

Obtaining the *I*/*Q* components occurs due to splitting the set scanned spectrum into sub-bands that are sequentially delivered to the analysis cycle. In the analysis cycle, the signal is received in the set sub-band, and its spectrum is obtained. Then, by spectrum values, one determines the maximal level of the noise component, and, where there is no radio signal, equates the spectral component amplitudes to 0. This is necessary to eliminate a false operation of the program in random noise. Then, by analyzing the width of the remained peaks in the sub-band, the algorithm selects a low-pass filter (LPF) to each signal and to the initial heterodyne frequency in the frequency converter [8], thus breaking the spectrum into signal segments. These segments are delivered then to the frequency converter that, with the set step, "runs" the carrier through the entire segment. As long as the program, on a certain heterodyne frequency, recognizes the modulation type (i.e., the modulation signatures are fulfilled), the frequency is determined as the signal carrier. Upon searching for all the segments, the analysis cycle performs the same within this sub-band *m* times, herewith, accumulating the obtained data. The accumulated data on the sub-band are compared with one another, after which the algorithm provides the ultimate result about the presence of these or those signals at the found frequencies. Afterward, the algorithm transfers to the next sub-band, and the cycle iterates. Figure 8 presents the described algorithm [9].

Such a principle to analyze the spectrum enables to implement:

- Algorithm correct operation at a varying noise level.
- Minimization of false operation due to accumulation of data and their comparison.

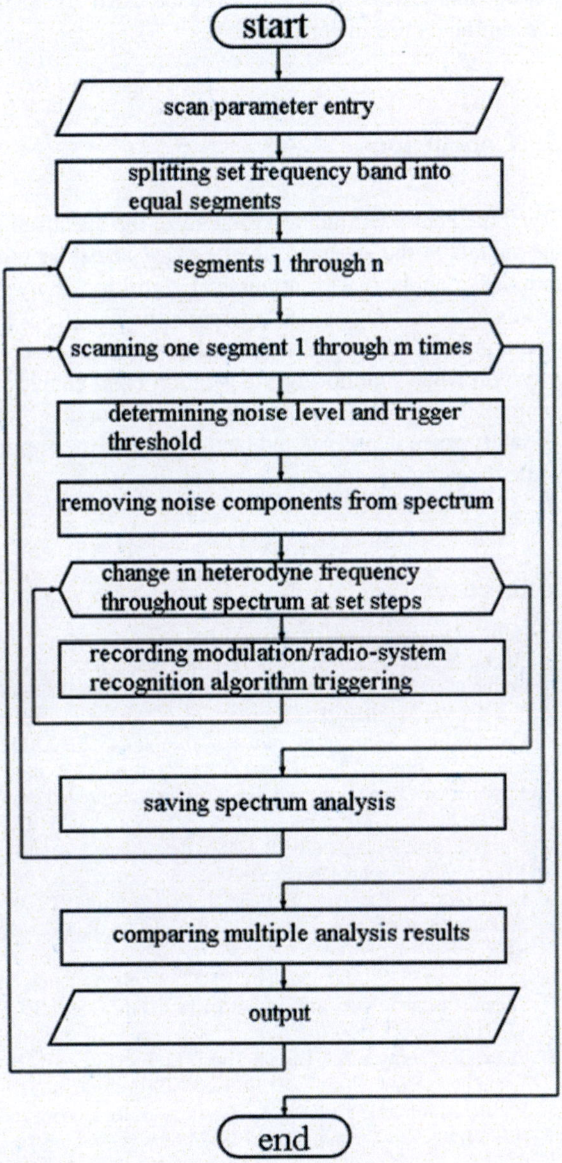

Fig. 8 Radio-spectrum analysis algorithm

- Recovery of the carrier and detection of spectrum band occupied by the signal, as long as it is impossible to accurately recognize the signal.
- Record of periodic signals and determination of their occurrence period.

However, such an algorithm is not perfect. In particular, its main disadvantage is the impossibility to recognize some digital signals, for example, OFDM ones [10–15]. The signals with specific spectra are better recognized by conventional methods, through analyzing their spectrum shape. Also, the addressed algorithm yields to the conventional ones in performance.

3 Conclusions

Thus, the addressed method to analyze the spectrum enables to exactly determine the signals in the set band, their carrier, and their modulation type. It is the most rational to apply it with a classical algorithm analyzing the signal spectra shape. In this case, as long as one interprets the data from both algorithms correctly, the result of studying the spectrum will be the most exact and complete. Applying such a system when monitoring the aviation band enables to determine the unintended jamming source as quickly and exactly as possible, which essentially will increase air safety when taking off and landing, where the communication is most important, while the jamming emergence is most probable.

References

1. Zhang R, Liu G, Liu J, Nees JP (2017) Analysis of message attacks in aviation data-link communication. Res Gate. https://doi.org/10.1109/ACCESS.2017.27670
2. Yue M, Wu X (2010) The approach of ACARS data encryption and authentication. Int Conf Comput Intell Secur. https://doi.org/10.1109/CIS.2010.127
3. Zhijun W, Jihai L, Jun H (2012) Research on ECC digital certificate in ATN. Computer, Informatics, Cybernetics and Applications. Lecture Notes in Electrical Engineering. Springer, Dordrecht. https://doi.org/10.1007/978-94-007-1839-5_97.
4. Pan W, Feng Z, Wang Y (2012) ADS-B data authentication based on ECC and X. 509 Certificate. J Electron Sci Technol 10(1):51–55
5. Berczyński S, Kravtsov Y, Pejaś J, Surovyatkina E (2004) Secure data transmission via modulation of the chaotic sequence parameters. Res Gate. https://doi.org/10.1007/0-387-23484-5_1
6. Mohammed RK, Abdullah HA (2020) Implementation of digital and analog modulation systems using FPGA. Indonesian J Electr Eng Comp Sci 18(1):485–493. https://doi.org/10.11591/ijeecs.v18.i1.pp485-493
7. Pham Q-V, Nguyen NT, Huynh-The T, Le Bao L, Lee K, Hwang W-J (2021) Intelligent radio signal processing: a survey. IEEE Access. https://doi.org/10.1109/ACCESS.2021.3087136
8. Wiesler A, Jondral FK (2002) A software radio for second- and third-generation mobile systems. IEEE Trans Veh Technol 51(4):738–748. https://doi.org/10.1109/TVT.2002.1015347

9. Rykaczewski P, Pienkowski D, Circa R, Steinke B (2005) Signal path optimization in software-defined radio systems. IEEE Trans Microw Theory Tech 53(3):1056–1064. https://doi.org/10.1109/TMTT.2005.843510
10. Rosmaniza ARS, Ahmad N, Yusof SS (2019) Signal modulation techniques in non-orthogonal waveform for future wireless communication system. Indonesian J Electr Eng Comp Sci 15(3):1458–1465
11. Akmaykin DA, Bolelov EA, Kozlov AI, Lezhankin BV, Svistunov AE, Shatrakov YG (2021) Pseudo-ranging radio navigation systems. In: Theoretical foundations of radar location and radio navigation. Springer Aerospace Technology. Springer, Singapore. https://doi.org/10.1007/978-981-33-6514-8_14
12. Skrypnik ON, Arefyev RO, Arefyeva NG (2020) The assessment of positioning error characteristics of combined GLONASS/GPS receivers. Crede Experto: Transp Soc Educ Lang 1:44–58
13. Bolelov EA, Sbitnev AV, Shalupin SV (2014) Matematicheskaya model' signalov na vyhode bortovyh radionavigacionnyh sistem, uchityvayushchaya ih vnezapnye otkazy (the mathematical model of the signals at the output of onboard navigation systems, taking into account their sudden failures). Nauchnyj vestnik Moskovskogo gosudarstvennogo tekhnicheskogo universiteta grazhdanskoj aviacii. 210:160–162 (In Russian)
14. Ermakov AK, Portnova TY, Lezhankin BV, Erokhin VV (2021) Trajectory control algorithms for unmanned aircraft complexes flying in formation. Wave Electron Appl Inf Telecommun Syst. https://doi.org/10.1109/WECONF51603.2021.9470714
15. Erokhin V, Lezhankin B, Portnova T (2021) Bi-criteria aircraft trajectory optimization in implementing the area navigation concept. Int J Aeron Space Sci 22(4):948–962. https://doi.org/10.1007/s42405-021-00353-3

Attitude Indicators in Bank Angle Determination: A Study of Errors

Olga Arinicheva, Natalia Lebedeva, Aleksei Malishevskii, and Roman Arefyev

Abstract The article discusses the results of an experiment that was conducted in the spring of 2021 at Saint Petersburg State University of Civil Aviation. The experiment was aimed at analyzing errors in bank angle determination made when using two different types of attitude indicators: the direct one, in which the horizon bar moves relative to the aircraft symbol, and the reverse one, in which the aircraft symbol moves relative to the horizon bar. As the direct type of indication currently prevails on board of almost all aircraft in use, it seems important from the safety point of view to understand what individual attributes are characteristic of those who find this type of indication inconvenient. This can serve as a foundation for solving the problem of improving professional screening procedures in civil aviation. In the experiment, heat maps and gaze plots were analyzed that had been obtained with the help of the Tobii REX eye tracker. Also, psychological testing was conducted to gather information on such characteristics of the participants as temperament and mental strength. Relationships between the results of the experiment, the data produced by the eye tracker, and the results of temperament and mental strength tests were studied. The sample included 46 fifth-year students at Saint Petersburg State University of Civil Aviation who major in flight operations management.

Keywords Desired professional attributes · Psychodiagnostics · Attitude indicator · Eye tracker · Mental strength · Temperament

---

O. Arinicheva (✉) · N. Lebedeva · A. Malishevskii
Saint Petersburg State University of Civil Aviation, Saint Petersburg, Russia
e-mail: 2067535@mail.ru

R. Arefyev (✉)
Irkutsk Branch of Moscow State Technical University of Civil Aviation, Irkutsk, Russia
e-mail: seven7772009@yandex.ru

O. A. Gorbachev et al. (eds.), *Proceedings of 10th International Conference on Recent Advances in Civil Aviation*, Lecture Notes in Mechanical Engineering, https://doi.org/10.1007/978-981-19-3788-0_24

1 Introduction

When a pilot gets disoriented, it poses a serious threat to flight safety. As a rule, such a situation leads to serious consequences, including aircraft upset and a stall, which are classified as loss of control in flight (LOC-I).

A typical example of a LOC-I accident in this category is the crash in Perm in 2008 that involved a Boeing 737, in which the pilot incorrectly determined the direction of the aircraft's roll and, instead of compensating for it, overrode it to 76°. The reason for this error is that the pilot misinterpreted the data from the attitude indicator. The direct type of indication (the view from the cockpit to the ground; the horizon bar moves relative to the aircraft symbol) used on the Boeing 737 was misinterpreted as reverse (the view from the ground to the aircraft; the aircraft symbol moves relative to the horizon bar) [1]. Another example is the crash of a Ka-27 helicopter, a model which also has direct indication. The investigation of the disaster convincingly showed that imperfection in the design of the artificial horizon in terms of roll indication from the ergonometric point of view was one of the causes of the accident. As it was established, instead of compensating for the roll, the pilot overrode it three times. For the first two times, he was able to recover the helicopter's attitude by moving the cyclic stick, but the third one turned out to be fatal, with the helicopter being overridden from 70° to 110° [2].

2 Problem Statement

The accident in Perm instigated heated debates in the Russian segment of the Internet as to the advantages and disadvantages of both types of indication. The debates revealed that many pilots found it quite inconvenient to use one or the other. As Professor V. V. Kozlov, member of the Russian Advisory Board on Civil Aviation Issues writes [1], "there are people for whom the direct type of indication is more familiar so to speak, and they find it very easy to determine the attitude of the aircraft using such artificial horizons. On the contrary, it is more difficult for them to do this when using attitude indicators of the reverse type. Consequently, a conclusion can be made that in terms of spatial orientation, some people have innate mental mechanisms that are based on a geocentric coordinate system ('I am moving while the earth is motionless') rather than an egocentric one ('I am motionless while the earth is moving'). It is impossible to say whether it is good or not, but according to a number of studies, there are much fewer people with an egocentric coordinate system (only 15–20%). It becomes obvious that by designing artificial horizons with the direct type of indication, we go against the nature of those who have a geocentric coordinate system. At the same time, the use of artificial horizons with reverse indication creates difficulties for those pilots whose mental processes function in an egocentric coordinate system. Therefore, no matter what kind of indication is used, there will always be both supporters and opponents".

As the direct type of indication currently prevails on board of almost all aircraft in use, it seems important to understand what individual attributes are characteristic of those who find this type of indication inconvenient. This can serve as a foundation for solving the problem of improving professional screening procedures in civil aviation that we considered in [3, 4] and a number of other works.

It should also be noted that objective research methods (for example, using an eye tracker) are often used to study the operator's attention since it is eye movements that are the marker which makes it possible to identify whether the operator's attention is directed to a specific object, which is especially true in the case of overt orienting attention. There are a lot of studies using eye trackers. In particular, such works as [5–10] can be mentioned that rely on this technology to some extent.

3 Materials and Methods

In order to study the issues under consideration, O. V. Arinicheva supervised an experiment conducted by her graduate students A. D. Voitik and V. D. Knyazheva in the spring of 2021.

The participants in the experiment were fifth-year students at Saint Petersburg State University of Civil Aviation majoring in flight operations management. The sample size was 46.

A comparative study was carried out to identify individual differences among the participants in the perception and interpretation of the bank angle when using the two previously described types of indication, or artificial horizon formats (see Fig. 1). The participants were shown a sequence of slides with data from attitude indicators of the two types. A total of 20 slides were demonstrated (ten for each of the two types of attitude indicators), with bank angles of 10, 20, 30, 40, and 50° being presented (see Table 1).

Before the experiments, the participants were given the instructions emphasizing that the task must be completed as quickly and as accurately as possible. In the course of the experiment, eight parameters shown in Table 2 were assessed.

In addition, all participants in the experiment underwent psychodiagnostics. The following tools were used:

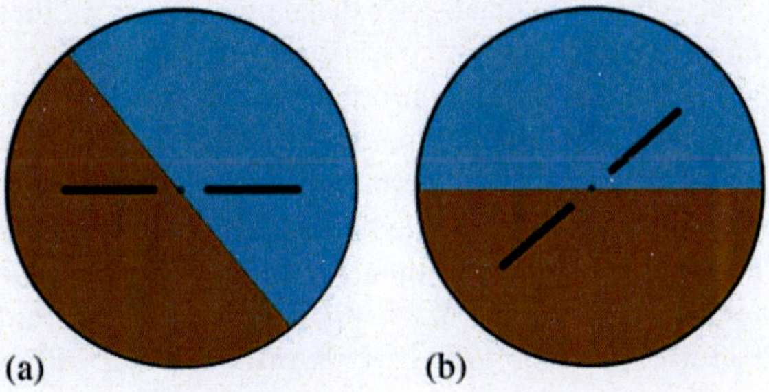

Fig. 1 Artificial horizon formants: **a** direct; **b** reverse

Table 1 The order of presenting slides with images of artificial horizons with different indication types

	Indication	Rolling motion	Bank angle		Indication	Rolling motion	Bank angle
1	Reverse	Left	20	11	Reverse	Right	20
2	Direct	Right	20	12	Direct	Left	40
3	Reverse	Right	40	13	Reverse	Right	50
4	Reverse	Left	40	14	Reverse	Left	10
5	Reverse	Right	30	15	Direct	Left	20
6	Direct	Left	30	16	Direct	Left	50
7	Direct	Right	10	17	Reverse	Right	10
8	Reverse	Left	50	18	Direct	Right	30
9	Direct	Left	10	19	Reverse	Left	30
10	Direct	Right	50	20	Direct	Right	40

Table 2 Parameters assessed in the experiment

Parameter	
T	Time taken to interpret one slide
$T_{\sum}$	Total time
γ	Bank angle, degrees
Δ_{γ}	Deviation of the bank angle from the set value
L/R	Roll direction
$m_{L/R}$	Error in roll direction determination
$S_{L/R}$	The sum of roll direction determination errors
$(S_{ERR})_{\Sigma}$	Total number of errors
E	Extraversion
n	Neuroticism
N_{MS}	Mental strength (scores)

- the Eysenck Personality Inventory (EPI) [11] adapted by A. G. Shmelev [12] for measuring two dimensions of personality—extraversion and neuroticism;
- the «Forecast» test [13] for assessing mental strength.

Also, the participants gave their subjective assessments on a 10-point scale of how convenient they found it to use each of the two types of indication.

To obtain additional information on attention distribution, the Tobii REX eye tracker was used in the experiment to analyze gaze plots and get more accurate data on the time spent on determining the roll direction and the bank angle [14, 15]. The eye tracker was installed on a IIYAMA ProLite T2250MTS 21.5" monitor. The characteristics of the computer were as follows: an Intel Core i5-3450 3.10 GHz processor, an ASUS P8Z77-V DELUXE motherboard, a Kingston DIMM DDR3 2048 MB PC10600 1333 MHz RAM, Windows 7 Professional.

For each of the participants, the equipment was adjusted and calibrated before the start of the test in accordance with the manual [15].

To process the data, we used a program developed by Doctor of Technical Sciences A. P. Plyasovskikh at the All-Russian Scientific Research Institute of Radio Equipment and designed specifically to analyze various aspects of eye movement when performing tests [14].

To analyze the results, the *R* programming language was used, which is a popular tool for data analysis and has become a standard choice for statistical programs [16] (available under the GNU GPL license [17]). Methods of correlation analysis were used in the study [18].

The experiment was carried out in accordance with the fundamental principles of bioethics [19] and on a voluntary basis.

4 Results and Discussion

Although all the participants in the experiment were familiar only with the direct type of indication in both simulation and real flight practice, the results of the experiment show that even for them, it is more convenient to work with artificial horizons of the reverse type. As can be seen from Fig. 2, the indicators showing the total number of errors made and the average time taken to interpret one slide are somewhat better for reverse indication. The difference is not too big, but it is a fraction of a second that can count in an emergency. Moreover, while the errors in bank angle determination are distributed almost equally for the direct and reverse indicators (51% and 49%, respectively), the errors in determining the roll direction (!!!), although rare, are accounted for by direct indication primarily, which was responsible for 94% (!!!) of errors. All this clearly confirms the opinion voiced by V. V. Kozlov [2].

The use of eye tracker in the experiment revealed a number of important aspects that seem to be promising for studying in the future. Of specific interest are the heat maps, which reflect the zones where the participant's gaze was fixed most often during the experiment (see Fig. 3). Most of these zones for both types of indications are located along the aircraft symbol (see Fig. 3a). However, in some cases, these zones have other locations, for example, perpendicular to the aircraft symbol (see Fig. 3b). These differences are especially pronounced on the heat maps showing how the whole task was performed.

It should also be noted that the results of the experiment confirm what we previously revealed concerning the focus of the participants' attention on the upper part of the screen [14].

To find differences in the participants' perception and interpretation of data from the two different types of attitude indicators, we checked whether there were correlations between temperament, mental strength and the results of the experiment.

As can be seen from the data presented in Fig. 4, 61% of the participants belong to the sanguine type, 26% are phlegmatic, 4% are choleric and none of the participants are melancholic. The remaining 9% have a mixed type of temperament. It should

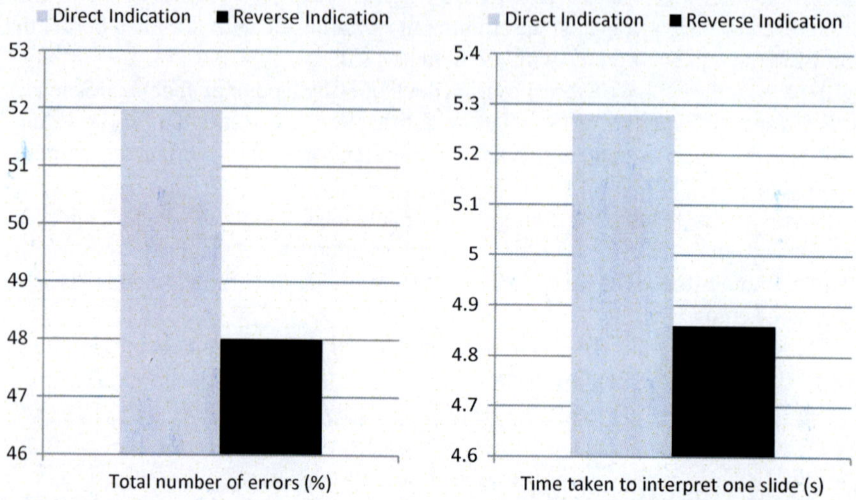

Fig. 2 Experimental results

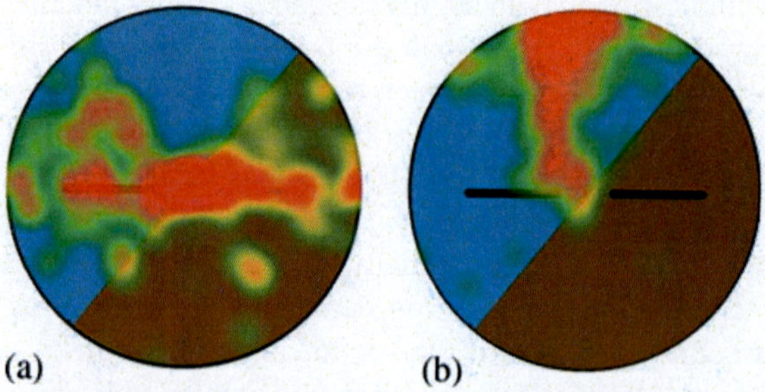

Fig. 3 Heat maps showing the results of completing the task: **a** typical, **b** non-typical

be noted that such a distribution of temperament types looks common and typical for aircraft personnel. As there are no participants with the temperament type that is deemed unfit for aviation (melancholic), it signals that the students were selected correctly for their future careers in aviation (at least based on one parameter).

The distribution of mental strength levels in the surveyed group of future pilots that is shown in Fig. 5 looks similar. This similarity might be the reason why the spread of the results is small and there were no correlations identified between the results of the experiment and the psychological parameters being tested. Most of the correlations turned out to be insignificant (see Table 3).

5 Conclusions

Based on the results of the experiment, several conclusions can be made.

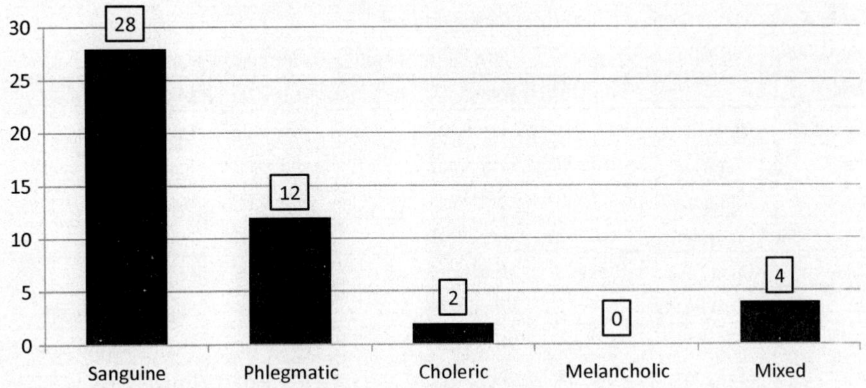

Fig. 4 Distribution of temperament types in the sample of future pilots

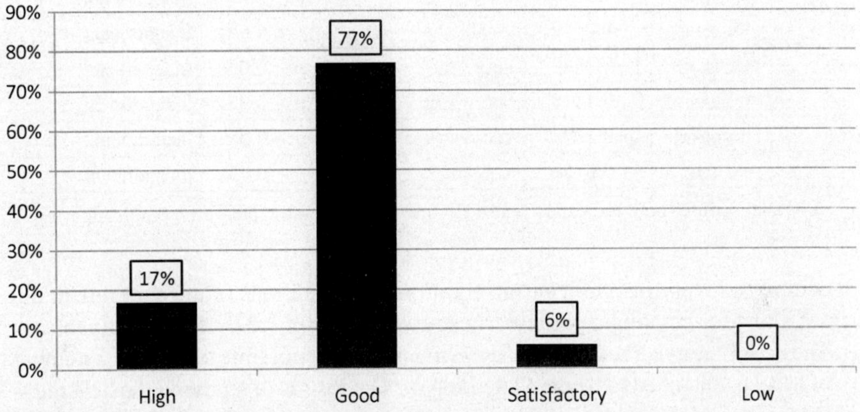

Fig. 5 Distribution of mental strength levels in the sample of future pilots

1. As a first approximation, the experiment revealed almost an equal ratio of errors (348/319, or 52/48% for direct and reverse types, respectively). Moreover, the participants' subjective assessments on a 10-point scale of the convenience of both indication types also show almost equal scores (7.5 and 7.3, respectively).
2. It should be noted that this study does not intend to retrain future pilots. All the participants in the experiment were familiar only with the direct type of indication in both simulation and real flight practice.
3. The experiment did not find any clear correlations between the results of performing the task, the psychological characteristics (extraversion and neuroticism), and the mental strength of the participants. All the correlations are predominantly very weak and insignificant. This is most likely due to the high sample homogeneity (63.8% of the participants belong to the sanguine type; 76.6% of them rank high in terms of mental strength), making the spread of data for

Table 3 Correlation data obtained from the experiment

1st value	2nd value	r_{CORR}	Correlation strength	Correlation significance	
E	n	+0.2551	Weak	$p \geq 0.05$	Insignificant
E	N_{MS}	+0.1050	Very weak	$p \geq 0.05$	Insignificant
E	$(S_{ERR})_{\Sigma}$	−0.1377	Very weak	$p \geq 0.05$	Insignificant
E	$(S_{ERR})_{DI}$	−0.1535	Very weak	$p \geq 0.05$	Insignificant
E	$(S_{ERR})_{RI}$	−0.0993	Very weak	$p \geq 0.05$	Insignificant
E	$(S_{L/R})_{\Sigma}$	-0.3159	Moderate	$p < 0.05$	Significant
E	$(S_{L/R})_{RI}$	−0.4775	Moderate	$p < 0.01$	Highly significant
n	N_{MS}	+0.4538	Moderate	$p < 0.01$	Highly significant
n	$(S_{ERR})_{DI}$	−0.0138	Very weak	$p \geq 0.05$	Insignificant
n	$(S_{ERR})_{RI}$	−0.0801	Very weak	$p \geq 0.05$	Insignificant
n	$(S_{L/R})_{\Sigma}$	−0.2139	Weak	$p \geq 0.05$	Insignificant
n	$(S_{L/R})_{DI}$	−0.2044	Weak	$p \geq 0.05$	Insignificant
n	$(S_{L/R})_{RI}$	−0.2439	Weak	$p \geq 0.05$	Insignificant
N_{MS}	$(S_{ERR})_{\Sigma}$	−0.1944	Very weak	$p \geq 0.05$	Insignificant
N_{MS}	$(S_{ERR})_{DI}$	−0.1823	Very weak	$p \geq 0.05$	Insignificant
N_{MS}	$(S_{ERR})_{RI}$	−0.1792	Very weak	$p \geq 0.05$	Insignificant
N_{MS}	$(S_{L/R})_{\Sigma}$	−0.0751	Very weak	$p \geq 0.05$	Insignificant

Note Σ—total sum for both indication types; RI—reverse indication; DI—direct indication

comparison very insignificant. In addition, the conditions of the experiment did not allow for creating a particularly stressful atmosphere for the participants.

4. The use of the eye tracker in this experiment made it possible to identify a number of aspects that need to be paid attention to. The results of studying the heat maps (i.e., zones of the focus of visual attention) seem to be of particular interest as most of the focus zones for both types of attitude indicators are located along the aircraft symbol. However, there are cases when they are located perpendicular to it. Consequently, we come to the conclusion that heat maps in some cases reflect pronounced differences between individuals in terms of their attention distribution.

All the above indicates that it looks promising to continue research in this direction and identify psychological characteristics that can serve as signals of a predisposition to making errors.

References

1. Final report B 737–505 VP-BKO (2021) Air accident investigation commission. Interstate Aviation Committee. https://reports.aviation-safety.net/2008/20080914-0_B735_VP-BKO.pdf. Accessed 07 Nov 2021

2. Kozlov VV (2017) Pilotu o pilote ot psihofiziologa letnogo truda (To the pilot about the pilot from the aviation psychophysiologist) Beliy veter, Moscow (In Russian)
3. Dzhapharadze TR, Malishevskii AV (2013) Problem a sovershenstvovaniya professional'nogo psixologicheskogo otbora pilotov grazhdanskoj aviacii (the challenge of improving professional psychological selection of pilots of civil aviation) Medico-Biol Socio-Psychol Prob Safety Emerg Situat 3: 66–70. https://doi.org/10.25016/2541-7487-2013-0-3-66-70 (In Russian)
4. Arinicheva OV, Lebedeva NA, Malishevskii AV (2020) Application of eye-tracking technology as a diagnostic tool for assessing flight operators. Part 2: a study of qualities important for operators in the aviation industry. Transp Prob 15(4, Part 1):5–18. https://doi.org/10.21307/tp-2020-043
5. Nistrom M, Holmqvist K (2010) An adaptive algorithm for fixation, saccade, and glissade detection in eyetracking data. Behav Res Methods 42(1):188–204. https://doi.org/10.3758/BRM.42.1.188
6. Wang Y, Cong W, Dong B (2015) Statistical analysis of air traffic controllers' eye movements. In: The 11th USA/Europe ATM R&D Seminar. EUROCONTROL, Lisbon
7. Lundberg J, Johansson J, Forsell C, Josefsson B (2014) The use of conflict detection tools in air traffic management: an unobtrusive eye tracking field experiment during controller competence assurance. In: Proceedings of the international conference on human-computer interaction in aerospace (HCI-Aero 2014). Association for Computing Machinery, New York, NY. https://doi.org/10.1145/2669592.2669655
8. Stanković A, Aitken MR, Clark L (2014) An eye-tracking study of information sampling and decision-making under stress: implications for alarms in aviation emergencies. Proc Human Factors Ergonomics Soc Ann Meet 58(1):125–129. https://doi.org/10.1177/1541931214581027
9. Martin C, Cegarra J, Averty P (2011) Analysis of mental workload during en-route air traffic control task execution based on eye-tracking technique. In: Proceedings of the 9th international conference on Engineering psychology and cognitive ergonomics (EPCE'11). Lecture Notes in Computer Science, vol 6781. Springer, Berlin, Heidelberg. https://doi.org/10.1007/978-3-642-21741-8_63
10. Kuravsky LS, Marmalyuk PA, Yuryev GA et al (2016) Flight crew diagnostic using aviation simulator training data. Experiment Psychol 9(3):118–137. https://doi.org/10.17759/exppsy.2016090310
11. Eysenck HJ, Eysenck SBG (1964) Manual of the eysenck personality inventory. University of London Press, London
12. Burlachuk LF (2008) Slovar'-spravochnik po psihodiagnostike (psychological testing handbook). Piter, St. Petersburg, p 685 (In Russian)
13. Prohorov AO (ed) (2004) Praktikum po psihologii sostoyanij (manual on mental states). Rech, St. Petersburg, p 480 (In Russian)
14. Arinicheva OV, Lebedeva NA, Malishevskii AV (2020) Application of eye-tracking technology as a diagnostic tool for assessing flight operators. Part 1: analise of flight operators' attention distribution and switching using eye-tracking. Transp Prob 15(3):167–179. https://doi.org/10.21307/tp-2020-042
15. Tobii Eye Tracker (2021) User's Guide. https://topics-cdn.dell.com/pdf/alienware-17-laptop_users-guide10_en-us.pdf. Accessed 07 Nov 2021
16. Research & Statistical Support Services (2021) University information technology. http://it.unt.edu/research. Accessed 07 Nov 2021
17. The Free Software Foundation (FSF) (2021) https://fsf.org/. Accessed 07 Nov 2021
18. Dowdy S, Wearden S, Chilko D (2004) Statistics for research, 2nd edn. John Wiley & Sons Inc., New York
19. Bioethics. Internet Encyclopedia of Philosophy. A Peer-Reviewed Academic Resource. https://www.iep.utm.edu/bioethic/. Accessed 07 Nov 2021

Research of Projection Algorithms for Solving Problems of Measuring Angular Coordinates of Low-Flying Radar Targets

Aleksandr Ermakov, Nikolay Povarenkin, and Nikolay Malisov

Abstract In the presented work, the urgent problem of detecting low-flying radar targets is touched upon. The problem has been known for a long time, but only now, there are computing capacities for testing algorithms for measuring angular coordinates. In this regard, the study of the ESPRIT super-resolution algorithm is carried out, and the error of measuring angular coordinates is estimated depending on the height of the surface irregularities. The presence and magnitude of this error are directly related to an increase in the power of the diffuse component of the signal, which in turn destroys correlations in the elements of the antenna array, on which the measurement of angular coordinates by the ESPRIT algorithm is based.

Keywords Low-flying radar targets · Kirchhoff method · Maxwell equations · Super-resolution algorithms · ESPRIT

1 Introduction

The problem of measuring angular coordinates is that at the receiving point, there is a superposition of the direct signal of the target and the signals re-reflected by the luminous points of the underlying surface [1, 2]. In this connection, a rugged phase front is observed at the receiving point, and the angular coordinates are measured with large errors. The problem is solved by lifting the phase center of the antenna system using balloons, quadcopters, and helicopters [3, 4]. However, solving this problem in this way is a special case and has a number of disadvantages. In this regard, it is necessary to solve the problem using mathematical algorithms.

A. Ermakov (✉) · N. Povarenkin
State University of Aerospace Instrumentation (SUAI), Saint Petersburg, Russia
e-mail: ermakov.alexandr.k@gmail.com

N. Malisov (✉)
Irkutsk Branch of Moscow State Technical University of Civil Aviation, Irkutsk, Russia
e-mail: malisovnik@mail.ru

O. A. Gorbachev et al. (eds.), *Proceedings of 10th International Conference on Recent Advances in Civil Aviation*, Lecture Notes in Mechanical Engineering, https://doi.org/10.1007/978-981-19-3788-0_25

2 Research of the ESPRIT Algorithm

When investigating ways to solve the problem of measuring the angular coordinates of low-flying radar targets in semi-natural conditions, first of all, it is necessary to determine the mathematical model of the signal reflected from the underlying surface. In a number of papers [5], various methods of calculating and modeling the signal reflected from the underlying surface were considered. Among them, an electrodynamic calculation method using the Kirchhoff method was chosen as optimal for this study. As the underlying surface, a cylindrical absolutely conductive sea surface was used, implemented using a Gaussian random process (Fig. 1).

The following conditions were determined for the task of studying the ESPRIT algorithm: A linear equidistant array with the number of elements N acts as a receiving antenna, a random Gaussian process acts as the underlying surface, since the normality of the distribution law of the sea surface has been repeatedly proven [6–8]. The time of signal accumulation by the ESPRIT algorithm is determined by the length of time during which the target does not change its position relative to the receiving antenna by more than 0.1°. The reflected signal model does not take into account the delay time, since it is proved that in the case of probing a low-flying radar target, the signal from the underlying surface and from the target that passed along a straight path have a time difference of less than a wavefront. The first stage of the study was to simulate the operation of the ESPRIT algorithm in order to obtain the dependence of the root-mean-square error of the measured angle on the root-mean-square deviation of the heights of the irregularities from the mirror surface [9–12]. The following values were set as simulation parameters: frequency f 3 GHZ, wavelength 10 cm, distance between the elements of the antenna array 5 cm,

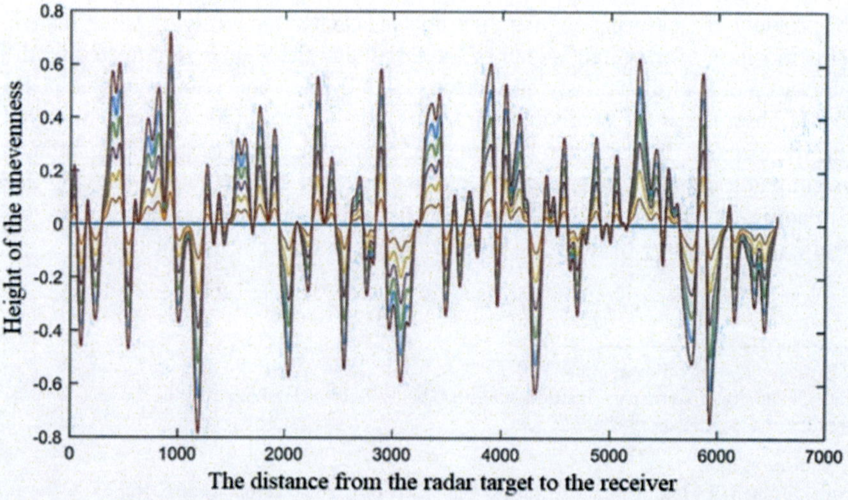

Fig. 1 Gaussian random process

respectively; the phase center of the antenna array was at a height of 11 m 25 cm, and the antenna aperture was 2.5 m, accumulation time t 0.3 s, pulse repetition period 0.1 s, target speed 100 m/s, the initial distance from the radar target to the receiving antenna 3 km. The signal-to-noise ratio and the intrinsic noise of the radio receiving equipment were not set; in this study, it is necessary to study the effect of only the diffuse component of the signal on the accuracy of the algorithm. For the study, 5 different surfaces were calculated: a mirror underlying surface and 4 surfaces with a mean square deviation of the heights of irregularities from the mirror surface from 0.1 to 0.4 m in increments of 0.1. The choice of surfaces with such standard deviation of heights is determined by the power of the mirror and diffuse components of the signals for the selected wavelength, the distance from the antenna to the radar target, the height of the antenna, and the flight altitude of the radar target [13–15]. As can be seen from Fig. 1, by choosing such surfaces, we cover all possible situations from the case of a fully known signal (the mirror component prevails) until the case when the signal is completely unknown (the diffuse component prevails). The theoretical curves in Fig. 2 are constructed according to the following formula coherent component:

$$|e_{zer}| = \exp(-p^2) \tag{1}$$

where $p = \sqrt{2} \cdot k \cdot s \cdot \sin\beta$ is the roughness coefficient;

S—is the standard deviation of roughness;

β—sliding angle.

Because, on those areas of the surface that satisfy the conditions:

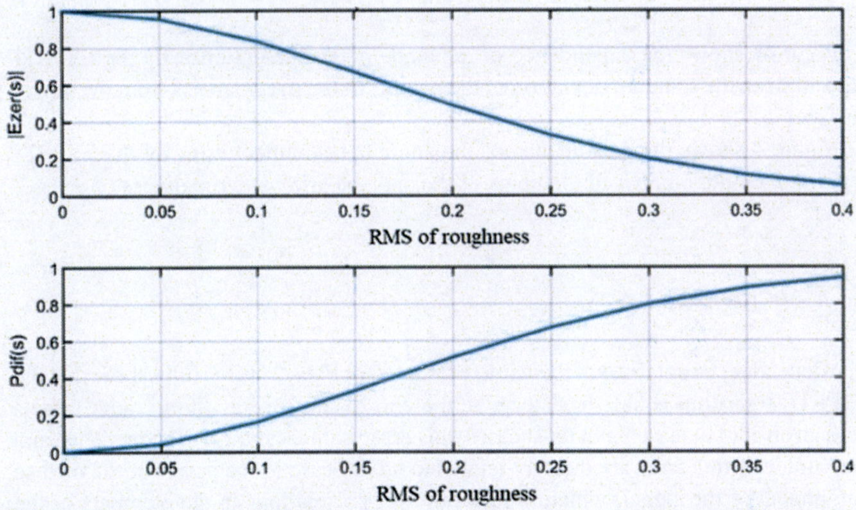

Fig. 2 Dependence of the diffuse and mirror components of the signal on the COE roughness

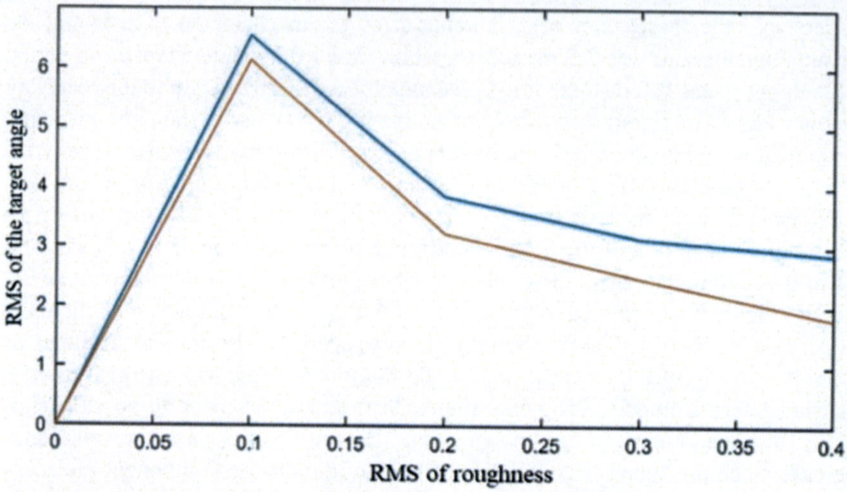

Fig. 3 Dependence of the radar target angle error on the surface roughness

$$\begin{aligned} l &>> \lambda, \\ h &\geq \lambda, \end{aligned} \tag{2}$$

The re-reflection is diffuse, and a random component begins to prevail in the signal, it is necessary to calculate a theoretical graph for it as well. The diffuse component is defined as $e_{dif} = e_{zer0} - e_{pr}$. The theoretical graph for the diffuse component is constructed according to the following formula:

$$P_d = \left(1 - \exp(-2 \cdot p^2)\right) \tag{3}$$

Figure 3 shows the dependence of the angle measurement error by the ESPRIT algorithm on the standard deviation of the heights of the irregularities from the mirror surface.

Figure 4 shows the dependence of the angle measurement error by the ESPRIT algorithm on the number of elements of the antenna array at a roughness of 0.1 m.

3 Conclusion

The dependences obtained in the work, presented in Figs. 2 and 3, fully show that the ESPRIT algorithm is able to determine the coordinates of low-flying radar targets with an error of at least 5°. In most cases, this error is unacceptable. The presence and magnitude of this error are directly related to an increase in the power of the diffuse component of the signal, which in turn destroys correlations in the elements of the

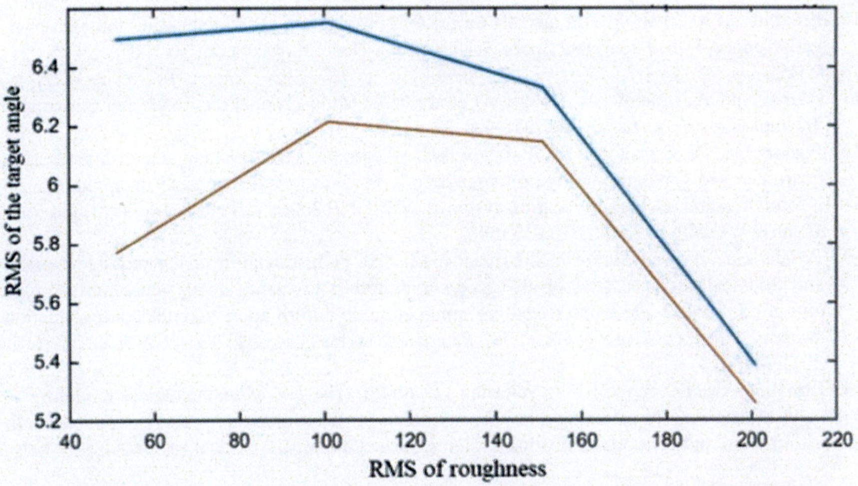

Fig. 4 Dependence of the radar target angle error on the number of antenna array elements

antenna array, on which the measurement of angular coordinates by the ESPRIT algorithm is based.

References

1. Galushko V, Vavriv DM (2017) Detection of radar targets moving with acceleration. In: IEEE microwaves, radar and remote sensing symposium (MRRS). https://doi.org/10.1109/MRRS.2017.8075043
2. Petrovic I, Kankaraš M, Cvetkovic K (2015) Significance and prospects of the development of air defence system. Vojno Delo 67(6):86–98. https://doi.org/10.5937/vojdelo1506086P
3. Kalenov EN (2015) Potential accuracy of measuring the angular coordinates of signal sources and accuracy of measuring them using optimal spatial filtration. Acoust Phys 61(2):205–212. https://doi.org/10.1134/S1063771015010054
4. Hussein MY (2014) Target tracking radar. IOSR J Eng 4(5):23–28. https://doi.org/10.9790/3021-04512328
5. Pedenko YuA (2018) Using the matrix pencil method for radar measurement of an elevation angle of low-altitude targets over the agitated sea. Radiofizika i elektronika 23(1):10–18. https://doi.org/10.15407/rej2018.01.010
6. Monakov A (2013) Asymptotically optimum estimation of signal waveform in the presence of uncertainties about the steering vector. Signal Process 93(12):3300–3305. https://doi.org/10.1016/j.sigpro.2012.12.007
7. Nesterov MY, Monakov AA (2014) Estimation of flight altitude in the aperture synthesizing mode for altimeter with continuous probing signal. Radioelectron Commun Syst 57(11):489–494. https://doi.org/10.3103/S0735272714110028
8. Mayorov BG (2018) Method for detecting and tracking low-flying targets. RU Patent 2713645C1
9. Monakov AA (2020) An Algorithm for estimating the parameters of a quadratic FM signal. In: Wave electronics and its application in information and telecommunication systems, WECONF 2020, St. Petersburg. https://doi.org/10.1109/WECONF48837.2020.9131159

10. Erokhin V (2021) Bi-criteria aircraft trajectory optimization in implementing the area navigation concept. Int J Aeronaut Space Sci 22(4):948–962. https://doi.org/10.1007/s42405-021-00353-3
11. Akmaykin DA, Bolelov EA, Kozlov AI et al (2021) Multi-channel radar systems. Aerospace Technol. https://doi.org/10.1007/978-981-33-6514-8_10
12. Povarenkin NV, Ermakov AK, Monakov AA, Velegev DS (2020) Using standard measuring instruments to simulate low-flying target signals. In: Wave electronics and its application in information and telecommunication systems, WECONF 2020, St. Petersburg. https://doi.org/10.1109/WECONF48837.2020.9131543
13. Povarenkin NV, Venedictov NA, Ermakov AK (2021) Justification of the required tactical and technical characteristics of air space survey radars to ensure timely detection of small aircraft. In: Wave electronics and its application in information and telecommunication systems, conference proceedings, St. Petersburg. https://doi.org/10.1109/WECONF51603.2021.9470614
14. Ermakov AK, Povarenkin NV, Velezhev DS (2020) The use of mathematical modeling to solve the problem of detecting a low-flying target. In: Wave electronics and its application in information and telecommunication systems. https://doi.org/10.1109/WECONF48837.2020.9131446
15. Kanevsky MB, Karaev VY, Balandina GN (2002) Model of a doppler spectrum of microwaves backscattered from a sea surface at small grazing angles. Radiophys Quantum Electron 45(8):619–622. https://doi.org/10.1023/A:1021776814301

Methods for Assessing the Influence of External Factors on Airplane Flight

Aleksey Verstiuk, Gennadii Kovalenko, Artem Fedorov, and Oleg Patrikeev

Abstract The article proposes three methods for predicting deviations from the specified parameters caused by an unfavorable environment in order to increase the flight safety level. The most effective method has been established for each stage of the flight. The corresponding functions with stationary coefficients for calculating the probable force impact of factors on the aircraft have been determined. During the research, it was observed that stationary coefficients in the function have a significant drawback. In a long flight, weather conditions can change a lot, and the coefficients will remain the same. The application of dynamic coefficients, the determination of which became possible thanks to the use of an artificial neural network, is most likely to solve this problem.

Keywords Forecasting · Deviation · External conditions · Flight safety · Information

1 Introduction

Flight safety is mainly influenced by three main groups of factors: human factor, technical factor, and adverse external conditions [1, 2]. Despite the fact that the number of air accidents caused only by the third group is about 3%, the reports also note that there is often a combination of the human factor as the main cause of the accident and adverse external conditions as a concomitant phenomenon. Primarily, these are weather conditions [3–5].

The crisis in the world in 2019–2020, caused by the pandemic of coronavirus infection, led to a significant decrease in air services. Accordingly, the number of aviation accidents by absolute measures also fell. Nevertheless, in 2020, the Federal

A. Verstiuk · G. Kovalenko (✉) · A. Fedorov
Saint Petersburg State University of Civil Aviation, Saint Petersburg, Russia
e-mail: kgvf@inbox.ru

O. Patrikeev (✉)
Irkutsk Branch of Moscow State Technical University of Civil Aviation, Irkutsk, Russia
e-mail: po_2010@mail.ru

O. A. Gorbachev et al. (eds.), *Proceedings of 10th International Conference on Recent Advances in Civil Aviation*, Lecture Notes in Mechanical Engineering, https://doi.org/10.1007/978-981-19-3788-0_26

Air Transport Agency of the Russian Federation and the IAC published documents on the state of flight safety, which paid special attention to the lack of due regard for weather conditions by aircraft crews [6, 7].

2 Problem Statement

From year to year, environmental factors, either on their own or in conjunction with other factors, lead to aviation accidents, which once again proves the need to find ways to reduce their impact on the flight of the aircraft. Therefore, methods are proposed to predict deviations in aircraft movement parameters from those specified for their advanced parading. They are based on data on deviations from previous time points.

For their development, data obtained from the flight information recorder of the flight navigation system Garmin 1000 of the aircraft—DA-40NG were used. Out of the 62 parameters recorded, 13 were used for research [8]. Only those points of time during which the flight was performed in manual mode were selected from the general data set.

3 Materials and Methods

At various stages of the flight, it is advisable to choose your own movement parameters that the crew withstood (Table 1), since only in this case it is possible to more objectively evaluate the results of predicted data during manual piloting. This makes it possible to track the influence of the human factor on the deviation of control surfaces.

In addition, not the deviations from the given parameters themselves are considered, but the rate of their increment, since they are proportional to the forces that cause them.

Control of runway axis direction after takeoff was taken into account according to data on touchdown zone coordinates (Buguruslan-Severny, Buguruslan-Main) and current aircraft coordinates. The required instrument airspeed, as well as the withstood altitude in horizontal flight, was determined in accordance with the flight plan. At the stages of descent and climb, a specific trajectory was not set but was determined taking into account the intention of the crew to change the flight altitude and return to the horizontal flight trajectory. However, when analyzing the second derivative by deviation from the theoretical vertical profile, this has no effect. As a result, the vertical flight profile can be represented, as shown in Fig. 1.

Denote the second derivative by deviation from the specified parameter—x_i a set of values from which the calculation is made—$\vec{x} = X = [x_0, x_1, x_2, x_3, x_4]$, and estimated value—$x_m$, where m—index of the method by which the prediction

Table 1 Movement parameters maintained by the aircraft crew

No.	Flight stage	Parameter controlled by the aircraft crew	Test value	Value to be analyzed
1	Set after take-off to $H = 60$ m	Flight directions along runway axis	Aircraft coordinates	Second derivative on deviation from runway axis
		(Angle of trajectory inclination)	Altitude	Second derivative on deviation from planned climbing trajectory
2	Climbing except for item 1	Aircraft instrument airspeed	IAS	Second derivative of the speed change from the assigned at this stage of flight
		(Angle of trajectory inclination)	Altitude	Second derivative on deviation from planned climbing trajectory
3	Horizontal flight	Flight altitude	Altitude	Second derivative by deviation from specified height
		Aircraft instrument airspeed	IAS	Second derivative of the speed change from the assigned at this stage of flight
4	Decrease except glide slope	Aircraft instrument airspeed	IAS	Second derivative of the speed change from the assigned at this stage of flight
		Angle of trajectory inclination	Altitude	Second derivative on deviation from planned descent trajectory
5	Glide slope	Vertical speed	Vertical speed from GPS	Second derivative on deviation from required vertical speed
		Flight directions along runway axis	Aircraft coordinates	Second derivative on deviation from runway axis

was carried out: *C*—1st method Cubic Spline; *U*—2nd method Univariate Spline; *F*—3rd method Flattop [3–13].

Regardless of the method, the accuracy of the x_m definition is not ideal. It is influenced by the error of measuring quantities and the random nature of atmospheric behavior. In order to take into account the effects of these phenomena, it is necessary to add coefficients in accordance with the equation:

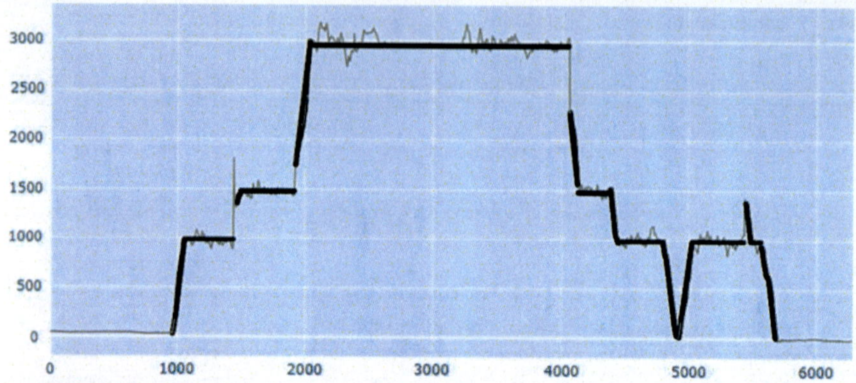

Fig. 1 Assigned vertical flight profile

$$x_m = k_1 * f_m(X)$$

Based on this function, you can select several additional ones, each of which increases the accuracy of the forecast:

- Simple definition $x_m = k_1 * f_m(X)$.
- Additional accounting of parameter change rate $x_m = k_1 * f_m(X) + k_2 * \dot{x}_{(m-1)}$.
- Additional accounting of trend of parameter change rate $x_m = k_1 * f_m(X) + k_2 * \dot{x}_{(m-1)} + k_3 * \ddot{x}_{(m-1)}$.

Estimation of the accuracy of the predicted data, as well as for determining the coefficients of the equations, was carried out by determining the standard deviation according to the graphic $\Delta x(t) = x_m(t) - x_a(t)$, where x_a—actual parameter value. In addition, it is important to take into account the landmark attitude. Indeed, if the values are located on both sides of zero, then applying the data in practice will become dangerous. In addition to the fact that external factors have a powerful effect on the aircraft, the forecast will also add an additional value to it.

The construction of all graphs, statistical analysis, as well as the implementation of the forecasting method was carried out using software—Anaconda, as well as the Jupyter Notebook web application. The program was written in the Phiton programming language. Prediction of acceleration of altitude deviation by method of cubic splines at the stage of descent, prediction of acceleration of deviation from runway axis by method of cubic splines in glide slope and prediction of acceleration of IAS deviation by method of Flattop at the stage of recruitment after $H = 60$ m are shown in Figs. 1, 2 and 3, respectively. The graph of the actual values of acceleration of deviation of the withstood parameter is highlighted in black, in gray—predicted, calculated on the basis of data on deviations for the last 5 s.

To identify the best method of three, two values were used with respect to the function describing the graph of the difference between the calculated and real value,

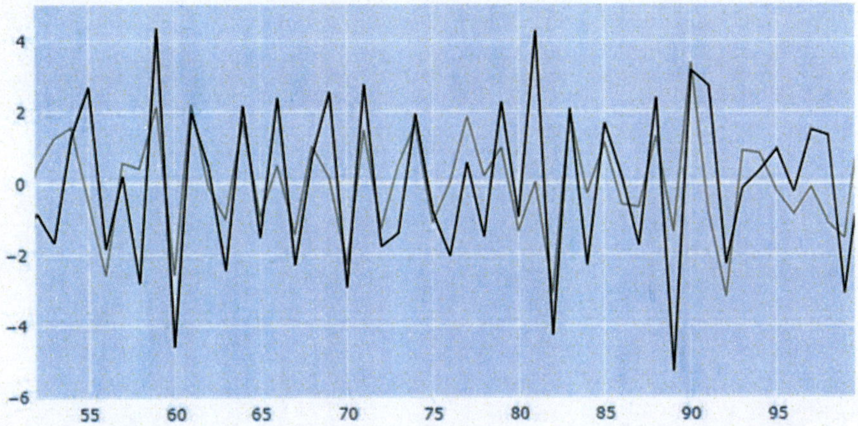

Fig. 2 Prediction of acceleration of altitude deviation by cubic splines at the stage of descent

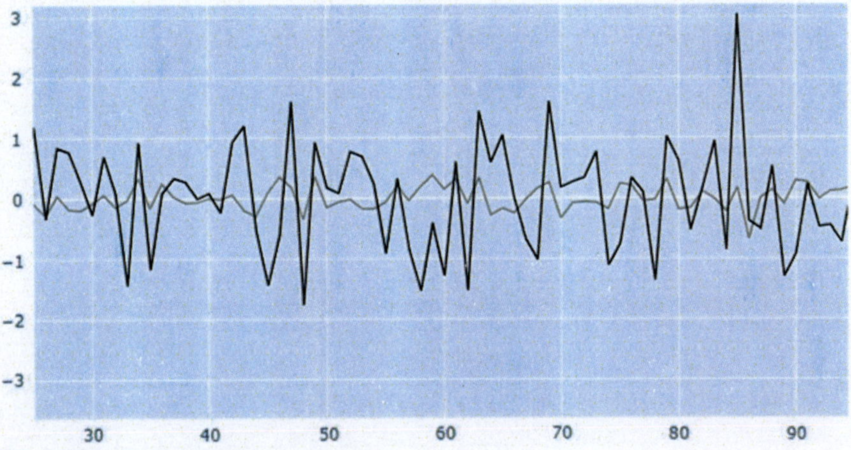

Fig. 3 Prediction of acceleration of deviation from runway axis by method of cubic splines in glide slope

for each stage and element of Table 1. First—standard deviation (StD) (σ), second—sign match index (Δ), which is calculated by formula [14, 15]:

$$\Delta = \frac{\sum_{i=4}^{n}\{x_{mi} * x_{ai} \geq 0\}}{n-4} * 100\%$$

Static data of forecasting functions by flight stages are presented in Table 2.

Part of the flight in the horizontal section was carried out in autopilot mode. Using methods at this interval allows you to determine the possibility of improving

Table 2 Static data of forecasting functions by flight stages

Flight stage	Parameter	Method 1 (cubic spline)			
		k_1	k_2	σ	Δ (%)
Set after take-off to $H = 60$ m	Direction of runway axis	−0.16	0.21	0.6996	73.3
	Altitude	0.16	−0.27	1.4237	75.75
Remaining climb	IAS	−0.77	0.95	1.9617	70.36
	Altitude	−0.21	0.18	3.6906	59.93
Horizontal flight	Altitude	0.29	−0.43	2.1905	74.29
	IAS	−1.13	1.51	1.6924	72.31
Decrease	IAS	−1.11	1.43	1.6411	75
	Altitude	−0.16	0.16	3.9392	59.02
Gglide slope	Altitude	−0.31	0.4	2.6049	63.8
	Direction of runway axis	−0.26	0.29	0.9029	56.73
Flight stage	Parameter	Method 2 (univariate spline)			
		k_1	k_2	σ	Δ (%)
Set after take-off to $H = 60$ m	Direction of runway axis	−0.04	−0.02	0.7033	70
	Altitude	0.11	−0.09	1.4234	72.72
Remaining climb	IAS	−0.49	0.33	2.1342	65.36
	Altitude	−0.15	0.06	3.6955	56.75
Horizontal flight	Altitude	0.17	−0.16	2.2055	71.58
	IAS	−0.46	0.27	1.8745	66.61
Decrease	IAS	−0.36	0.12	1.8708	70
	Altitude	−0.03	−0.04	3.9593	57.38
Gglide slope	Altitude	−0.2	0.15	2.6048	62.86
	Direction of runway axis	0	−0.1	0.9153	60.58
Flight stage	Parameter	Method 3 (flattop)			
		k_1	k_2	σ	Δ (%)
Set after take-off to $H = 60$ m	Direction of runway axis	0.21	−0.05	0.9799	66.6
	Altitude	−0.14	−0.02	1.4168	69.69
Remaining climb	IAS	0.09	−0.22	2.3118	62.85
	Altitude	−0.07	−0.12	3.7114	65.25
Horizontal flight	Altitude	0.04	0.02	2.407	66.79
	IAS	0.12	−0.22	2.0163	62.24
Decrease	IAS	0.15	−0.28	1.9218	66.66
	Altitude	−0.02	−0.07	3.9596	59.83
Gglide slope	Altitude	−0.12	−0.05	2.6331	71.43
	Direction of runway axis	0.18	−0.12	0.8949	61.54

the accuracy and quality of maintaining parameters by the automatic flight control system.

For automatic flight, the 2nd method showed the best prediction result in the vertical channel. The graphical representation at one of the intervals is shown in Fig. 4. By amplitude, the coincidence is too small, however, extremes are determined accurately, and the sign coincidence is 75.2%.

By the parameter of the instrument speed, method 1 turned out to be the most effective—Cubic Spline. The graph shows a pronounced similarity of shapes in amplitude (Figs. 5 and 6).

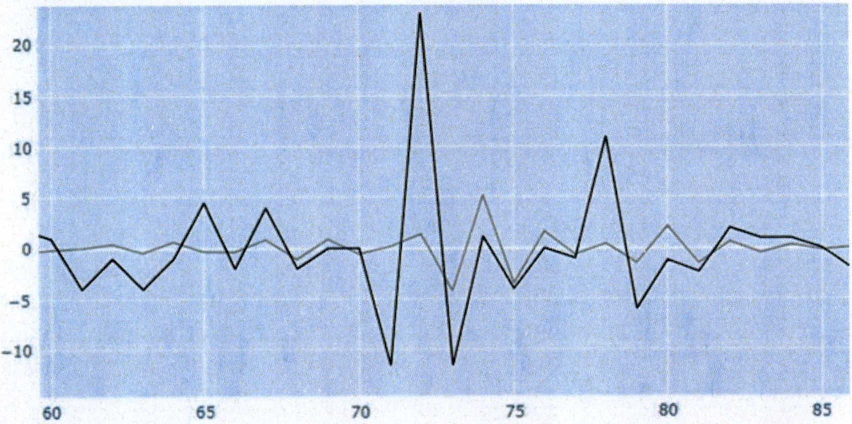

Fig. 4 Prediction of acceleration of IAS deviation by Flattop method at climbing stage after $H =$ 60 m

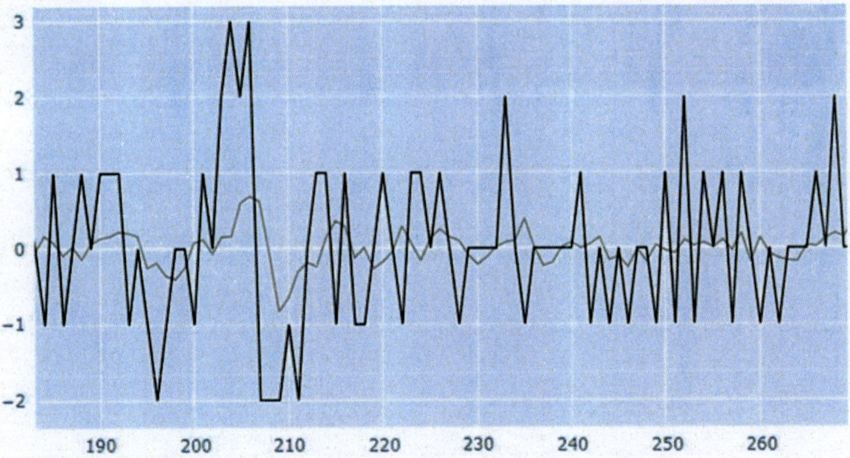

Fig. 5 Prediction of acceleration of altitude deviation in automatic mode

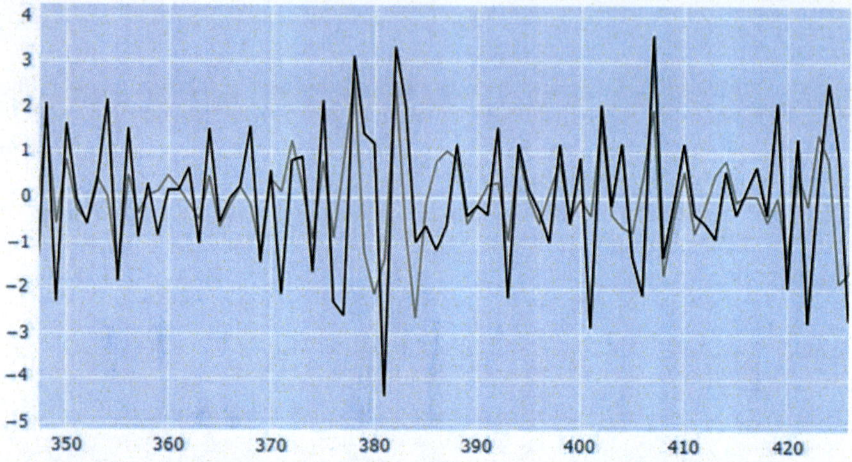

Fig. 6 Prediction of acceleration of deflection of instrument speed in automatic mode

The coefficients of functions, as well as statistics on all three methods, are presented in Table 3. The parameter ε demonstrates how much percent consideration of the prediction method can improve the generation of autopilot control signals. At that n—flight time using automatic control system, expressed in seconds. A limit of 1 m/c^3 for speed, introduced in connection with the accuracy of determining this parameter.

$$\varepsilon = \overline{\left\{\frac{|x_i| - |x_i - x_{mi}|}{|x_i|} * 100\right\}}, \quad \text{at} \quad \begin{cases} x_i \in (4, n]; \\ |x_i| \geq 1 \end{cases}$$

Table 3 Data on forecasting results during horizontal flight in automatic mode

Parameter	Method 1 (cubic spline)				
	k_1	k_2	σ	Δ [%]	ε [%]
Altitude	0.2	−0.33	1.1906	73.9	1.95
IAS	−1.06	1.43	1.4291	71.3	28.9
Parameter	Method 2 (univariate spline)				
	k_1	k_2	σ	Δ [%]	ε [%]
Altitude	0.19	−0.1	1.1784	75.2	4.09
IAS	−0.43	0.22	1.5798	65.6	15.2
Parameter	Method 3 (flattop)				
	k_1	k_2	σ	Δ [%]	ε [%]
Altitude	−0.06	−0.03	1.2018	74.7	1.2
IAS	0.07	−0.17	1.7056	63.6	7.4

4 Conclusions

The maximum value of the sign coincidence of actual deviations and deviations determined by the prediction methods was 75% (by maintaining the instrument speed at the descent stage), with a minimum value of 56.73% (when maintaining the direction of the runway axis on the glide path). For civil aviation, these indicators are unacceptable, however, if you do not take into account the results of the prediction function in the interval ± 0.05 from 0, then the sign match indicator will range from 96 to 98%, depending on the flight stage and parameter.

The standard deviation by the difference function of the calculated and actual acceleration of the deviation from the given height is high at all stages. It is most important to control this parameter on take-off and glide slope, for this reason, to consider the obtained forecasting methods, to warn crews in advance about deviation in the vertical profile, is not permissible.

During the research, it was observed that stationary coefficients in the function have a significant drawback. In a long flight, weather conditions can change a lot, and the coefficients will remain the same. The application of dynamic coefficients, the determination of which became possible thanks to the use of an artificial neural network, is most likely to solve this problem.

The use of the methods presented in automatic flight showed the theoretical possibility and benefit of their implementation since it does not require material investments in additional equipment for the aircraft. However, the data obtained are not sufficient to objectively assess the safety of their use. For this purpose, it is advisable to study more flight data.

References

1. Luzik E, Akmaldinova A (2006) Psychological aspects of ensuring flight safety in civil aviation. Aviation 10(1):25–35. https://doi.org/10.3846/16487788.2006.9635924
2. Sharov VD, Vorobyov VV, Zatuchny DA (2021) Methods of safety risk management. In: Risk management methods in the aviation enterprise. Springer Aerospace Technology. Springer, Singapore. https://doi.org/10.1007/978-981-33-6017-4_1
3. Kim CY (1996) Study on the type and compatibility of flight crew personalities. J Korean Soc Aeronaut Flight Oper, pp 39–62, 4 Dec
4. Boeing, statistical summary of jet airplane accidents. Worldwide Operations 1959–2002
5. Akmaikin DA et al (2021) Theoretical foundations of radar location and radio navigation. Springer Aerospace Technology
6. Smurov MY, Arinicheva OV, Kovalenko GV et al (2017) Possible solutions to reduce the negative impact of human factors on flight safety. J Ind Pollut Control 33(2):1195–1201
7. Sharov VD, Vorobyov VV, Zatuchny DA (2021) Analysis of navigational and meteorological risks that influence civil aviation safety. In: Probabilistic-statistical methods for risk assessment in civil aviation. Springer Aerospace Technology. Springer, Singapore. https://doi.org/10.1007/978-981-16-0092-0_2
8. Flight Data Logging—Garmin G1000 Manual. https://www.manualslib.com/manual/1396612/Garmin-G1000.html?page=634. Accessed 30 Jan 2021

9. Natural Cubic Spline Interpolation (2019) https://codetobuy.com/downloads/natural-cubic-spline-interpolation/. Accessed 30 Jan 2021
10. Wolberg G, Alfy I (2002) An energy-minimization framework for monotonic cubic spline interpolation. J Comput Appl Math 143:145–188
11. Sablonnière P (2005) Univariate spline quasi-interpolants and applications to numerical analysis. Rendiconti del Seminario Matematico 63(3)
12. D'Antona G, Ferrero A (2006) The digital signal processing. In: Digital signal processing for measurement systems. Information technology: transmission, processing and storage. Springer, Boston, MA. https://doi.org/10.1007/0-387-28666-7_1
13. D'Antona G, Ferrero A (2006) Digital signal processing for measurement systems theory and applications. Springer Science+Business Media. https://link.springer.com/book/10.1007%2F0-387-28666-7#toc. Accessed 30 Jan 2021
14. Kuklev EA, Shapkin VS, Filippov VL, Shatrakov YG (2019) Solving the rare events problem with the fuzzy sets method. In: Aviation system risks and safety. Springer Aerospace Technology. Springer, Singapore. https://doi.org/10.1007/978-981-13-8122-5_3
15. Erokhin V (2021) Bi-criteria aircraft trajectory optimization in implementing the area navigation concept. Int J Aeronaut Space Sci 22(4):948–962. https://doi.org/10.1007/s42405-021-00353-3

Assessment of Aircraft Conditions in Flight

Gennadii Kovalenko, Sergey Lobar, Ivan Muravyev, and Sergey Turintsev

Abstract When flight conditions become more difficult, the pilot needs to control a greater number of parameters than in normal modes. This results in higher workload on the pilot and an increase in the probability that some mistakes will be made when operating highly automated aircraft. The problem lies in the fact that in such situations, the pilot is to perform operations safely regarding the use of both the flight control system and the air traffic control system, even if the number of information signals exceeds that which can be processed by the pilot. The article presents the results of a study on the assessment of different methods by which the aircraft status can be determined and its control systems can be monitored. The study was conducted due to the need to meet stringent requirements in flight safety regarding pilots' ability to serve as backups when flying highly automated aircraft. In order to do this, pilots need to develop an integral skill in processing static and dynamic information coming to them from various sources.

Keywords Information · Efficiency · Status assessment · Piloting · Automation

1 Introduction

When flying a modern highly automated aircraft, the pilot does not know the position of the flight control surfaces or the aircraft's attitude. The pilot can only evaluate the position of the control stick or the side stick [1–3]. This is explained by the fact that control operations are performed by many intelligent systems that work according to the algorithms embedded in them. It is impossible for the pilot to directly monitor the operation of all these systems. However, an assessment of the state of the systems

G. Kovalenko (✉) · S. Lobar · I. Muravyev
Saint Petersburg State University of Civil Aviation, Saint Petersburg, Russia
e-mail: kgvf@inbox.ru

S. Turintsev (✉)
Irkutsk Branch of Moscow State Technical University of Civil Aviation, Irkutsk, Russia
e-mail: basek@rambler.ru

O. A. Gorbachev et al. (eds.), *Proceedings of 10th International Conference on Recent Advances in Civil Aviation*, Lecture Notes in Mechanical Engineering, https://doi.org/10.1007/978-981-19-3788-0_27

involved in aircraft control by indirect indicators is necessary in order to back up the automated flight systems, or for the pilot to take over the control from the autopilot.

In cases when flight conditions are difficult, the pilot needs to monitor and control a lot of different parameters. This results in higher workload on the pilot and an increase in the probability that mistakes will be made when operating highly automated aircraft [4–9]. However, the pilot needs to perform operations safely regarding the use of both the flight control system and the air traffic control system, even if the number of information signals exceeds that which can be processed by the pilot in terms of both physical and mental resources [10–12]. To solve such a problem, it is necessary to identify the number of information signals necessary for performing manual control by the crew, which, on the one hand, would satisfy the safety of the aircraft as a technical system, and, on the other hand, would not exceed the amount of information that the pilot is able to process per unit time.

Since the pilot manages his attention in a closed loop consisting of a certain group of information sources, an important part of this process is the choice of exactly that group of information sources that most fully describes the current flight phase and the status of the aircraft [13–15]. Some of the sources should display the status of the automation systems, and the other part of the information sources should display the parameters of the flight independent of the automation systems. Thus, when evaluating how well pilots control aircraft in flight, an information assessment method can be used that uses the principle of hierarchical consistency between the selected groups of instruments and the principle of controllability of the parameters included in the groups of instruments selected by the pilot. Let us consider how these two principles of selecting information coming to the pilot can work within one method. It should be noted here that in order to implement the two principles of information selection and processing in flight, the pilot needs to use cognitive converters of activity algorithms. They are the processes of regulating the algorithms of the actions performed by pilots by means of information resources selected according to the criteria of consistency and controllability of flight parameters. Using the information resources of the crew-aircraft-environment system, the algorithms of the pilot's actions are regulated by the pilot through the process of combining the sources being monitored in the algorithm. These sources are long-period and algorithmic flight parameters. The algorithms of the pilot's actions are also regulated by means of gradual strengthening of the control of information sources used in the algorithm. To increase the probability that the pilot will make the right decision concerning whether to take over the autopilot or not, the processes of regulating the algorithms are modeled. Mathematically, the decision-making process is modeled as a comparison between what the pilot should see in a situation and what he can actually see [10, 16–18]. Next, we will demonstrate how the use of cognitive converters of activity algorithms affects the efficiency of how the pilot assesses the status of the crew-aircraft-environment system.

2 Materials and Methods

Let us consider aircraft control on Boeing 737 (see Table 1) in three different flight situations (see Table 1):

1. Mode III (Autopilot, Autothrottle/Flight Guidance): climbing after takeoff, flaps extended;
2. Mode III (Autopilot, Autothrottle/Flight Guidance): activating the Vertical Speed mode (VS) and setting the vertical rate of climb with flaps extended;
3. Mode IV (Autopilot, Autothrottle, data from the Flight Management System (FMS)): the moment of being below or above the attitude profile during descent. In this case, the Autopilot Flight Director (AFDS) tries to maintain the profile parameters without using forward and vertical speed in the calculations.

In the first situation, the crew needs to monitor four information sources: Flight Mode Annunciator (FMA), which checks whether the autopilot operation mode corresponds to the specified one; Primary Engine Display (PED), which demonstrates the key parameters of the engine and is used to compare the operation of the engines with the parameters set in FMS; FMS compares the thrust displayed on PED with the pre-calculated climb thrust; kinesthetic control of the thrust levers to check the change in thrust. However, by applying cognitive converters of activity algorithms to combine the kinesthetic control of the thrust levers and thrust control using the data displayed on PED, the pilot will need to work with three rather than four parameters. Mathematically, the control process will be expressed as a ratio of one to three (see Table 3), and not one to four (see Table 2).

In the second situation, AFDS and the autothrottle (A/T) work separately, maintain their own parameters. AFDS maintains the vertical rate of climb by changing the pitch, and A/T maintains forward speed by changing the mode of operation of the engines. Naturally, in the V/S mode, a conflict situation may arise between the autopilot and the autothrottle, which will lead to the aircraft going beyond critical

Table 1 Types of control modes of a highly automated aircraft for a Boeing 737

I	Hand flown/raw data	Manual piloting with flight directors off/FMA and directors off
II	Hand flown/flight guidance	Manual piloting with flight directors on/FMA control while flying; flight modes are set on the mode control panel (MCP) by the pilot monitoring (PM)
III	Autopilot, Autothrottle/flight guidance	Autopilot mode using AFS with AFDS and autothrottle (A/T) on/mandatory FMA control, where the pilot flying (PF) himself sets the flight mode on the MCP
IV	LNAV, VNAV/flight guidance	Autopilot mode with AFDS and autothrottle (A/T) on/using data entered into the Flight Management Computer (FMC)

Table 2 Parameters of incorrect (F) and correct (T) information presented to the pilot during the specified flight phases without the pilot using cognitive converters

Aircraft status	Flight situation No. 1	Flight situation No. 2	Flight situation No. 3
F	1/4	1/3	1/5
T	3/4	2/3	4/5

Table 3 Parameters of incorrect (F) and correct (T) information presented to the pilot during the specified flight phases with the pilot using cognitive converters

Aircraft status	Flight situation No. 1	Flight situation No. 2	Flight situation No. 3
F	1/3	1/2	1/3
T	3/4	2/3	4/5

flight speeds. According to an expert survey conducted among instructors, the pilot in such situations only controls the compliance of the FMA mode with the specified piloting mode, which usually includes a long-period parameter that depends on an algorithm (for example, the airspeed parameter on the flight display and the airspeed parameter on the autopilot panel). However, if the pilot understands that conflict situations may arise between the control systems of the aircraft and what they can lead to, then he or she will understand that the flight situation may become more complicated and the application of cognitive converters is necessary. Therefore, the pilot needs to use such operations as combination and step-by-step control in order to maintain awareness of the aircraft status in flight and not exceed the limits on the amount of information being processed.

Thus, the probability that the pilot is presented with information requiring intervention if cognitive converters are not used will be one in three (see Table 2), and if they are used, it will be one in two (see Table 3).

In the third situation, the crew needs to control five information sources: wind direction and speed on the navigation display (ND); compliance of the current piloting mode with the one set by FMA; control of the forward speed for increase according to the commands on the FMS display (THRUST REQUIRED or DRAG REQUIRED); kinesthetic control of the position of the thrust levers; monitoring the atmospheric situation based on the comments of the crews in the air. The probability that the information received by the pilot in this mode without the use of cognitive converters does not require intervention in the control will be equal to one in five (see Table 2). And when using cognitive converters, the pilot will control the other four modes after moving to the current flight phase and making sure that the FMA displays the correct autopilot mode. Thus, the probability that the information coming to the pilot in this mode requires intervention in the control will be equal to one in three when the pilot uses cognitive converters (see Table 3).

The parameters of incorrect (F) and correct (T) information presented to the pilot during the specified flight phases without the pilot using cognitive converters and with the use of these converters are shown in Figs. 2 and 3.

There is an equation for calculating the probability of the expected gain of any process related to human activity in the control loop of a technical system [16]:

$$E = p(T)\big[p(t|T)V_t - p(f|T)C_f\big] + p(F)\big[p(f|F)V_f - p(t|F)C_t\big] \quad (1)$$

Let us assume that the probability of gain for the study of the control of a highly automated aircraft will be the efficiency of the pilot's assessment of the aircraft status in flight when controlled in automatic mode. Then the values of the components in Eq. (1) will be as follows:

- $p(T)$—a priori probability that the aircraft will be in the correct status, i.e., not in the status that the pilot set through automation;
- $p(F)$—a priori probability that the aircraft will be in the wrong status, i.e., in the status that the pilot set through automation;
- $p(t|T)$, $p(f|T)$, $p(f|F)$, $p(t|F)$—conditional probabilities f (control intervention) and t (no control intervention) for the corresponding statuses of the aircraft F (different from the set one) and T (set one);
- V_t—gain with the correct non-intervention in the current piloting situation;
- V_f—gain in the case of necessary intervention in the automatic control in the current piloting situation;
- C_f—loss in the case of unnecessary intervention in the automatic control in the current piloting situation;
- C_t—loss due to the inaction of the pilot in the case when it is necessary to intervene in the control.

3 Discussion

In order to find the probabilities $p(f|T)$, and $p(f|F)$, it is necessary to perform a number of transformations with the data presented in Tables 2 and 3. We will first discuss the data in Table 2. Let us reduce all the fractions in the table to a common denominator, which is $g = 60$. Next, we will compile a new table (see Table 4), which will contain data on the values of the numerators of all fractions previously reduced to the common denominator, each written in its own row and column but without a denominator. This table also presents the ratios of these observed values or the so-called cutoff $L(x)$.

The next step is to search for the values of the likelihood ratio criterion (K). To do this, it is necessary to know a priori probabilities $p(T)$ and $p(F)$, as well as gains and losses V_t, V_f, C_f, C_t. Since the pilot's behavior is unknown before any action is performed, it is assumed that $p(T) = p(F) = 0.5$. Gains and losses are distributed in this case as follows: $V_t = 1$, $V_f = 4$, $C_f = 1$, $C_t = 1$. This distribution is explained by the fact supported by the results of expert surveys that when the autopilot system is on, pilots rely on automatic controls when making decisions on performing control actions [19]. Therefore, the gain V_f in the case when there is necessary intervention in the control in the current piloting situation is the key criterion for the safe control

Table 4 Observed values of conditional probabilities for various flight modes, with corresponding flight modes presented in Table 2

Conditional probabilities	Flight situation No. 1	Flight situation No. 2	Flight situation No. 3
$P(x\|F)$	15	20	12
$P(x\|T)$	45	40	48
$L(x) = \frac{P(x\|F)}{P(x\|T)}$	0.33	0.5	0.25

of the aircraft for the pilot. In this case, as in all subsequent cases, reliance on the visual channel is inevitable, and all actions that pilots perform in flight are checked by monitoring instruments and other information sources. Therefore, intervention in control follows only after the pilot has processed information by means of the visual channel. This is why V_f has such a high value.

According to [15], the value of the K-criterion will be as follows:

$$K = \frac{p(T)}{p(F)} \times \frac{V_t + C_f}{V_f + C_t} = \frac{0.5}{0.5} \times \frac{1+1}{4+1} = 0.4.$$

The next step in our calculations is to find the values $p(t|T)$, $p(f|T)$, $p(f|F)$, $p(t|F)$:

$$p(f|F) = \frac{\sum p(x|F)}{g} = \frac{20}{60} = 1/3,\ p(f|T) = \frac{\sum p(x|T)}{g} = \frac{40}{60} = 2/3,$$

$$p(t|T) = 1 - \frac{2}{3} = 1/3,\ p(t|F) = 1 - \frac{1}{3} = 2/3.$$

It should be noted here that the sum of all values $\sum p(x|F) u \sum p(x|T)$, It should be noted here that the sum of all values $L(x) = \frac{P(x|F)}{P(x|T)}$, that is greater than or equal to the value of the K-criterion (see Fig. 1).

Next, let us find the efficiency of assessing the aircraft status by the pilot when performing a flight in the automatic mode using Eq. (1):

$$E = 0.5\left[\frac{1}{3} \times 1 - \frac{2}{3} \times 1\right] + 0.5\left[\frac{1}{3} \times 4 - \frac{2}{3} \times 1\right] \cong 0.20.$$

Let us find the efficiency of the pilot's assessment of the aircraft status when flying in the automatic mode when the pilot uses the processes of cognitive converters of activity algorithms (see Table 5 and Fig. 2).

In accordance with [16], the value of the likelihood criterion will be as follows:

$$K = \frac{p(T)}{p(F)} \times \frac{V_t + C_f}{V_f + C_t} = \frac{0.5}{0.5} \times \frac{2+1}{4+2} = 0.5.$$

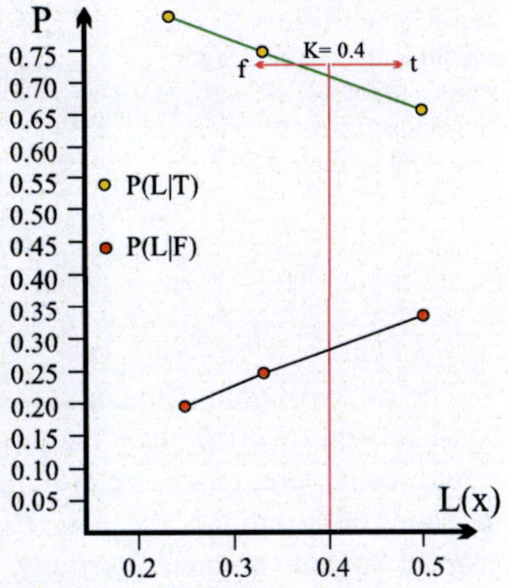

Fig. 1 Distribution of conditional probabilities for likelihood ratios without the pilot's reliance on cognitive converters of activity algorithms

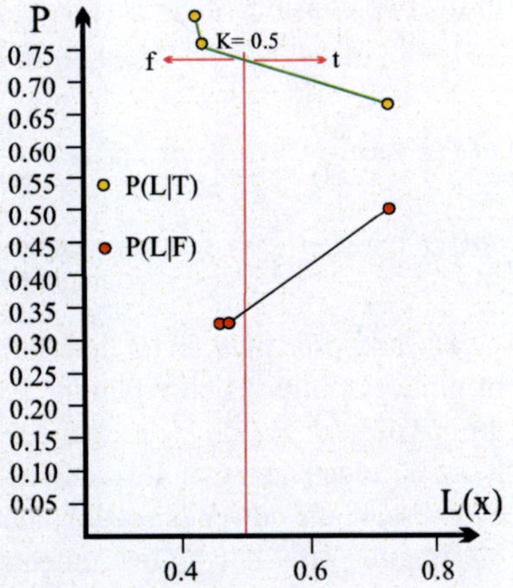

Fig. 2 Distribution of conditional probabilities for likelihood ratios with the pilot's reliance on cognitive converters of activity algorithms

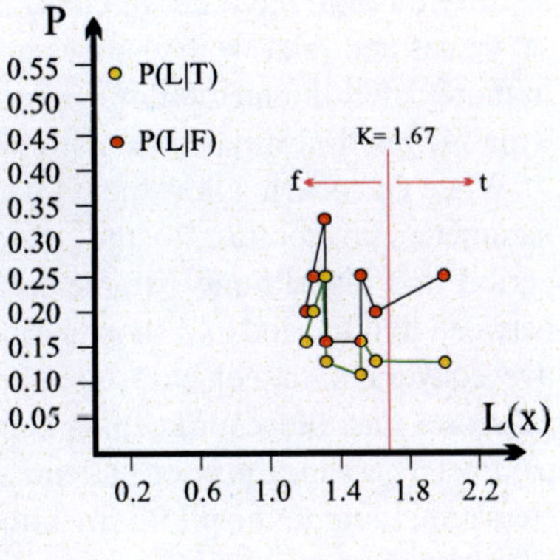

Fig. 3 Distribution of conditional probabilities for likelihood ratios with the pilot's reliance on cognitive binding to current information

Table 5 Observed values of conditional probabilities for various flight modes, with corresponding flight modes presented in Table 3

Conditional probabilities	Flight situation No. 1	Flight situation No. 2	Flight situation No. 3
$P(x\|F)$	20	30	20
$P(x\|T)$	45	40	48
$L(x) = \frac{P(x\|F)}{P(x\|T)}$	0.44	0.76	0.41

The values of gain with correct non-intervention and those of loss with incorrect non-intervention grow since the processes of cognitive converters of activity algorithms, as it were, return the pilot to the control loop. In this case, when assessing the status of the aircraft, the pilot relies not only on the results provided by the visual channel but also on mental activity, which involves a forecast of how the situation will develop. Therefore, in accordance with [19], the risks of non-intervention in any of the two considered cases (C_f, C_t) increase.

Further calculations produce the following result:

$$p(f|F) = \frac{30}{60} = 1/2,\ p(f|T) = \frac{40}{60} = 2/3,\ p(t|T) = 1 - \frac{1}{2} = 1/2,$$
$$p(t|F) = 1 - \frac{2}{3} = 1/3,\ E = 0.5\left[\frac{1}{2} \times 2 - \frac{2}{3} \times 1\right] + 0.5\left[\frac{1}{2} \times 4 - \frac{1}{2} \times 1\right] \cong 0.50.$$

The proposed method for determining the efficiency of assessing the aircraft status by a pilot in flight is based on the static regularity of attention distribution between groups of information sources and certainly affects the quality of piloting. However, it can be added here that the method does not explain how to maintain the pilot's skill of assessing the current situation, although it reflects the laws of interaction between information provided by instruments and other information received in flight.

There is also a pattern in the dynamic behavior of the aircraft, which the pilot can identify through the order of change of certain flight parameters both relative to the set values and relative to changes in other related flight parameters. Let us consider in more detail the method of assessing of the aircraft status taking into account the dynamic relationship between aircraft flight parameters.

When considering how the pilot assesses the aircraft status by means of dynamic parameters in contrast to the static assessment, the relationship between long-period and algorithmic parameters is no longer considered, but the relationship between long-period and short-period parameters is considered. In this case, cognitive converters cannot be used. Therefore, the pilot needs to use other cognitive processes that allow for keeping control over the flight parameters when any of the parameters escape the pilot's attention by comparing the rate of change of the parameters remaining in the pilot's attention focus. Such processes are called processes of cognitive binding to current information.

The dynamic approach is based on the pilot's assessment of the status of a highly automated aircraft through monitoring the relationship of the rate of change in aircraft flight parameters. There are two such rates of parameter change for each source of information. One of them is to the difference between the current change in the observed parameter and its set value, and the other is the change in the observed value of the parameter relative to another dependent parameter. For example, the rate of speed change depending on the amount of power supplied to the engines can have different values. Therefore, if the pilot sets aircraft engine power in the manual mode, he or she expects that there will be a corresponding rate of increase in speed. However, in automatic modes with kinesthetic control, it is difficult for the pilot to track whether the change in speed is as predicted because the pilot's processes of perception are distorted due to overload and control is performed by the automation system. Therefore, the control in the autopilot mode should be guided by information coming from indirect indicators. For monitoring speed, such indicators apply as the rate of change of speed on the PFD depending on the engine speed value demonstrated on the PED and the rate of change of pitch (or angle of attack) on the PFD depending on the change in speed.

Thus, the processes of cognitive binding to current information are reduced to the detection of signals coming to the pilot and are based on determining and comparing two parameters:

- the rate of change of the parameter being monitored relative to its own set value;
- the rate of change of the parameter being monitored relative to another, which is most closely related to it.

Let us consider several flight operations: maintaining speed and altitude in level flight, maintaining speed and altitude in climb or descent, and maintaining heading at various flight phases. The conditions for maintaining the flight speed at a constant for the case of determining the correct aircraft status (F) are as follows: (1) constant pitch; (2), (3) change in pitch to the side contributing to an increase or decrease in speed that tends to a given value; (4), (5) increase or decrease in engine speed in the direction corresponding to the specified speed value. The conditions for maintaining the flight speed at a constant for the case of determining the incorrect aircraft status (T) are as follows: one more parameter is added to the parameters for determining the correct aircraft state (F)—a constant value of engine speed (see Table 6). The remaining parameters of the modes considered in the table are determined according to the same principle. In order to develop a mathematical model of the conditions for assessing the aircraft status using cognitive converters, it is first necessary to represent the process as a fraction whose numerator is one and whose denominator is the number of conditions that determine the aircraft status at the current time. The unity in the numerator means that the pilot determines the aircraft status by one of the selected parameters in the denominator of the fraction (see Table 6).

Further calculations are performed according to Eq. (1). The results of reducing the values of the characteristics of incorrect (F) and correct information (T) to a common denominator and finding the definition of the cutoff $L(x)$ are presented in Table 7. The distribution of conditional probabilities for likelihood ratios in the case of using the processes of cognitive binding to current information is shown in Fig. 3. The a priori probabilities of both incorrect and correct states $p(F)$, $p(T)$ were chosen

Table 6 Parameters of incorrect (F) and correct (T) information presented to the pilot when performing specified flight operations using cognitive binding to current information

Aircraft status	Maintaining speed in level flight		Maintaining altitude in level flight		Maintaining speed in climb or descent		Maintaining altitude in climb or descent		Maintaining heading at various flight phases	
	V const	V ↓ or ↑	H const	H ↓ or ↑	V const	V ↓ or ↑	H const	H ↓ or ↑	hdgl const	hdg ↓ or ↑
F	1/5	1/4	1/5	1/4	1/6	1/4	1/6	1/4	1/4	1/3
T	1/6	1/5	1/6	1/6	1/9	1/8	1/9	1/8	1/5	1/4

equal to 0.25 due to the high load on the visual channel when assessing changes in flight parameters. Gains and losses, in this case, are distributed as follows: $V_t = 3$, $V_f = 2$, $C_f = 2$, $C_t = 1$. This distribution is explained by the fact that the loss in the case of unnecessary intervention in automatic control is equal to the gain in the case of necessary intervention in automatic control since the dynamic assessment of the current information equates these two processes as they do not require an element of prediction.

On the contrary, the gain from correct non-intervention implies a forecast of changes in the parameter being monitored, which makes it bigger than the loss from non-intervention. Therefore, the likelihood criterion is:

$$K = \frac{p(T)}{p(F)} \times \frac{V_t + C_f}{V_f + C_t} = \frac{0.25}{0.25} \times \frac{3+2}{2+1} = 1.67.$$

Further calculations produce the following result:

$$p(f|F) = \frac{180}{360}, p(f|T) = \frac{90}{360}, p(t|T) = \frac{180}{360}, p(t|F) = \frac{270}{360}$$

$$E = 0.25\left[\frac{270}{360} \times 3 - \frac{90}{360} \times 2\right] + 0.25[0.5 \times 2 - 0.5] \cong 0.81.$$

Table 7 Conditional probabilities and likelihood ratios for various flight situations, with their mode correspondence probabilities presented in Table 6

Conditional probabilities	Maintaining speed in level flight		Maintaining altitude in level flight		Maintaining speed in climb or descent		Maintaining altitude in climb or descent		Maintaining heading at various flight phases	
	V const	V ↓ or ↑	H const	H ↓ or ↑	V const	V ↓ or ↑	H const	H ↑ or ↑	hdg cons t	hdg ↓ or ↑
P(x|F)	72	90	72	90	60	90	60	90	90	120
P(x|T)	60	72	45	60	40	45	45	45	72	90
L(x)	1.20	1.25	1.6	1.5	1.5	2	1.33	2	1.25	1.33

4 Results

It follows from the calculations that the efficiency of the method for assessing the aircraft status by the pilot when piloting in automatic mode without relying on cognitive converters of activity algorithms is approximately 20%. The efficiency of the pilot's assessment of the aircraft status when piloting in automatic mode using the processes of cognitive converters of activity algorithms increases by approximately 30% relative to the situation in which cognitive converters are not used. The efficiency of the pilot's assessment of the aircraft status when piloting in automatic mode using the processes of cognitive binding to current information increases by approximately 30% relative to when cognitive converters are used.

5 Conclusion

The results of the theoretical study presented in this article show that the efficiency of assessing the status of the aircraft and its automated control systems by means of indirect indicators increases significantly when using methods that include the processes of cognitive converters of activity algorithms and processes of cognitive binding to current information. This allows for combining the static and dynamic parameters of the assessment of the current flight situation.

However, the application of cognitive converters and cognitive binding processes requires that pilots have an integral skill needed to operate modern highly automated aircraft. In order to develop such an integral skill in the area of processing static and dynamic information in flight, it is necessary to design and apply new flight training methods.

References

1. Miroshnichenko AV (2018) Podderzhanie navykov ruchnogo pilotirovaniia VS A320 pri vkliuchennom avtopilote (Maintaining manual piloting skills A320 with autopilot enabled) (In Russian). Izdatel'skie resheniia, Moskva, 15 p
2. Megson T (2021) Aircraft structures for engineering students. Burlington, USA. https://doi.org/10.1016/C2019-0-03113-5
3. Watkinson J (2004) Art of the helicopter. Butterworth-Heinemann, UK. https://doi.org/10.1016/B978-0-7506-5715-0.X5000-5
4. Guoqing W, Wenhao Z (2020) The principles of integrated technology in avionics systems. Academic Press, Shanghai. https://doi.org/10.1016/C2018-0-00489-2
5. Stewart M, Mueller J (2018) Are we safe enough? Measuring and assessing aviation security. Elsevier, UK, USA. https://doi.org/10.1016/C2016-0-01215-9
6. Rosenbaum D (2009) Human motor control. Academic Press, USA. https://doi.org/10.1016/B978-0-12-374226-1.X0001-0

7. Kletz T, Amyotte P (2019) What went wrong? Case histories of process plant disasters and how they could have been avoided. Butterworth-Heinemann, UK. https://doi.org/10.1016/B978-0-12-374226-1.X0001-0
8. Price J, Forrest J (2017) Practical aviation security. Predicting and preventing future threats. Butterworth-Heinemann, UK
9. Modern cloud services: key trends, models and tools for interactive education. In: Integrating engineering education and humanities for global intercultural perspectives proceedings of the conference "integrating engineering education and humanities for global intercultural perspectives", 25–27 March 2020, St. Petersburg, Russia. Conference proceedings IEEHGIP 2020, pp 883–890
10. Muravyev IS, Kovalenko GV (2017) Metod obucheniia pilotov vertoletov navykam bezopasnoi posadki vne aerodroma v usloviiakh neopredelennosti (Helicopter pilots safe landing outside the airfield in conditions of uncertainty skills training methods) (In Russian). Nauka, St. Petersburg
11. Fadali M, Visioli A (2013) Digital control engineering. Analysis and design. Academic Press, New York
12. Latash M (2016) Biomechanics and motor control. Defining central concepts. Academic Press, USA. https://doi.org/10.1016/B978-0-12-374226-1.X0001-0
13. Datta B (2004) Numerical methods for linear control systems. Design and analysis. Academic Press, UK, USA.https://doi.org/10.1016/B978-0-12-203590-6.X5000-9
14. Hetherington S (2006) Aspects of knowing. A volume in perspectives on cognitive science. Elsevier Science, Sydney. https://doi.org/10.1016/B978-0-12-203590-6.X5000-9
15. Wiener E, Nagel D (1988) Human factors in aviation. A volume in cognition and perception. Academic Press, USA
16. Sheridan T, Ferrell W (1974) Man-machine systems: information, control, and decision models of human performance. The MIT Press, London
17. Wilson M (2015) Implementation of robot systems. An introduction to robotics, automation, and successful systems integration in manufacturing. Butterworth-Heinemann, USA. https://doi.org/10.1016/B978-0-12-203590-6.X5000-9
18. Jaulin L (2015) Mobile robotics. ISTE-Elsevier, USA. https://doi.org/10.1016/C2014-0-004743-0
19. Fishburn P (1970) Utility theory for decision making. Research analysis. New York, London, Sydney, Toronto

Integrated Piloting Skills Training

Gennadii Kovalenko, Yuriy Mikhal'chevskiy, Ivan Muravyev, and Sergey Turintsev

Abstract Operating highly automated aircraft requires transitioning from automatic to manual control and back. In this process, problems can be caused due to information overload, changes in the operation mode and the pilot's being exposed to forces in excess of 1 G (Kletz and Amyotte in What went wrong? Case histories of process plant disasters and how they could have been avoided. Butterworth-Heinemann, UK, 2019 [1]; Megson in Aircraft structures for engineering students. Burlington, USA, 2021 [2]; Guoqing and Wenhao in The principles of integrated technology in avionics systems. Academic Press, Shanghai, 2020 [3]; Muravyev and Kovalenko in Helicopter pilots safe landing outside the airfield in conditions of uncertainty skills training methods. Nauka, St. Petersburg, 2017 [4]). The existing method for training pilots using cognitive converters of activity algorithms does not take into account the way how pilots control the dynamic characteristics of the aircraft in flight. The article proposes a new method that allows the pilot to develop an integral skill that takes into account both static and dynamic parameters of the aircraft. The relevance of the study stems from the fact that it is vital for pilots to know how to backup automated systems in modern aircraft.

Keywords Learning method · Skill · Information processing · Stimulus · Pilot command

1 Introduction

Despite the fact that more and more automation technologies are introduced in the aviation sector as a response to more complex operating conditions and the need to ensure flight safety, whether landing will be successful or not still largely depends on

G. Kovalenko (✉) · Y. Mikhal'chevskiy · I. Muravyev
Saint Petersburg State University of Civil Aviation, Saint Petersburg, Russia
e-mail: kgvf@inbox.ru

S. Turintsev
Irkutsk Branch of Moscow State Technical University of Civil Aviation, Irkutsk, Russia

O. A. Gorbachev et al. (eds.), *Proceedings of 10th International Conference on Recent Advances in Civil Aviation*, Lecture Notes in Mechanical Engineering, https://doi.org/10.1007/978-981-19-3788-0_28

pilots and their skills. As the tasks that the crew-aircraft system has to deal with are constantly becoming more difficult, the process of fitting this system with necessary cognitive and information resources does not keep up with this pace. There are two main reasons why aircraft operation is now significantly more difficult than it used to be.

The first reason is that the functions assigned both to pilots and to automated aircraft systems are becoming more complex and incorporated into the algorithms of machine actions. This, in turn, requires aircraft crews to perform more complex mental operations. Pilots also have to perform a wide range of manual operations [5].

The second reason is that not all operations can be entrusted to the automation system. It means that pilots have to switch to manual control quite often.

To assess the efficiency of the crew-aircraft system, to analyze its performance, and to ensure safe flights, it is necessary to analyze manual control and automatic manual in terms of the interaction between them.

An analysis of air accidents in global aviation shows that the two most common types of accidents are controlled flight into terrain (CFIT) and loss of control inflight (LOC-I). These are the weak points in which air safety could be improved [6].

Modern approaches to the study of human-machine systems provide for a deeper understanding than before of the forms and structures of the controls performed by the operator in the aviation sector. However, there is a lack of both algorithms for pilots and methods of engineering and cognitive research in issues connected with the crew-aircraft system. As a result, it creates a subjective perception of the principles of control in today's aircraft and results in flawed training methods. This, in turn, leads to conflicts between the crew and onboard systems due to the inconsistency of their actions in emergencies.

Despite the high level of automation on board of modern aircraft, the backup function remains with the pilot [7], and the task for researchers is to find ways to maintain it at a higher level when the pilot is out of the control loop. To do this, when training pilots, it is necessary to develop their cognitive functions, which will enable them to backup intelligent control systems more efficiently when being out of the control loop. In order for the pilot to successfully perform this operation, it is necessary to use level F of brain cognitive-affective organization, which is one of the six levels within the Grand Design model [4]. According to this model, mental activity can be divided into elements of different levels. Elements of levels A, B, and C are the simplest. They are traditionally called sensations. Level D and partly level C refer to the perceptions of space (C) and actions (D). These mechanisms of regulating the subject's consciousness are responsible for its dynamic localization and identification of the objects surrounding it. Subsequent levels D, E, and F require using memory and its multiple subsystems. The highest levels of cognitive regulation (E and F) mainly involve thinking and imagining. They do not operate on their own but are based on a combination of contributions from the lower five levels. For us, it is of interest how consciousness functions according to this model because the pilot who controls the automated aircraft is not only a control element of the system but also a backup element. Also, any pilot of an automated aircraft should be able to assess the performance of its systems not only by its individual parameters (at levels A, B, and C), but also by forming a comprehensive understanding of the complex interaction between both intra-system factors and factors affecting the system from the outside. The regulation of such activity is possible only at levels E and F. The problem is that in this model, the range of functional states is wider than that of their forms that are used to try to solve cognitive problems. As already noted, levels E and F do not exist on their own but send to the subject's consciousness an image of the situation which is based on a combination of contributions from mechanisms located at different levels (A, B, C, D). By utilizing the pilot's metacognitive function during training, new skills can be developed for operating aircraft. In this case, the pilot will solve the problem better than when using any other level of regulation of consciousness (D, E, etc.), because when there are random equipment failures, pilots need to not only know how to act in such a situation but also be creative in their thinking to find the fastest and most effective way out of an emergency. This ability to avoid an emergency can be formed in a pilot during training when triggering level F.

The decrease in the amount of information coming from the pilot to the automated aircraft [8] did not reduce the load on the pilot himself. At the moment, the understanding of the motor-reflex principle of aircraft control as a fundamental one is now being reconsidered. With aircraft moving up the levels of automation all the time, it is required to develop new methods for modern aircraft controlling in a visual-cognitive manner.

Extrapolating the results of cognitive research in related areas onto the aviation sector is associated with some problems. The following are the key ones:

- the extreme complexity and irreversibility of the process of professional training of pilots and the development of aircraft control skills;
- the need to combine skills in controlling various types of aircraft;
- the individuality of each pilot in terms of the cognitive parameters of their perception and information processing both in the process of professional training and in flight.

Based on the study of the experience of official flight tests and the introduction of automated aircraft in the process of professional training both in state and civil aviation, it was concluded that the methods used do not fully take into account the features of the functioning of an automated aircraft and do not make it possible to shape a full picture of how pilots act in modern aircraft [4–8].

The imperfections of methods and tools for predicting the behavior of the crew-aircraft system hinder training and retraining of crews for using modern types of aviation equipment. It can be concluded that not enough attention is paid to cognitive models for managing automated systems in aviation.

There are various methods that help to develop piloting skills to be used in highly automated aircraft. One of these methods is using cognitive converters of activity algorithms [4, 9–12]. This method allows for developing the skill of understanding the interaction between the parameters of the instrument panel and other information without losing the ability to assess the current situation. The method is as follows.

When controlling flight modes, the use of cognitive converters of activity algorithms relies on the following:

- repeated information checks using the instrument panel;
- gathering parameters from various sources of information and repeating information checks using the instrument panel;
- combining information without repeatability while controlling information from the instruments;
- consolidating information coming to the pilot.

The method we developed for teaching flight crews how to make decisions on pilot controls includes several instruments used sequentially. The method itself is designed as follows. Pilots being trained need to be presented with pairs of images reflecting different flight situations. In one situation for each pair of images, the pilot needs to recognize that he or she has to perform some controls. In another situation for the same pair of images, the pilot needs to recognize that no intervention is required. First, it is necessary to present the pilot with paired images with two sources of information during three tests. Next, it is necessary to present paired images with three sources of information during four tests. Then it is necessary to present paired images with four sources of information for five tests, and then present paired images with five sources of information for fourteen tests.

Such an arrangement of sources from test to test allows the pilot as a subsystem in the aircraft to learn the regularity of the distribution of parameters in the selected groups of instruments and correctly analyze the proposed situation, as suggested in [10–16]. This, in turn, makes it possible to make the processes of information control repeatable. With information consolidated in various combinations, the pilot develops the skill of identifying the kind of situation presented. Using a mathematical model and the method of cognitive converters of activity algorithms shows that such training can improve the efficiency of controlling a highly automated aircraft up to approximately 40%. To do this, it is necessary to conduct 25 tests during the training course.

The method under consideration helps to understand how different parameters interact, but it does not help pilots improve aircraft control. This method is suitable for teaching the pilot how to monitor the functioning of the automation during flight, but it is not well suited for the pilot to perform a successful return to manual control in the event of an automatic failure or in the event of a sudden need to pilot the aircraft in manual mode. This requires a deep assessment by the pilot of the process itself, which means not only understanding the relationship between parameters, but also the process of managing these parameters. In other words, the method for determining the laws of interaction between the parameters of the instrument panel and other information is a static parameter of the pilot's activity. However, for a safe transition to manual piloting, a dynamic parameter of the pilot's activity is necessary. It can be added that the method does not determine how to maintain the pilot's ability to assess the current situation, although it determines the laws of interaction between instrument data and other information in flight.

The method also shows how it is necessary to arrange operating procedures based on the resources of the pilot as a biological system performing technical tasks. The method also determines the principles of understanding the processes occurring around the pilot and the procedure for the pilot to perceive instrumental data and other information. This method is more focused on the controller. However, for the safe and efficient management of such a complex system as the aircraft-environment system, it is the object being controlled that needs to be analyzed. This method does not offer tools for such an analysis.

Therefore, it is necessary to consider another method that will improve the efficiency and safety of controlling a highly automated aircraft through the use of the dynamic parameters that instruments and other information sources present to the pilot.

2 Materials and Methods

Any human activity, be it flight control or any other, consists of periods that end with the adoption of decisions and subsequent actions or inactions. Let us assume that the pilot's decision-making process at each of the minimum periods of his activity is characterized by the states in which he can be and the probabilities that precede the occurrence of these states. Each minimum period of the pilot's activity, which is shown in Fig. 1, consists of the following states (see Fig. 1):

- $C_{i,n}$—the state in which the pilot's activity is at the present moment;
- $A_{1,n}$—the state of the pilot after making a decision to perform some pilot control;
- $A_{2,n}$—the state of the pilot after making a decision not to intervene;
- $E_{1,n}$—reinforcing event showing that it was necessary to intervene ($A_{1,n}$);
- $E_{2,n}$—reinforcing event showing that it was not necessary to intervene the automatic control and that the flight mode corresponded to the specified one;
- $C_{i-1,n+1}$—the state in which the pilot's activity was during the previous minimum period;
- $C_{i-1,n+1}$—the state in which the pilot's activity is during the next minimum period.

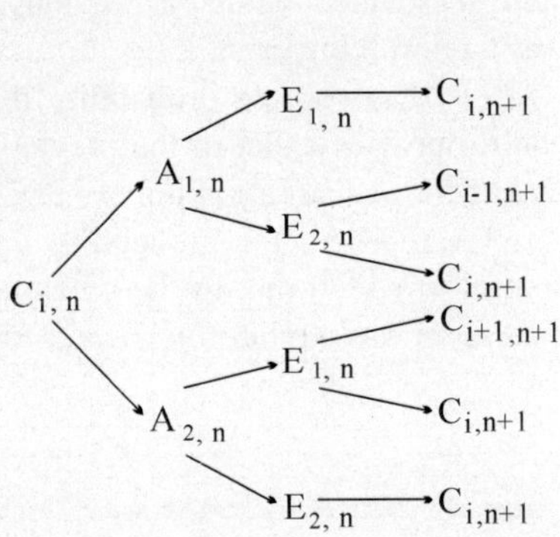

Fig. 1 Tree of pilot activity states at minimum segment

Also, all states of the minimum period of the pilot's activity are characterized by the following probabilities of occurrence:

- $\frac{i}{N}$—the probability that the pilot will make a decision to perform some control (i is the number of elements influencing the pilot's intervention in control, N is the total number of stimulus elements in the form of sources of information presented to the pilot);
- $\frac{1-i}{N} - N$—the probability that the pilot will decide not to interfere with the control ($1 - i$ is the number of elements that affect this decision);
- π is the probability that a reinforcing event E_1 will occur;
- $1 - \pi$ the probability that a reinforcing event E_2 will occur;
- c the probability that the reinforcing event E_1 will change the state of the process E_1 (into C_i, C_{i+1} or C_{i-1});
- $1 - c$—the reciprocal of c.

There is a mathematical model for predicting successive statistics for the final tests in the sequence [14]:

$$P(A_1|E_1\ A_1) = \pi(1 - 1/N) + 1/N, \tag{1}$$

$$P(A_1|E_2\ A_1) = \pi(1 - 1/N) + (1 - c)/N, \tag{2}$$

$$P(A_1|E_1\ A_2) = \pi(1 - 1/N) + c/N, \tag{3}$$

$$P(A_1|E_2\ A_2) = \pi(1 - 1/N) \tag{4}$$

The interpretation of the presented model in the framework of studying how pilots make decisions will be as follows:

$P(A_1|E_1A_1)$—the probability that the pilot will make a decision to perform control provided that in the previous period the pilot intervened, and after that, there was a reinforcing event E_1;

$P(A_1|E_2A_1)$—the probability that the pilot will make a decision to perform control provided that in the previous period the pilot intervened, and after that, there was a reinforcing event E_2;

$P(A_1|E_1A_2)$—the probability that the pilot will make a decision to perform control provided that in the previous period the pilot did not intervene, and after that, there was a reinforcing event E_1;

$P(A_1|E_2A_2)$—the probability that the pilot will make a decision to perform control provided that in the previous period the pilot did not intervene, and after that, there was a reinforcing event E_2.

3 Discussion

When the autopilot system is on, the pilot has two types of reactions: A_1—to intervene, and A_2—not to intervene. Also, there are N stimulus elements in flight, one of which the pilot relies on to make a decision in the minimum period of his activity. Also in flight, after making a decision, the pilot receives a reinforcement based on visual feedback: either E_1 that it was necessary to use the reaction A_1, or E_2 that it was necessary to use the reaction A_2. This reinforcement for the pilot in flight will be a change (or no change) in the parameter that he chose as the key one.

If π is the probability of occurrence of a reinforcing event E_1, then this probability will consist of several elements. Let us consider the automatic flight control mode. We shall assume that the total number of stimuli N is ten. Imagine the pilot has performed some actions and is now in a new minimum period of activity at this stage of the flight. The probability that the pilot in the next minimum period of activity will make a decision to intervene will be expressed by one of the Eqs. (1–4). On the way to the state of making a decision shown in Fig. 1, the pilot faces two steps: reinforcement recognition E_1 or (E_2), occurring with probability π or $\pi - 1$, and system state recognition, C_i, C_{i+1} or C_{i-1}, occurring with probability c or $c - 1$. The first step, or the probability of the transition process in the considered minimum period of the pilot's activity, is expressed through cognitive converters of activity algorithms. The use of these converters at this stage makes it possible to correctly select a group of information sources that need to be monitored, and to select the order and sequence of control of these sources. Mathematically, this will be expressed in determining the number of sources that are in the denominator and the numerator of the fraction $\pi = \frac{l}{n}$, where l is the number of sources that the pilot must control, and is the total number of sources in the group selected by the pilot. The second step, or the probability of a transient process in the considered minimum period of the pilot's activity, is expressed through the use by the pilot of the processes of cognitive binding to current information. Mathematically, this will be expressed in determining the number of sources that are in the denominator and the numerator of the fraction $c = \frac{k}{d}$, where k is the number of flight parameters that the pilot must monitor, and d is the total number of sources in the group selected by the pilot. Cognitive binding occurs in the group of information sources selected through the use of cognitive converters, and this binding allows the pilot to track changes in flight parameters, which will further affect the decision. The integral piloting skill means combining these processes in accordance with the mathematical model (1–4).

It is necessary to determine how to develop this skill in a pilot being trained. To do this, it is necessary to determine how many sources of information (N, l, n, k, and d) and in what ratio, in accordance with the model (1–4), should interact in the pilot's field of active attention during the training process.

The study of how pilots process various amounts of incoming information in its various combinations showed the following results:

1. With a general (or individual) increase in the amount of information coming to the pilot N, the number of sources in the group selected by the pilot, the number

of parameters the pilot monitors in flight, and if there is a direct relationship between them, the probability of the pilot making a decision to intervene in any of the conditions (E_1A_1, E_2A_1, E_1A_2, E_2A_2) decreases.

2. The maximum value of the probability of the pilot making a decision to intervene in any of the conditions (E_1A_1, E_2A_1, E_1A_2, E_2A_2) occurs with an increase in the number of information sources that the pilot includes in processing by means of cognitive converters and which is excluded from the process when forming control groups with a static assessment of the current situation ($\pi = 2/5$) (see Fig. 2).
3. The probability that the decision to interfere will be effective is the greatest when the graph changes its direction (see Fig. 3), where under the conditions E_1A_1 and E_2A_1 the probability of making the decision to interfere has maximum values, with minimum values at E_2A_1 and E_2A_2. This occurs when the number of parameters between which the pilot identifies a direct relationship increases, and this relationship is taken into account when making a decision.

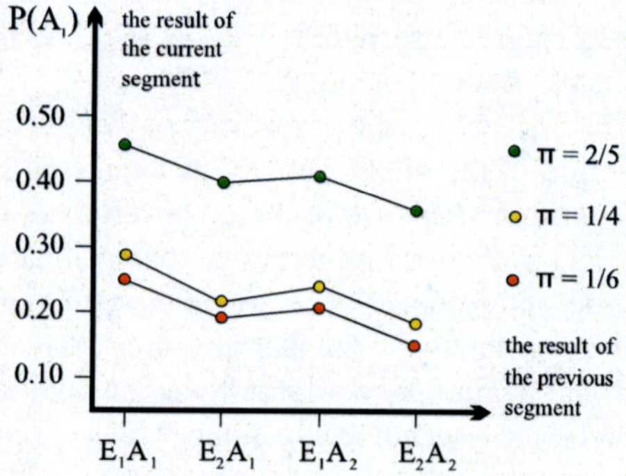

Fig. 2 Predicted values of sequential statistics for the minimum segments of the pilot's activity with a variable value of π, with $N = 10$ and $c = 3/5$

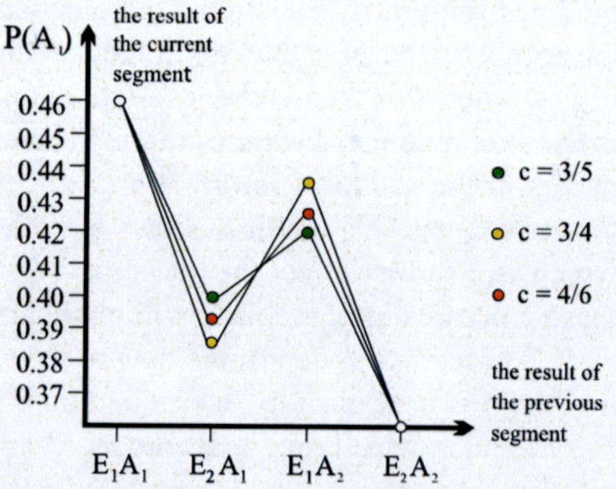

Fig. 3 Predicted values of sequential statistics for the minimum segments of the pilot's activity with a variable value of c, with $N = 10$ and $\pi = 2/5$

4 Results

The results of the study allow for making several conclusions. For piloting modern highly automated aircraft successfully, the pilot must have an integral skill in processing the information that comes in flight. The essence of this skill lies is using dynamic information processing along with static information processing. The method of processing dynamic information consists in mastering the understanding of the relationship between long-period flight parameters and short-period flight parameters. The method for developing the integral piloting skill consists of two parts and includes going through two steps in the process of training. At the first step of training, the pilot needs to master the following processes connected with the cognitive converters of activity algorithms:

- repeated information checks using the instrument panel;
- gathering parameters from various sources of information and repeating information checks using the instrument panel;
- combining information without repeatability while controlling information from the instruments;
- consolidating information coming to the pilot.

At the first step of training, the pilot needs to master the following processes of cognitive binding to current information:

- processes of interconnection between the parameters of long-period sources of information and the parameters of short-period sources of information;
- processes of replacing one parameter with another from short-period sources of information when the same parameter is influenced by a long-period source of information.

5 Conclusion

In order to use cognitive converters of activity algorithms and cognitive binding to current information to the greatest efficiency possible, it is necessary to apply them consistently in such a way that the probability of the decision being made is greatest; the inflection of the graph (see Fig. 3) has maximum values under the conditions of the decision made in the previous minimum period of the pilot's activity to intervene with positive reinforcement or the decision made not to intervene with negative reinforcement; it has minimum values if the decision in the previous period of activity was to intervene with negative reinforcement, or not to intervene with positive reinforcement in the previous periods of activity. Such efficiency, as shown in Fig. 1, is achieved through the use of the following features of cognitive converters of activity algorithms and the cognitive binding processes:

- exclusion from the process of information processing a number of sources that do not affect the course of assessing the state of the aircraft in flight when forming control groups during a static assessment of the current situation;
- increasing the number of aircraft flight parameters between which the pilot identifies a direct relationship and takes it into account when making a decision.

Thus, in order to develop an integral skill among pilots for processing incoming information when flying modern highly automated aircraft, it is necessary to apply the following method in the learning process, which includes two steps. In the first step of training, the pilot needs to master the following processes connected with cognitive converters of activity algorithms:

- repeated information checks using the instrument panel;
- gathering parameters from various sources of information and repeating information checks using the instrument panel;
- combining information without repeatability while controlling information from the instruments;
- consolidating information coming to the pilot.

At the first step of training, the pilot needs to master the following processes of cognitive binding to current information:

- processes of interconnection between the parameters of long-period sources of information and the parameters of short-period sources of information;
- processes of replacing one parameter with another from short-period sources of information when the same parameter is influenced by a long-period source of information.

References

1. Kletz T, Amyotte P (2019) What went wrong? Case histories of process plant disasters and how they could have been avoided. Butterworth-Heinemann, UK. https://doi.org/10.1016/C2016-0-0118-3
2. Megson T (2021) Aircraft structures for engineering students. Burlington, USA.https://doi.org/10.1016/C2019-0-03113-5
3. Guoqing W, Wenhao Z (2020) The principles of integrated technology in avionics systems. Academic Press, Shanghai. https://doi.org/10.1016/C2018-0-00489-2
4. Muravyev IS, Kovalenko GV (2017) Metod obucheniia pilotov vertoletov navykam bezopasnoi posadki vne aerodroma v usloviiakh neopredelennosti (Helicopter pilots safe landing outside the airfield in conditions of uncertainty skills training methods). Nauka, St. Petersburg (In Russian)
5. Salas E, Marino D (2010) Human factors in aviation. Academic Press, USA. https://doi.org/10.1016/C2009-0-01731-0
6. Price JC, Forrest JC (2017) Practical aviation security. Predicting and preventing future threats. Butterworth-Heinemann, USA
7. Kanki BG, Jose A, Chidester TR (2019) Crew resource management. Academic Press, USA. https://doi.org/10.1016/C2016-0-03953-0

8. Sgobba T, Kanki B, Clervoy J, Sandal GM (2018) Space safety and human performance. Butterworth-Heinemann, USA. https://doi.org/10.1016/C2016-0-00319-4
9. Miroshnichenko AV (2018) Podderzhanie navykov ruchnogo pilotirovaniia VS A320 pri vkliuchennom avtopilote (Maintaining manual piloting skills A320 with autopilot enabled). Izdatel'skie resheniia, Moskva (In Russian)
10. Kovalenko GV, Smurov MY, Arinicheva OV, Balyasnikov VV, Chepiga VE (2017) Possible solutions to reduce the negative impact of human factors on flight safety. J Ind Pollut Control 33(1):1195–1201
11. Kovalenko G, Barabanov M, Balyasnikov V, Smurov M, Chepiga V (2017) Experimental validation for the training method and mathematical model of the pilot skill formation in maintenance of attitude orientation. Transp Prob Int Sci J 12(4):127–140
12. Miroshnichenko AV (2017) Pilotirovaniie samoleta A320 (A320 aircraft piloting). Izdatel'skie resheniia, Moskva (In Russian)
13. Fadali M, Visioli A (2013) Digital control engineering. Analysis and design. Academic Press, New York
14. Modern cloud services: key trends, models and tools for interactive education. Integrating engineering education and humanities for global intercultural perspectives proceedings of the conference "integrating engineering education and humanities for global intercultural perspectives", 25–27 March 2020, St. Petersburg, Russia. Conference proceedings IEEHGIP 2020, pp 883–890
15. Atkinson RC, Bower GH, Crothers EG (1965) An introduction to mathematical learning theory. Stanford University, New York, London, Sydney
16. Erokhin V (2021) Bi-criteria aircraft trajectory optimization in implementing the area navigation concept. Int J Aeronaut Space Sci 22(4):948–962. https://doi.org/10.1007/s42405-021-00353-3

Emergency Performance Assessment in Air Traffic Control

Aleksei Malishevskii, Igor Krivoborsky, Alexander Khumorov, and Sergey Vorobyov

Abstract The article discusses the influence of the human factor in air traffic control on flight safety. Errors made by air traffic controllers in emergencies are explained primarily by lack of professionalism or low stress resistance. Since psychological testing is the first step in screening out candidates unfit for the job, the task of identifying candidates with low performance in emergencies is important and relevant. In order to study methods for assessing emergency performance, an experiment was conducted using the expert and navigator simulators. Also, to analyze the innate component of stress resistance in the participants, personality assessment tests were carried out in order to determine their neuroticism scores. The acquired component of emergency performance was assessed indirectly by comparing how accurately the participants evaluated the risks of potential conflict situations with expert assessments. An analysis was performed of the correlations between experimental results. The experiment was conducted on a sample of 20 fifth-year students at Saint Petersburg State University of Civil Aviation who major in air traffic management.

Keywords Desired professional attributes · Psychodiagnostics · Emergency performance · Temperament · Air traffic control · Stress resistance

1 Introduction

It is well known [1, 2] that the human factor is the key reason negatively affecting flight safety. It is the human factor that is the cause of the majority of accidents and incidents (about 80%). It is primarily associated with aircraft crews, but in many cases, air traffic controllers (ATCs) are also responsible. "One of the reasons of aircraft proximity and the resulting conflicts, violations of separation standards, near

A. Malishevskii (✉) · I. Krivoborsky · A. Khumorov
Saint Petersburg State University of Civil Aviation, Saint Petersburg, Russia
e-mail: 9909395@bk.ru

S. Vorobyov
Irkutsk Branch of Moscow State Technical University of Civil Aviation, Irkutsk, Russia

O. A. Gorbachev et al. (eds.), *Proceedings of 10th International Conference on Recent Advances in Civil Aviation*, Lecture Notes in Mechanical Engineering, https://doi.org/10.1007/978-981-19-3788-0_29

midair collisions, and collisions in the air is the presence of points of intersection and convergence of both aircraft traffic flows and individual aircraft trajectories. In these points, aircraft proximity situations can occur randomly, creating dangers" [3].

"Another reason for conflicts, violations of separation standards, near midair collisions, and collisions in the air are accidental deviations of the aircraft from the flight path. The third reason is aircraft deviations from flight paths due to errors made by air traffic controllers and aircraft crews, aircraft failures, or abnormal circumstances in flight" [3].

First of all, errors made by ATCs in emergencies result from their lack of professionalism and/or low stress resistance.

There are a lot of examples of such errors. In 2001, the Boeing 747-446D Domestic was operating a flight from Tokyo to Naha at flight level 390. Meanwhile, the DC-10-40D aircraft was flying across its flight path at flight level 370, performing an international flight from Busan to Narita. The aircraft was to separate at a safe altitude of 2000 ft. At the Tokyo Area Control Center in Tokorozawa, a 26-year-old trainee who was also working with a dozen other flights saw the intersecting trajectories of Flights 907 and 958 on the screen and instructed Flight 958 (DC-10) to descend to FL 350. However, he made a mistake and called the wrong number–907 (B747). A little later, noticing that Flight 958 was still moving at the same altitude, the trainee gave the command to this crew to turn right. On board Flight 958, this command was not heard. Realizing the danger of the situation, the supervisor of the trainee gave the command "957, start ascent", although no Flight 957 was seen on the radars at that moment. Meanwhile, the TCAS aboard the B747 (Flight 907) began commanding the climb while commanding the DC-10 (Flight 958) pilots to descend. However, the B747 pilot ignored the TCAS commands and continued to execute the descent command from the ATC. Later, in 2002, the pilot of the Tu-154 over Lake Constance did the same, but with much more tragic consequences [4]. Following the TCAS commands, the DC-10 pilot also began descent. Thus, both aircrafts were then flying across each other's flight paths at the same altitude. Then, the command was given to Flight 907, which at that moment was at an altitude of 36,200 ft, to climb and ascend to level 390. However, the B747 pilot did not have time to complete it, because after only a few seconds he saw the DC-10 flying toward him. He quickly sent the plane diving under the DC-10. The planes separated at a distance of less than 100 m. The B747 pilot himself stated that the distance was only 35 ft (11 m). An unknown passenger later said in one of the TV programs on the topic: "I have never seen a plane fly so close. I thought we were going to crash" [5, 6]. In this situation, both the first and third reasons named in [3] are present. ATCs have to deal with very heavy workloads, due to which they may start making one mistake after another in an emergency when there are intersecting or converging flight paths. In the situation discussed above, is was only by a lucky chance that no tragedy happened.

In accordance with the well-known document by Eurocontrol [7], flight safety statistics takes into account not only violations of separation standards, near midair collisions, and collisions, but also such situations as short-term conflict alert (STCA), minimum safe altitude warning (MSAW), area proximity warning (APW), ground proximity warning system (GPWS), airborne collision avoidance system (ACAS), and advanced surface movement guidance and control systems (A-SMGCS). In other words, the safety statistics of the Eurocontrol member countries take into account not only situations with serious consequences, such as violations of separation standards, near midair collisions, and collisions, but also events with relatively negligible, or insignificant consequences. This is the reason why in 2019, Eurocontrol recorded 62,333 incidents related to air traffic control [8], while Russia recorded only 20 incidents [9]. That is, it can be reasonably assumed that the number of incidents related to ATC in the Russian Federation is unlikely to be fundamentally different from that in Europe.

Speaking about Reason's model [10], A. P. Plyasovskikh rightly notes that "this model clearly shows that a sequence of more serious situations, in fact, is formed from a sequence of less serious situations. Thus, less serious situations (e.g., short-term conflicts) are precursors, or a litmus test, of the impending onset of more serious situations (near midair collisions, or even collisions). A large number of STCAs that did not end in violations of separation standards is an indicator that sooner or later, more serious situations will arise where there are only short-term conflicts" [3].

Therefore, the issue of emergency performance in air traffic control is very important and urgent.

2 Problem Statement

Psychological testing is the first step in screening out candidates unfit for the job of a pilot or ATC. The problem of improving screening practices in the aviation sector is discussed in many works, including the following ones [2, 11–14].

In order to improve psychological tests, it is necessary to know what psychological characteristics are important for an ATC. Based on the key points discussed in the introduction to this article, it seems quite obvious that such a quality is potential emergency performance (PEP) [15], or the ability of the ATC to do his or her job well in an emergency.

Since PER is a quality that depends on innate factors and skills acquired in the process of work [15, 16], a direct measurement of this quality in an applicant is not possible even theoretically. An ideal solution to this problem would be to test the candidate for, first, the innate psychological qualities that determine PEP, and, second, those necessary psychological qualities (NPQs) that will help the person develop high emergency performance in the course of their professional activity.

In order to determine such NPQs, it is necessary to find a connection between the emergency performance that a student or worker already demonstrates and his or her psychological qualities. Even this task is extremely difficult. Also, it is greatly complicated by such a fundamental problem as how to evaluate PEP. This article is devoted to finding ways to solve the problem of assessing the performance of an ATC in an emergency.

3 Materials and Methods

To test it, A. V. Malishevskii supervised an experiment using the navigator and expert air traffic control simulators that was conducted by I. Yu. Krivoborsky and students K. A. Aleksanyan and G. I. Baklanov (see Fig. 1).

20 fifth-year students at Saint Petersburg State University of Civil Aviation took part in this experiment. They had to perform a rather difficult task. The task included dealing with five potential conflict situations. The instructors assessed the participants on a 10-point scale using five criteria: following the rules of radio communications phraseology (V_1); interaction with other air traffic control rooms (V_2); conducting procedural control and performing control panel operations (V_3); ensuring efficient use of airspace (V_4); ensuring air traffic safety (V_5). Then, based on these criteria, the total score (V_Σ) was derived. The scores were accompanied with a detailed description of the mistakes made by the participants to further assess the quality of their performance.

After that, the participants were tested to determine their potential emergency performance (Ω), which is a function of their innate stress resistance and the emotional experience acquired in the process of professional development [15, 16].

The innate component can be measured directly as it depends on the parameters of the nervous system, or, to be precise, on a personality trait dimension called neuroticism, which was assessed using the Eysenck Personality Inventory (EPI) [17] adapted by Shmelev [18].

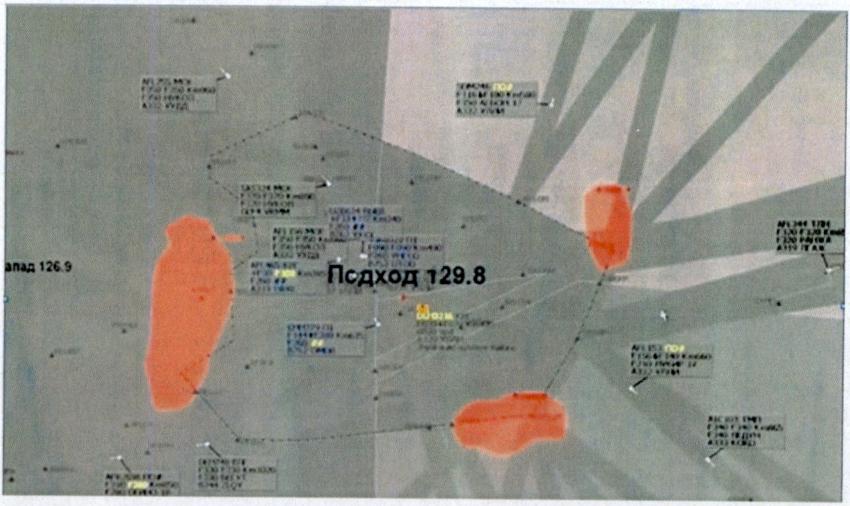

Fig. 1 Example of a developing air situation that was assessed as part of the experiment

Measuring emotional experience is a much more difficult problem. As noted earlier, measuring emotional experience directly is impossible in principle. Therefore, an indirect approach was used that is proposed in [15]. It was assumed that participants with more developed emotional experience should manifest more adequate risk assessments than those with less developed emotional experience (which, of course, is a very serious assumption). To assess the danger of a situation in the experiment, potential conflict situations were used that were previously described in [15, 19] to have a benchmark for comparison. An example of such a potential conflict situation is shown in Fig. 2. The participant in the experiment was given the following instruction explaining the situation:

"Situation 1. Four aircraft (85751, 85774, BAW8, JAL 408) are moving in the same direction to the Ukhta radio beacon. Two of them (85774 and BAW8) are on the same flight level 10,600 and calculate to move past the Ukhta radio beacon at the same time. JAL 408 is moving at 11,600, and there is no safe interval between them. Changing the flight level of Board 85774 is problematic, because if it is raised to a passing flight level, it will conflict with JAL 408 in the future. By lowering it, we create a conflict situation with 85751 that is following 9600. In addition to this situation: from KEDORA to POLOON and from BURKI to POLOON, there are two foreigner flights at 10600 (VIR 901 and FIN 52). Give your risk assessment" [15, 19].

To assess how dangerous a potential conflict situation is, a scale from zero to one was used, with zero meaning there is no danger, and no action is required and one

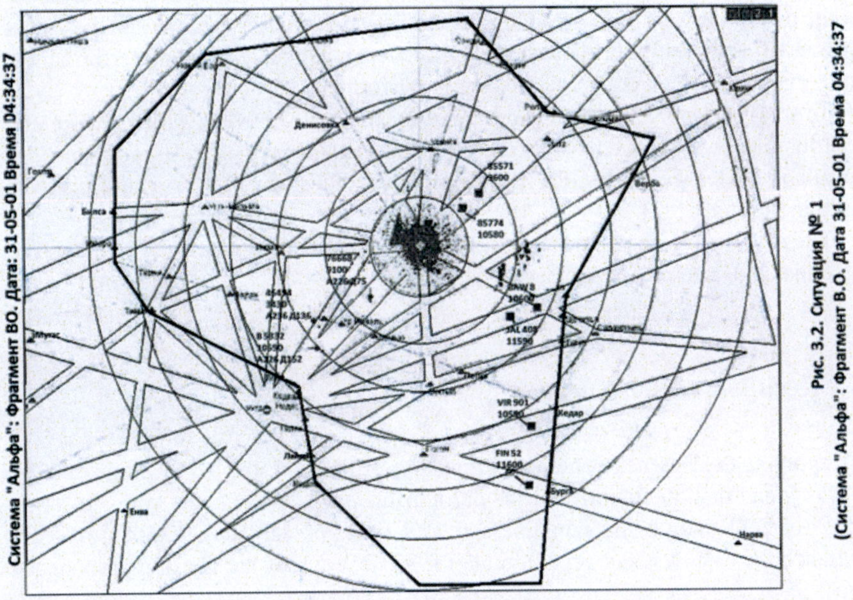

Fig. 2 Example of a potential conflict situation described in [15, 19] (the first situation in the experiment)

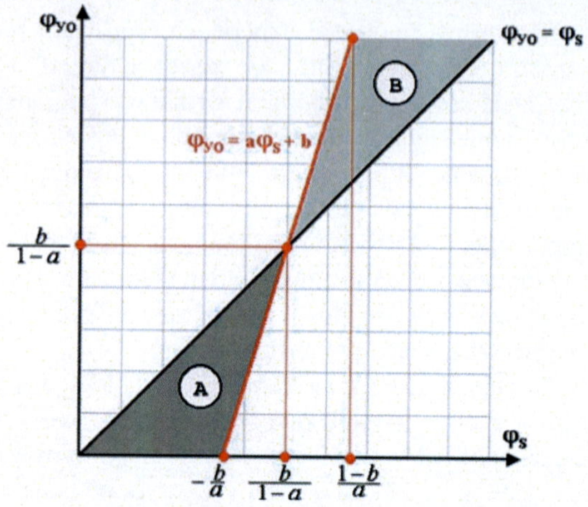

Fig. 3 Example of a graph for determining whether an individual risk assessment of a potential conflict situation is adequate (φ_S is the subjective risk assessment given by the participant, and φ_{yo} is the notionally adequate risk assessment that is found using a special procedure)

meaning that the situation is catastrophic, rendering any action performed by the ATC controller useless.

The emotional experience of each participant was assessed using graphs that can serve as a foundation for determining whether individual risk assessments are adequate. To do this, the values of α and β were calculated, which are functions of the area of triangles A and B, which determine the deviation of the individual risk assessment of an emergency from the fully adequate one (see Fig. 3).

Then, by using the method described in [15], the potential emergency performance of the participant (Ω) was determined as a function of neuroticism (n) and the values of α and β. The value of Ω ranges from 1 (excellent potential emergency performance) to 6 (extremely weak potential emergency performance).

To analyze the results, the R programming language was used, which is a popular tool for data analysis and has become a standard choice for statistical programs [20] (available under the GNU GPL license [21]). Methods of correlation analysis were used in the study [22].

The experiment was carried out in accordance with the fundamental principles of bioethics [23] and on a voluntary basis.

4 Results and Discussion

By applying the Eysenck Personality Inventory (EPI) [17] adapted by Shmelev [18], it was found that the majority of the participants have temperament types fit for an ATC (65% of them being sanguine and 25% being phlegmatic). Another 10% are melancholic, which is not very suitable for ATCs, but they are not of a pronounced type.

Subjective (individual) assessments of the risks given by the participants in the experiment were compared with previously obtained expert assessments (Table 1).

Table 1 Objective risk assessments of the emergencies analyzed in the experiment [15, 19]

Average risk assessment by	Total	φ_{o1}	φ_{o2}
Experts	8	0.425	0.088
ATCs	59	0.431	0.211
Both categories	67	0.431	0.196

For example, previously published works [15, 19] provide data on the same potential conflict situations, where assessments of their risks were given by experts and air traffic controllers. The results demonstrated by the participants in the experiment are shown in Table 2. The previously encountered problem of inadequate risk assessment in assessing potential conflict situations by individual subjects was revealed.

While the assessments of the first situation given by both experts and participants were relatively uniform, risk assessments for the second situation differed: the experts assigned it an average score of 0.088, whereas the participants in the experiment assessed the risk at the initial stage of the development of the situation (which is relatively safe) as 0.13, with the score reaching 0.585 for the final stage. Some respondents even ranked the risk as 0.8, which corresponds to an emergency.

Table 2 Subjective risk assessments in the experiment

Participant	Situation 1	Situation 2	Situation 2 after 3 min	Situations 2 after 20 min
K.A.	0.5	0.3	0.3	0.7
A.M.	0.6	0	0.3	0.6
S.T.	0.6	0.1	0.3	0.7
P.D.	0.4	0	0.2	0.2
A.G.	0.3	0	0	0.7
A.O.	0.3	0	0	0.5
D.F.	0.3	0.2	0.2	0.7
V.L.	0.8	0.4	0.2	0
Sh.S.	0.5	0.5	0.4	0.7
G.B.	0.5	0.1	0.1	0.3
S.L.	0.3	0	0	0.5
T.M.	0.5	0.3	0.3	0.8
M.A.	0.5	0	0.4	0.7
A.P.	0.5	0.1	0.3	0.7
N.K.	0.5	0	0.3	0.5
V.G.	0.3	0	0.3	0.7
M.S.	0.3	0	0	0.5
A.G.	0.4	0.2	0.3	0.7
D.K.1	0.5	0.4	0.6	0.7
D.K.2	0.5	0	0.5	0.8

Table 3 Summary of the results

Participant	V_1	V_2	V_3	V_4	V_5	V_Σ	Ω	E	n
K.A.	10	9	10	8	10	8	1	10	5
A.M.	8	9	10	8	5	6	1	14	6
S.T.	6	6	8	6	8	7	1	12	9
P.D.	9	9	8	8	10	9	6	8	14
A.G.	6	8	8	7	8	7	1	6	15
A.O.	6	7	10	7	10	7	1	18	5
D.F.	4	7	10	10	10	4	1	6	2
V.L.	10	8	10	9	10	9	1	15	4
Sh.S.	7	9	10	6	10	7	1	13	2
G.B.	6	8	10	10	10	7	1	9	10
S.L.	7	10	10	6	10	7	1	14	6
T.M.	10	10	10	7	10	8	1	17	5
M.A.	10	6	10	9	10	7	1	17	5
A.P.	10	10	10	10	10	10	1	15	5
N.K.	10	10	10	6	10	7	1	16	6
V.G.	9	7	10	7	3	5	1	11	2
M.S.	4	7	10	8	5	5	2	15	6
A.G.	7	8	10	6	5	4	1	12	2
D.K.1	10	9	10	8	10	9	1	18	8
D.K.2	10	8	10	7	6	5	1	18	3

Note E denotes extraversion measured by the Eysenck Personality Inventory

Table 3 summarizes the results of the experiment. As can be seen from the data given in the table, most of the participants are characterized by high potential emergency performance.

Table 4 shows the most significant correlations between potential emergency performance, and the results of the experiment carried out using the ATC simulators. The significant correlation between potential emergency performance and the assessment of conducting procedural control and performing control panel operations is most likely due to some random deviation as these assessments are almost homogeneous (seventeen participants have a score of 10, and three participants have a score of 8), but it is difficult to make an unambiguous conclusion.

Table 4 Most significant correlations obtained as a result of the experiment

Values		r_{CORR}	H_F	H_{CR}	Correlation	
1st	2nd				Strength	Significance
V_Σ	V_3	0.7555	3.293	2.960	Strong	Very highly significant
V_3	n	−0.7828	3.412	2.960	Strong	Very highly significant
V_3	ω	−0.5800	2.528	2.450	Average	Highly significant
n	E	−0.3825	1.667	1.650	Moderate	Reliable
ω	n	0.5938	2.588	2.450	Average	Highly significant

Note ω is an intermediate value (ranging from 0 to 100), which is used to determine the potential emergency performance Ω

5 Conclusions

Obviously, 20 participants make a relatively small sample, which does not allow for making any far-reaching conclusions. Nevertheless, the experiment produced several results.

1. The total score in the simulator experiment was mainly influenced by how well the participant performed in terms of ensuring flight safety. Almost no influence was exerted by the quality of conducting procedural control and performing control panel operations.
2. The results of assessing potential emergency performance by using this method showed that the majority of the participants rank high in this domain, at least according to the results of this experiment. In addition, it should be taken into account that even a student who has demonstrated insufficient PEP has the potential to develop it in the process of professional activity, since it is professional activity that is the key source of the formation of a high PEP.
3. Potential emergency performance and the actions performed by the ATC in a potential conflict situation are significantly influenced by temperament. Most of the participants have temperament types that are fit for their jobs. Only two of them are classified as melancholic, which is a type that is unfit for ATCs. Their scores are quite low, which corresponds to theoretical assumptions.
4. When assessing potential emergency performance, it is also necessary to take into account the ATC's socionics type, since this job implies processing a large amount of incoming information.
5. As a methodology for assessing potential emergency performance of an ATC, the method proposed at Saint Petersburg State University of Civil Aviation that assesses such parameters as neuroticism and the adequacy of risk assessment was analyzed. This method seems to need refinement and requires additional tests.

References

1. Human Factors Training Manual. Doc. 9683-AN/950 (1998) First edition. ICAO, Montréal, Canada
2. Malishevskii AV, Vlasov EV, Kajmakova EM (2015) Vozmozhny'e puti resheniya problemy' snizheniya negativnogo vliyaniya chelovecheskogo faktora v chrezvy'chajny'x situaciyax na transporte (Possible ways to reduce the negative impact of human factor in transport emergencies) Medico-biological and socio-psychological problems of safety in emergency situations 1:108–114. https://doi.org/10.25016/2541-7487-2015-0-1-108-114 (In Russian)
3. Plyasovskikh AP (2020) Ot reagirovaniya na aviacionnye sobytiya k ih uprezhdeniyu (From responding to emergencies to prevention) Aviapanorama 6(144):22–29 (In Russian)
4. Investigation Report AX001-1-2/02 (2004) German Federal Bureau of Aircraft Accidents Investigation. BFU, Braunschweig
5. Blame pinned on air traffic controllers (2001) The Japan Times. Saturday February 3. https://www.japantimes.co.jp/news/2001/02/03/national/blame-pinned-on-air-traffic-controllers/. Accessed 07 Nov 2021
6. Aircraft accident investigation report Japan Air Lines flight 907 Boeing 747-446D, JA8904, Japan Air Lines flight 958 Douglas DC-10-40, JA8546. A near midair over the sea of Oyaizu city, Shizuoka prefecture, Japan at about 15:55 JST, January 31 (2001) https://www.mlit.go.jp/jtsb/eng-air_report/JA8904.pdf. Accessed 07 Nov 2021
7. ESARR 2 Reporting and assessment of safety occurrences in ATM (2009). https://www.eurocontrol.int/sites/default/files/2019-05/esarr2-bil30.pdf. Accessed 07 Nov 2021
8. PRR 2019 Performance review report an assessment of air traffic management in Europe during the calendar year 2019. EUROCONTROL, Brussels 2020. https://www.eurocontrol.int/sites/default/files/2020-06/eurocontrol-prr-2019.pdf. Accessed 07 Nov 2021
9. Analiz sostoyaniya bezopasnosti poletov v grazhdanskoj aviacii Rossijskoj Federacii v 2019 godu (The analysis of flight safety in the civil aviation of the Russian Federation in 2019). Federal Air Transport Agency, Moscow, 2020 (In Russian)
10. Reason JT (1997) Managing the risks of organizational accidents Aldershot, Hants, Brookfield, Vt., Ashgate, cop.
11. Dzhapharadze TR, Malishevskii AV (2013) Problema sovershenstvovaniya professional'nogo psixologicheskogo otbora pilotov grazhdanskoj aviacii (The challenge of improving professional psychological selection of pilots of civil aviation) Medico-biological and socio-psychological problems of safety in emergency situations 3:66–70. https://doi.org/10.25016/2541-7487-2013-0-3-66-70 (In Russian)
12. Arinicheva OV, Lebedeva NA, Malishevskii AV (2019) Intellectual functioning in students and conflict management strategies. Revista Espacios 40(44):29
13. Arinicheva OV, Lebedeva NA, Malishevskii AV (2020) Socionics aspects of the human factor in aviation. Revista Espacios 41(21):391–407
14. Arinicheva OV, Lebedeva NA, Malishevskii AV (2020) Socionic models of a person and their application in aviation. Revista Espacios 41(43):129–144
15. Leichenko SD, Malishevskii AV, Mikhailik NF (2006) Chelovecheskij faktor v aviacii (Human factor in aviation). In: State University of civil aviation, vol 2, St. Petersburg, State Flight Academy of Ukraine, Kirovograd (In Russian)
16. Mikhailik NF, Tsitser VA (1994) Issledovaniya vozmozhnosti ispol'zovaniya teorii katastrof dlya ocenki rabotosposobnosti pilota v osoboj situacii (The researches of the opportunity of the crashes the theory usage for the pilot work capacity estimation in an abnormal case. In: Proceedings of the Academy of Civil Aviation: "The optimization of flight operation and the air staff professional training. Psychophysiological problems of professional work capacity of the civil aviation specialists"). St. Petersburg. In Russian
17. Eysenck HJ, Eysenck SBG (1964) Manual of the Eysenck personality inventory. University of London Press, London
18. Burlachuk LF (2008) Slovar'-spravochnik po psihodiagnostike (Psychological testing handbook) Piter, St. Petersburg (In Russian)

19. Arinicheva O, Dalinger Ya, Malishevskii A, Sukhikh N, Khoroshavtsev Yu (2018) Objectification of subjective estimation of abnormal cases danger in air traffic control. Transp Prob 13(1):5–18. https://doi.org/10.21307/tp-2018.13.1.1
20. Research & Statistical Support Services. University Information Technology. http://it.unt.edu/research. Accessed 07 Nov 2021
21. The Free Software Foundation (FSF). https://fsf.org/. Accessed 07 Nov 2021
22. Dowdy S, Wearden S, Chilko D (2004) Statistics for research, 2nd edn. Wiley, New York
23. Bioethics. Internet encyclopedia of philosophy. A peer-reviewed academic resource. https://www.iep.utm.edu/bioethic/ Accessed 07 Nov 2021

Analysis of the Accelerometer Signal for Gait Asymmetry Detection

Roman Pirozhkov, Aleksandr Ermakov, Danil Muzafarov, and Vitaliy Karachentsev

Abstract The analysis of informative signal parameters is popular with researchers in various scientific fields. This scientific paper presents a method for detecting gait asymmetry in aircraft crews by analyzing the vertical acceleration signal generated during walking from the accelerometer of a mobile device. This method allows detecting asymmetry by a number of features that are revealed in the signal structure in the time domain, and does not require large computational resources. In the presence of gait asymmetry, the durations of one and the other leg steps are determined, followed by a comparison of their numerical values, and the largest average difference between the extremes of the vertical acceleration of the steps is determined.

Keywords Gait asymmetry · Aircraft crew · Accelerometer signal · Step structure · Segmentation · Step duration · Musculoskeletal system

1 Formulation of the Problem

Human gait analysis is one of the methods to identify and diagnose various pathologies of the human body during the rehabilitation period, such as scoliosis, Parkinson's disease and other diseases of the musculoskeletal system and the central nervous system, manifested in gait disturbance. These diseases are quite common nowadays, they affect people of different professions, including aircraft crews: pilots, navigators, flight engineers, flight mechanics, flight attendants and so on. At the same time these diseases are difficult to detect and diagnose in the early stages, which is important for effective treatment. The difficulty lies in the need to use expensive diagnostic equipment, as well as hospitalization of patients. This creates additional motivation for the development of methods to detect such diseases using publicly available mobile

R. Pirozhkov · A. Ermakov (✉) · D. Muzafarov
Saint Petersburg State University of Aerospace Instrumentation, Saint Petersburg, Russia
e-mail: ermakov.alexandr.k@gmail.com

V. Karachentsev (✉)
Irkutsk Branch of Moscow State Technical University of Civil Aviation, Irkutsk, Russia
e-mail: kara1126@mail.ru

O. A. Gorbachev et al. (eds.), *Proceedings of 10th International Conference on Recent Advances in Civil Aviation*, Lecture Notes in Mechanical Engineering, https://doi.org/10.1007/978-981-19-3788-0_30

devices. Therefore, it is an urgent task to develop methods for human gait analysis using various tools and means, including computer modeling [1–4].

2 Analysis of Known Solutions

It stands to mention that human gait analysis is a relatively young area in biomechanics—a science at the intersection of biology, medicine and physics. Today there is a large number of gait analysis methods. They can be divided into the following groups: methods of registration of general, kinematic, dynamic parameters, sub-dynamometry, accelerometry and some others. The proposed algorithm will relate to the accelerometry method [5–7].

Special attention should be paid to [8–11]. The proposed method uses a reference step model, which is the dependence of the vertical acceleration on time in the interval of one step of a healthy person. This model is generated once and then used to calculate its two-dimensional correlation function (CF) with the analyzed vertical acceleration signal of the current subject. The first area of the CF is time (the shift of the model relative to the beginning of the analyzed signal allows to isolate individual steps), the second area is the time scale of the model (allows to estimate the difference in the duration of individual steps, and also provides invariance to the subject's walking pace). By analyzing the local maxima of the KF, the model scales corresponding to each step are determined, and after averaging the scale differences for the left and right legs over the entire duration of the analyzed signal (60–90 s), the desired value of the relative difference in step duration is formed.

Objective of this work is to identify signal signs of asymmetry in the gait of the subject by analyzing the accelerometer signal from a mobile device.

Tasks to achieve the stated objective:

1. to analyze the general structure of the vertical acceleration signal during walking, considering all steps as a whole;
2. to analyze the structure of the accelerometer signal between individual steps;
3. to compare signal structures of the steps with one leg and the other.

3 Carrying Out the Experiment

An accelerometer is hung on the subject's belt. An accelerometer is a device that measures the projection of apparent acceleration (the difference between the true acceleration of an object and gravitational acceleration) on three orthogonal axes. Data from the accelerometer should be recorded when placing the mobile device with the sensor on the strap in the middle of the body (front or back) so that the local maximum accelerations generated by the left and right legs have the same values. When the device is placed on the side, the vertical acceleration fluctuations for the step of the farthest leg from the accelerometer have a lower intensity, as a result of which the local maximum of such a step becomes comparable in level with the internal acceleration fluctuations within the duration of the step. This leads to the incorrect allocation of step boundaries: the boundaries are indicated only by the local maxima of one leg (the one on the side of the attached accelerometer), and instead of the sequence of individual steps necessary for further processing, an unacceptable sequence of paired steps is formed in this case. To simulate an uneven gait, additional insoles are placed in one of the subjects' boots [12]. The subject walks along the corridor for 30 s. At first they do it without additional insoles, which corresponds to a relatively even gait, then with each new pass they add an additional insole to the boot. After that, the accelerometer data is transferred to the computer and analyzed using the MATLAB program [13–16].

4 Algorithm of the Proposed Method

The algorithm of the proposed method will consist of several successive actions:

1. highlighting the steps in the accelerometer signal;
2. finding the durations between steps for the right and left legs, as well as finding the average value of the step duration for each leg;
3. analysis of the internal structure of the steps, i.e. the accelerometer signal between the selected steps;
4. finding individual extrema of the accelerometer signal in the internal structure of the steps, as well as finding the average value of each extremum for each leg.

The first two points can be attributed to the analysis of the overall structure of the steps, and the second two to the analysis of the internal structure of the steps.

1. Analysis of the general structure of the accelerometer signal.

Let us analyze the situation when the subject walks along the corridor and in one of his boots there are five additional insoles, which corresponds to a total height of about 5 mm. In this case, the signal from the vertical axis of the accelerometer will look as shown in the Fig. 1. After removing the constant component (acceleration of gravity), this signal is segmented into separate steps—shown by vertical lines.

Let us calculate the step durations separately for both legs and compare the average values. To do this, we will build a graph of dependence, where we indicate the values of the step durations with dots: the dot is placed at the intersection of the step end moment along the horizontal axis and the duration of the step along the vertical axis. The values for one leg are highlighted in red, and the values for the second leg are highlighted in blue (at the current stage of the study, it does not matter which of them is the right leg and which is the left one). The horizontal lines indicate the average step length for one leg and the other. This graph is shown in the Fig. 2.

Obviously, the middle lines for one leg and the other have a discrepancy in duration. Significant discrepancy may indicate gait asymmetry.

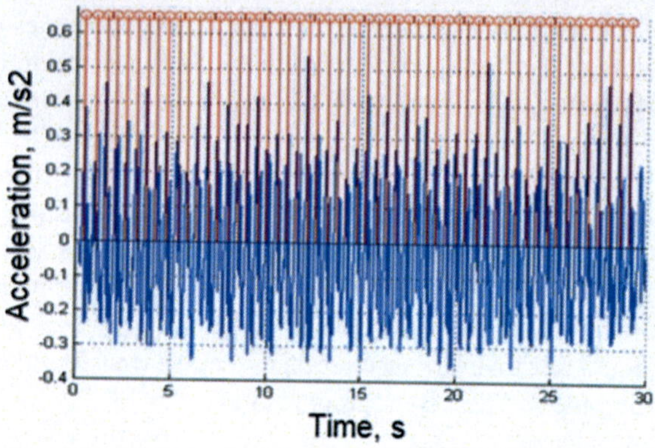

Fig. 1 Signal from the vertical axis of the accelerometer, segmented into steps, with the removed constant acceleration component

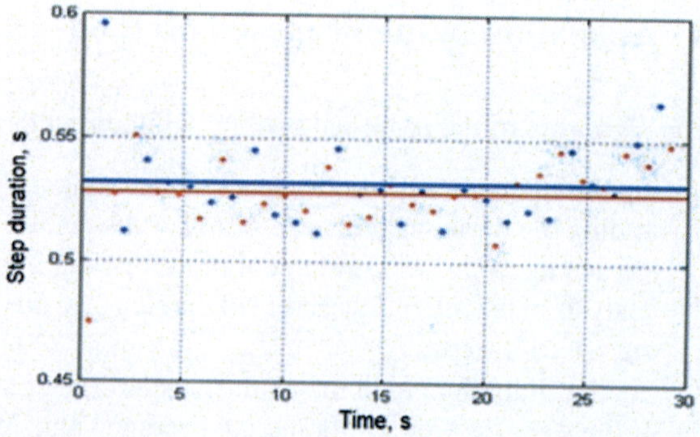

Fig. 2 Step duration dependence on time graph

The numerical value of the average step duration with one leg is 0.528 s, with the other leg-0.531 s. The difference between the mean values is, respectively, 0.003 s.

2. Analysis of the structure of the accelerometer signal between individual steps.

For further analysis of the gait, we will move from considering a set of steps to analyzing the internal structures of each step (segment). An example of the internal structure of a step is shown in Fig. 3.

Plot the acceleration values of individual steps on a graph and place them on one graph. The result is shown in Fig. 4.

Analyzing Fig. 4, we can say that the structure of the steps is similar and has a general pattern, but we can see a scattering of values relative to the average. For further comparison of the structures of the steps of the right and left legs, we will construct a graph, in which the values for one leg will be highlighted in red and the values for the other leg in blue. The corresponding graph is shown in Fig. 5.

Significant difference in the behavior of these dependencies is easily visible. For a numerical estimation, consider the temporal location of some extrema relative to the beginning of the step (shown in Fig. 5) for each curve. Let us build a graph of the dependence of the number of the reading, which corresponds to the extremum, on

Fig. 3 Example of the internal structure of a step

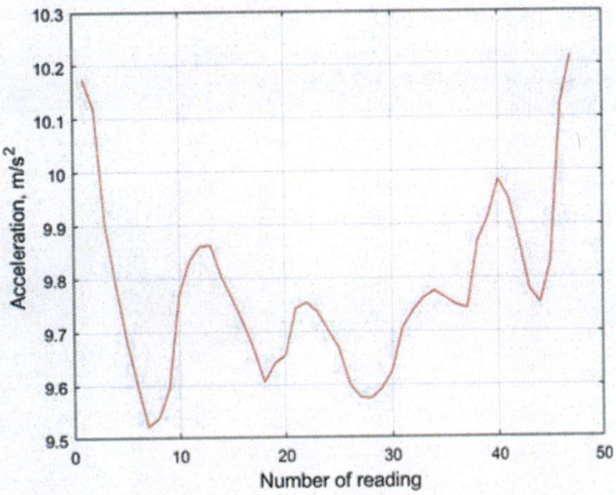

Fig. 4 Plot of acceleration of individual steps against array value number and step number: **a** Three-dimensional plane view; **b** "top" view

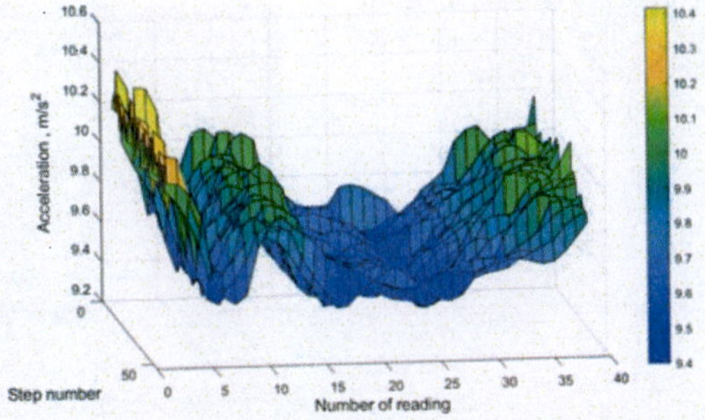

(a)

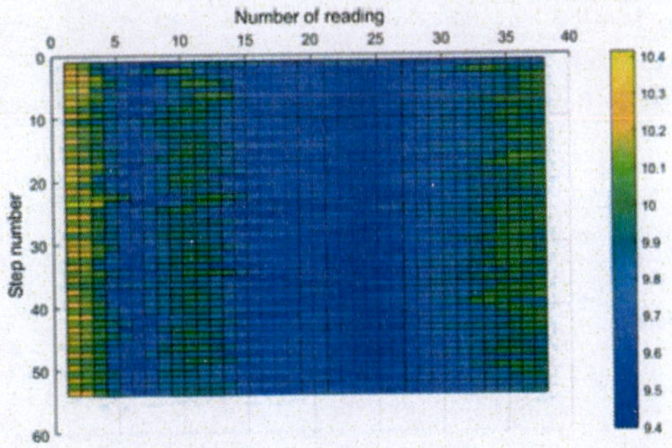

(b)

the step number. The graph is shown in Fig. 6. The values for one leg are highlighted in red, and the values for the second one are highlighted in blue. The middle lines on the chart represent the average values of the extrema of all steps. Considering that only 3 extrema were taken for consideration, there will be 6 middle lines.

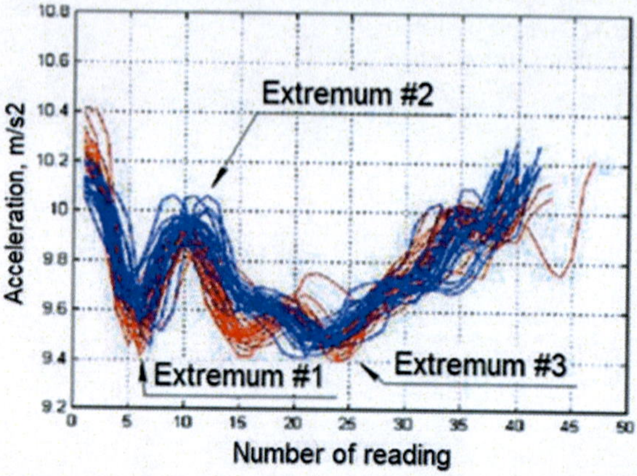

Fig. 5 Dependence of the individual steps acceleration on the number of the array value graph

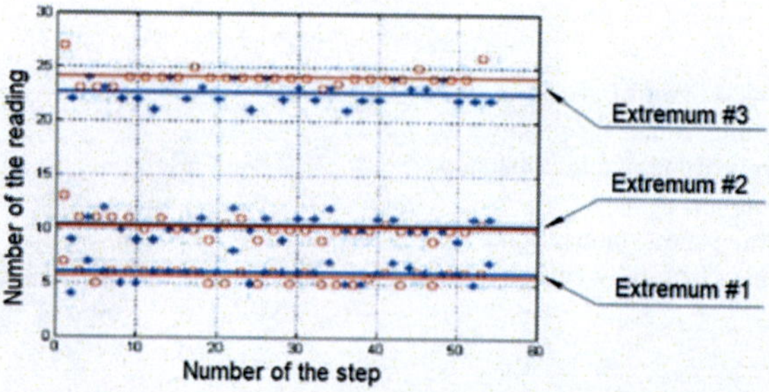

Fig. 6 Graph for dependence of the number of the reading corresponding to the extremum on the step number

The numerical value of the difference between the middle lines is presented in Table 1.

Table 1 Comparison of the average values of the extrema of all steps

	For the first leg	For the second leg	Difference
Average value of the first extremum for all steps, readings	5.648	5.944	0.296
Average value of the second extremum for all steps, readings	10.240	10.270	0.03
Average value of the third extremum for all steps, readings	24.090	22.560	1.53

5 Assessment of the Analysis Results

The results obtained should be evaluated both graphically (visually) and numerically. Analyzing the graphs and tables, in each case it can be noted that there are discrepancies in the values.

Considering Fig. 2, we can say that the average step durations for the left and right legs differ, but this difference is not significant (only 0.003 s). This order of numbers can be attributed to measurement errors.

The most interesting are Figs. 5 and 6 and Table 1. In Fig. 5, the difference in the behavior of the dependencies is obvious. In the case of a symmetrical gait, the appearance of the curves should repeat each other. In this case, this is not so. Therefore, it is already possible to assume the presence of asymmetry in the gait.

Considering Fig. 6 and Table 1, it can be noted that the average values of the first two considered extrema for the right and left legs are similar, and the average values of the third extremum differ significantly. Numerically, the discrepancy is equal to 1.53 of the reading, which can also signal the presence of asymmetry in gait.

6 Conclusion

A method for detecting the asymmetry of a person's gait is shown based on the analysis of the temporal structure of the vertical acceleration of individual steps in the process of walking. The presented method works in the time domain based on the analysis of local signal extrema and does not require large computational resources. The authors recommend the use of this method for medical examinations of flight personnel in order to detect pathologies at an early stage of development. According to the results of the measurements, in the presence of asymmetry, the largest average difference of 0.3–2 ms is detected between the extrema indicated in the article by numbers 3 and 1. Further development of the proposed technique will largely depend on the results of the analysis of a large number of statistical data with imitation of varying degrees of asymmetry. Existing methods will be improved and new ones proposed, which will make it possible to more accurately identify signs and quantify the degree of gait asymmetry.

Acknowledgements The authors are grateful to the Director of the Institute of Information Systems and Information Security of the SUAI A. M. Tyurlikov and the student of the Department of Infocommunication Systems of the SUAI M. D. Savinova for useful ideas on the research topic and provided test materials.

References

1. Ren B, Liu J, Chen J (2020) Simulating human-machine coupled model for gait trajectory optimization of the lower limb exoskeleton system based on genetic algorithm. Int J Adv Rob Syst 17(1):15. https://doi.org/10.1177/1729881419893493
2. Zhang L et al (2021) Modeling and simulation of a **human** knee exoskeleton's assistive strategies and interaction. Front Neurorobot. https://doi.org/10.3389/fnbot.2021.620928
3. Park H, Ri Y, Jehee L (2020) Multi-segment foot for human modelling and simulation. Comput Graph Forum, Feb. pp 637–649, https://doi.org/10.1111/cgf.13896
4. Ganea D et al (2017) Analysis of reaction forces in human ankle joint during gait. In: MATEC web of conferences. https://doi.org/10.1051/matecconf/20171120701
5. Skvortsov DV (1996) Klinicheskij analiz dvizhenij. Analiz pohodki (Clinical analysis of movements. Gait analysis). Izdatel'stvo NPC - "Stimul" (In Russian)
6. De Marsico M, Mecca A (2017) Biometric walk recognizer: gait recognition by a single smartphone **accelerometer**. Multimedia Tools Appl 76(4):4713–4745. https://doi.org/10.1007/s11042-016-3654-1
7. Moore ST, Yungher DA, Morris TR et al (2013) Autonomous identification of freezing of gait in Parkinson's disease from lower-body segmental accelerometry. J NeuroEng Rehab. https://doi.org/10.1186/1743-0003-10-19
8. René S et al (2020) Intra- and interobserver reliability comparison of clinical gait analysis data between two **gait** laboratories. Appl Sci. https://doi.org/10.3390/app10155068
9. Chen J, Mu X, Du F (2017) Biomechanics analysis of human lower limb during walking for exoskeleton design: J Vibroeng 19(7):5527–5539. https://doi.org/10.21595/jve.2017.18459
10. Xiong Y, Tang X, Shi C, Yang Y (2021) A design of stretcher with auxiliary functions of lateral positioning and transferring for immobilized patients. In: MATEC web of conferences. https://doi.org/10.1051/matecconf/202133602014
11. Songhai MO et al (2018) Control strategy of rehabilitation lower extremity exoskeleton based on gait state machine. China Sciencepaper 13(16):1889–1895. https://doi.org/10.1051/matecconf/202133602014
12. Borisova MD, Tomchuk KK, Turlikov AM (2020) Correlation analysis of the data from a mobile device accelerometer to detect the difference in step duration. In: 2020 Wave electronics and its application in information and telecommunication systems (WECONF). https://doi.org/10.1109/WECONF48837.2020.9131479 (In Russia)
13. Borowska-Terka A, Strumillo P (2017) Algorithms for head movements' recognition in an electronic human-computer interface. In: Przeglad Elektrotechniczny 93(8):131–134. https://doi.org/10.15199/48.2017.08.35
14. Nandy A, Chakraborty P (2017) A study on human gait dynamics: modeling and simulations on OpenSim platform. Multimedia Tools Appl 76(20):21365–21400. https://doi.org/10.1007/s11042-016-4033-7
15. Wu H et al (2017) Study on the control algorithm for lower limb exoskeleton based on ADAMS/Simulink co-simulation. J Vibroeng 19(4):2976–86. https://doi.org/10.21595/jve.2017.17303
16. Ren B et al (2020) Simulating human–machine coupled model for gait trajectory optimization of the lower limb exoskeleton system based on genetic algorithm. Int J Adv Rob Syst. https://doi.org/10.1177/1729881419893493

Comparison of the Effectiveness of the MUSIC and ESPRIT Superresolution Algorithms

Timur Tagaev, Aleksandr Ermakov, Danil Mokhort, and Nikolay Malisov

Abstract In radar, it is of great importance to distinguish targets separated by a small angular distance, for example, two close-flying aircraft located at a great distance from the radar station or distinguishing a low-flying target and its antipode, arising due to signal reflection from the underlying surface. This article compares the efficiency of the MUSIC and ESPRIT superresolution algorithms for the problem of distinguishing two targets separated by a small angular distance. According to the simulation results, the ESPRIT algorithm is faster than MUSIC for the same antenna array and signal parameters, ESPRIT has a smaller RMS deviation for the same parameters, which suggests that the ESPRIT superresolution algorithm is more efficient.

Keywords Superresolution algorithms · ESPRIT · MUSIC · Direction of arrival · Angular resolution

1 Introduction

To increase the resolving power of the antenna, algorithms of secondary or tertiary signal processing—superresolution algorithms are used [1, 2]. They by means of mathematical processing of the received signals allow to distinguish the targets located at a smaller angular distance than physically allowed by the antenna array. Thus it is possible, without changing the parameters of the radar station itself, to increase its resolving power. The most used superresolution algorithms are MUSIC and ESPRIT [3].

The MUSIC algorithm is used to process signals obtained by gratings with an arbitrary, predetermined configuration of elements and response. The essence of

T. Tagaev · A. Ermakov (✉) · D. Mokhort
State University of Aerospace Instrumentation (SUAI), Saint Petersburg, Russia
e-mail: ermakov.alexandr.k@gmail.com

N. Malisov (✉)
Irkutsk Branch of Moscow State Technical University of Civil Aviation, Irkutsk, Russia
e-mail: ermakov.alexandr.k@gmail.com

O. A. Gorbachev et al. (eds.), *Proceedings of 10th International Conference on Recent Advances in Civil Aviation*, Lecture Notes in Mechanical Engineering, https://doi.org/10.1007/978-981-19-3788-0_31

MUSIC is to construct a pseudospectrum to estimate the angular coordinates of targets, but it needs an additional algorithm that selects in the pseudospectrum the maximums corresponding to the targets [4].

The ESPRIT algorithm is used for equidistant antenna arrays. The ESPRIT algorithm divides the antenna array into two sublattices, offset relative to each other by a distance multiple of the distance between the elements of the array. The direction of signal arrival from the target is determined by the phase shift in the elements of the antenna sublattices [5].

This paper compares the performance of the MUSIC and ESPRIT algorithms by basic parameters: speed, dependence of the standard deviation on the signal-to-noise ratio (SNR), and dependence of the SNR on the number of antenna array elements [6]. We consider a linear equidistant antenna array receiving signals from targets located at an angular distance of 0.028 degrees.

2 Mathematical Model of the Signal Received by the Antenna Array

On the linear equidistant antenna array, consisting of M elements, a plane wave from N sources falls [7]. The direction of the signal coming from the sources is given by the direction matrix A, the columns of which are N direction vectors:

$$A = \begin{pmatrix} 1 & 1 & \cdots & 1 \\ e^{-\frac{2\cdot\pi\cdot i\cdot d\cdot\sin(\theta_1)}{\lambda}} & e^{-\frac{2\cdot\pi\cdot i\cdot d\cdot\sin(\theta_2)}{\lambda}} & \cdots & e^{-\frac{2\cdot\pi\cdot i\cdot d\cdot\sin(\theta_N)}{\lambda}} \\ \vdots & \vdots & \ddots & \vdots \\ e^{-\frac{2\cdot\pi\cdot i\cdot d\cdot(M-1)\cdot\sin(\theta_1)}{\lambda}} & e^{-\frac{2\cdot\pi\cdot i\cdot d\cdot(M-1)\cdot\sin(\theta_2)}{\lambda}} & \cdots & e^{-\frac{2\cdot\pi\cdot i\cdot d\cdot(M-1)\cdot\sin(\theta_N)}{\lambda}} \end{pmatrix} \quad (1)$$

where θ_i—angle of arrival i-th signal,

d—distance between neighboring elements of the antenna array,

λ—signal wavelength.

The signals falling on the antenna array are given by the matrix S, the rows of which are the signals falling on each element of the antenna array, and the columns are the time samples:

$$S = \begin{pmatrix} S_1(t_1) & S_1(t_2) & \cdots & S_1(t_K) \\ S_2(t_1) & S_2(t_2) & \cdots & S_2(t_K) \\ \vdots & \vdots & \ddots & \vdots \\ S_M(t_1) & S_M(t_2) & \cdots & S_M(t_K) \end{pmatrix} \quad (2)$$

where K is the number of time stamps.

The power of the signal from the i-th direction is given by the matrix P:

$$P = \begin{pmatrix} P_1 & 0 & \cdots & 0 \\ 0 & P_2 & \cdots & 0 \\ \vdots & \vdots & \ddots & \vdots \\ 0 & 0 & \cdots & P_N \end{pmatrix} \tag{3}$$

The matrix of signals taken from the output of the antenna array, $x(t)$ has the form:

$$x(t) = A \cdot P \cdot S + n = \begin{pmatrix} 1 & 1 & L & 1 \\ e^{\frac{2\cdot\pi\cdot i\cdot d\cdot\sin(\theta_1)}{\lambda}} & e^{\frac{2\cdot\pi\cdot i\cdot d\cdot\sin(\theta_2)}{\lambda}} & L & e^{\frac{2\cdot\pi\cdot i\cdot d\cdot\sin(\theta_N)}{\lambda}} \\ M & M & 0 & M \\ e^{\frac{2\cdot\pi\cdot i\cdot d\cdot(M-1)\cdot\sin(\theta_1)}{\lambda}} & e^{\frac{2\cdot\pi\cdot i\cdot d\cdot(M-1)\cdot\sin(\theta_2)}{\lambda}} & L & e^{\frac{2\cdot\pi\cdot i\cdot d\cdot(M-1)\cdot\sin(\theta_N)}{\lambda}} \end{pmatrix} \times$$

$$\begin{pmatrix} P_1 & 0 & L & 0 \\ 0 & P_2 & L & 0 \\ M & M & 0 & M \\ 0 & 0 & L & P_N \end{pmatrix} \cdot \begin{pmatrix} S_1(t_1) & S_1(t_2) & L & S_1(t_k) \\ S_2(t_1) & S_2(t_2) & L & S_2(t_k) \\ M & M & O & M \\ S_M(t_1) & S_M(t_2) & L & S_M(t_k) \end{pmatrix} + \begin{pmatrix} n_1(t) \\ n_2(t) \\ M \\ n_M(t) \end{pmatrix} \tag{4}$$

where $n_1(t)$—additive receiver noise with dispersion σ^2, received by the i-th antenna element.

The choice of signal frequency is determined by its attenuation in the atmosphere. Since the targets are at a considerable distance, the signal must propagate well in the medium, so there must be a low attenuation coefficient, which is achieved at a frequency of 10 GHz [8]. Graphs of the dependence of the attenuation coefficient on the frequency are shown in Fig. 1.

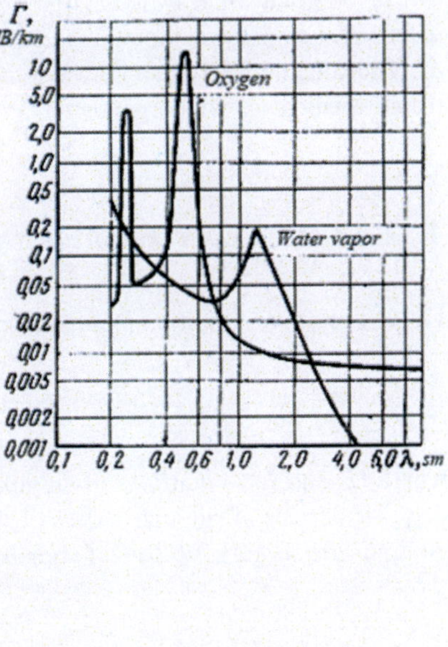

Fig. 1 Frequency dependences of the linear attenuation in oxygen and water vapor

3 Determining the Number of Targets

Some superresolution algorithms, including ESPRIT, require to know in advance the number of targets from which the signal comes [9]. This number can be determined by a singular decomposition of the correlation matrix of the received signal.

Correlation matrix R:

$$R = \frac{1}{T} \cdot \int_0^T x(t) \cdot x^H(t)dt \tag{5}$$

Or for counting:

$$R = \frac{1}{K} \cdot \sum_{t=t_1}^{K} x(t) \cdot x^H(t) \tag{6}$$

where $x^H(t)$ is the hermitian-conjugate matrix to the matrix $x(t)$,

t_1—the initial moment of time.

Singular decomposition of the correlation matrix (6):

$$R = U \cdot \Lambda \cdot U^H \tag{7}$$

where U—matrix, the columns of which are the singular vectors of the correlation matrix,

$\wedge$—diagonal matrix of singular values of the correlation matrix, $\wedge = \mathrm{diag}\{\lambda_1, \lambda_2, ..., \lambda_M\}, \lambda_1 \geq \lambda_2 \geq ... \geq \lambda_M \geq 0$.

The singular values of the signal vectors will be greater than σ^2, the singular values of the noise vectors will be equal to its variance: $\lambda_{N+1} = \lambda_{N+2} = \ldots = \lambda_M = \sigma^2$. Thus, the number of sources will be equal to the number of singular numbers of the correlation matrix, large than σ^2.

4 ESPRIT Superresolution Algorithm

Let us decompose matrix (7) into orthogonal signal and noise subspaces:

$$R = U_s \Lambda_s U_s^H + U_n \Lambda_n U_n^H \tag{8}$$

where U_s and U_n—matrices of singular vectors of signal and noise subspaces, Λ_s и Λ_n—matrices of singular values of signal and noise subspaces. Matrix U_s, whose columns are singular vectors of signals has the form:

$$U_s = \begin{pmatrix} U_{s_{11}} & U_{s_{21}} & \cdots & U_{s_{N1}} \\ U_{s_{12}} & U_{s_{22}} & \cdots & U_{s_{N2}} \\ \vdots & \vdots & \ddots & \vdots \\ U_{s_{1M}} & U_{s_{2M}} & \cdots & U_{s_{NM}} \end{pmatrix} \tag{9}$$

Matrices U_{s_x} and U_{s_y} are the matrices of the singular vectors of the first $M-1$ elements of the antenna array and the last $M-1$ elements, respectively:

$$U_{s_x} = \begin{pmatrix} U_{s_{11}} & U_{s_{21}} & \cdots & U_{s_{N1}} \\ U_{s_{12}} & U_{s_{22}} & \cdots & U_{s_{N2}} \\ \vdots & \vdots & \ddots & \vdots \\ U_{s_{1(M-1)}} & U_{s_{2(M-1)}} & \cdots & U_{s_{N(M-1)}} \end{pmatrix} \tag{10}$$

$$U_{s_y} = \begin{pmatrix} U_{s_{12}} & U_{s_{22}} & \cdots & U_{s_{N2}} \\ U_{s_{13}} & U_{s_{23}} & \cdots & U_{s_{N3}} \\ \vdots & \vdots & \ddots & \vdots \\ U_{s_{1M}} & U_{s_{2M}} & \cdots & U_{s_{NM}} \end{pmatrix} \tag{11}$$

Matrix Ψ is the solution to the matrix equation $U_{s_x} \cdot \Psi = U_{s_y}$:

$$\Psi = U_{s_y} \cdot U_{s_x}^{-1} = U_\Psi \cdot \Lambda_\Psi \cdot U_\Psi^{-1}, \tag{12}$$

$$\varphi = \mathrm{diag}\{\Lambda_\Psi\} \tag{13}$$

Matrix $\varphi = diag\{\Lambda_\Psi\}$ – vector string of singular numbers Ψ.
The arrival angles of the signals are determined by the expression:

$$\theta_i = \arcsin\left(\frac{-|\varphi_i|}{2 \cdot \pi \cdot d}\right) \tag{14}$$

One of the disadvantages of the ESPRIT algorithm is that the viewing sector decreases when the d_s—sublattice offsets relative to each other [10–13].

5 MUSIC Superresolution Algorithm

Based on the singular decomposition of matrix R into orthogonal signal and noise subspaces, a pseudospectrum is constructed $P_{\mathrm{MU}}(\theta)$:

$$P_{MU}(\theta) = \frac{1}{\left|a^H(\theta) U_n U_n^H a(\theta)\right|} \tag{15}$$

where a—directional vector.

The main disadvantage of the MUSIC algorithm is the need for an additional algorithm to extract the maxima of the pseudospectrum corresponding to the targets. This requires significant computational cost, especially when the number of targets is large [14].

6 Comparison of Algorithms

To compare the efficiency of the algorithms, we measure the standard deviation (SD), depending on the signal-to-noise ratio, and the number of elements of the antenna array. The SD is determined by the formula:

$$\sigma = \sqrt{\frac{\sum_{i=1}^{n} (x_i - M[x])^2}{n}} \tag{16}$$

where n—number of measurements, x_i—i-th value of a random variable, $M[x]$—mathematical expectation.

The dependence of the SD on the ratio of signal power to noise power for the ESPRIT algorithm is presented in Fig. 2. The dependence of the SD on the ratio of the signal power to the noise power for the MUSIC algorithm is presented in Fig. 3. The graphs show that the ESPRIT algorithm is unable to discriminate targets located at an angular distance of 0.028 degrees with signal-to-noise ratio below 25 dB. For the MUSIC algorithm to work, the required signal-to-noise ratio is higher than 40 dB.

The dependence of SD on the number of elements of the antenna array for the ESPRIT algorithm is presented in Fig. 4. The dependence of SD on the number of elements for the MUSIC algorithm is presented in Fig. 5. From the graphs we can see that the ESPRIT algorithm requires a much smaller number of antenna array

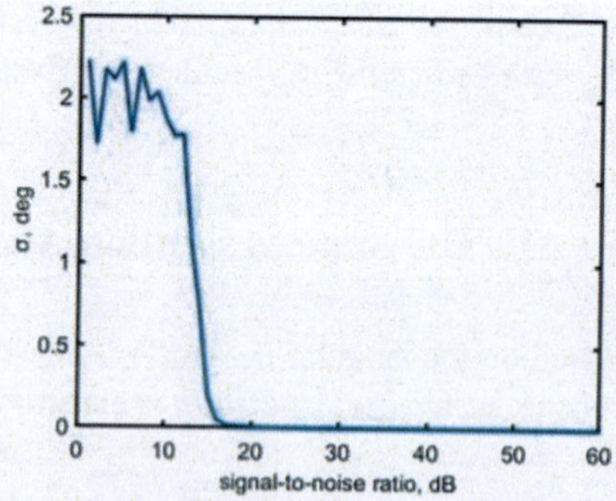

Fig. 2 The dependence of the SD on the signal-to-noise ratio for the ESPRIT algorithm

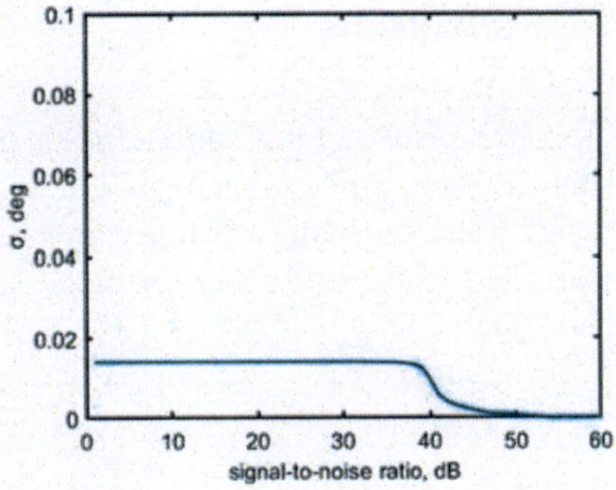

Fig. 3 The dependence of the SD on the signal-to-noise ratio for the MUSIC algorithm

elements to distinguish targets, that is, it can provide the SD needed to resolve two targets at smaller dimensions of the antenna array.

An analysis of the speed of the algorithms shows that ESPRIT is on average 12 times faster than MUSIC. The execution of the MUSIC algorithm includes an additional algorithm for selecting pseudospectrum maxima. Excluding the sampling of maxima corresponding to targets, ESPRIT is on average 10 times faster than MUSIC.

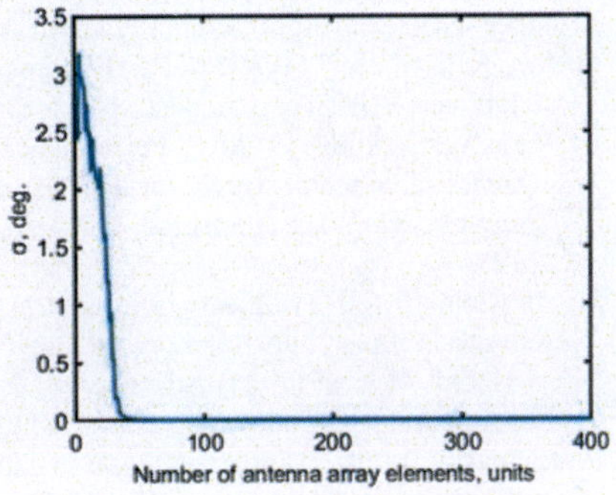

Fig. 4 Dependence of SD on the number of antenna array elements for the ESPRIT algorithm

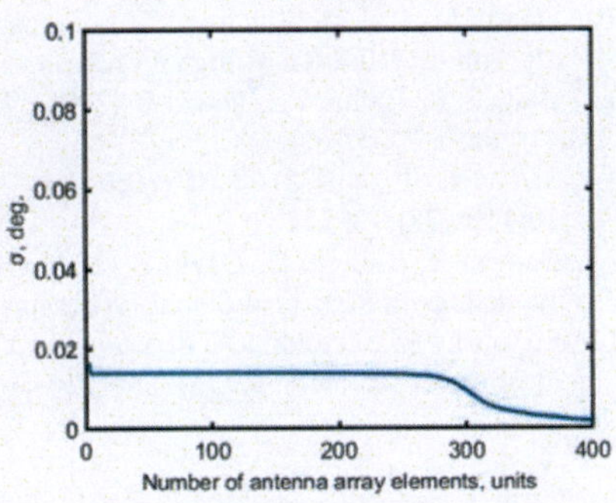

Fig. 5 Dependence of SD on the number of antenna array elements for the MUSIC algorithm

7 Conclusion

Thus, the results of our simulations suggest that the ESPRIT superresolution algorithm is more efficient than MUSIC in resolving the problem of two targets separated by a small angular distance. In the general case, ESPRIT is several times faster than MUSIC, so it is better for processing signals received by the antenna array in real-time.

References

1. Erdyneev ZT, Manokhin GO, Velikanova EP et al (2015) An formation algorithm of the synthetic aperture in an automotive radar with use of the MUSIC algorithm. In: Progress in Electromagnetics Research Symposium, pp 1834–1838
2. Lagovsky BA (2017) Superresolution in signal processing using smart antenna. In: Progress in Electromagnetics Research Symposium, St. Petersburg, https://doi.org/10.1109/PIERS.2017.8261787
3. Manokhin GO, Erdyneev ZT, Geltser AA, Monastyrev EA (2015) MUSIC-based algorithm for range-azimuth FMCW radar data processing without estimating number of targets. Mediterranean Microwave Symposium. https://doi.org/10.1109/MMS.2015.7375471
4. Liao W, Fannjiang A (2016) MUSIC for single-snapshot spectral estimation: Stability and super-resolution. Appl Comput Harmon Anal 40(1):33–67
5. Roy R, Kailath T (1989) ESPRIT-estimation of signal parameters via rotational invariance techniques. IEEE Trans Acoust Speech Signal Process 37(7):984–995
6. Vikas's B, Vakula's D (2017) Performance comparision of MUSIC and ESPRIT algorithms in presence of coherent signals for DoA estimation. In: International Conference of Electronics, Communication and Aerospace Technology (ICECA). https://doi.org/10.1109/ICECA.2017.8212844
7. Apostolov P (2011) Linear equidistant antenna array with improved selectivity. IEEE Trans Antennas Propag. https://doi.org/10.1109/TAP.2011.2163743
8. Al Hadidi M et al (2016) Adaptive regulation of radiated power radio transmitting devices. Contemporary Eng Sci 9(10):473–485. https://doi.org/10.12988/ces.2016.629
9. Lagovsky B (2017) Superresolution in signal processing using a priori information. Progress in Electromagnetics Research Symposium—Spring (PIERS). https://doi.org/10.1109/PIERS.2017.8261879
10. Haard M (1997) Structured least squares to improve the performance of ESPRIT-type algorithms. IEEE Trans Signal Process 45:792–799
11. Badeau R, Richard G, David B (2005). Fast adaptive ESPRIT algorithm. https://doi.org/10.1109/SSP.2005.1628608
12. Strobach P (1998) Fast recursive subspace adaptive ESPRIT algorithms. IEEE 10(1109/78):709531
13. Badeau R, Richard G, David B (2005) Fast adaptive ESPRIT algorithm. In: IEEE/SP 13th Workshop on Statistical Signal Processing. https://doi.org/10.1109/SSP.2005.1628608
14. Lagovsky B, Samokhin A, Shestopalov Y (2021) Angular superresolution based on A Priori information. Radio Sci 56(3). https://doi.org/10.1029/2020RS007100

Development of a Multifactorial Flight Safety Level Assessment Methodology in the Russian Federation Civil Aviation

Vyacheslav Besogonov, Anatoly Kostylev, and Mikhail Ushakov

Abstract Currently, the main factor in assessing the flight safety level in Russia is the number of aviation occurrences—incidents and accidents, while other multiple parameters directly affecting or powerful to affect flight safety are not taken into consideration and remain in the shadows. The research article proposes a variant of developing a methodology for a multi-factor assessment of the flight safety ensuring effectiveness applicable to operators acting in different segments of civil aviation and incorporating a variety of performance indicators. Use of multifactorial flight safety level assessment methodology with account of all operators' features make it possible to create a comparative assessment of flight safety level for various operators, which will make it possible to take the necessary measures to improve the civil aviation flight safety level.

Keywords Flight safety · Aviation accident · Aviation incident · Civil aviation · Civil aviation operator

1 Introduction

It is definitely necessary to take into consideration multifactoriality, developing recommendations for the application of a proactive flight safety management strategy (FS) by operators, in terms of flight safety level assessing methodology and its comparative assessment. First of all, in a historical context, the multifactoriality is related to the peculiarities of the structure of the aviation transportation system (ATS) of our country.

V. Besogonov · A. Kostylev (✉)
Saint Petersburg State University of Civil Aviation, Saint Petersburg, Russia
e-mail: kag-guga@yandex.ru

M. Ushakov (✉)
Irkutsk Branch of Moscow State Technical University of Civil Aviation, Irkutsk, Russia
e-mail: ushakov-an@yandex.ru

O. A. Gorbachev et al. (eds.), *Proceedings of 10th International Conference on Recent Advances in Civil Aviation*, Lecture Notes in Mechanical Engineering, https://doi.org/10.1007/978-981-19-3788-0_32

This point is being dictated by the fact that after the transformation of our country ATS in 1991, there was not only a change in the ownership form but also a change of operated aircraft (ACFT).

This period is known to be the time range when domestic aircraft were being replaced by foreign equipment, but this process did not take place at once and simultaneously, just like the adaptation of domestic operators when it comes to flight personnel training, maintenance, and creation of operational documentation. As well, the regulatory framework, governing the activities of civil aviation, has been changed. This justifies the value of the past 20 years, let's call it the operating period.

2 Problem Statement

When developing a multi-factor flight safety assessment method, it is supposed to divide the parameters affecting the flight safety level into two basic groups:

- 1st group—relative and absolute flight safety indicators;
- 2nd group—operational activity factors, general and specific.

Whereas general factors without exceptions are associated with all civil aviation operators, specific factors are known to be advisable and possible to take into account only for one type of activity operators, this is dictated by different technologies of the operators' activities and certification requirements, e.g., factors inherent in operators performing aviation work are not inherent in operators performing commercial air transportation [1–6].

Normally, in assessing the aircraft, flight safety relative and absolute quantitative indicators are used, e.g., the number of events that occurred for a certain number of flight hours or the number of aviation events that occurred for a certain period [7, 8]. However, it is extremely important to consider the age of occurrence. Hence by analogy with the aviation transportation system changeability thesis, operator is claimed to not remain with the same parameters set and conditions in which operational activities are carried out. Therefore, the impact of one or another aviation event occurring 20 years ago on the flight safety level and the event occurring 3 years ago, cannot be the same. This is the recommendation for taking into account the age of the aviation event, let's talk about them.

Currently, the Russian Federation of civil aviation consists of two general segments: civil aviation, providing services for the implementation of passenger air transportation service, baggage, cargo, and mail transportation, which refers to commercial civil aviation, and general aviation [1].

Operators operating in various civil aviation segments are to meet with the relevant federal aviation regulations requirements.

If the operator carries out air transportation of passengers, baggage, cargo, mail, performs aviation work, and performs flights as a general aviation airline, it must have several aircraft operator's certificates.

The point is to consider the operator and its flight safety level, the influence of a certain factor on the flight safety level is different than when the operator is considered for each available certificate separately. But the flight safety level is incorrectly determined by dividing the operator, taking into consideration its activities in separate segments of aviation activity. The safety effectiveness ensuring level isn't distorted only when the operator as a whole is taken into account, since only this reflects the real state of affairs in the field the effectiveness ensuring of the operator's safety [9].

3 Materials and Methods

Assessment of the current flight safety level with a data, with the level that must be achieved, that also known as the target flight safety level, may correspond to the acceptable flight safety level set by the state, or maybe better than it.

When setting the target safety level, it is necessary to take into account such factors as the prevailing level of risk to flight safety, the acceptability of the risk to flight safety, and expectations regarding the level of safety in a particular area of aviation activity [10, 11]. The correctness and effectiveness of the target levels will only be met when realistically achievable levels are set. It is also worth considering the dynamics of these indicators in recent years.

It is important to take into consideration the age of the operator since only the newly emerging operators have not yet been able to demonstrate the flight safety level, at a time when the operators who have been operating for a long time could already accumulate certain negative parameters that negatively affect the operator's flight safety index, but this will not necessarily mean that that the newly created carrier has a better flight safety level. It is especially recommended to take into account the experience of the flight employees since, at the moment in the existing methods for assessing the operator's safety level, such an important factor as the experience of the crew is not taken into account.

While creating a flight safety level assessing methodology, it is definitely necessary to take into consideration not only the aforementioned conventional flight safety indicators, but also the risks accompanying the operator's operational activities and other factors that can potentially affect this level, because in accordance with federal aviation regulations safety is known to be a state in which risks are reduced to an acceptable level and are held on to it. Risks, developing a methodology taking into account these recommendations, will mostly relate to the second group of parameters to particular factors, since this or that operator has its own characteristics of operating activities.

To consider and concretize the parameters is necessary—flight safety indicators, general and particular factors for the development of subsequent recommendations, as well as the basis of the methodology. While assessing the power supply level, it is proposed to start with a reference level of power supply efficiency [12], which is recommended to be expressed in so-called safety units.

To express flight safety level take the following L_{fs}. The reference flight safety level is 100 safety units $L_{fs} = 100e\partial$. Then each parameter is going to affect the final level L_{fs}.

There are the following indicators for flight safety level assessing: the relative aviation incidents number indicator (AI) and serious incidents (SAI) per 1000 flight hours; the number of aviation accidents (AA) for 20 years; the number of emergencies (EM) per 1000 flight hours; death toll in 20 years [13–15].

The relative aviation incidents number indicator (AI) and serious incidents (SAI) per 1000 flight hours is supposed to be calculated as:

$$i_{1000} = \left(\frac{n_{AI \times k} + n_{SAI \times k}}{T}\right) \times 1000 \tag{1}$$

where, i_{1000}—the relative aviation incidents number indicator (AI) and serious incidents (SAI) per 1000 flight hours; T—total flight hours; n_{AI}, n_{SAI}—amount of AI and SAI in certain period; k—correlation coefficient.

Moreover, the AI and SAI will have different correlating coefficients, since the difference between these two aviation events in the degree of probability of an accident is determined in international and domestic regulations. The correlating coefficient for the AI will be taken as equal to 1, for the SAI the correlated coefficient will be taken as equal to 1.3. This indicator should be noted to make a significant contribution to the final flight safety level assessment, so there is a serious flight safety impact—accident (Table 1).

Another indicator, called the number of emergency situations per 1000 flight hours, in mathematical terms is the similar to the previous one.

There is the number of emergencies (EM) per 1000 flight hours:

$$E_{1000} = \left(\frac{n_{EM}}{T}\right) \times 1000 \tag{2}$$

where, E_{1000}—EM amount per 1000 h; T—total flight hours; n_{EM}—amount of EM in certain period.

There is the suggestion to turn the E_{1000} to safety units, refers to Table 2.

Table 1 i_{1000} conversion to safety units

i_{1000}	0–0.0056	0.0057–0.012	0.013–0.021	0.022–0.033	More than 0.033
i_{1000}(units)	0	3	5	8	10

Table 2 E_{1000} conversion to safety units

E_{1000}	0–0.0010	0.0011–0.0015	0.0016–0.0025	0.0026–0.0037	More than 0.0037
E_{1000}(units)	0	1	2	4	7

Table 3 Accounting crash age

Age of crash	0–3	4–5	6–10	10–15	15–20
Additional coefficient	13	10	5	3	1

Table 4 Accounting accident age

Age of accident	0–3	4–5	6–10	10–15	15–20
Additional coefficient	18	14	10	5	3

The next flight safety indicator is the number of accidents that have occurred over 20 years, this is an absolute indicator. What is remarkable, it is advisable to use this indicator in absolute terms, since accidents occur rarely within the same operator. However, when taking into consideration this indicator, as in the case of the previous one, it is imperative to take into account the accident type, an accident brings the least consequences than a disaster. It should be noted that this indicator makes the greater negative contribution, the greater its value. And as mentioned earlier, it is imperative to take into consideration the age of the accident. There are the recommendations with an example of taking into consideration the age of the accident are presented in Tables 3 and 4.

Therefore this indicator is calculated as:

$$I = n_{1c \times K_i} + n_{2c \times K_i} + n_{ic \times K_i} + n_{1acc \times K_i} + n_{2acc \times K_i} + n_{iacc \times K_i} \tag{3}$$

where, I—aviation accidents amount for 20 years; n_c—crash; n_{acc}—accident; k_1—additional coefficient.

The indicator reflecting the number of death on board over 20 years is associated with the plane accidents, but as mentioned earlier the emergency may also be the result of death.

Therefore, this indicator, although not the main one, is taken into consideration in world practice and allows more fully to take into consideration the severity of emergency and accidental accidents. It is advisable to make this indicator relative, taking into consideration such a production indicator as the number of passengers carried over the period under consideration. The representation of it is in formulas below.

$$V = \left(\frac{nn_{\text{Death}}}{p}\right) \times 10^6 \tag{4}$$

where, V—death onboard for 20 years; nn_{death}—death onboard in certain period; p—amount of passengers with transportation service in certain period.

The resulting value must be recalculated. Recalculation example from value to safety units is represented in Table 5.

Table 5 V indicator conversion to safety units

V value	0	0.1–0.47	0.48–1	1–2	More than 2
Safety units	0	5	8	15	20

There are the second group of safety parameters – general factors:

- history of the operator's certificate;
- the age of the operator;
- ACFT type;
- the age of the operator's ACFT;
- the average total flight time of the operator's flight crews;
- topography and climatic conditions of the flight region;
- flight region aviation complex infrastructure.

For example, let's look at some of them.

Factor of topography and climatic conditions of the flight region. According to the analysis of FS, the share of adverse meteorological conditions accounts for only 3% of the AA. However, do not forget that, as a rule, AA is a combination of active failures and hidden conditions, that is, in special situations in flight, the external environment, whether it is meteorological conditions or difficult terrain, can predetermine the outcome of a special situation. Operators perform flights along different routes, with different climatic, topographical, meteorological features. Different regions have their own characteristics and carry certain risks for FS. For example, flights in mountainous areas, in accordance with the Federal Aviation Regulations, refer to flights in special conditions. Flights in polar conditions, in the conditions of the far north, high-altitude airfields, long-haut flights over the water space, and flights in adverse conditions also pose risks to the FS. To take into account this factor, it is recommended to divide flight routes into risk classes for five categories of conditions that determine risk—Table 6.

This factor will be designated as Rvu. Each category contains a route characteristic, the compliance of which will be considered as the presence of a hazard with its

Table 6 Accounting for the Topography and climatic conditions of the flight region

Characteristics of the route	Risk index r_{By}	
Flights to high-altitude airfields	5	
Flights in mountainous areas	4	
Flights with frequent local adverse weather conditions	Frequency r/month	Risk index r_{By}
	More than 5	5
	3–5	3
	0–3	2
Flying in the far North	4	
Flights over the water area	4	

inherent risk index, this indicator is evaluated in points (safety units) and is defined as:

$$R_{By} = r_{By} + r_{2By} + r_{iBy} \tag{5}$$

where, r_{By}—risk index.

This indicator is expressed in safety units.

The next safety parameter, which is a common factor, is the infrastructure of the aviation complex of the flight region. As of 01.01.2019, 241 units are registered in the State Register of Airfields and Heliports of the Civil Aviation of the Russian Federation. First of all, when we talk about infrastructure for aviation, we mean technical equipment with navigation aids, light-signal equipment, and the state of the runway. At the moment, the most used precision approach system is the ILS instrumental approach system. This system provides guidance in the horizontal and vertical plane and provides high accuracy, which increases with increasing approach category. Of the 241 airfields in the Russian Federation: 38 airfields are operated in category I; 12 airfields are operated under category II; four airfields are operated under the III category of accurate approach. There are also less accurate approach systems for approaching inaccurate systems, such as: VOR and NDB. The VOR system provides higher accuracy than the NDB system, however, it still does not provide guidance in the vertical plane. The basis of the navigation field of the Russian Federation when flying along the routes is created by 130 sets of drive radio stations (NDB). Also, navigation is carried out using 67 VOR beacons, usually combined with the DME rangefinder. One of the most modern methods of navigation is zonal navigation, which is set out in the concept of PBN, the concept provides for the gradual abandonment of ground infrastructure in favor of zone navigation. In some regions of Europe, this concept is already being fully implemented. Of the 241 airfields, 180 have an artificial hard surface, 60 airfields have a dirt runway and one airfield has a hydraulic runway.

This factor, we denote it as A, is recommended to be taken into account in terms of the equipment of the airfields to which the operators fly. An example of parameter A values in safety units is specified in Table 7.

This indicator is taken into account according to the maximum value and is summed up only within the framework of different categories of conditions. Under

Table 7 Parameter A value under different conditions

Condition categories	Conditions	Value of A (unit)
1	Precise approach	0
	Inaccurate approach (except zonal navigation)	2
	Zonal navigation	0
2	Only Dirt runway or water area	3
	Artificial runway	0

category 1 conditions, accounting is made according to the prevailing value, for example, if the operators perform flights to aerodromes of 55% or more of which have the worst value A, the worst value is taken into account. Within the framework of the conditions of category 2, the calculation is made according to the worst value, that is if at least one airfield has the worst value A.

The next common factor is the age of the operator's aircraft. Each aircraft approved for flight meets the airworthiness standards established by the state. The recommendation to take this factor into account does not carry a declaration that the older the aircraft, the more unreliable it is, but there are studies in the field of reliability, survivability, and fault tolerance that conduct, by a logical and probabilistic method, a relationship between the age of the aircraft and the frequency of failures. In these studies, the fault flow parameter is taken as the main indicator of the reliability of systems and units. The dynamics of changes in the parameter of the flow of failures is determined depending on the operating time of the aircraft in flight hours or flights, for our tasks the most suitable is the dependence of the failure flow parameter on the age of the aircraft. And depending on the time calendar duration of operation, the task of assessing the operational level of reliability of the units is solved. The dependence of the failure flow parameter on the age of the aircraft is shown in Fig. 1.

This factor, we denote it as Y, is expressed in safety units and is taken into account by the maximum value from the proposed table of values—Table 8.

If the aircraft fleet of the operator is made up of aircraft of different ages, in this case, it is recommended to use the following rule: If 50 percent or less of the aircraft by age fall into the category with the lowest (best) value of Y, and the rest in the worst category, the calculation is made according to the largest (worst) value of factor Y; If between 51 and 88% of the suns by age fall into the category with the lowest Y value, and the rest in the worse category, the calculation is made according to the average value of Y; If 88% or more of the aircraft by age fall into the category with the smallest value Y, and the rest in the worse category, the calculation is made according to the best (lowest) value Y. For example, if the operator has 10 aircraft

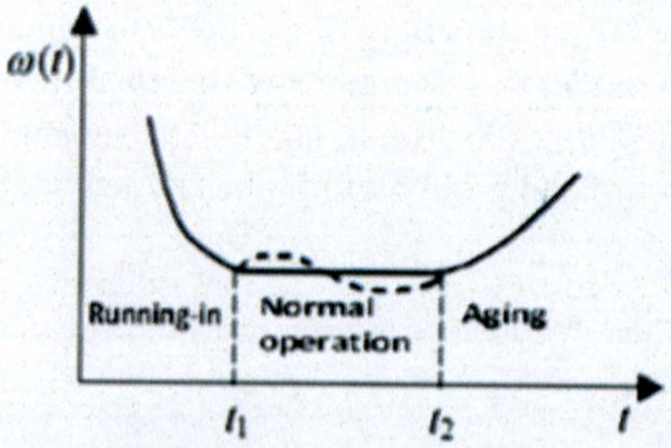

Fig. 1 Dependence of the parameter of the flow of failures depending on age

Table 8 Value of parameter Y depending on the age of the aircraft

Aircraft age (years)	0–2	2–19	19–28	More than 28
Value of Y (units)	4	0	2	15

Table 9 Safety parameters applicability

L_{fs} value	L_{fs} characteristics	Recommendations
100–85	Acceptable	Acceptable level, no corrective actions required
84–80		An acceptable level can be a trigger. Corrective action may be required due to a specific safety parameters
79–60	Admissible	May be allowed on the basis of measures to improve the flight safety level. Acceptance of this level requires a comprehensive analysis of safety parameters
Up to 59	Unacceptable	Immediately take action to improve flight safety or cease operations. Implement priority safety risk mitigation actions to move to the acceptable L_{fs} value

and the age of three of them is from 5 to 12 years, and the age of seven of them is more than 28 years, then the parameter Y will be equal to 15 safety units based on the values of Table 8.

While performing flight safety assessing method, there is the simple flight safety level formula L_{fs}.

$$L_{fs} = 100 - i_{1000} - E_{1000} - I - \mathrm{V} - \mathrm{K}_{\text{общ}} - R_{\text{ву}} - \mathrm{A} - \mathrm{Y} - \mathrm{C} - \mathrm{T} - \mathrm{M} - \mathrm{P} - \mathrm{U} - W - S \tag{6}$$

If there are the inapplicable private factors, these points turn to zero.

Table 9 below proposes a final operator's safety level assessment, depending on the obtained L_{fs} value, according to which recommended actions are proposed to improve the operator's safety level.

4 Conclusions

Use of multifactorial flight safety level assessment methodology with all operators' operation features makes it possible to create a comparative flight safety level assessment of various operators, which will make it possible to take the necessary measures to improve the civil aviation flight safety level.

References

1. Sharov VD, Vorobyov VV, Zatuchny DA (2021) Concept of risk and safety: Analysis of aviation safety regulations. In: Probabilistic-statistical methods for risk assessment in civil aviation. Springer Aerospace Technology. Springer, Singapore. https://doi.org/10.1007/978-981-16-0092-0_1
2. Yakovlev AV, Istomin AS, Zatuchny DA, Shatrakov YG (2021) Analysis of the problem functioning modeling ergatic air traffic management information system. In: Conditional function

control of aircraft. Springer Aerospace Technology. Springer, Singapore. https://doi.org/10.1007/978-981-16-1059-2_1
3. Prikaz Mintransa Rossii ot 19.11.2020 N 494 "Ob utverzhdenii federal'nyh aviacionnyh pravil trebovaniya k yuridicheskim licam, individual'nym predprinimatelyam vypolnyayushchih aviacionnye raboty, vklyuchennye v perechen' aviacionnyh rabot, predusmatrivayushchih poluchenie dokumenta, podtverzhdayushchego sootvetstvie trebovaniyam federal'nyh aviacionnyh pravil, yuridicheskogo lica, individual'nogo predprinimatelya. Forma i poryadok vydachi dokumenta" (Order of the Ministry of Transport of Russia dated 19.11.2020 N 494 "On approval of federal aviation rules requirements for legal entities, individual entrepreneurs performing aviation work included in the list of aviation works, providing for the receipt of a document confirming compliance with the requirements of federal aviation rules, a legal entity, an individual entrepreneur. Form and procedure for issuing the document"). https://normativ.kontur.ru/document?moduleId=1&documentId=380611. Accessed 05 Nov 2021 (in Russian) (2021)
4. Prikaz Mintransa Rossii ot 31.07.2009 N 128 (red. ot 22.04.2020) "Ob utverzhdenii federal'nyh aviacionnyh pravil «Podgotovka i vypolnenie poletov v grazhdanskoj aviacii Rossijskoj Federacii" (Order of the Ministry of Transport of Russia dated 31.07.2009 N 128 (as amended on 22.04.2020) "On approval of federal aviation rules "Preparation and performance of flights in civil aviation of the Russian Federation"). https://docs.cntd.ru/document/902172421. Accessed 05 Nov 2021 (in Russian) (2021)
5. Prikaz Ministerstva transporta RF ot 13 avgusta 2015 g. N 246 «Ob utverzhdenii Federal'nyh aviacionnyh pravil "Trebovaniya k yuridicheskim licam, individual'nym predprinimatelyam, osushchestvlyayushchim kommercheskie vozdushnye perevozki. Forma i poryadok vydachi dokumenta, podtverzhdayushchego sootvetstvie yuridicheskih lic, individual'nyh predprinimatelej, osushchestvlyayushchih kommercheskie vozdushnye perevozki, trebovaniyam federal'nyh aviacionnyh pravil" (Order of the Ministry of Transport of the Russian Federation of August 13, 2015 N 246 "On approval of the Federal Aviation Rules"). https://base.garant.ru/71216992. Accessed 01 Nov 2021 (in Russian) (2021)
6. Xiong S-H, Chen Z-S, Chang J-P, Chin K-S (2019) On extended power average operators for decision-making: a case study in emergency response plan selection of civil aviation. Comp Industrial Eng 130:258–271, April. https://doi.org/10.1016/j.cie.2019.02.027
7. International Civil Aviation Organization (2013) Appendix 19 to the convention on International Civil Aviation. Flight Safety Management, 1st edn. ICAO, Quebec. https://www.icao.int/Meetings/a40/Documents/10004_ru.pdf. Accessed 7 September 2021 (in Russian) (2021)
8. Doc 9859 AN/460 "Safety Management Manual (RUBP)" (2018) International Civil Aviation Organization, 4th edn. http://www.scac.ru/ru/wp-content/uploads/2016/DOC_9859_3.pdf. Accessed 7 Sept 2021 (in Russian) (2021)
9. Besogonov VV, of flight Kostylev A.G. O neobhodimosti razrabotki mnogofaktornoj metodologii ocenki urovnya bezopasnosti poletov (On the need to develop a multifactorial methodology for assessing the level safety). Problemy letnoj ekspluatacii i bezopasnosti poletov (Problems of flight operation and flight safety) – St. Petersburg State University of Civil Aviation, 2021, 192–193 (In Russian)
10. Shi D, Guan J, Zurada J, Manikas AA (2017) Data-mining approach to identification of risk factors in safety management systems. J Manag Inf Syst 34(4):1054–1081. https://doi.org/10.1080/07421222.2017.1394056
11. Rybalkina AD, Trusova EI, Sharov VD (2018) Metodika ocenki riska predstoyashchego poleta dlya vertoletov s uchetom neblagopriyatnyh meteouslovij (Methods of risk assessment of the upcoming flight for helicopters taking into account adverse weather conditions). Nauchnyj vestnik MGTU GA. https://doi.org/10.26467/2079-0619-2018-21-6-124-140. (In Russian)
12. Karanikas N, Chionis D (2019) Tracing new safety thinking practices in safety investigation reports. MATEC Web of Conferences. https://doi.org/10.1051/matecconf/201927301001
13. Zhao Q, Li Q, Wen J (2018) Construction and application research of knowledge graph in aviation risk field. MATEC Web of Conferences. https://doi.org/10.1051/matecconf/201815105003

14. Thomas J, Davis A, Samuel MP (2020) Integration-in-totality: The 7th system safety principle based on systems thinking in aerospace safety. Aerospace (MDPI Publishing) 7(10):149–149. https://doi.org/10.3390/aerospace7100149
15. Rose RL, Puranik TG, Mavris DN (2020) Natural language processing based method for clustering and analysis of aviation safety narratives. Aerospace (MDPI Publishing) 7(10):143–143. https://doi.org/10.3390/aerospace7100143

Application of ADS-B for Providing Surveillance at Civil Aviation Regional Aerodromes

Andrey Kalintsev, Evgeniy Rubtsov, and Nikolay Povarenkin

Abstract Automatic Dependent Surveillance-Broadcast is an inexpensive promising technology that can ensure the required surveillance quality. This technology, however, has some shortcomings and vulnerabilities due to which the ICAO insists on confirming the ADS-B data by other surveillance tools, including radars and multi-position systems, which can increase surveillance infrastructure costs. For regional Russian aerodromes, particularly those located in the Arctic zone, the economic component is crucial. The task to conduct observations using the ADS-B stations and alternative coordinate confirmation methods is urgent. The article provides an overview of methods employed in the algorithmic confirmation of ADS-B data reliability based on estimating mathematical expectation and standard deviation (for different levels of reliability) of normally distributed aircraft positioning errors. The gating method, which involves selecting an area around the extrapolated aircraft coordinate, is proposed. The methodology can be applied to describe the ground and air traffic in the working area of aerodromes and the aerodrome area.

Keywords Flight safety · ADS-B · Data confirmation · Gate · Airfield area · Surveillance equipment

A. Kalintsev
Arkhangelsk ATM Center of the North-West Air Navigation Branch of the Federal State Unitary Enterprise State ATM Corporation in the Russian Federation, Mezen, Russia

E. Rubtsov (✉)
Saint Petersburg State University of Civil Aviation, Saint Petersburg, Russia
e-mail: rubtsov.spb.guga@rambler.ru

N. Povarenkin
Saint Petersburg State University of Aerospace Instrumentation, Saint Petersburg, Russia

O. A. Gorbachev et al. (eds.), *Proceedings of 10th International Conference on Recent Advances in Civil Aviation*, Lecture Notes in Mechanical Engineering, https://doi.org/10.1007/978-981-19-3788-0_33

1 Introduction

According to the Decree of the President of the Russian Federation No. 645, "On the Development Strategy for the Arctic Zone of the Russian Federation and National Security Insurance until 2035" of October 26, 2020, threats to the development of the Arctic Zone of the Russian Federation (AZRF) include a poorly developed transport infrastructure at small aerodromes. The Strategy entails improving and developing aviation infrastructure. The cost of facilities is a crucial factor.

The Automatic Dependent Surveillance-Broadcast (ADS-B) technology implemented at regional aerodromes in the Russian Arctic meets the objectives set by the Strategy since modern surveillance equipment is not expensive.

The ICAO documents refer to ADS-B as a promising technology. The ICAO documents emphasize the economic benefits of implementing ADS-B ground systems, which are more efficient than expanded radar observation zones in remote and hard-to-reach regions [1]. Thus, the ADS-B technology can reduce infrastructure and maintenance costs.

However, the ICAO Aeronautical Surveillance Operations Manual contains a list of ADS-B vulnerabilities [2]. It is recommended that coordinates received via the 1090ES data link are confirmed by radar stations (RS) or multi-position surveillance systems (MPSS).

It is more difficult to install radar and MPS than ADS-B ground stations, located in small settlements, remote and hard-to-reach areas. Financially, implementing these systems at regional civil aviation aerodromes is economically ineffective and runs contrary to the Strategy. Large distances between settlements and the seasonal nature of transport links are characteristic of the Russian Arctic.

Thus, it is necessary to develop algorithmic ADS-B data confirmation methods. The methods will allow the use of the ADS-B system as the primary tool for controlling aircraft in the airfield zone and the working area of regional airfields, including those located in the Russian Arctic.

The problems associated with the accuracy and reliability of ADS-B data are described in sufficient detail [3–6].

The ADS-B system validates data based on the analysis of message subfields (NIC, NACP, NUC, SIL): navigation data's integrity, accuracy and reliability, and source integrity levels [7].

In [6], the method of ADS-B data reliability estimation was proposed. It can be used in conducting observations and controlling the aircraft. The method includes two stages. The ADS-B message subfields (NIC, NACP, SIL, NUC) are estimated at the first stage. In the second stage, a digital trajectory is constructed, and deviation parameters are estimated.

The methods proposed in [6] involve the accumulation of measurements of aircraft coordinates and the statistical analysis of data received. They are developed to ensure ADS-B data reliability in conducting observations and controlling the aircraft. The present study proposes a method based on the one described in [6]. It can be used to conduct observations in the airfield area based on the gating method, widely used in radar systems [8, 9].

2 Materials and Methods

Following [2], positioning errors have a normal distribution. The parameters of a normally distributed random variable such as the mathematical expectation and the positioning error variance concerning the digital (nominal) aircraft trajectory can be estimated. The digital trajectory is a sequence of points (coordinates) of an aircraft trajectory in the plane, including the permissible deviation width and other parameters. Two criteria are checked. Criterion 1 is given by the following inequality [6].

$$|\tilde{m}| < {}^1\!/_4 W_{\mathrm{RW(TW)}} + \varepsilon_m, \tag{1}$$

where $\tilde{m}$ is the mathematical expectation of the lateral deviation from the digital trajectory, obtained from the ADS-B data processing results;

${}^1\!/_4 W_{\mathrm{RW(TW)}}$ is the permissible deviation from the digital trajectory;

$W_{\mathrm{RW(TW)}}$ is the runway (taxiway) width;

ε_m is the lateral deviation estimation accuracy.

The mathematical expectation is based on the ADS-B observations—coordinates received from the aircraft. The following formula calculates the mathematical expectation [10]:

$$\tilde{m} = \frac{\sum_{i=1}^{n} x_i}{n}, \tag{2}$$

where n is the number of observations.

An additional condition for the lateral deviation estimation accuracy is

$$\varepsilon_m \leq {}^1\!/_4 W_{RW(TW)} \tag{3}$$

This condition ensures that the estimated random variable falls within half the runway width. The lateral deviation estimation accuracy is calculated by the formula [10].

$$\varepsilon_m = \sigma[\tilde{m}] \arg\Phi\left(\frac{\beta_\varepsilon}{2}\right), \tag{4}$$

where $\arg\Phi(x)$ is the reverse function of $\Phi(x)$;

$\sigma[\tilde{m}]$ is the mean square deviation (MSD) of the mathematical expectation estimate for the lateral deviation from the digital trajectory;

β_{ε} is the reliability of the estimate $\tilde{m}$ of the lateral deviation.

Reliability values close to unity are scommonly used (e.g., 0.95; 0.99 or 0.999, for which the Student's coefficient values $t_{\beta} = \arg \Phi\left(\frac{\beta_2}{2}\right)$ are equal to 1.960; 2.576 and 3.290, respectively). The following formula calculates the MSD [10].

$$\sigma[\tilde{m}] \approx \sqrt{\frac{\tilde{D}}{n}}, \tag{5}$$

where $\tilde{D}$ is the lateral deviation variance estimation.

The variance is estimated based on the ADS-B data processing results by the following formula [6]

$$\tilde{D} = \left(\sum_{i=1}^{n} (X_i - \tilde{m})^2\right) / (n-1), \tag{6}$$

$$\tilde{m} = \frac{1}{n} \sum_{i=1}^{n} X_i \tag{7}$$

Criterion 1 is given by the following inequality

$$\tilde{D} < D^{l} + \varepsilon_D, \tag{8}$$

where $\tilde{D}$ is the lateral deviation variance estimate obtained from the ADS-B data processing results;

D^{l} is the limit variance estimate value X_i;

ε_D is the variance estimation accuracy.

Condition (8) means that the lateral deviation variance calculated using ADS-B data with reliability β_{ε} does not exceed the sum of estimation accuracy ε_D and given limit value D^{l}.

The following expression gives the additional accuracy condition for the variance estimate:

$$\varepsilon_D = \sigma\left[\tilde{D}\right] t_{\beta} \leq D^{l}, \tag{9}$$

where $\sigma\left[\tilde{D}\right]$ is the MSD of a variance estimate.

Application of criteria 1 and 2 assumes the calculation of confidence intervals for a random variable based on n ADS-B observations. The variance estimate $\left(\tilde{D}\right)$ is calculated by formulas (6, 7), while the standard deviation is calculated by the following formula [6]:

$$\sigma\left[\tilde{D}\right] \approx \sqrt{\frac{2}{n-1}}\tilde{D} \tag{10}$$

Additional condition (10) can be written as

$$\sqrt{\frac{2}{n-1}}\tilde{D}t_{\beta} \leq D^{\mathrm{l}} \tag{11}$$

The variance is estimated using the accumulated ADS-B data. The number of measurements n required to fulfill the additional condition is expressed in condition (11) and calculated as [6]

$$n \geq 1 + 2\left(\frac{\tilde{D}t_{\beta}}{D^{\mathrm{l}}}\right)^2 \approx 1 + 2t_{\beta}^2 \tag{12}$$

The gating method was applied to estimate the reliability of ADS-B data in the airfield area. The gate is an area around the extrapolated aircraft mark. The method of algorithmic verification of ADS-B data involves the estimation of positioning error parameters. According to Russian Certification requirements (Basis) for wide-area multi-sensor surveillance system (approved by the Department of Development Programs of the Ministry of Transport of the Russian Federation №08–04/5228-IS of March 15, 2018), the gate area dimensions are based on the standard deviation of the surveillance system. The membership condition for estimating the mathematical expectation can be written as [11]

$$|\tilde{m} + \varepsilon_M| \leq \frac{R}{2}, \tag{13}$$

where ε_M is the positioning error, $\varepsilon_M < \frac{R}{2}$;

$\tilde{m}$ is the estimate of the mathematical expectation of the lateral aircraft deviation from the extrapolation line obtained from the ADS-B data;

R is the gate area radius.

Formula (2) calculates the mathematical expectation of the lateral aircraft deviation from the extrapolation line obtained from the ADS-B data. Formula 6 calculates the unbiased variance that is based on the ADS-B data. To estimate the variance, the membership condition was introduced

$$\left|\tilde{D} + \varepsilon_D\right| \leq D_l, \tag{14}$$

where ε_D is the accuracy of estimation of the aircraft deviation variance;

D_l is the limit variance estimate, $D_l = \left({}^{R}/_{2}\right)^2$.

The additional condition is

$$\varepsilon_D = \sigma_{\tilde{D}} t_\beta \leq D_l, \tag{15}$$

where $\sigma_{\tilde{D}}$ is the MSD of the variance estimate.

Formula (6) calculates the variance. Formula (10) [6] calculates the standard variance deviation by taking the observation as a normally distributed random variable [2]. Taking into account additional conditions (15), the expression can be written as

$$\sqrt{\frac{2}{n-1}}\tilde{D}t_\beta \leq D_l \tag{16}$$

Formula (12) calculates the required number of observations.

The mathematical expectation and the positioning error variance are calculated with a given degree of reliability. After the parameters are calculated, the membership condition is checked at each iteration. To calculate the mathematical expectation, condition (13) must be fulfilled; to calculate the variance, condition (14) must be fulfilled. If the membership condition is fulfilled, the ADS-B data are considered reliable; if the condition is not fulfilled, the ADS-B data are considered unreliable and have to be validated in an alternative way, or other air traffic control methods should be used.

The ADS-B data confirmation method in the aerodrome area is applied according to the following algorithm:

1. When the aircraft is approaching the maneuver point, the ATC AS initiates the extended Mode *S* request by the secondary radar.
2. In response to the request, the airborne transponder transmits a Mode *S* message with its current coordinates and additional intent information.
3. The aircraft trajectory is extrapolated while checking that the maneuver scheme coordinates received from the onboard transponder belong to the constructed trajectory.
4. The mathematical expectation and the standard deviation of positioning errors are calculated to the constructed trajectory.
5. Based on the calculation results, the membership conditions are checked for various reliability values β.
6. If the conditions are fulfilled, the ADS-B data are considered reliable. If the conditions are not fulfilled, the coordinates are considered unreliable, and the controller is informed about the impossibility of using the ADS-B data as the main means of observation and the need to confirm the data.

3 Results

In accordance with the Russian Certification requirements (Basis) for wide-area MPSS, the horizontal positioning error should not exceed 350 m for the route zone. In accordance with the requirements of Eurocontrol, for providing joint surveillance and separation of 3 NM, the horizontal error should not exceed 300 m [12].

Take the radius (R) of the gate area equal to 300 m, which meets the Eurocontrol and certification requirements. The gate is a circle whose center coordinates are maneuver scheme points. The membership of the mathematical expectation and the positioning error variance are determined from the conditions $|\tilde{m} + \varepsilon_M| \le {}^{R}\!/_{2}$ and $\left|\tilde{D} + \varepsilon_D\right| \le D_l$, where $D_l = \left({}^{R}\!/_{2}\right)^2$.

The fulfillment of these conditions means that the estimated values are within the gate with a given probability (reliability β).

Estimates of the mathematical expectation and the standard positioning error deviation were simulated in MATLAB (the mathematical expectation and standard deviation estimation graphs are shown in Figs. 1, 2, 3 and 4). In determining the membership condition, the limit value was 150 m. The initial data were: 50–110 m for the ME and 50–70 m for the MSD.

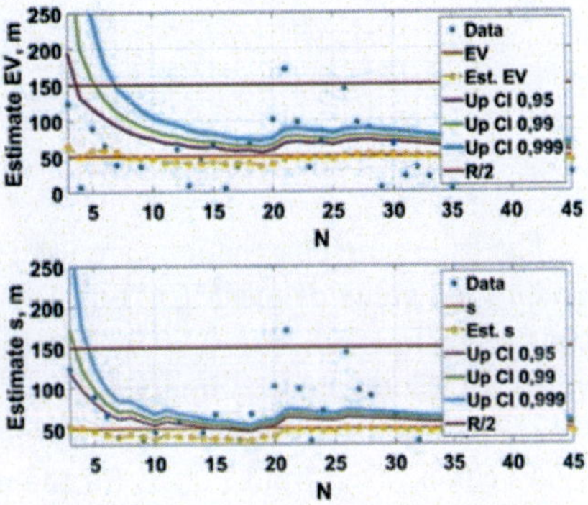

Fig. 1 Simulation results with the ME value of 50 m and the MSD value of 50 m

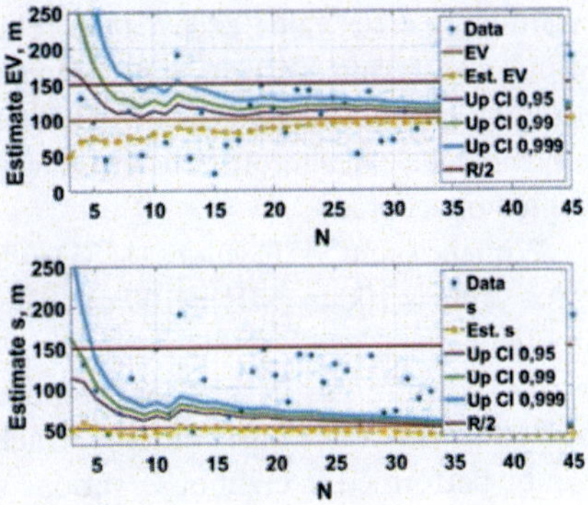

Fig. 2 Simulation results with the ME value of 100 m and the MSD value of 100 m

Fig. 3 Simulation results with the MSD value of 100 m and the MSD value of 100 m

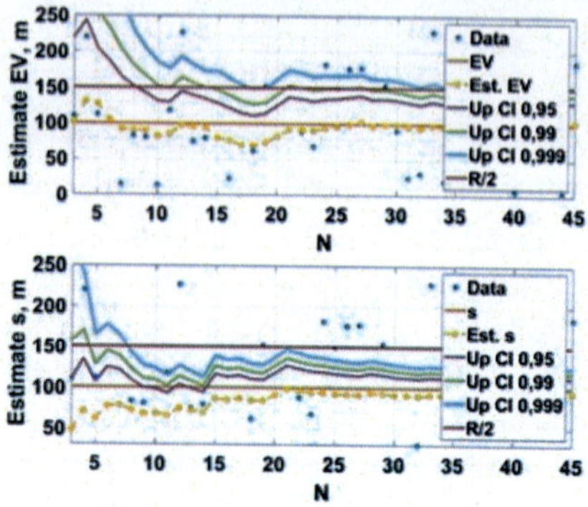

Fig. 4 Simulation results with the ME value of 110 m and the MSD value of 70 m

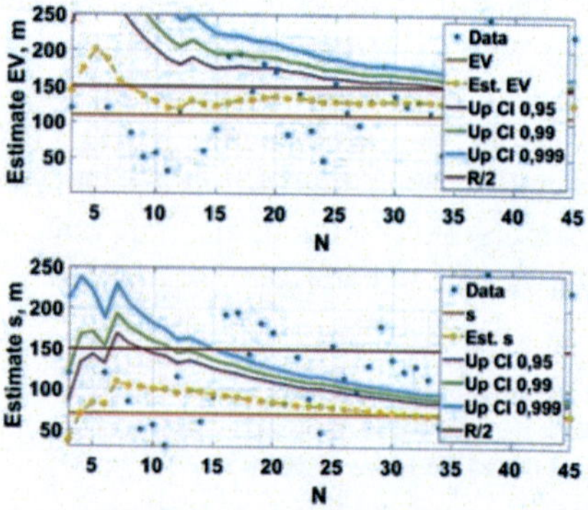

The following designations are used in the figures: "Data"—input data, "EV"—established value of the ME, "Est. EV"- ME estimate, "Up CI 0.95"- upper 95% confidence interval, "Up CI 0.99"- upper 99% confidence interval, "Up CI 0.999"- upper 99.9% confidence interval, "R/2"- data validity condition, "s"—given MSD value, "Est. s"—MSD estimate, "N"- number of observations.

As can be seen from Fig. 1, for the upper confidence interval (Up CI) of 0.999, seven observations (4 s) are required to estimate the ME (provided that two messages per second are received). To estimate the MSD, five observations are required. The positioning error value exceeds the reliability threshold of the GLONASS system, which is given in Russian State Standard GOST 32,454–2013 "Global navigation satellite system. Parameters of radio navigation field. Technical requirements and test methods". Therefore, this error will be detected and identified with a required degree of accuracy.

With the same MSD values (MSD = 50 m), consider the case when the systematic error value is 100 m (ME = 100 m). The estimation simulation results are presented in Fig. 2.

As shown in Fig. 2, the MSD estimation with a confidence interval of 0.999 can be performed in five observations, which require three seconds. The ME estimation can be performed in eight observations which require at least four seconds.

Figure 3 shows the simulation results for the MSD of 100 m and the MSD of 100 m.

As can be seen from Fig. 1, for the upper confidence interval (Up CI) of 0.999, seven observations (4 s) are required to estimate the ME (provided that two messages per second are received). To estimate the MSD, five observations are required. The positioning error value exceeds the reliability threshold of the GLONASS system, which is given in Russian State Standard GOST 32,454–2013 "Global navigation satellite system. Parameters of radio navigation field. Technical requirements and test methods". Therefore, this error will be detected and identified with a required degree of accuracy.

As shown in Fig. 3, the ME estimation can be performed in 18 observations; however, immediately after the fulfillment of the membership condition, outliers are observed, which indicates the unstable operation of the algorithm in the presence of a positioning error approaching the upper limit. The MSD estimation is performed in seven observations which require at least 4 s.

Of interest is when the systematic error is close to half the radius of the gate area, and the sum of ME and MSD values exceeds the size of half the radius of the gate area. Figure 4 shows the simulation results for the ME of 110 m and the MSD of 70 m.

Figure 4 shows that for the limit MSD and ME value equal to 70 m and 110 m, respectively, the membership condition is not fulfilled. The MSD estimate with Up CI 0.999 fulfills the condition after 15 observations, which require 8 s (provided that two messages per second are received). The ME estimate for the upper confidence intervals Up CI 0.999 Up CI 0.99 does not fulfill the membership condition even for a larger number of observations.

In this scenario, the data will not be confirmed. The controller should be warned of the impossibility of using ADS-B as the main monitoring tool and the need to confirm the data.

4 Discussion

The number of observations is of theoretical and practical interest. In [6], it was revealed that for different reliability values, the most significant required number of observations is used when validating data following Criterion 2. When calculating the required number of observations, it was assumed that $D^{\mathrm{l}} = 49$, and $\tilde{D} \approx D^{\mathrm{l}}$. Thus, for reliability values β_{ε} 0.95; 0.99, or 0.999, the number of observations was 9, 15, and 22, respectively [6].

The simulation revealed that the number of observations required at different MSD and ME values vary depending on random variable values. For small MSD (50 m) and ME (50 m) values, the random variable can be estimated in seven observations, which require 4 s. With an increase in the values, the number of observations will increase.

If there is a significant systematic error of 100 m for the ME and 100 m for the MSD, outliers are observed after 18 observations, i.e., the membership condition is not fulfilled. Thus, if there is a positioning error approaching the upper limit, the algorithm may be unstable.

The estimated input data are not confirmed at the ME value of 110 m and the MSD value of 70 m; information on the impossibility of using ADS-B data should be forwarded to the controller.

The simulation revealed that when the systematic error and standard deviation values approach the limit, the algorithm may have outliers. With these values, the number of observations increases and can exceed 34 (Fig. 3), which requires 17 s (provided that two messages per second are received). When controlling the aircraft in the terminal area, estimating the parameters for an acceptable number of observations is necessary.

In order to reduce the required number of observations, an additional measuring channel can be used. The automatic radio direction finder (ARDF) can be used as a channel for measuring the angular coordinate of aircraft using the ADS-B signals. Estimating the required number of observations in the joint processing of data received from the ATM and the ADS-B ground station is an avenue for further research.

If there is no systematic error, the membership condition can be determined using the Rayleigh distribution. If there is a systematic error with equal MSD values along the *Ox, Oy* axes of the rectangular Cartesian coordinate system, the probability of remaining in the gate area of radius *R* can be determined using the Rice distribution [13].

In practice, the most probable situations are the positioning errors along the *Ox, Oy* axes differ. With the ME equal to zero and different MSDs along the *Ox, Oy* axes of the rectangular Cartesian coordinate system ($m = 0, \sigma_x \neq \sigma_y$), the probability of remaining in the gate area can be determined using the Hoyt distribution (Nakagami-q) [14, 15]:

$$w\{r\} = \frac{r}{\sigma_x \sigma_y} \exp\left[-\frac{r^2}{4}\left(\frac{1}{\sigma_x^2} + \frac{1}{\sigma_y^2}\right)\right] I_0\left[\frac{r^2}{4}\left(\frac{1}{\sigma_x^2} - \frac{1}{\sigma_y^2}\right)\right] \tag{17}$$

If there is a systematic error (the ME is not equal to zero) and unequal MSDs $\sigma_x \neq \sigma_y$, the probability of remaining in the gate area can be determined using the Beckmann distribution. The Rayleigh, Rice, and Hoyt distribution laws are exceptional cases of the Beckmann distribution [16]. Thus, the Beckmann distribution includes all possible typical situations of the aircraft positioning error distribution and can calculate the probability of aircraft remaining in the gate area.

Thus, it is necessary to develop a method that can estimate the probability of aircraft remaining in the gate area using the Rayleigh, Rice, Hoyt, and Beckmann distributions, which will make it possible to identify typical situations that occur during the flight.

5 Conclusions

The present paper described the methods used to check the validity of ADS-B data to assess the possibility of using ground stations as the primary observation tool. The methods can be applied to conduct observations at regional aerodromes to estimate the reliability of the information in the aerodrome area. Information about unreliable ADS-B data can be displayed in automatic mode, which will reduce the load on the controller controlling the ground and air traffic.

The ME and MSD of normally distributed aircraft positioning errors were estimated in MATLAB. The simulation revealed that under normal operating conditions of the satellite navigation systems (the ME value of 10 m and the MSD value of 20 m), it is possible to calculate the required estimates and confirm data in no more than five observations, which require three seconds or less.

With the ME and MSD values of 50 m, the required estimates can be calculated with a confidence interval of 0.999 in 7 observations, which require 4 s. The simulation revealed that the algorithm might be unstable with a significant standard deviation (after 18 observations, outliers were observed on the graph) (Fig. 3). In this case, the membership condition is not fulfilled, and the controller should be warned of the need to confirm ADS-B data. Thus, the method can identify an increase in the positioning error. The membership condition is not fulfilled for the ME value of 110 m and the MSD value of 70 m. The controller should be warned of confirming the aircraft coordinates received from the ADS-B ground station or using other procedural air traffic control methods.

The method can be used at aerodromes with a low air traffic level, characteristic of aerodromes located in the Russian Arctic.

References

1. Global air navigation plan for 2016–2030 (2016) https://www.icao.int/publications/Documents/9750_5ed_en.pdf. Accessed 12 November 2021
2. Aeronautical Surveillance Manual (2020) https://web.archive.org/web/20170824134230, http://www.aerohelp.ru/data/432/Doc9924.pdf. Accessed 29 November 2021
3. Soto A, Merino P, Valle J (2011) ADS-B integration in the SESAR surface surveillance architecture. In: Tyrrhenian International Workshop on Digital Communications—Enhanced Surveillance of Aircraft and Vehicles (TIWDC/ESAV), pp 13–18
4. Ali BS (2013) A study of ADS-B data evaluation and related problems. In: Proceedings of the International Technical Meeting of The Institute of Navigation, San Diego, California, January, pp 444–455
5. Zhao W, Liu D (2020) Integrity verification of ADS-B navigation data source. J Phys: Conf Ser 1570:012081. https://doi.org/10.1088/1742-6596/1570/1/012081
6. Plyasovskih AP, Rubtsov EA (2020) Teoreticheskoe obosnovanie podtverzhdeniya dostovernosti informacii o mestopolozhenii ob"ekta na rabochej ploshchadi aerodroma (Theoretical substantiation of confirmation of the validity of information about the location of the object on the work area of the aerodrome). T-Comm: Telekommunikacii i transport 14(3): 32–40. https://doi.org/10.36724/2072-8735-2020-14-3-32-40. (in Russian)

7. RTCA DO-260B. Minimum operational performance standards for 1090 MHz extended squitter automatic dependent surveillance – broadcast (ADS-B) and traffic information services—broadcast (TIS-B) (2009). https://standards.globalspec.com/std/14356196/rtca-do-260. Accessed 12 November 2021
8. Bar-Shalom Y (2001) Estimation with applications to tracking and navigation
9. Vovk V, Prokopenko I, Prokopenko K (2014) Sequential target tracking based on local trajectory parameters estimation in Rayleigh clutter. In: Proceedings International Radar Symposium, pp 1–5. https://doi.org/10.1109/IRS.2014.6869249
10. Polyanin A, Manzhirov A (2006). Handbook of mathematics for engineers and scientists. https://doi.org/10.1201/9781420010510
11. Kalintsev AS, Rubtsov EA, Plyasovskih AP (2021) Podtverzhdenie dannyh AZN-V v aerodromnoj zone metodom strobirovaniya (Confirmation of ADS-B data in the aerodrome traffic zone by gating method). T-Comm: Telekommunikacii i transport 15(6):39–49. https://doi.org/10.36724/2072-8735-2021-15-7-39-49 (in Russian)
12. Eurocontrol Specification for ATM Surveillance System Performance (2015) https://www.eurocontrol.int/sites/default/files/publication/files/201509-eassp-specification-vol1-v1.1.pdf. Accessed 20 November 2021
13. Rajan D (2017). Probability, random variables, and Stochastic processes. https://doi.org/10.1201/9781351105668-7
14. Mitra R, Mishra AK, Choubisa T (2012) Maximum likelihood estimate of parameters of Nakagami-m distribution. In: 2012 International Conference on Communications, Devices and Intelligent Systems (CODIS), pp 9–12. https://doi.org/10.1109/CODIS.2012.6422123
15. Wolfram. Nakagami Distribution. https://reference.wolfram.com/language/ref/NakagamiDistribution.html. Accessed 12 November 2021
16. Beckmann P (1962) Statistical distribution of the amplitude and phase of a multiply scattered field. J Res NBS 66:231–240. https://nvlpubs.nist.gov/nistpubs/jres/66D/jresv66Dn3p231_A1b.pdf. Accessed 12 November 2021

Effects of Electromagnetic Fields on Aviation Personnel, Their Behavior, and Erroneous Actions

Vladimir Tsetlin, Galina Stepanova, Nikolay Nikolaykin, and Natalia Korepina

Abstract The cause of many adverse aviation events is the erroneous actions of the flight crew on board the aircraft and aviation personnel performing maintenance of aircraft on the ground. As part of the study, the biotropic effects of fluctuations in non-ionizing radiation, helio-geomagnetic background, electromagnetic radiation, and the surrounding ionosphere were assessed based on changes monitoring in the hydrogen parameter (pH) and redox potential of the aquatic environment in the measurement zone (radius up to 100 m). The indices of the heart rate autonomic regulation were estimated. The circadian rhythms of heart rate variability indicators were studied. It has been shown that the stressful state of the organism, explaining the non-standard behavior and errors of aviation personnel, can also be associated with abrupt changes in the power of electromagnetic radiation in the workers' environment surrounding their workplaces. It is justified that the effect of electromagnetic radiation on humans occurs through a change in the redox properties of the body's aquatic environment. Some measures have been developed to prevent erroneous personnel actions and improve flight safety.

Keywords Non-ionizing radiation · Redox potential of the aqueous medium · pH value · Human behavior errors · Flight safety

V. Tsetlin · G. Stepanova
Institute of Medical and Biological Problems of the Russian Academy of Sciences, Moscow, Russia

N. Nikolaykin (✉)
Moscow State Technical University of Civil Aviation, Moscow, Russia
e-mail: nikols_n@mail.ru

N. Korepina
Irkutsk National Research Technical University, Irkutsk, Russia

O. A. Gorbachev et al. (eds.), *Proceedings of 10th International Conference on Recent Advances in Civil Aviation*, Lecture Notes in Mechanical Engineering, https://doi.org/10.1007/978-981-19-3788-0_34

1 Introduction

In the first three quarters of 2021 in Russia, there were almost three dozen military and civil aircraft disasters (excluding hang gliders and homemade aircraft). One missile carrier of the country's military space forces was seriously damaged, resulting in the death of crew members. More than 80 people died as a result of these disasters. The general trend of flight safety changes [1] in the Russian Federation is not comforting yet [2].

So, in particular, the following adverse aviation events, aviation accidents, and incidents have already occurred in 2021:

- January 8: in the Leningrad Region, in a collision with another plane, the RA-28 airliner crashed, three people were killed;
- June 24: the MI-8 helicopter crashed in the Leningrad Region, three people were killed;
- June 30: in the Lytkarino area (Moscow region), a 2-seater sports plane Harmony LSA RA-2086G crashed, two people were injured;
- August 17: in the Moscow region, an Il-112 V plane crashed on a test flight, three people died;
- August 23: in the Astrakhan region, after the flight, the fighter burned down on the ground due to a malfunction.

Several works recently published by the researchers of the Institute of Medical and Biological Problems of the Russian Academy of Sciences (IMBP RAS) and the Moscow State Technical University of Civil Aviation (MSTU CA) set forth the scientific results confirming the relationship between some water properties and the helio-geomagnetic characteristics of the environment (cosmophysical disturbances). An assumption about the mechanism of the dangerous impact of cosmophysical disturbances on living organisms was made [3]. The hypothesis about the environment electromagnetic background changes influence on the changes in the functional state of a person, that is, on the so-called "human factor" (from here on - HF) was substantiated [4]. Moreover, as shown in [5–7], the HF is recognized as the cause of a significant part of adverse aviation events in civil aviation, such as incidents and aviation accidents [1]. At the same time, it was revealed that the HF is the cause of malfunctions (erroneous actions) among the flight personnel onboard the aircraft and among the aviation personnel engaged in the aircraft maintenance on the ground [6].

Among the constantly acting environmental factors, electromagnetic radiation (EMR) is often called "non-ionizing". It is created near the earth's surface by radiations of the radio frequency range, mainly by the radiation of ions and electrons of the ionospheric plasma [8]. A significant number of such radiation sources are man-caused, including relay station generators, antennas of cellular communication systems operators, and many others located in dangerous proximity to buildings (both residential and industrial) [3]. The total number of such EMR technogenic sources is steadily increasing. They are often located near places of human activity.

Significantly, many similar man-caused EMR sources are on civil aviation facilities: airports, and airfields [6]. Modern aircraft are literally "stuffed" with electronic equipment for navigation, communication, control, and other systems. This complex of computerized onboard radio-electronic equipment cannot help but influence the parameters of natural geomagnetic fields and their disturbances.

As shown in [9], the absorption of high-frequency radio range electromagnetic radiation in liquid media causes "oxidative stress" in the human body, associated with the formation of hydrogen complexes in an aqueous medium, primarily hydrogen ions. The authors of another study [10] argue that a slight change in the hydrogen parameter, pH, within the range of 0.02–0.2 transforms blood parameters, leads to an increase in erythrocyte adhesion, promotes thrombus formation, a decrease in human immunity, performance, and causes psychophysiological reactions as well as hypoxia.

Erroneous actions of aviation personnel caused by changes in their health, subsequently entailing adverse aviation events, further lead to environmental problems of one scale or another [11]. Therefore, the study of the impact of the solar and geomagnetic environment changes on the human body and possible reactions of its heart rate [12] is currently relevant not only for astronauts' training but also for eliminating the HF influence on the accidence during air transportation [7].

The present research aims to study the impact of the environmental electromagnetic fields on a human, operator of complex technical systems and devices, through changes in the redox properties of water.

Research objectives:

- To determine the dynamics of the influence of the environmental electromagnetic fields on the redox properties of water.
- To investigate the human cardiovascular system (CVS) response to changes in the electromagnetic background of the environment.
- To determinate the parameters of a complex visual-motor reaction of a human, operator of complex technical systems and devices, when the electromagnetic background of the environment changes.

2 Research Tools and Methods

A technique for monitoring the water redox properties in the experimental setup was used during the experiments. The central part of the installation consisted of weighing bottles—cylindrical vessels made of Pyrex glass with a ground-in stopper and two electrodes inside, immersed in specially purified (bidistilled) water. Laboratory ionomers of the I-160 MI type measured the characteristics of water around the clock and recorded all deviations in the values of electric currents, pH, and redox potential (ORP). Thus, the electrochemical cell is operated in the mode of a galvanic cell, a device in which the chemical energy of a redox reaction is converted into an electric current.

The reference resistance voltage (which made it possible to estimate the change in the current value between the electrochemical cell electrodes) was amplified, converted from an analog signal to a digital code, and recorded using a personal computer (laptop). Then, it was displayed on the screen as a graph of the dependence of a particular value on the current time. The program provided a continuous automatic recording of readings.

Psychophysiological testing was carried out on three men aged 31–44 years and three women aged 29–33 years twice a month under realistic conditions, twice in the background period. A device of the UPFT-1/30 type (Medicom MTD Ltd, Research and Development Company, Taganrog, Russia) was used. Analyzing the psychophysiological state using the method of complex visual-motor response (CVMR), the following indicators were selected: the mean-square deviation of the average response time (MSD), the integral reliability indicator (IRI), the mean response time *M* (ms), and the response error.

A 24-h (Holter) electrocardiogram (ECG) tracing was carried out using a HOLTERLIVE monitor (Sovtest ATE, Ltd., Kursk, Russia). For the analysis of heart rate variability (HRV), the most common indicators were chosen, as follows:

- SDNN is the standard deviation of normotopic (NN) intervals;
- RMSSD is a square root of the average sum of the squares of the differences between adjacent NN intervals;
- SI is a stress index;
- TP or "Total Power" is the total power of the heart rate spectrum, which reflects the general activity of regulatory systems in the area of harmonic frequencies;
- HF is the power of the spectrum of high-frequency heart rate fluctuations;
- LF is the power of the spectrum of low-frequency oscillations of the heart rate;
- VLF is the power of the spectrum of very low-frequency heart rate fluctuations.

The daily dynamics of heart rate variability were calculated using the software "Iskim 6.2". The measurements were performed twice a month (with an interval of 15–16 days), mainly at noon. Eight measurement cycles were run; their numbers are used on the abscissa axis of the graphs discussed below.

3 Results

Throughout the four-month experiment, ORP experienced a noticeable general multi-day trend. By the end of the first month, ORP increased to 300 mV. By the middle of the second month, the isolation values of ORP were already 329–340 mV. Subsequently, a gradual decrease in the redox potential was observed up to the end of the experiment. In our opinion, despite the ORP fluctuations insignificance during the middle of the second, third and fourth months of the experiment (Fig. 1), there is a pronounced fluctuation in ORP during the day. As can be seen from the figure, the moments of reaching the extreme values of pH and ORP do not coincide. The minimum pH values are reached 2–4 h later than the minimum ORP value.

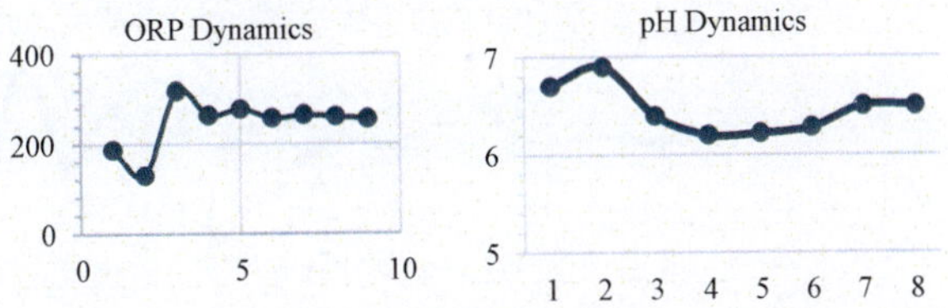

Fig. 1 Changes in ORP (mV) and pH during the experiment

The data of daily (hourly average) values of heart rate (HR) as a whole represent a typical [13] chronobirhythmological "picture" for healthy individuals, men, and women aged 20–40 years. It is customary to calculate the circadian index (CI), which does not have significant gender and age differences in healthy subjects and is in the range from 1.24 to 1.44 relative units (on average 1.32 ± 0.08 [10, 13]). There were no significant differences in individual CI values in the subjects.

The HRV frequency analysis shows the power increase in all components, with the maximum increase on weekdays of the second month of testing. At the same time, in the second month of isolation, the HRV data on weekends (both temporal and frequency indicators) are inverted, accompanied by a decrease in overall variability and a noticeable drop in the RMSSD (Root Mean Square of Successive Differences) link [14] against the background of a relative increase in CI. In this case, the indicators of the HRV spectrum lose power values both in the total (TP) and in other ranges—HF, LF, and VLF. However, in the third month, to a lesser extent, temporal indicators and very pronounced indicators of spectral analysis of HRV have an inverse, almost mirror, data inversion (weekdays/weekends): a significant decrease in the HRV spectrum on weekdays compared to weekends.

For a more in-depth analysis of the effect of the electromagnetic radiation of the ionospheric plasma, the subjects' psychophysiological state data were evaluated. A rapid assessment of the performance level according to the method "Complex visual-motor response" (CVMR), based on the analysis of stability and the level of visual-motor responses to light stimuli, made it possible to estimate the reaction errors of the subjects and the average reaction time *M* (ms). The mean-square deviation of the average response time (RMSD) was used to reveal the stability of nervous processes. The results are illustrated in Fig. 2, where the horizontal axis shows the number of days on which the subjects' health was assessed using the Holter (daily) ECG monitoring method [15, 16].

Thus, against the ORP background drop from 220 to 150 mV in the first testing month, a sharp drop in the circadian index (CI) from 1.27 to 1.17 was noted. By the end of the first month, when the ORP increased to 300 mV, the CI increased to 1.29. By the middle of the second month of the experiment, the ORP values were 329–340 mV (Fig. 3).

At the same time, in the second month of the experiment, HRV data, both temporal and frequency indicators, are inverted. There is a slight activation of the sympathetic section, accompanied by a decrease in overall variability and a noticeable drop in the parasympathetic link activity [14]. In the third month, a gradual decrease in the redox potential of water continued. Until the end of the tests, a violation of the decline

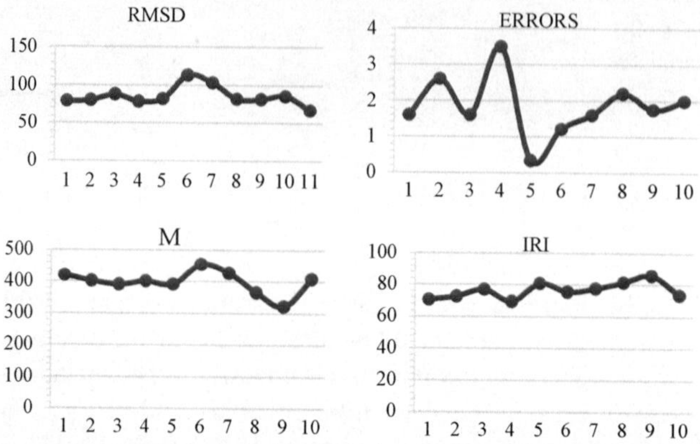

Fig. 2 Average daily indicators of complex visual-motor reaction on the days of measurements

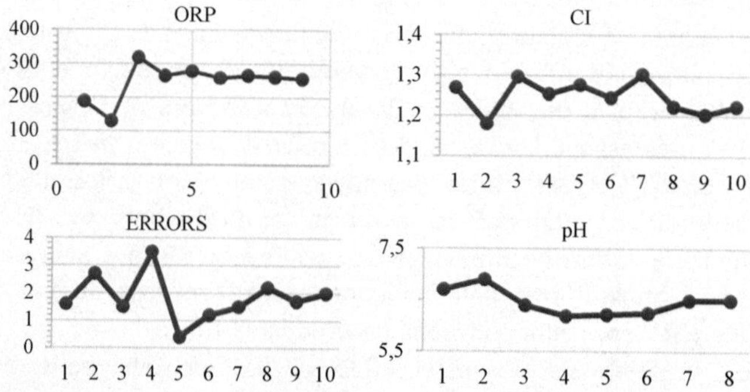

Fig. 3 Average daily circadian index, number of response errors, ORP (mV), and pH on the days of measurements

monotonicity was noted in the last month. By the end of the test, the circadian index decreased to 1.20. When comparing the graphs of changes in pH and the number of response errors with the ORP and circadian index graphs, their mirror similarity was registered.

The data analysis monitoring the redox potential (ORP) changes and water acidity (pH) in control water cells, carried out continuously in 2021, showed that during the days of accidents and incidents given as examples at the beginning of this article, there were noticeable deviations of the ORP and pH values (see Table 1).

Table 1 Ranges of ORP and water pH changes under control conditions on the days of adverse aviation events in 2021, according to the monitoring data under IBMP RAS conditions

Date: YYYY/MM/DD	ORP, mV		pH	
	min	max	min	max
2021/01/08	164.2	167.7	–	–
2021/06/24	184.34	195.72	–	–
2021/06/30	186.19	193.97	5.80	5.95
2021/08/17	150	172	6.39	10
2021/08/23	153	169	6.45	6.6

4 Discussion

The model of confined space conditions, humidity, illumination, and some other elements, embracing the peculiarities of the duties distribution between crew members, essentially reproduces the work of people in a confined space, including the flight crew's activities in the aircraft cockpits. On the other hand, electromagnetic radiation oscillations should be considered one of the possible and independently acting environment factors (if not the leading).

An earlier conditions analysis for performing flight AA 973 en route New York—Rio de Janeiro on a Boeing 767–300 liner [7, 17] revealed that the effect of a magnetic "pseudo-storm" can be observed on board the aircraft. Its amplitude-frequency characteristics are more than a thousand times higher than the corresponding values of the extreme magnetic storm (class G5). According to the scientists from Ufa State Aviation Technical University (USATU), a magnetic environment is observed onboard the aircraft that does not meet the sanitary rules and regulations in force in Russia. Thus, a particular response of "biological objects" is possible.

The rooms where the subjects participating in the described experiment were located were not strictly isolated (shielded) from the external EMR effects. Moreover, in the aircraft's cockpit, the conditions for weakening the earth's magnetic field influence on a person have not been created.

Under the artificially created aircraft environment conditions, it is necessary to monitor vital human functions during the flight, for example, indicators of heart rate variability (HRV). In doing this, the simultaneous registration of pH and ORP changes in the aquatic environment serves as indirect indicators in the EMR external background since it has been experimentally proven that these changes are interrelated [18, 19]. The experimental results are consistent with studies [9, 10], which showed that an increase in HRV and parasympathetic activity is associated with ORP increase. The findings support the hypothesis about the environment's influence, including the electromagnetic radiation of the radio range, on psychophysical processes.

While in flight, the number of potential hazards, according to the expert assessment of the flight crew, is three times higher than the number of accounted accidents. Pilots reflect the impact of the external environment by stressing their functional state, their performance changes, the risk of erroneous decisions and actions increases.

5 Conclusion

The crew led by the captain perform their professional tasks in a highly complex environment onboard. The study confirms that the human psychophysiological well-being and their cardiovascular system state can manifest themselves as a regulatory response to solar activity disturbances, as well as the impacts of the environment, including bursts of radio and electromagnetic radiation of man-caused nature.

Time-synchronous changes in the human functional state and changes in the electromagnetic background are registered (sometimes with a physiological delay). The delay duration depends on the type of body autonomic regulation. In average daily data, the activation of the parasympathetic division of the autonomic nervous system is traced in response to the disturbing effect.

To reduce the risk of the aviation personnel erroneous actions, primarily, flight crews, and, thereby, improve the level of aircraft flight safety in civil aviation, it is advisable:

- to test the impact of radio-electronic equipment and computerized systems onboard aviation equipment on the possible evolution of the characteristics of the geomagnetic disturbances at the stages of the take-off and landing cycle and during cruising flight;
- to check the dependence of the quality of solving various problems, both psychological and professionally oriented, under varying conditions of electromagnetic radiation for all types of examinations of the health status of aviation personnel and, first of all, of the flight crew;
- to constantly control the electromagnetic environment characteristics, in which aviation personnel work, for which purpose, equip aviation enterprises, primarily aircraft, with appropriate means of hardware control;
- to immediately inform the heads of units performing aircraft maintenance upon detecting geomagnetic disturbances and give them the right to take immediate short-term breaks (up to 10 min) carrying out critical operations.

Thus, as mentioned earlier, all those will contribute to the elimination of erroneous actions and other failures in the behavior of aviation personnel caused by sharp disturbances in the parameters of electromagnetic fields while performing critical production operations. The so-called "human factor" is the dominant cause of events in the modern statistics of adverse aviation events. The exclusion of its influence helps reduce accidents and improve air travel safety.

The work is within the framework of fundamental research on the primary topics of the Russian Academy of Sciences No. 64.1 and 65.1.

References

1. IATA Safety report 2018 (2019) 55th edn, IATA, pp 56–61
2. Ministry of Transport of the Russian Federation. Federal Agency of Air Transport. Rosaviatsiya Homepage. https://favt.gov.ru/dejatelnost-vozdushnye-perevozki-osnovnye-proizvodstvennye-pokazateli-ga/, Accessed 09 Sept 2021.
3. Nikolaykin NI, Tsetlin VV, Savchukov SA, Pozheluyeva ZV, Starkov EY (2017) O neobhodimosti i vozmozhnosti snizheniya vozdejstviya chelovecheskogo faktora na bezopasnost' polyotov (Concerning the Need and Possibility of Decreasing the Human Factor Influences on Flight Safety). Crede Experto: Transport, Society, Education, Language 2:201–218 (in Russian)
4. Nikolaykin NI, Pozheluyeva ZV, Starkov EY, Tsetlin VV, Savchukov SA (2018) Geogeliomagnitnoe vliyanie na organizm cheloveka i bezopasnost' aviaperevozok (Geo-Heloimagnetic Influence of a Human Body and Air Transportation Safety). Crede Experto: Transport, Society, Education, Language 2:26–54 (in Russian)
5. Sharov VD, Vorob'ev VV, Nikolaikin NI, Kuznetsova VL, Tolstykh SA (2020) Methodology for estimating the safety and quality of the aviation service provider activities using the principal component analysis. Russ Aeronaut 63:575–585. https://doi.org/10.3103/S1068799820040030
6. Ivanov AI, Kuznetsov AA, Nikolaykin NI, Sharov VD (2017) Sovershenstvovanie vzaimodejstviya rabotnikov, obsluzhivayushchih aviatekhniku, putyom podbora sostava brigad dlya snizheniya chisla oshibok (Interaction Improvement of the Workers Serving Aviation Equipment by Selection of Crews Structure For Decrease in Mistake Number). XXI vek: itogi proshlogo i problemy nastoyashchego plyus (XXI Century: Resumes of the Past and Challenges of the Present plus) 1(35):41–47 (in Russian)
7. Nikolaykin NI, Tsetlin VV, Stepanova GP, Koroleva EK, Starkov EY (2020) Kosmofizicheskie vozmushcheniya i chelovecheskij faktor v aviacii kak prichina negativnyh sobytij (Cosmophysical Disturbances and the Human Factor in Aviation as the Cause of Negative Events). Nauchnyj vestnik UI GA (Scientific Bulletin of UI CA) 12:56–61 (in Russian)
8. Budko N, Zaitsev A, Karpachev A, Kozlov A, Filippov B (2006) Space environment around us. TROVANT, Troitsk
9. Tsetlin VV, Belisheva NK, Muravyev VS, Martynova AA, Pryanichnikova SV, Mikhailov RE (2016) The conjugacy of the water properties dynamics with the person's psychophysiological state and variations of cosmophysical agents—the basis for the detection of flight conditions through the water sensory properties. In: Proceedings of the XVI Conference on Space Biology and Medicine with international participation. pp 254–256, Moscow
10. Tsetlin VV, Stepanova GP (2019) Study of the electromagnetic environmental factors' impact on water and the internal environment of living organisms. Aerospace Environ Med 53(6):70–76
11. Nikolaykin NI, Nikolaykina NE, Sekerin VD, Gorokhova AE (2017) Environmental and economic model of an aircraft accident evaluation. J Environ Manag Tourism 8(5):1128–1135
12. Alabdulgader A, McCraty R, Atkinson M, Dobyns Y, Vainoras A, Ragulskis M, Stolc V (2018) Long-term study of heart rate variability responses to changes in the solar and geomagnetic environment. Sci Rep 8(1):1–14. https://doi.org/10.1038/s41598-018-20932-x
13. Bersenev EY, Rusanov VB, Chernikova AG (2011) The study of body adaptive reactions in conditions of long-term isolation (experiment "Mars-500"). In: 9th International Proceedings on Donozology-2011 "Healthy lifestyle and harmful factors for health", pp 663–665. St. Petersburg
14. Risdiana N (2017) Comparison of Root Mean Square of Successive Differences (RMSSD) among adolescent smokers and nonsmokers in Yogyakarta. Adv Sci Lett 23(12):12660–12664
15. Holter Monitor. Johns Hopkins Medicine Homepage https://www.hopkinsmedicine.org/health/treatment-tests-and-therapies/holter-monitor. Accessed 23 October 2021
16. Baevskii RM, Nikulina GA (2000) Holter monitoring in space medicine: Analysis of heart rate variability. Journal Arrhythmology 16:6–16
17. Vorobiev AV (2017) Meteoinformatics: geomagnetic variations and space weather. Innovative Mechanical Engineering, Moscow

18. Demidova MM, Tikhonenko VM (2001) Circadian dynamics of the indices of heart rate variability in healthy pesrsons. Journal Arrhythmology 23:52–58
19. Vinnichenko MB, Belisheva NK, Zhirov VK (2009) Modulation of water properties by cosmic ray variations. Reports Academy Sci 429(6):816–820

Application of Remotely Piloted Aircraft Systems

The papers are devoted to urgent issues and promising areas of developing remotely piloted aircraft systems: systems of automatic control, flight and navigation equipment, computer-aided designs, closed-circuit television systems, data collection and processing systems, aerial survey equipment, high-speed wireless data transmission systems, software for simulating flights of remotely piloted aircraft systems, component parts and materials for remotely piloted aircraft systems, maintenance of remotely piloted aircraft systems, maintenance documentation for remotely piloted aircraft systems, personnel's training system.

Keywords Remotely piloted aircraft system · UAS · Automatic control system · GCS · НПУ · Unmanned aerial vehicle · UAV · Remote pilot · UAV operator · Aerial photography · Aerial monitoring

Decision-Making Related to UAV Control Under Uncertainty

Nikolay Kim and Nikolay Bodunkov

Abstract The article is devoted to the decision-making onboard of the unmanned aerial vehicles (UAVs) under complex or, in some cases, even uncertain operating conditions, for example, automatic and even unplanned landing on improvised sites. We have proposed applying decision-making to specific tasks, e.g., decisions to "land" or to "continue flight" within the "Game Against Nature" framework. In addition, decisions are made, while a UAV performs an autonomous flight, which is why various scenarios should be taken into account, based on uncertainty of a current situation. To identify nature's strategies, situation awareness method has been used in this work to assess the state of the current situation and to predict the consequences of decisions made and controls implemented. The current technology for situation awareness has been implemented, based on the use of databases built in advance and containing information about possible objects of interest, including area maps, knowledge bases containing products that describe cause-and-effect relationships between objects (phenomena, processes), as well as a set of models describing current statuses. The UAV and its subsystems that safeguard efficient task execution—namely, the UAV performance, including range, speed, and maneuverability under current conditions; the operating spectral range of the surveillance system, etc. The environment, e.g., time of day, weather conditions, etc., includes effects of destabilizing factors. Objects of interest specified for a task assigned. We have proved that making decisions about landing or continuing flight through implementing onboard algorithms implies taking into account possible nature's strategies, the status of the UAV, and the expected pessimism index of the Hurwitz criterion.

Keywords Unmanned aerial vehicle · Situation awareness · Environment description model · Hurwitz criterion

N. Kim (✉) · N. Bodunkov
Moscow Aviation Institute (National Research University), Moscow, Russia
e-mail: nkim2011@list.ru

O. A. Gorbachev et al. (eds.), *Proceedings of 10th International Conference on Recent Advances in Civil Aviation*, Lecture Notes in Mechanical Engineering, https://doi.org/10.1007/978-981-19-3788-0_35

1 Introduction

Broadening application areas and increasing sophistication of tasks executed by unmanned aerial vehicles (UAVs) call for improving onboard software and hardware to safeguard autonomous decision-making under complex operation conditions. An autonomous emergency landing of a UAV is one of the most complicated situations a UAV may face while being airborne [1].

Autonomous landing of an unmanned helicopter is similar to the landing process of other types of UAVs, but imposes a number of additional restrictions on the properties of the landing pad and its environment [2, 3]. Choosing a landing site that is suitable in terms of the location, dimensions, the surface quality, the environment, and the approach area adds to the complicity of emergency landing:

- The size of the landing area is not less than the diameter of the helicopter propeller;
- The slope of the landing zone should not exceed 2%, the slope of the approach zone—no more than 7%;
- The supporting characteristics of the ground must withstand the impact of the main rotor and the load when landing the helicopter.

Thus, the complex algorithm of an emergency landing of an unmanned helicopter should include the stages of analyzing the current situation and making a decision about landing, choosing a site, and implementing a landing.

The current action planning methods are based on using mathematical models and formal criteria that take into account pre-set factors and operation conditions [4]. Any autonomous implementation of these approaches is complicated by the impact of the factors listed below:

- Uncertain conditions under which a task assigned (TA) is being executed, e.g., uncertain criteria, operation conditions, the UAV status, etc.;
- No tools to reduce or to eliminate the uncertainties;
- No procedures to match situations with their respective efficient controls.

Currently, a task assigned can only be accomplished with assistance from a human operator if being executed under the above uncertainty conditions. Hence, developing approaches based providing descriptions to the above situations and their classification to adapt a UAV to the above factors would significantly increase the UAV operational autonomy while executing various real-life tasks [5, 6].

Now, let us consider how situation awareness method can be applied to a UAV autonomous landing.

2 Situation Awareness Method

Autonomous control of an UAV is essentially choosing between control strategies, depending on a current situation parameters (nature's strategies), where a situation can be defined as a set of models and descriptions of objects of interest and factors that are involved in the process [7]. A specific strategy is selected, based on a criterion specified in the task assigned.

It is the aim of situation awareness [8–10] to build situation descriptions to provide more profound understanding of interactions between factors that are essential to accomplishing a task assigned (TA). Situation awareness formulates classifications of both current and forecasted situations, which allows selecting the most efficient controls, while taking into account a task being executed.

A situation awareness technology we have proposed is based on using.

1. Databases (DB) that are built in advance and contain data of possible objects, phenomena, and processes of interest and their attributes, including area maps;
2. Knowledge bases (KB) that are built in advance and contain rules that describe cause-and-effect relationships between the objects, phenomena, and processes of interest;
3. Current situation models.

The complicity of building the above models is in describing uncertain and changeable factors, the impossibility of altering multiple parameters, uncertain relations (and reciprocal impact) between various factors, including problematic tracing of cause-and-effect relationships, etc. The model that forms a situation description is a hierarchic model and integrates various type models [11].

Generally, a situation model is an aggregate of the three models

$$M = \langle O, \text{Env}, \text{Task} \rangle$$

where O is the UAV model, Env is the environment model, and Task is the task assigned (TA) model.

As a TA description is being built, criteria for and limitations to be imposed on the solution are formalized. The model of a UAV and its subsystems determine the efficiency and a probability of accomplishing the task assigned, while the environment model describes the objects and the factors that impact the UAV subsystems.

There exist various model representation methods, semantic descriptions being best suited to uncertain situations. Examples of semantic descriptions and rules for selecting an aircraft-type UAV landing site were provided in the research [12]. An abstract from a description of a TA was provided and can be as shown below

$$G((\text{SO}(1) : \text{Type}(\text{Surface}(n)) : \quad /* \text{ TA is to find a specific Surface, shape } n */$$

An abstract from the object attribute descriptions

$$\text{Atr}M(H(15)\hat{}L(50)\hat{}W(20)) \ldots \quad /* \text{ Shape } */$$

$$\text{AtrObs}(B(\text{Not})\hat{}C(G)\hat{}T(\text{Not}))\hat{}F(\text{Not}))\hat{} \ldots \quad /* \text{ Observability } */$$

Descriptions of both the current and the forecasted situations are formed, based on the above models. For example, an abstract from a description of a forecasted situation having a code (150) can be presented as

$$Pr\_S(N(150), CS(i(8), j(12)) : P(0,85)\hat{}T(40)\hat{}\ldots$$

with the codes of strategies used by the UAV (*i*) and the adversary (*j*), a probability of accomplishing the task assigned (*P*), the process time (*T*), etc., provided.

Many of situation model parameters can be fuzzy, stochastic, or uncertain. Any uncertainty of the model parameters, in turn, causes uncertainty of the selection of UAV strategies. Any uncertainty of a situation can be assessed quantitatively by using information entropy [3].

By way of illustration, let us consider emergency landing of a UAV. As landing on an improvised site is being made [2] under adverse flight and observability conditions, the landing situations can be viewed as a game against nature. In the latter case, choosing between decisions either to land (P1) or to continue the flight (P2) is the controllable strategies (P*i*, $i = 1,\ldots, I$) of the UAV, while the weather conditions, the earth surface types [13] that have not been specified clearly enough by onboard algorithms, gusts of wind, which are hard to predict, etc., are the nature's strategies (Env *j*, $j = 1,\ldots, J$). Various combinations of the factors may result in various consequences, including a crash.

3 Criterion for Strategy Selection

We propose using the Hurwitz criterion to select a compromise decision as to a UAV control strategy, which takes into account a possibility of the best and the worst of nature's behaviors, in terms of a task assigned and being executed. A decision is made, based on combining the minimal and the maximal payoffs $\propto_{ij}$. The best strategy is that for which the value of the criterion *S* is the highest

$$S = \max_i[(1-a)\max_j \propto_{ij} + a\min_j \propto_{ij}],$$

where $\propto_{ij}$ is the optimism degree index, the payoff, when the UAV executes the *i*th strategy, while the adversary, nature, executes the *j*th strategy.

To generate descriptions and obtain estimates of $\propto_{ij}$,we suggest using the situation analysis approach. During the analysis, fuzzy rules are checked sequentially from a pre-prepared knowledge base, for example:

IF *Weather.conditions* = *"Good"* ***and*** *Surface* = *"Bad"* , ***then*** $\propto_1$ = *"High"*.

IF *Weather.conditions* = *"Bad"* ***and*** *Surface* = *"Bad"* , ***then*** $\propto_1$ = *"Very High"*.

The basic rules link the environment in a certain area (e.g., landscape) and the search object to the reliability of its presence in it. However, as shown in the example above, rule parameters can be fuzzy. The result of the rules is also fuzzy (it takes the values "*Very Low*", "*Low*", " *Average*", "*High*", and "*Very High*"). The conversion to confidence values occurs according to known fuzzy inference rules.

When $a = 1$, the Hurwitz criterion transits into the maximal Wald criterion, the extreme pessimism; when $a = 0$, it is no different from the maximal criterion—the extreme optimism.

In other words, selecting a value a, we can find the criteria for all of the UAV strategies, and then, the maximum payoff strategy is chosen among these strategies.

Selecting an index a is a problem that is solved subjectively in most cases, which can hardly be done when a UAV performs an autonomous flight.

Within the framework of the problem being solved, that is, making a decision whether to land or to continue the flight, the pessimism index a denotes the "crash expectation" when landing the UAV. The higher the value a is, the lower is a probability that the landing will be safe, and the lower is the possible payoff value. As the optimism index increases and a tends to zero, an expected payoff increases, and the landing strategy P1 becomes more competitive.

There exist various methods of selecting an index a. The most complex, yet the most accurate method is founded on the situation awareness technology that implies building models of both current and forecasted situations. Specifying the environment model parameters, including the landing surface parameters, provides higher probability assessment of landing safety and, accordingly, allows for more efficient decisions about UAV control.

Pessimism index values are selected by comparing respective consequences of decisions taken. In simpler cases, the index a value can be set, based on, e.g., the cost of the UAV. Please refer to Fig. 1 for (C) dependence cases, where C is the cost of the UAV.

Figure 1 suggests that using various $a(C)$ models may yield various a values. For example, setting the relative price of the UAV $C = 0.6$ following Model 1 (the solid line) yields the pessimism index $a = 0.6$, while Model 2 (the dashed line) yields $a = 0.8$.

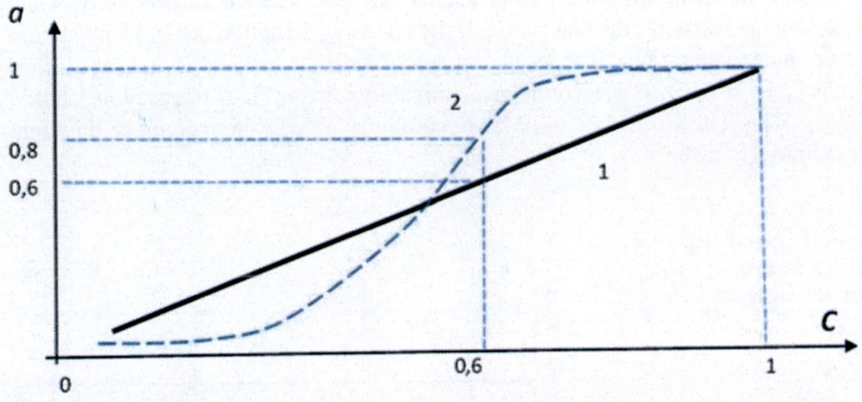

Fig. 1 $a(C)$ dependence cases

Using this approach, one is able to find the relative price of the UAV and to find the index a from the dependence $a(C)$ set.

4 Choosing Strategy Under Uncertainty

A payoff matrix derived from choosing a specific strategy α_{ij} is the basis under making a decision with the Hurwitz criterion taken into account. Whenever a necessity of landing arises, the payoff equals to safe landing under the current conditions (Strategy $P1$) or to continuing the flight as a safer option (Strategy $P2$). Let safety be set by an index value within the range between 0 and 10. When $\alpha_{ij}= 0$, the UAV will certainly be lost. When $\alpha_{ij}= 10$, will either certainly land safely or continue the flight. It is obvious that Strategy $P1$ depends on landing conditions, namely the landing site parameters (e.g., the landing surface type) and the environment conditions (e.g., weather conditions)—the nature strategies, while Strategy $P2$ depends on the status of the UAV, the flight conditions, and the task assigned parameters.

Let us consider the four nature's strategies.

- Env1—Weather conditions: Good; Surface: Good;
- Env2—Weather conditions: Good; Surface: Bad;
- Env3—Weather conditions: Bad; Surface: Good;
- Env4—Weather conditions: Bad; Surface: Bad.

When both the weather conditions and the landing surface quality are "Good", the landing safety is maximal ($\alpha_{ij}= 10$). Bad weather conditions may worsen the landing site quality, which reduces landing safety.

Let the payoff from Strategy $P2$ depends on the UAV status, the TA parameters, and the flight conditions, and unrelated to the nature's strategies listed above. In the latter case, the payoff from selecting this strategy will be constant.

The payoff matrix for the case in question can be arranged in a Table 1.

A suitable string of Table 1 can selected whenever there is full certainty, while selecting the Hurwitz criterion is essentially choosing an option that has a maximum safety index value. E.g., whenever the current weather conditions and the surface quality are bad, with the confidence probability being 1, Strategy 4 is chosen ($P_{\text{Env4}} = 1$). Then, $\alpha_{14}= 3$, $\alpha_{24}= 5$. Accordingly, a decision to continue the flight is definitively taken.

Table 1 Payoff matrix for UAV strategies depending on nature's strategies

	$P1$	$P2$
Env1	10	5
Env2	4	5
Env3	5	5
Env4	3	5

Table 2 Payoff matrix for full uncertainty in nature's strategies

	*P*1	*P*2
Env1	2.5	1.25
Env2	1	1.25
Env3	1.25	1.25
Env4	0.75	1.25

Seeing as the very concepts of "good" and "bad" weather conditions are fuzzy and soft, the nature's strategy confidence distribution $P_{\text{Env}j}$ should be considered. Whenever descriptions of the conditions contain any uncertainty, the indexes must be recalculated

$$\widetilde{\propto_{i,j}} = P_{\text{Env}j} * \propto_{i,j}$$

In the event there is no information, the options for the conditions are considered equally probable. Here, $P_{\text{Env}j} = 1/4$. Therefore, the safety index table will look like shown below (Table 2).

Let us assume the Hurwitz index being $a = 0.7$. Table 2 above suggests that.

$$\max \widetilde{\propto_{1,j}} = 2.5;\ \max \widetilde{\propto_{1,j}} = 0.75$$

$$\min \widetilde{\propto_{2,j}} = 1.25;\ \min \widetilde{\propto_{2,j}} = 0.75$$

Thus, from the Hurwitz criterion, we find that

$$S_1 = 2.5 * 0.3 + 0.75 * 0.7 = 1.275$$

$$S_2 = 1.25$$

Therefore, whenever a situation is fully uncertain, the decision made is based immediately on the optimism index.

The uncertainty in the condition descriptions may change dynamically. E.g., the environment parameters (including weather conditions or the landing surface quality) can be specified and altered. The distribution of losses will change accordingly and, eventually, so will the decision to be made.

Assuming we know that the confidence distribution of nature's strategies is $P_{\text{Env1}} = 0.5$; $P_{\text{Env2}} = 0.14$; $P_{\text{Env3}} = 0.14$; $P_{\text{Env4}} = 0.22$, and the payoff matrix looks like shown below.

Let us assume the index of the Hurwitz criterion being $a = 0.7$. Table 3 above suggests that.

$$\max \widetilde{\propto_{1,j}} = 5;\ \min \widetilde{\propto_{1,j}} = 0.56$$

Table 3 Payoff matrix for known distribution of nature's strategies

	$P1$	$P2$
Env1	5	2.5
Env2	0.56	0.7
Env3	0.7	0.7
Env4	0.66	1.1

$$\max \widetilde{\propto_{2,j}} = 2.5;\ \min \widetilde{\propto_{2,j}} = 0.7$$

Thus, from the Hurwitz criterion

$$S_1 = 5 * 0.3 + 0.56 * 0.7 = 1.892$$

$$S_2 = 1.24$$

Consequently, a decision about landing is made, based on the Hurwitz criterion.

Thus, reducing uncertainty of a situation model reduces the "subjectivity" of a decision made.

5 Conclusions

A task of autonomous decision-making related to choosing between landing a UAV or continuing the flight under uncertainty has been considered in this article. We have proposed implementation of decision-making within the framework of the "Game Against Nature" method. UAV's strategies (either "land" or "continue flight") and the nature's strategies (an aggregate of factors that impact decision-making) have been considered.

Situation awareness and description methods have been used for assessment of current situations and forecasting consequences of decisions made and controls implemented to identify nature's strategies. Parameters of descriptions, and, consequently, those of nature's strategies, can be generally fuzzy or uncertain.

We have proved that possible nature's strategies, the UAV status, outcomes expected, and optimism indexes taken into account by the Hurwitz criterion should be provided by onboard algorithms implemented and related to decision-making about landing.

A method of selecting an optimism index has been proposed, and its impact on a decision being made has been described for various uncertainty levels of nature's strategies.

This research has been carried out with financial support from the Russian Foundation for Basic Research (RFFI) as part of a science project No. 19-08-00613 A.

References

1. Tan L (2019) Research on optimal landing trajectory planning method between an UAV and a moving vessel. Appl Sci 9(18)
2. Kim N (2020) Avtomatizacija posadki bespilotnogo vertoleta na neoborudovannuju ploshhadku (Automation of the landing of an unmanned helicopter on an unequipped site). Izvestija JuFU. Tehnicheskie nauki 1:17–25
3. Kim NV, Mikhaylov NA (2019) An Entropy approach in solving the search problem by a UAV group. Stanki i Instrument 9:28–31
4. Tulum K (2009) Situation aware UAV mission route planning. In: IEEE Aerospace Conference Proceedings. https://doi.org/10.1109/AERO.2009.4839602
5. Kim N (2018) Automatic decision-making by the on-board system of an unmanned aerial vehicle during the traffic monitoring. Aerospace MAI J 25:99–108
6. Kim NV, Chervonenkis MA (2015) Situation control of unmanned aerial vehicles for road traffic monitoring. Mod Appl Sci 9(5):1–13
7. Amiri S, Shirazi MS, Zhang S (2020) Learning and reasoning for robot sequential decision making under uncertainty.In: AAAI 2020—34th AAAI Conference on Artificial Intelligence, pp 2726–2733
8. Endsley M, Bolte B, Jones D (2003) Designing for situation awareness: an approach to user-centered design. Taylor & Francis, New York p 312
9. Kim N (2015) Adaptive surveillance algorithms based on the situation analysis. In: M Favorskaya, L Jain (Eds) Computer vision in control systems. Springer, New York, Chapter 7 pp 169–200
10. Mutzenich C (2021) Updating our understanding of situation awareness in relation to remote operators of autonomous vehicles. Cognitive Res: Principles Implications 6:1
11. Gaeta A (2021) A comprehensive model and computational methods to improve situation awareness in intelligence scenarios. Appl Intell 51(9):6585–6608. https://doi.org/10.1007/s10489-021-02673-z
12. Bodunkov N (2018) Semantic descriptions in the problem of autonomous landing of an unmanned aerial vehicle. Tekh Zrenie 1(12):21–25
13. Vazaev AV, Noskov VP, Rubtsov IV (2019) Neural network module for selection of standard for recognition of the types of support surface. In: Conference "Advanced Systems and Control Problems", Rostov-on-Don: Yuzh. Fed. University, pp 29–33

Environmental Monitoring in the "Land–Water" Contact Zone of Water Bodies with the Help of Small Unmanned Aerial Vehicles

Denis Efimov, Alexandr Shablov, and Elena Shavalieva

Abstract The paper presents the results of using UAV, small unmanned aerial vehicles, for the purposes of environmental monitoring of coastal ecosystems of large continental bodies of water. Large water bodies of the Lake Baikal system in the south of Siberia were used as test sites. The survey was carried out using small unmanned aerial vehicles—the Geoscan 101 aircraft type and DJI multi-rotor UAV—during the warm season. From June to September, more than 6,000 high-resolution aerial photographs were obtained, which were used to compile orthophotomaps. As a result of the analysis of the aerial photographs, a classification of objects for environmental monitoring of coastal ecosystems of a specific area was developed, as well as a technical algorithm for the stages of shooting. The results of the study demonstrate the advantages of using small unmanned aerial vehicles as an additional control tool in monitoring the state of environmental objects, which significantly expands the range of factors of potential and real impact on the state of ecosystems, helps to identify the location of impact sources and to obtain quantitative estimates of a wide spatial coverage.

Keywords Aerophotometry · Contact zone · Environmental monitoring · Unmanned aerial vehicles · Water bodies

D. Efimov
Papanin Institute for Biology of Inland Waters, Russian Academy of Sciences, Borok, Yaroslavl Region, Russia
e-mail: dnsfmv@gmail.com

A. Shablov (✉)
Irkutsk Branch of Moscow State Technical University of Civil Aviation, Irkutsk, Russia
e-mail: avshablov@mail.ru

E. Shavalieva
Irkutsk State University, Irkutsk, Russia
e-mail: allona2005@yandex.ru

O. A. Gorbachev et al. (eds.), *Proceedings of 10th International Conference on Recent Advances in Civil Aviation*, Lecture Notes in Mechanical Engineering, https://doi.org/10.1007/978-981-19-3788-0_36

1 Introduction

Unmanned aerial systems (designation of the technology and the vehicles) are a technology that, along with methods of remote sensing of the Earth, has recently been increasingly used to obtain geospatial data [1, 2]. At the same time, small unmanned aerial vehicles, or small unmanned aircraft, as an aerial monitoring tool have a number of significant advantages in obtaining detailed large-scale data of small areas of the earth's surface [3], are comparatively not expensive and more accessible to the population. Significant advantages of using small unmanned aerial vehicles are manifested in those areas where mobility and high accuracy of information is required, for example, in industrial areas, in mining, in agriculture and other large spheres of human activity, as well as in assessing the risks of natural disasters, etc. [4], especially against the background of anthropogenic climate change in the world.

Remote sensing data of aquatic anthropogenic complexes and their coasts are actively used in monitoring and mapping the diversity of natural ecosystems, protected, technogenic, and hard-to-reach territories [5, 6]. Small unmanned aerial vehicles are already used to assess the state of wetland ecosystems, monitor erosion phenomena [7], and measure the consequences of environmental pollution [8].

In vegetation monitoring, UAV is used to estimate biomass [9] and assess the incidence of disease in agricultural crops [10], count and determine the size of plants [11], map plant communities [12] and ecosystems [13], and detect weed plant species [14]. At the same time, in the analysis of vegetation objects on aerial photographs obtained using UAV and subsequent modeling, various methods and techniques are used, which are based on RGB, NDVI, multispectral sensors, and their combinations [15].

The coastal zone of water bodies and watercourses is a very vulnerable element in the "land–water" contact system [16]. This concerns the variability of parameters not only of small inland water bodies, but of large objects—sea and oceanic coasts [17, 18]. The use of SUAV makes it possible to obtain real-time data on the state of environmental objects, which are necessary, among other things, for the purposes of environmental monitoring at the local, regional, and global levels [4]. However, while vegetation indices have been developed and are successfully used for vegetation objects, not enough of such indices have been developed for objects of non-plant origin, and especially those of technogenic nature, the shape, structure, and composition of which can be very diverse.

Currently, the coastal ecosystem of Lake Baikal, associated streams and reservoirs, is exposed to significant anthropogenic pressure. Until now, the coastal zone of many large continental lakes, including Lake Baikal, has been poorly explored [18–20]. Therefore, our work was aimed at: 1) assessing the possibilities of using small UAV for environmental monitoring of the coastal zone of the water bodies of southern Siberia and identifying natural and man-made objects, 2) identifying real and potential sources of environmental threats to the ecosystem in the coastal zone.

2 Materials and Methods

The coastal areas of two water bodies in the south of Siberia were selected as test objects. The first section (I) was located on the Angara River, in the lower reaches of the Irkutsk reservoir (Lake Baikal basin). The second section (II) was located in the central part of the Irkutsk reservoir. The two sections are large water bodies in the south of Siberia, which are connected elements of one hydraulic system, differing from each other in terms of the parameters of the hydrological regime (see Fig. 1).

In 2021, aerial photography was carried out using the Geoscan 101 aircraft type UAV (see Fig. 1d) and an amateur multi-rotor UAV according to the DJI classification (see Fig. 1e) DJI Mini, DJI Mini 2, DJI Mavic Air 2, DJI Mavic 2 Zoom. The testing areas covered the coastal strip, including shallow waters. The Geoscan 101 UAV was used with a Sony A6000 camera; the horizontal flight speed was about 30 m/s; the flight altitude was 100–150 m. DJI multi-rotor UAVs were used in standard configuration. Aerial photography with multi-rotor UAV was carried out from heights from 1 to 100 m, horizontal, vertical, and perspective low-altitude aerial photography.

When using the Geoscan 101 UAV at the preparatory stage, the flight task is designed: the shooting parameters, the model of the UAV and the camera, parameters of areal or linear aerial photography, flight routes, waiting points, landing route are entered. After starting, the UAV operates automatically, without the participation of the operator. Aerial photography from a height of 110 m gives a resolution of 2.2 cm/pix, from a height of 150 m–2.5 cm/pix.

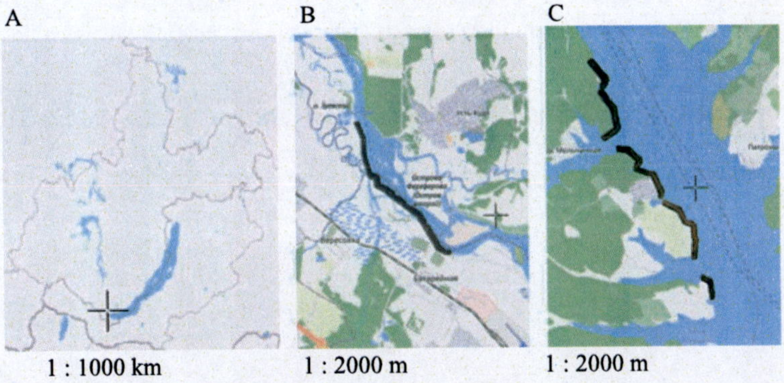

Fig. 1 Area of work and flights: **a** Southern section of Baikal Siberia (Russia, Irkutsk region); **b** I—section in the area of the village of Ust-Kuda on the Angara River; **c** II—section in the area of the village of Novogrudinino on the Irkutsk reservoir

3 Results and Discussion

The array of aerial photographs, after 6 cycles of shooting, amounted to 6100 items. After each cycle, the images were processed in the Agisoft Metashape program in order to compile an orthophotomap with subsequent processing in the QGIS program (see Fig. 2).

The developed algorithm for aerial photography using small unmanned aircraft as applied to the objects of research roughly includes three sequential blocks of work (Fig. 3).

The first "preparatory" block includes a series of sequential stages, such as identifying the features of objects to be monitored, choosing locations for flights, drawing up a flight plan, ending with coordination with ATM, and obtaining appropriate permits for flights. The second block of "collection of aerial photo data" includes the stages of flight operations, aerial photography, and the accumulation of a photo

Fig. 2 Small unmanned aircraft used and examples of aerial photographs: **a** Geoscan 101 UAV, preparation for launch; **b** DJI Mavic 2 Zoom UAV; **c** A snapshot of the site of a spontaneous dump of household waste in the floodplain of the Angara River in the district of the village of Ust-Kuda; **d** A snapshot of an area of active erosion of the coastal slope and destruction of forest vegetation in the area of the village of Novogrudinino

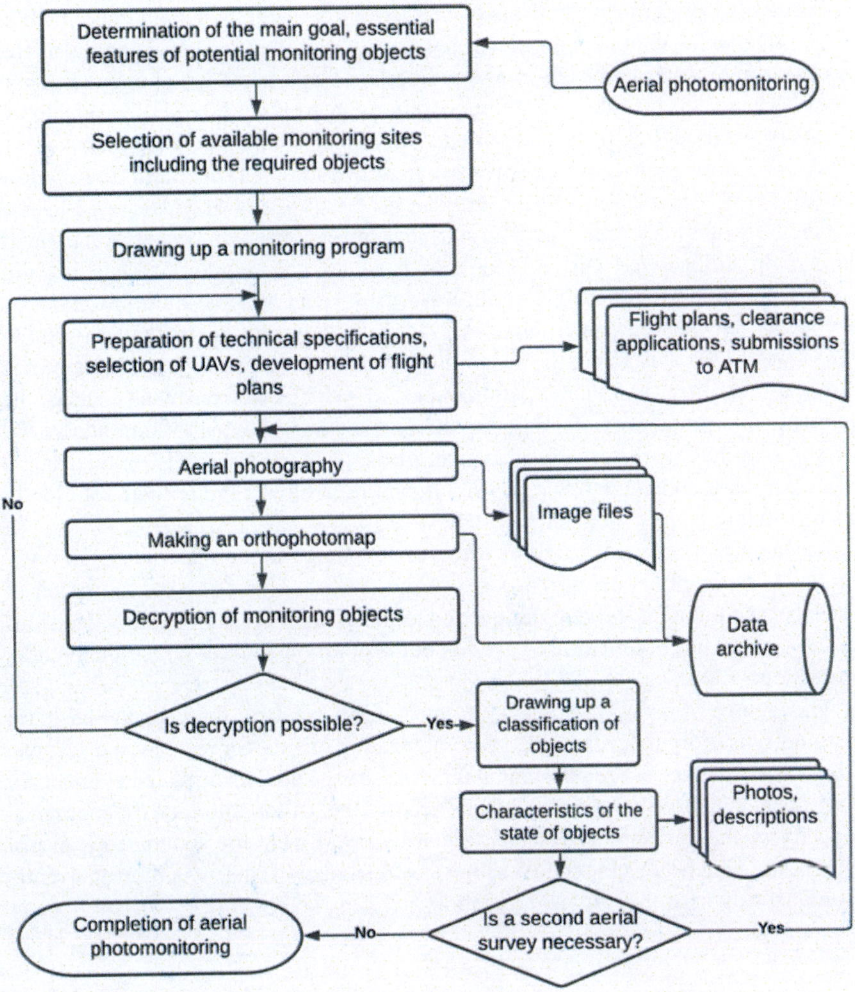

Fig. 3 Algorithm for environmental monitoring using an unmanned aerial vehicle

archive. The block is completed by drawing up an orthophotomap. The third final block of "processing" includes procedures for decoding orthophotomaps, compiling a classification of the monitored objects, creating an archive of descriptions of objects.

The nature of further work with archives of images and descriptions of objects is determined by the detailed tasks of specific environmental studies, for example, the creation of large-scale thematic maps (of vegetation, recreational load, technogenic pollution of the territory, etc.), operational assessment of the state of ecosystem parameters (area of land occupied by agricultural crops, residential buildings, recreational load, etc.), assessment of dynamic phenomena (monitoring of the condition of river banks, vegetation, animals, birds, etc.).

Based on a preliminary analysis of the aerial photography data, a classification of environmental monitoring objects has been developed on the example of the coastal strip of water bodies in southern Siberia. The prepared list of environmental monitoring objects includes: by origin/condition—natural (forests, shrubs, meadows, aquatic vegetation) and man-made (roads, buildings, landfills, degraded territories); by shape and location—leveled surfaces, coastal slopes, splash zones, and shallow water.

The use of small unmanned aircraft has revealed a number of some operational difficulties that currently do not have an unambiguous technical solution. В частности, погодные условия, влияют не только на возможность осуществления полетов и качество фотоматериалов. In particular, weather conditions do no only affect the possibility of flights or the quality of photographic materials. Another significant operational effect, which also somewhat reduces the quality of interpretation, is the shooting time since different angles of illumination by the sun significantly distort and shade part of the territory, especially on the water's edge. All these and some other less significant difficulties require their solution in the near future.

An important practical result of using aerial photographic materials is an operational assessment of the current state of the ecosystem and the concentration of efforts of the public, local population, and regional authorities on making decisions on the optimal use of territories, carrying out restoration work or taking them under protection.

The archive of aerial photo data, in our opinion, can be a fundamental basis for planning a territory for recreation, i.e., organization of sites for recreation of people, placement of vehicles, fire pits, containers for garbage and other related elements of the recreational infrastructure, taking into account the real situation and the ecological capacity of the territory. This leads to a withdrawal from the spontaneous use of the coasts, which are potentially subject to degradation and which are currently spontaneously used for rest and recreation.

4 Conclusion

The results of aerial photo monitoring using small unmanned aerial vehicles to assess the state of the coastal zone of water bodies in the south of Siberia show ample opportunities for the purposes of environmental monitoring. The developed algorithm ensures high-quality aerial monitoring of water bodies in the south of Siberia. The resulting array of geospatial data makes it possible to identify objects of various origins—natural (forests, bushes, meadows) and anthropogenic (roads, landfills, etc.), assess their current state, track the seasonal and annual dynamics of the parameters of objects, determine quantitative, qualitative and spatial indicators. Assessment of the state of land and water bodies at the boundaries of environments using small unmanned aerial vehicles is an additional tool for environmental monitoring at the local, regional, and global levels.

As well as being a source of operational assessment of the current state of the ecosystem and its components, aerial photographs represent an important foundation for the consolidation of society and regional authorities in terms of the optimal use of territories and the provision of high-quality environmental services.

Acknowledgements The work was supported by the grant competition of environmental projects En+ Group, project № БПП/ГК-En-ЦСП-Д-21-350. The work of D. Yu. Efimov was also held within the state assignment of IBIW RAS (theme 121051100099-5).

Author Contribution DE, AS conceived the idea, analyzed the data, wrote and edited the manuscript; ES, DE, AS translated the text into English.

Declaration of Competing Interest
The authors declare that they have no conflict of interest.

References

1. Elvidge CD, Ziskin D, Baugh KE et al (2009) A fifteen year record of global natural gas flaring derived from satellite data. Energies 2(3):595–622. https://doi.org/10.3390/en20300595
2. Mulac B, Storvold R, Weatherhead EC (2011) Remote Sensing in the Arctic with Unmanned Aircraft: Helping Scientists to Achieve Their Goals. Int Sym Rem Sen Env, Sydney, Australia, (ICRSE), April 10–15. https://www.isprs.org/proceedings/2011/isrse-34/211104015Final00863.pdf
3. Romankevich AP, Kachanovskaya D, Chernyakov G (2017) Primenenie bespilotnyh letatel'nyh apparatov s cel'yu krupnomasshtabnogo kartografirovaniya i sozdaniya cifrovoj osnovy dlya monitoringa rastitel'nosti (The use of unmanned aerial vehicles for the purpose of mapping and creating a digital basis for monitoring vegetation). Land Bel 3:46–48. http://elib.bsu.by/handle/123456789/209976 (In Russian)
4. Skudneva OV (2016) Bespilotnye letatel'nye apparaty v sisteme lesnogo hozyajstva Rossii (Unmanned aerial vehicles in the forestry system of Russia). News high educ For J 6(342):150–154 (In Russian)
5. Frater T, Juzsakova T, Lauer J (2015) Unmanned aerial vehicles in environmental monitoring—An efficient way for remote sensing. J Env Sci Eng A4:85–91. https://doi.org/10.17265/2162-5298/2015.02.004
6. Alvear O, Zema NR, Natalizio E et al (2017) Using UAV-based systems to monitor air pollution in areas with poor accessibility. J Adv Transp 2017:1–14. https://doi.org/10.1155/2017/8204353
7. d'Oleire-Oltmanns S, Marzolff I, Peter KD et al (2012) Unmanned aerial vehicle (UAV) for monitoring soil erosion in Morocco. Rem Sen 4:3390–3416. https://doi.org/10.3390/rs4113390
8. Leifer I, William J, Lehr WJ et al (2012) State of the art satellite and airborne marine oil spill remote sensing: Application to the BP deepwater horizon oil spill. Rem Sen Env 124:185–209. https://doi.org/10.1016/j.rse.2012.03.024
9. Fu Y, Yang G, Song X et al (2021) Using multiscale textures extracted from UAV-based digital images and hyperspectral feature analysis. Rem Sen 13(4):581. https://doi.org/10.3390/rs13040581
10. Guo A, Huang W, Dong Y (2021) Wheat yellow rust detection using UAV-based hyperspectral technology. Rem Sen 13(1):123. https://doi.org/10.3390/rs13010123
11. Machefer M, Lemarchand F, Bonnefond V et al (2020) Mask R-CNN refitting strategy for plant counting and sizing in UAV imagery. Rem Sen 12(18):3015. https://doi.org/10.3390/rs12183015

12. Chemeris EV, Kutuzov AV, Efimov DY et al (2020) Changes in the vegetation cover of the lake Pleshcheyevo (Yaroslavl region) from 1899 to 2017. Proc IBIW 90(98):33–52. https://doi.org/10.24411/0320-3557-2020-10011
13. Bhatnagar S, Gill L, Ghosh B (2020) Drone image segmentation using machine and deep learning for mapping raised bog vegetation communities. Rem Sen 12(16):2602. https://doi.org/10.3390/rs12162602
14. Sivakumar ANV, Li J, Scott S (2020) Comparison of object detection and patch-based classification deep learning models on mid- to late-season weed detection in UAV imagery. Rem Sen 12(13):2136. https://doi.org/10.3390/rs12132136
15. Ashapure A, Jung J, Chang A (2019) A comparative study of RGB and multispectral sensor-based cotton canopy cover modelling using multi-temporal UAS data. Rem Sen 11(23):2757. https://doi.org/10.3390/rs11232757
16. Cendero A (1989) Land-use problems planning and management in the coastal zone. Ocean and Shor Manag 12(5/6):367–381. https://doi.org/10.1016/0951-8312(89)90019-2
17. Hays GC, Richardson AJ, Robinson C (2005) Climate change and marine plankton. Tre Eco Evo 20:337–344. https://doi.org/10.1016/j.tree.2005.03.004
18. Afanasyeva EL, Shimaraev MN (2006) Long-term zooplankton variations in the pelagial of Lake Baikal under global warming. Aquatic ecology at the dawn of the XXI century. KMK, Moscow, pp 253–265
19. Hampton SE, Izmest'eva LR, Moore MY et al (2008) Sixty years of environmental change in the world's largest freshwater lake—Lake Baikal, Siberia. Glo Ch Bio 14(8):1947–1958. https://doi.org/10.1111/j.1365-2486.2008.01616.x
20. Timoshkin OA, Suturin AN, Bondarenko NA (2011) Biology of the coastal zone of Lake Baikal. 1. Overview of the current knowledge on the splash zone, first results of interdisciplinary investigations, monitoring as a basic tool in ecological research. Irk St Univ Bul 4(4):75–110

A Complex of Ground Equipment for an Unmanned Search Aircraft

Sergey Stukalov, Vladimir Kostenkov, and Roman Gavryushin

Abstract Approaches to the creation of a complex of ground equipment for an unmanned search aircraft are considered. To ensure stable control and information acquisition, it is proposed to use the main and additional channels for control and reception of video information with a combined antenna system as part of the complex. For stable operation of the main channel, it is proposed to use a system of two circular polarization antennas: spiral and clover. The technical characteristics of the antennas are presented. The results of practical application of two circularly polarized antennas are presented. It is shown that the inclusion of two different antennas makes it possible to compensate for the deficiencies in the directivity characteristics and to ensure the stable operation of the transceiver devices of the complex. Experiments were carried out with flights of an unmanned search aircraft along the route. Recommendations are given for the practical application of the ground equipment complex for an unmanned search aircraft.

Keywords Ground equipment complex · Unmanned aerial vehicles · Circular polarization antennas · Directional patterns

1 Introduction

In various areas of civil aviation, unmanned aerial vehicles (UAVs) are increasingly used, which have the ability to perform search tasks, en-route flights, flights at specified points in automatic mode, etc. One of the tasks of the UAV flight can be search tasks: viewing the terrain, analyzing the surface for the purpose of examining the places of flight accidents and finding typical for these situations, areas, and objects. In such cases, an important advantage of the UAV (Fig. 1) is the speed of finding the

S. Stukalov (✉) · V. Kostenkov · R. Gavryushin
Moscow State Technical University of Civil Aviation, Moscow, Russia
e-mail: s.stukalov@mstuca.aero

R. Gavryushin
e-mail: tu-204-200@mail.ru

O. A. Gorbachev et al. (eds.), *Proceedings of 10th International Conference on Recent Advances in Civil Aviation*, Lecture Notes in Mechanical Engineering, https://doi.org/10.1007/978-981-19-3788-0_37

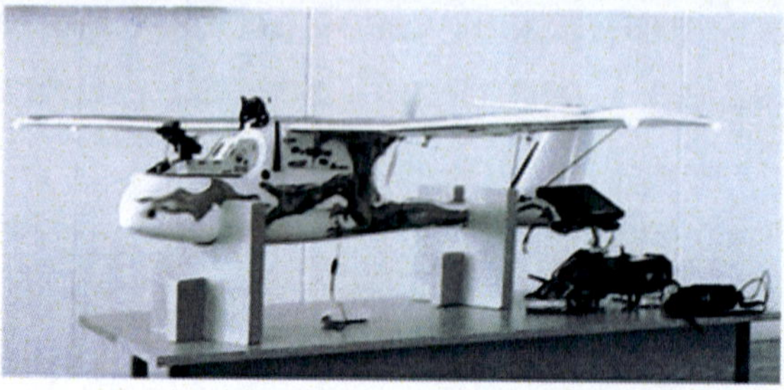

Fig. 1 An experimental sample of an unmanned aerial vehicle with control equipment

coordinates of places and objects of flight accidents, as well as the ability to fly and perform tasks in the daytime and at night.

2 Selection and Construction of the Complex Structure

A complex of ground equipment has been developed to control this type of UAV. In order to ensure stable control and information acquisition, it was proposed to introduce a main control channel, an auxiliary control channel, a main channel for receiving video information, and a backup channel for receiving video information into the composition of the ground equipment of the complex (Fig. 2).

To control the flight of the UAV, a main control channel was created using the J-FPV system, operating at a frequency of 900 MHz, and an auxiliary control channel, operating at a frequency of 2400 MHz. A feature of the technical solution is controlled using the steering column (Fig. 3).

As part of the main channel for receiving video information, the Head Tracker equipment (control of turning the operator's head) is used with the output of information to the operator's video glasses FPV (First-Person View - first-person view) (Fig. 4).

The flight control of the UAV is carried out using the operator's video glasses together with the information displayed on the display. This makes it possible to realize the advantages of applying the idea of automatic dependent surveillance systems-radio broadcasting (ADS-B) [1]. As indicated in [2, 3], the use of this type of operator's video glasses improves visual sensations when analyzing important visual information. And collaboration with displays becomes one of the modes in the specifics of pilot interfaces. In [4] it is shown that such solutions for visualization systems reduce the visual stress of the operator and increase productivity at work.

The auxiliary control channel is based on the FRSKY DJT transmitting module operating at a frequency of 2400 MHz. The control is carried out using the Turnigy 9XR joystick. Two reception channels provide information presentation separately to the monitor of ground equipment and the operator's video glasses, which makes it possible to quickly control the complex and make it available for analysis by ground means. It also helps to implement a UAV control strategy based on a stable vision

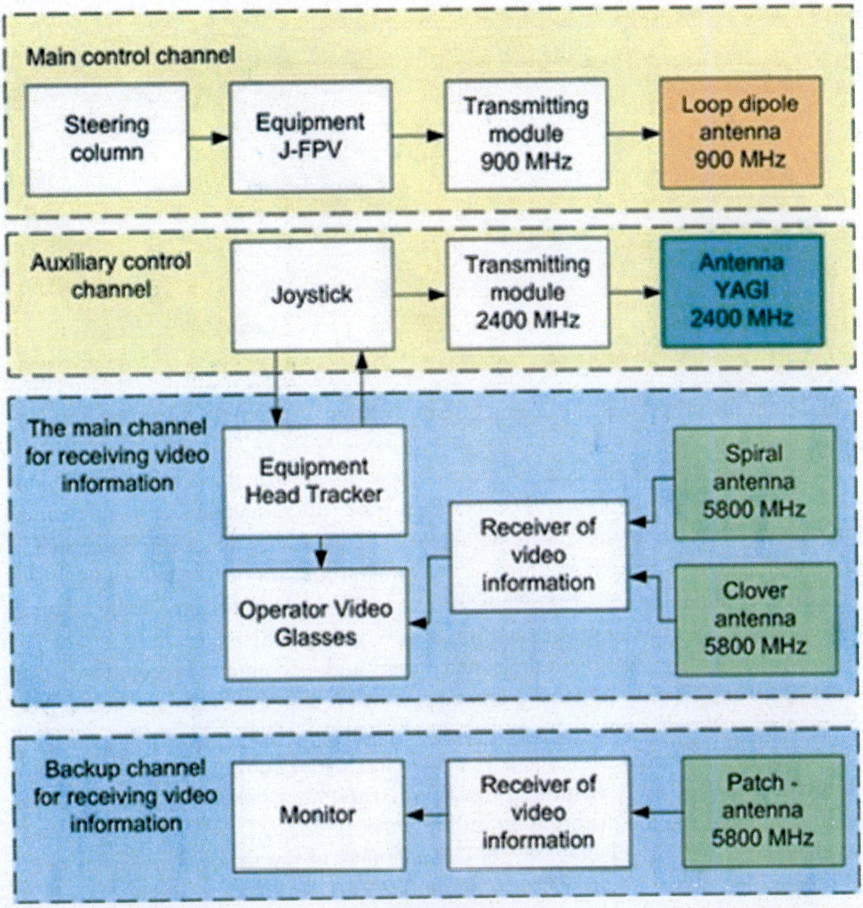

Fig. 2 Structural diagrams of the ground equipment complex

Fig. 3 Main control channel

Fig. 4 Video glasses of the operator

to highlight the properties of evading possible obstacles when flying at low altitudes [5].

A feature of the antenna system of the complex is the use of a combined antenna-mast system, which includes five antennas: a loop dipole antenna, a wave channel antenna, also known as the Yagi-Uda antenna, or Yagi antenna, a spiral antenna, a clover antenna, and a patch antenna. The unique properties of these antennas make it possible to find solutions for establishing a stable operation of the radio channel between the UAV and the ground equipment complex [6–8].

To ensure stable operation of the main channel and reception of video information of the complex, it is proposed to use antennas with circular polarization. Another type of antenna with linear polarization sends a direct signal that can be reflected from objects in the path of the signal [9]. The reflected signal, hitting the receiving point simultaneously with the transmitted signal, can cause interference, i.e., the quality of the video image will be distorted by interference, lines [10, 11]. Circular-polarized antennas lack this disadvantage [12]. The antennas of the receiver and transmitter of linear polarization are very demanding to the mutual arrangement [13, 14]. Both antennas should be located in the same plane (vertically or horizontally). When maneuvering the UAV, there may be a mismatch of polarization and deterioration of the received video information [15]. Antennas with circular polarization are less demanding on the relative position of the antennas [16]. This allows UAVs to fly, performing various maneuvers without losing the quality of the information signal.

3 Discussion of Results and Recommendations

For stable operation of the main channel for receiving video information, it was proposed to use a combined system of two circular polarization antennas: spiral and clover (Fig. 5). The spiral antenna belongs to the class of circularly polarized antennas. It is advisable to use it for multiple amplification of video information. The antenna is more complex than typical dipole whip antennas. characterized by good directional properties and high gain. When using such an antenna in a complex,

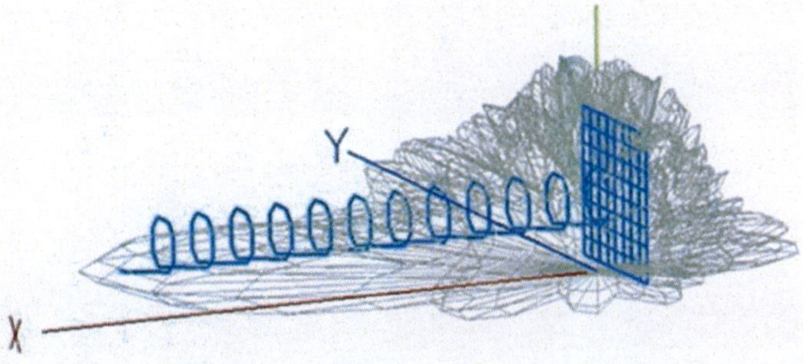

Fig. 5 The radiation pattern in the horizontal and vertical planes and the designed antenna in the MMANA program

the range of video information reception, the range of action increases many times in the direction where the antenna is oriented. The directional pattern of the helical antenna calculated in the MMANA program is shown in Fig. 5.

The obtained characteristic shows that the spiral antenna of the complex will provide directional reception of information signals.

The clover antenna is a circularly polarized antenna. Such an antenna consists of flexible copper conductors connected by soldering and a coaxial cable (Fig. 6).

An important advantage of this antenna design is that the direct and reflected signals from the transmitter do not cancel each other out. The antenna radiation pattern calculated in the MMANA program is shown in Fig. 7.

The obtained characteristic shows that the clover antenna of the complex will provide an omnidirectional reception of information signals. A plastic case was used to protect the antenna from external factors (Fig. 8).

Such a housing is made of radio-transparent material and does not attenuate the signal. When the signal of one of the channels deteriorates, it automatically switches to another antenna. The use of two different antennas makes it possible to compensate

Fig. 6 Clover antenna

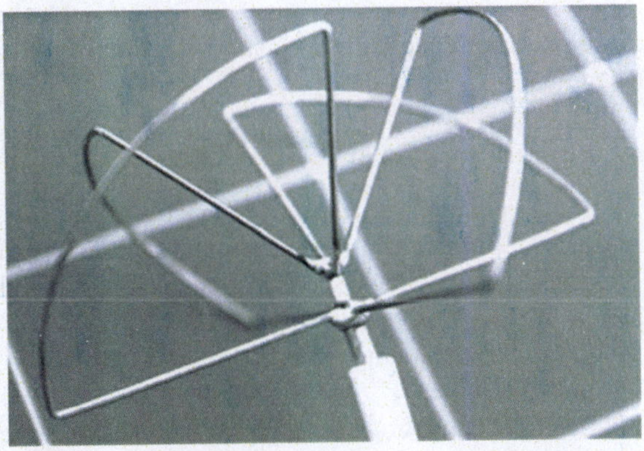

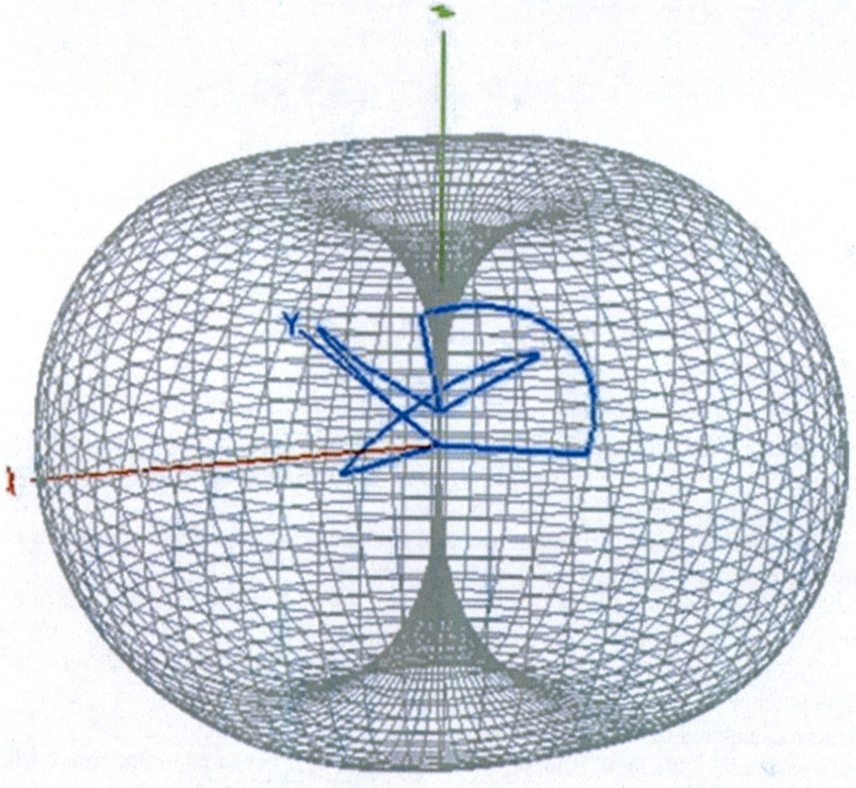

Fig. 7 The radiation pattern designed in the MMANA program

for the deficiencies in the directivity characteristics of each of the antennas separately and to ensure the stable operation of the transceiver devices of the complex (Fig. 9).

To evaluate the operation of the ground equipment complex of the unmanned search aircraft, experiments were carried out with the UAV flight. The flight was carried out along the route with automatic return via the satellite navigation system (Fig. 10).

Based on the experiments carried out, it is possible to give recommendations on the use of the complex in tasks of a research and exploratory nature:

1. In the operation of the antenna-mast system of the main channel for receiving video information for stable ground control, it is advisable to use a combined system of two circular polarization antennas: spiral and clover.
2. To improve the stability of the UAV flight in search tasks, it is necessary to eliminate positioning errors from the satellite navigation system and to correct the flight route.
3. The task of correcting the UAV flight route can be performed automatically based on the results of registration of video images of the terrain.

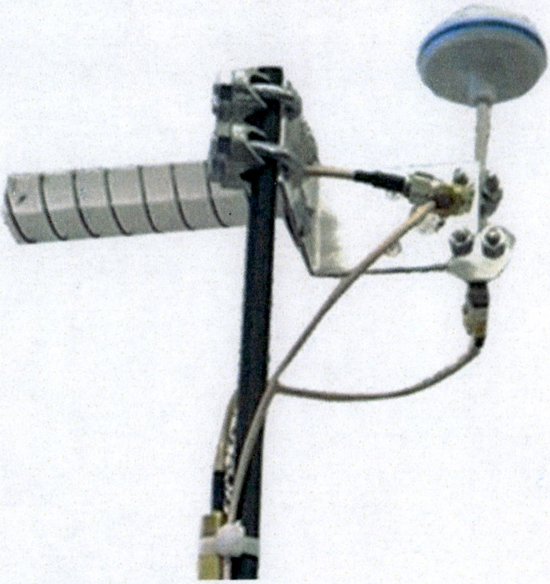

Fig. 8 Circular-polarized antenna system

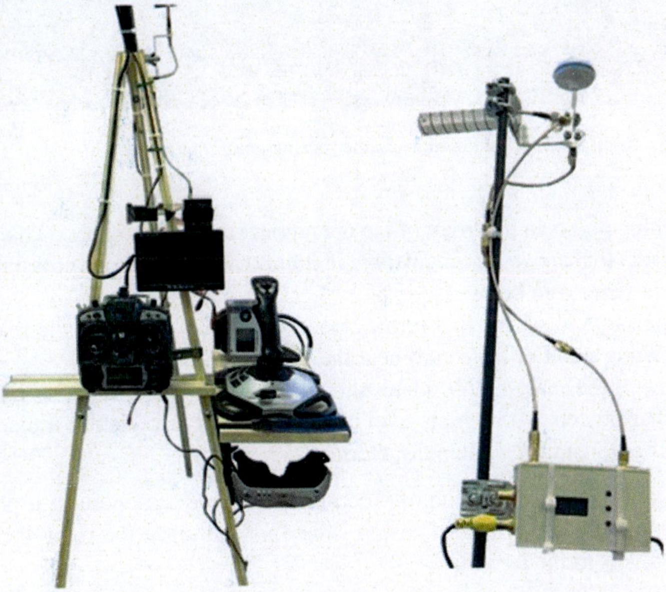

(a) (b)

Fig. 9 Receiving and transmitting devices (**a**) and antennas of the main channel with a video information receiver (**b**) of the ground equipment complex

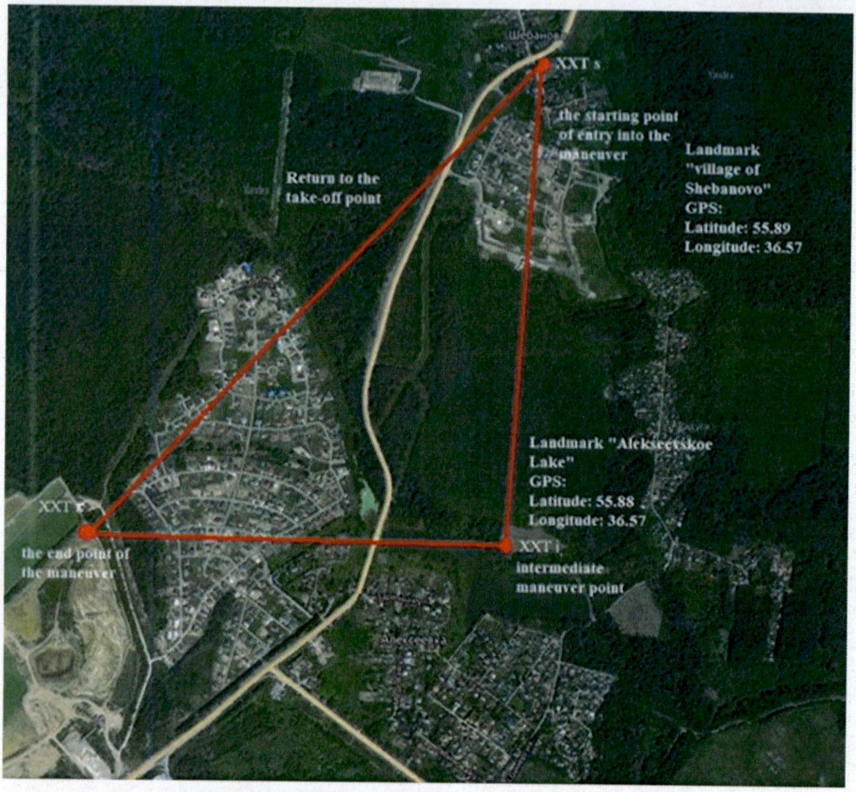

Fig. 10 UAV flight trajectory with return to the starting point

4. As turning points in the flight of the operational research complex, characteristic landmarks that are well recognizable in flight (river bends, road crossings, single buildings, etc.) can be used.
5. It is advisable to set the first turning point of the flight route of the search UAV (the starting point of the route) near the starting point.
6. The line of the search UAV's path should not pass near high-power power lines and other objects with a high level of electromagnetic radiation (radar stations, receiving-transmitting antennas, etc.).

The results of the registration of video images of the terrain make it possible to identify objects characteristic of search situations and make the right decisions on search problems in the terrain.

References

1. Naganawa J, Miyazaki H (2021) Theory of automatic dependent surveillance–broadcast position verification using time difference of arrival. IEEE Trans Aerospace Electronic Syst 57(3):1387–1404. https://doi.org/10.1109/TAES.2020.3043536
2. Gorbunov, A.L. (2014) Stereoscopic augmented reality in visual interface for flight control. Aerospace Science and Technology 38:116–123. https://doi.org/10.1016/j.ast.2014.08.002
3. Gigilashvili D, Thomas JB, Hardeberg JY, Pedersen M (2021) Translucency perception: a review. J Vision 21(8):1–41. https://doi.org/10.1167/jov.21.8.4
4. Pladere T, Luguzis A, Zabels R, Smukulis R, Barkovska V, Krauze L, Konosonoka V, Svede A, Krumina G (2021) When virtual and real worlds coexist: visualization and visual system affect spatial performance in augmented reality. J Vision 21(8):1–18. https://doi.org/10.1167/jov.21.8.17
5. Agrawal P, Ratnoo A, Ghose D (2017) Inverse optical flow based guidance for UAV navigation through urban canyons. Aerospace Sci Tech 68:163–178, September. https://doi.org/10.1016/j.ast.2017.05.012
6. Harl L, He Y, Papapolymeru J (2017) 3D printing of the 77 GHz Yagi-Uda planar antenna. In: IEEE International Symposium on Antennas and Radio Wave Propagation and the USNC/URSI National Radioscience Meeting, San Diego, USA, pp 13–14. https://doi.org/10.1109/apusncursinrsm.2017.8072049
7. Pei T, Zhu L, Wang J, Wu V (2021) Low-profile decoupling structure for mutual coupling suppression in a MIMO patch antenna. IEEE Trans Antennas Prop 69(10):6145–6153. https://doi.org/10.1109/tap.2021.3098565
8. Li M, Luk KM (2015) Wideband magnetoelectric dipole antennas with dual polarization and circular polarization. IEEE Antennas Prop Magazine 57(1):110–119. https://doi.org/10.1109/map.2015.2397091
9. Wei D, Li Yang J (2017) Vortex radio waves realized by spiral antennas. In: IEEE International Symposium on Antennas and Their Propagation, San Diego, USA, pp 175–176. https://doi.org/10.1109/apusncursinrsm.2017.8072130
10. Slomyan I, Odrobina S, Grushchinsky S, Vincha K (2017) Integrated two-beam antenna array with double circular polarization. In: IEEE International Symposium on Antennas and Radio Wave Propagation and the USNC/URSI National Radioscience Meeting, San Diego, USA, pp 2311–2312. https://doi.org/10.1109/apusncursinrsm.2017.8073198
11. Rappoport TS, Sin Y, McCartney GR, Malysh AF, Melis E, Zhang J (2017) Overview of millimeter-band communications for fifth-generation (5g) wireless networks—with emphasis on propagation models. IEEE Trans Antennas Prop 65(12):6213–6230. https://doi.org/10.1109/tap.2017.2734243
12. Hoang H, John M, McAvoy P, Ammann MJ (2021) Near-field propagation analysis for vivaldi antenna design: understanding the propagation process to optimize directivity, signal transmission integrity and efficiency. IEEE Antennas Prop J 63(5):46–60. https://doi.org/10.1109/map.2020.3003217
13. Gryte K, Sollie M, Johansen T (2021) Control system architecture for automatic recovery of fixed-wing unmanned aerial vehicles in a moving arrest system. J Intelligent Robotic Syst 103(4):2–20. https://doi.org/10.1007/s10846-021-01521-z
14. Mesa F, Valerio G, Rodriguez-Berral R, Quevedo-Teruel O (2021) Efficient computation of the dispersion diagram of periodic structures using modeling: an exhaustive review of applications to filters, radiating wave antennas, and metasurfaces. IEEE Antennas Prop Magazine 63(5):33–45. https://doi.org/10.1109/map.2020.3003210
15. Duan H, Xin L, Shi Y (2021) A self-guided autonomous navigation system for unmanned aerial vehicles inspired by homing pigeons. IEEE Trans Aerospace Electronic Syst 57(4):2218–2224. https://doi.org/10.1109/beret.2021.3054060
16. Karim R, Malkovati P (2021) Embedded antennas: the next milestone in the big world of small satellites—an overview of opportunities, challenges and future directions. IEEE Aerospace Elect Syst Magazine 36(1):46–60. https://doi.org/10.1109/MAES.2020.3016751

Transport Economics

The chapter deals with various aspects of organizing passenger and freight air transportation. The papers consider general and specific issues of optimizing the financial, economic and productive activities of airports and airlines, air safety, digitalization, improving business processes. Special attention is paid to the issues of state support of the industry.

Keywords Regional air transportation · Improving business processes of air transport · Air safety · Air transportation subsidies · Airport modernization

Methodology for Evaluating Transport Accessibility in the Arctic Zone: Organization of Passenger Transportation

Irina Poleshkina

Abstract The article defines the concept of transport accessibility of the territory, examines the existing methods for assessing transport accessibility. The reasons that do not allow using the existing methods of assessing transport accessibility in relation to the areas of the Arctic zone were identified. Statistical methods such as transport network density per 1000 square kilometers of territory, per 1000 inhabitants; Engel, Goltz, Uspensky, Vasilevsky coefficients; methods of topological analysis of transport networks; methods of spatial analysis and delimitation of transport networks; methods of potentials and balance methods have been considered as existing methods of transport accessibility assessment. Based on the analysis of the state of networks of different types of transport and the survey of the local population, the main reasons that limit transport accessibility of these territories have been identified. These reasons include: the lack of railway service, the lack of roads for year-round use, the high cost of passenger intra-regional air transportation, the low regularity of flights, and the limited regular route network of existing airlines. Taking into account these factors, the paper proposes a methodology for assessing the transport accessibility of the regions of the Arctic zone in terms of the organization of passenger transportation, which is based on the calculation of the full cost of all correspondence from a particular area to all other considered areas of the region per 1 km of distance between them. The full cost of transportation is calculated as the direct costs of purchasing tickets for transportation, the cost of waiting time for connecting flights, and the waiting time for flight delays due to weather conditions. The cost of waiting time is estimated based on the calculation of the average cost of accommodation for one day in Yakutsk.

Keywords Transport availability of the territory · The Arctic zone regions · Air transport · Passenger transportation · Transportation cost · Transportation time

I. Poleshkina (✉)
Moscow State Technical University of Civil Aviation, Moscow, Russia
e-mail: i.poleshkina@mstuca.aero

O. A. Gorbachev et al. (eds.), *Proceedings of 10th International Conference on Recent Advances in Civil Aviation*, Lecture Notes in Mechanical Engineering, https://doi.org/10.1007/978-981-19-3788-0_38

1 Introduction

Scientific literature uses a large number of concepts reflecting the degree of transport connectivity of a particular territory with other territories of the region and the transport system of the country. Among them, we can distinguish the following terms: transport accessibility, transport availability, transport development, transport connectivity of territories. They all have a close meaning in one way or another. The Transport Strategy of the Russian Federation defines that "accessibility of transport services and their volume determine the completeness of connections within the country and outside it, as well as the ability to move all segments of the population to meet production and social needs" [1]. In this paper, we propose to consider the term transport accessibility of the territory, by which we mean the ability of the population and economic entities to access a transport service to move from a certain point of origin to the required destination using transport networks of all available modes of transport within a certain period of time. Thus, the term transport accessibility can be applied to a territorial unit of any level (settlement, district, region, allocated territory). Most often in the scientific literature the term transport accessibility is used to determine the degree of transport connectivity of a region (its administrative center) with other regions of the country (their administrative centers). The indicator of transport availability reflects the ability of the transport system to meet the entire demand for transport services, i.e., it allows assessing the existing transport capacity. Transport development and transport connectivity of the territory, in our opinion, are practically synonymous and reflect, as a rule, the level of development of the transport system within the region, i.e., between its settlements and/or economic centers. We propose to use the term "transport accessibility" in relation to the regional level as well, with a clear definition of which territory this indicator refers to.

The regions of the Arctic zone have pronounced characteristics that impose significant restrictions on the development of transport networks of all modes of transport. In addition, these features do not allow the application of standard methods for assessing the transport accessibility of territories within the Arctic zone. The main volume of cargo delivery is carried out by long-term complex multimodal schemes with seasonal involvement of river transport (in summer) and road transport (in winter). The average period of cargo delivery within the Arctic zone is more than one year [2]. Local and regional passenger transport in the Arctic areas is mainly carried out by air transport due to the long distances involved and the lack of year-round road connections. At the same time, the share of air transport in the organization of cargo transportation is negligible due to the high cost of delivery.

We propose to assess the transport accessibility of the Arctic zone regions separately in terms of the organization of passenger and freight transportation. The purpose of this paper is to develop a methodology for assessing the transport accessibility of the Arctic zone in terms of the organization of passenger transportation using the example of the Arctic zone of Eastern Siberia of the Russian Federation.

2 Research Methods

This study was based on the analysis of existing methods for assessing the transport accessibility of territories and the possibility of their use in relation to the regions of the Arctic zone. For this purpose, at the first stage, the peculiarities of the state of networks of different types of transport in the Arctic zone of Krasnoyarsk Krai, the Republic of Sakha (Yakutia) and Chukotka Autonomous Okrug were identified.

Statistical methods such as the density of the transport network per 1000 square kilometers of territory, per 1000 inhabitants; coefficients of Engel, Goltz, Uspensky, Vasilevsky [3–8] were considered as existing methods to assess transport accessibility. In addition, the methods of topological analysis of transport networks proposed by Professor Tarkhov [9, 10]; methods of spatial analysis and delimitation of transport networks [11]; potential methods and balance methods were considered [12]. The article considers the work on the assessment of transport accessibility of European countries on a regional scale, carried out as part of the project ESPON 2013 [13]. In these works, transport accessibility is evaluated according to three traditional indicators: access to regional centers, daily accessibility of jobs, and regional potential accessibility of the population by means of road and public transport. Access to regional centers is evaluated on the basis of the time it takes to travel from the settlement to the administrative district center by road and public transport. Daily accessibility of jobs is estimated by the number of jobs that can be reached by road or public transport in 60 min. Regional potential accessibility is estimated by the number of people that can be reached by automobile or public transportation. This indicator is used to estimate the size of the sales market for individual companies, determine the size of the labor market, etc.

Papers [13] also propose the use of three non-classical indicators of transport accessibility: access to healthcare facilities, availability of upper secondary schools, and potential accessibility to basic health care. The access to healthcare facilities is measured by the time it takes to travel by road and public transportation to the nearest hospital. The availability of upper secondary schools is measured by the number of high schools located within a 30-min transportation distance. The potential accessibility to basic health care is measured by the number of doctors of various specialties that can be reached by car and public transport from the locality in question.

With regard to each method, limitations were identified that do not allow us to comprehensively assess the transport accessibility of the Arctic zone areas in terms of organizing passenger transportation. In order to identify the problems of transport accessibility of remote settlements in the Arctic zone, a survey of the local population living in 13 Arctic regions of the Republic of Sakha (Yakutia) was conducted. The survey data showed that the main transportation problems are caused by low regularity of air flights in intra-regional communication, high flight costs, and the limited route network of existing airlines. These indicators were taken into account when developing the methodology for assessing the transport accessibility of northern territories. Based on the main target indicators reflecting the level of transport accessibility of the territories, such as time and cost of transportation, the methodology for assessing the transport accessibility of the regions of the Arctic zone of Eastern Siberia of the Russian Federation was developed.

3 Methodology for Assessing the Transport Accessibility of the Arctic Zone Districts

Statistical indicators of transport accessibility of territories allow calculating the ratio of the length of transport networks either relative to the territory area or relative to population, or simultaneously relative to the territory area and population. The Goltz coefficient allows calculating the ratio of the length of the transport network relative to the population size and number of settlements, the Uspensky coefficient—relative to the territory area, population size, and volume of transported goods, the Vasilevsky coefficient—relative to the territory area, population size, and volume of goods produced. However, all these indicators do not allow taking into account the seasonal limitations of the use of modes of transport in the northern territories, the participation of air transport in transportation, time and cost of delivery.

The methods of topological analysis of transport networks make it possible to determine the shape and structure of transport networks, their connectivity and isolation, cyclicality and branching. However, this tèchnique is of little relevance for areas of the Arctic zone, which usually have a single transport axis, the main characteristics of which are its capacity and seasonal limitations in the use of individual sections. Methods of spatial analysis and delineation of transport networks are quite convenient in terms of assessment of network configuration, so we propose to use them in combination with methods of quantitative assessment of transport accessibility, which will be proposed below. Methods of potentials and balance methods are more suitable for the assessment of transport availability of territories in terms of freight traffic organization, as in the conditions of low population density in the northern areas air transport has no capacity limitations. The methods, that assess transport accessibility by the amount of time spent on the trip, in our opinion, are the most objective for the areas of the Arctic zone. These methods allow us to take into account seasonal limitations and low regularity of transportation.

Analysis of the state of the road transport network in the Arctic regions of Krasnoyarsk Territory, the Republic of Sakha (Yakutia) and Chukotka Autonomous District showed that these territories have roads of seasonal use—winter roads. In the harsh climatic conditions of the Arctic winter and the extremely limited road infrastructure, long-distance passenger transportation by road is extremely dangerous. On winter roads, there is a lack of gas stations, repair shops, and places to eat and get warm. Therefore, road transport accounts for a small share of passenger traffic. There is no railway transport in the areas under consideration. River transportation is seasonal, taking into account the winter freeze-up and the period of summer shallowing of the rivers. Navigation along the channels of northern rivers opens gradually, starting from the upper reaches. Transportation of cargo between river beds is carried out along the Northern Sea Route during an even more limited navigation period. Delivery of cargoes in this way takes more than one year [14].

Air transport is the main in the provision of passenger transportation services in Arctic zone due to the objective impact of natural and climatic factors. This is confirmed by the experience in the development of transportation systems in Alaska and Northern Canada [15, 16]. The accessibility of the northern territories in terms of passenger transportation is determined by two criteria: the cost of transportation and the time it takes to travel.

Surveys of the local population of settlements in the Arctic zone of the Republic of Sakha (Yakutia) have shown that due to the high cost of air transport tickets this service is becoming unaffordable for the local population. For example, the cost of a flight from Moscow to Yakutsk for a distance of 4883 km is on average 17,500 rubles. And the cost of an intraregional flight from Yakutsk to Chokurdakh for a distance of 1245 km is on average 30,000 rubles. The cost of a flight from Yakutsk to Tiksi is about 27,000 rubles. The regular route network of Polar Airlines, which serves local passenger traffic in the Republic of Sakha (Yakutia), connects only administrative district centers with the capital of the Republic. There is no direct regular air service between the republic's Arctic districts. High flight costs and a limited route network result in a reduced demand for air transportation on the part of the local population.

The high cost of air transport tickets makes this service unaffordable for the local population. Reduced demand for air travel leads to underutilization of flights and loss of profit for airlines, which in turn reduce the number of regular flights and increase the cost per ticket. As a result, the affordability of the transport service decreases and the waiting time for transport increases, reducing its physical accessibility. A similar approach to assessing transport accessibility of territories of the country was proposed in the works of Lavrinenko et al. [17]. As applied to the northern regions, we propose to extend this methodology due to the high level of impact of natural and climatic factors and low regularity of passenger flights.

We propose to assess the transport aviation accessibility of the Arctic zone area by the formula:

$$a_i = \frac{\sum_{j \ni R} c_{ij}}{\sum_{j \ni R} l_{ij}};\ j \ni R \tag{1}$$

where c_{ij} is the cost of moving passengers from district i to all other districts j of the region in question, expressed in total transport cost, travel time cost, and waiting time cost; l_{ij} is the distance between district i and the rest of district j of the region.

In fact, this indicator will reflect the total cost of transporting passengers from the selected area i to all other considered areas j per 1 km of distance between these areas.

We propose to calculate the total cost c_{ij} of passenger transportation from district i to all other districts j of the region in question using the formula:

$$c_{ij} = \sum_{j \ni R} p_{ij} + \sum_{j \ni R} t_{ij} \times s_i;\ j \ni R \tag{2}$$

where p_{ij} is the direct monetary cost of traveling from area i to all other areas j with which no transport alternative is available, determined at a specific date; t_{ij} is the total time cost of traveling the distance between area i and all other areas j under consideration; s_i is the cost of waiting time, which is calculated as the average cost of one day's accommodation in Yakutsk.

The cost of money for the flight is estimated by the real cost of tickets for the selected date because the cost of the flight depends on a large number of parameters, starting from the cost of aviation fuel, the amount of airport charges and ending with the occupancy of the flight. Therefore, it is not reasonable to take it into account on the basis of calculation of these indicators in the formulation of this problem.

The amount of total time t_{ij} to cover the distance between area i and all other areas under consideration j is calculated by the formula:

$$t_{ij} = \sum_{j \ni R} t_{ij}^{t} + \sum_{j \ni R} t_{ij}^{w};\ j \ni R \tag{3}$$

where t_{ij}^{t} is the time it takes to travel from region i to all other regions j under consideration; t_{ij}^{w} is the waiting time for the transport, which is made up of the waiting time for the connecting flight and the waiting time for the departure due to adverse weather conditions from region i to all other regions j under consideration.

Total waiting time for transport t_{ij}^{w} we propose to calculate according to the formula:

$$t_{ij}^{w} = \sum_{j \ni R} t_{ij}^{w1} + \sum_{j \ni R} t_{ij}^{w2} \times r;\ j \ni R \tag{4}$$

where t_{ij}^{w1} is waiting time for connecting flights when flying from district i to all other districts j through the region's central airport due to lack of direct air connectivity between the northern districts; t_{ij}^{w2} is time delay of flight from airport i to all other considered areas j; r is probability of flight delay due to weather conditions from area i to all other considered areas j on selected departure date.

This methodology will make it possible to assess the transport aviation accessibility of the Arctic areas, taking into account the cost of transportation, the existence of a route network between the areas in question, the regularity of existing flights and the risk of flight cancellation due to weather conditions. However, the application of this methodology requires the development of an Arctic transport accessibility scale. This scale can be developed in two ways. The first way is to assess the transport accessibility of all the Arctic regions of Eastern Siberia and rank them in ascending order. Take the smallest index value as one. Then, calculate the deviation of the index values of the other districts from the benchmark. And convert this value into a relative value in the range from 0 to 1. The second way is to determine the amount of money that can be spent on travel, depending on the level of income of the population, and calculate the normative value of average expenses per 1 km of travel. This value should also be taken as a unit, and all indices should be converted by analogy into a relative value between 0 and 1. This methodology reflects the main features of passenger transport organization in the northern regions. To calculate this transport accessibility index, it is necessary to determine the probability of flight delays at each of the airports in the selected areas under consideration. The probability of flight delays can be determined on the basis of analysis of data on flight delays departing from these airports in previous years.

A further line of research is to define a methodology for assessing the transport accessibility of the Arctic areas in terms of organizing cargo delivery through multi-stage multimodal delivery schemes, taking into account the seasonal limitations of using each mode of transport.

4 Conclusions

The proposed methodology for assessing transport accessibility can be used not only for the administrative district centers but also for all the settlements of the Arctic zone. The ranking of settlements by the level of aviation transport accessibility will allow to make a more objective decision on the implementation of transport projects, reconstruction, modernization and construction of airports, runways and landing sites, the distribution of budget funds to subsidize socially important routes of air transport.

References

1. Decree of The Government of The Russian Federation of 27.11.2021 N3363-r "Transportnoj strategii Rossijskoj Federacii do 2030 goda s prognozom na period do 2035 goda" (Transport Strategy of the Russian Federation until 2030 with a forecast for the period up to 2035), https://docs.cntd.ru/document/727294161?marker=65A0IQ. Accessed 20 Dec 2021. (in Russian)
2. Poleshkina I, Gorbunov V (2021) Development of air transport network in the Arctic Zone of Eastern Siberia. Transportation Res Procedia 57:443–451
3. Bolshakov NM (2015) Strategiya razvitiya dorozhno-transportnoj seti regiona: metodologiya i praktika (Strategy for the development of the region's road transport network: methodology and practice). SPbGLTU, Saint Petersburg (in Russian)
4. Tsyganov VV (2021) Kompleks modelej strategicheskogo upravleniya transportnoj infrastrukturoj Sibiri, Dal'nego Vostoka i Rossijskoj Arktiki (A set of strategic management models for transport infrastructure in Siberia, the Far East and the Russian Arctic). ITNOU: Informatsionnyye tekhnologii v nauke, obrazovanii i upravlenii 1(17):3–8 (in Russian)
5. Tsyganov V (2019) Development of infrastructure in Siberia, the Far East and the Arctic Zone of Russia. In: Proceedings of 2019 12th International Conference "Management of Large-Scale System Development", MLSD, 8910968

6. Dabiev DF, Dabieva UM (2015) Ocenka transportnoj infrastruktury makroregionov Rossii (Assesment of the infrastructure of transport of Russia). Mezhdunarodnyy zhurnal prikladnykh i fundamentalnykh issledovaniy 11–2:283–284 (in Russian)
7. Malygin IG (2019) Infrastruktura Sibiri, Dal'nego Vostoka i Arktiki. Sostoyanie i tri etapa razvitiya do 2050 g (Infrastructure of Siberia, the Far East and the Arctic. Status and three stages of development until 2050). Solomenko Institute of Transport Problems of the Russian Academy of Sciences, Saint Petersburg. (in Russian)
8. Tsukerman VA, Goryachevskaya ES (2021) Transport system of Northern and Arctic regions: Assessment and development problems. IOP Conference Series: Materials Science and Engineering 1079:062090
9. Tarkhov SA (2018) Transportnaya osvoennost' territorii (Transportation development of territories). Moscow University Bulletin. Series 5. Geography 2:3–9 (in Russian)
10. Tarkhov SA (2018) Izmenenie aviatransportnoj svyaznosti gorodov Rossii v 1990–2015 gg. (Changes in air transport connectivity of Russian cities in 1990–2015). Izvestiya rossiiskoi akademii nauk. Seriya geograficheskaya 2:5–26 (in Russian)
11. Neretin AS, Zotova MV, Lomakina AI, Tarkhov SA (2019) Transportnaya svyazannost' i osvoennost' Vostochnyh regionov Rossii (Transport connection and development of the eastern regions of Russia). Izvestiya rossiiskoi akademii nauk. Seriya geograficheskaya 6:35–52 (in Russian)
12. Horak J (2006) Transport Accessibility Evaluation. Geografie 111:115–132
13. Biosca O, Spiekermann K, Stępniak M (2013) Transport accessibility at regional scale. Europa XXI 24:5–17
14. Poleshkina IO (2018) Ocenka effektivnosti prodovol'stvennogo obespecheniya rajonov Krajnego Severa Rossii (Problems of food security in the regions of the Far North of Russia). Economy of Region. 14(3):820–835 (in Russian)
15. Alaska Aviation System Plan, Phase III, CFAPT00484 | AIP 3–02–0000–024–2018. Alaska Department of Transportation & Public Facilities Statewide Aviation, January (2021)
16. Civil Aviation Infrastructure in the North. Transport Canada. Spring Reports of the Auditor General of Canada to the Parliament of Canada (2017), https://www.oag-bvg.gc.ca/internet/English/parl_oag_201705_06_e_42228.html. Accessed 20 Dec 2021
17. Lavrinenko PA, Romashina AA, Stepanov PS, Chistyakov PA (2019) Transport accessibility as an indicator of regional development. Studies Russian Econ Develop 30(6):694–701

Modeling of Sustainable Business Processes of Solid Municipal Waste Removal in Civil Aviation

Alexander Sukhorukov, Nikolay Koryagin, Ekaterina Bogdanova, and Elena Zakharova

Abstract The article deals with the problems of organizing business processes for the removal of solid municipal waste (MSW) in civil aviation, taking into account the construction of a closed-cycle economy. Particular attention is paid to reducing the overall consumption of natural resources by recycling waste into secondary resources and goods. A typical scheme of the main business processes of accumulation, transportation and disposal of MSW in civil aviation, developed using a business process modeling system, is given. Based on the conducted functional and cost analysis and ABC-analysis of business processes on the Pareto principle, economically sound recommendations are given for the organization of sustainable business processes for the export of MSW at air enterprises, which consist in the need to increase the share of separate waste accumulation and recycling.

Keywords Business processes · Civil aviation · Solid municipal waste · Closed-loop economy

1 Problems of Sustainable Development of Business Processes of Solid Municipal Waste Removal in Civil Aviation

1.1 Circular Economy in the Field of Solid Municipal Waste Management

Every year, the problem of increasing the consumption of natural resources is becoming more obvious in the world, which exacerbates economic, environmental

A. Sukhorukov (✉) · N. Koryagin · E. Bogdanova
The Moscow State Technical University of Civil Aviation, Moscow, Russia
e-mail: sukhorukov@mstuca.aero

A. Sukhorukov · E. Zakharova
Plekhanov Russian University of Economics, Moscow, Russia

O. A. Gorbachev et al. (eds.), *Proceedings of 10th International Conference on Recent Advances in Civil Aviation*, Lecture Notes in Mechanical Engineering, https://doi.org/10.1007/978-981-19-3788-0_39

and social contradictions. One of the ways to solve this problem is the transition from a linear model of the economy to a closed-cycle, where the principles of sustainable development defined in 2015 by the UN General Assembly are implemented [1]. Russia also pays great attention to the principles of sustainable development and the gradual transition to a circular economy. This is reflected in the national project «Ecology», the terms of implementation of which are prescribed from 01.10.2018 to 31.12.2024, and a budget of 4041 billion rubles has been allocated for it. Of these, 296.2 billion rubles were allocated for the project «Integrated Solid Municipal Waste Management System». At the same time, a clear task has been set to implement the principles of sustainable development in this area by bringing the level of MSW processing to 36% by 2024.This will reduce the total consumption of natural resources in a circular economy (Formula 1) [2].

$$Na = Nr + Ns + Nw \tag{1}$$

where:

Na—total consumption of natural resources;

Nr—rational consumption of natural resources in a circular economy;

Ns—«structural» consumption (waste, excess consumption, overconsumption) of natural resources in production processes;

Nw—waste at the stage of consumption, which accumulate natural resources.

The problem of handling MSW is associated with both excessive consumption of resources and environmental pollution. Air, reservoirs and soils are polluted. For industrial waste and hazardous products of large man-made accidents, there are various remote monitoring methods that allow you to track environmental anomalies [3]. In the case of MSW, which belong to the lowest hazard classes, it is advisable to focus on the organization of sustainable business processes instead of expensive remote monitoring systems.

1.2 *Problems of Solid Municipal Waste Management in Civil Aviation*

Civil aviation, as a high-tech transport industry, implementing its main business processes (transportation of passengers and cargo) [4], contributes a large share to the consumption of natural resources. Of course, this primarily concerns the consumption of petroleum products, which are non-renewable natural resources. For airlines, the transformation of business processes should proceed under the principles of environment, social, governance (ESG), which include concern for the environment, society and transparent corporate governance [5].

At first glance, in comparison with the consumed petroleum products, the sustainable management of MSW at airports is not a primary task in the field of environmental management. It should be borne in mind that major airports are located in large cities and megacities, where the problem of handling MSW is already acute. Therefore, the considerable volumes of MSW generated in the processes of transportation of passengers and cargo further aggravate the problem of waste in these regions. In some regions, such as Moscow and the Moscow region, the handling of MSW is a critical problem for the population and management. In these regions there is an impressive air transport hub with the ability to transport more than 100 million passengers per year. Figure 1 shows the dynamics of the number of passengers transported by year for one of the largest airports in Moscow Domodedovo [6]. Despite the doubling of these figures since 2019 due to the COVID-19 pandemic, the number of people transported looks quite impressive—more than 15 million passengers.

There is also a reverse impact of MSW on the safety of civil aviation. It is worth remembering here how on August 15, 2019, a passenger plane made an emergency landing on a corn field due to seagulls that took off from a landfill getting into the engines. In this case, the processes of handling MSW in the Moscow region revealed not only environmental, economic and social problems, but also the problem of flight safety.

The main business processes in civil aviation are quite strictly regulated by various «green» regulations on pollution of the atmosphere, soil, reservoirs, noise. However, additional business processes, such as meals for passengers on board, comfortable waiting in the terminal halls, which entail the formation of a considerable amount of MSW, are not yet sufficiently regulated. As a result, various conflicts arise between airlines, airports and regional operators on the handling of MSW.

Firstly, in civil aviation, every aircraft participates in the business processes of handling MSW. According to the Federal Classification Catalog of Waste 2021 [7], waste generated from passengers of an aircraft with hazard class 4 belongs to MSW.As a result of catering for passengers and shopping in duty-free on board the aircraft, a certain standard set of MSW is formed, usually consisting of plastics, polyethylene terephthalates (PET), glass, aluminum, paper, food waste [8].

Fig. 1 Annual passenger traffic at Domodedovo Airport

Secondly, numerous public catering outlets and industrial goods stores in airport terminal buildings are involved in the accumulation of MSW, which also form waste with a similar composition.

2 Process Model of Sustainable Solid Municipal Waste Management in Civil Aviation

2.1 A Typical Scheme of Business Processes of Accumulation, Transportation and Disposal of MSW in Civil Aviation

An important task in such large regions as Moscow and the Moscow region is the development of adequate electronic territorial schemes for the management of MSW, which should take into account all waste flows from accumulation sites to disposal sites. Currently, the process model of such a territorial scheme can be presented in general form as a sequence of business processes of accumulation, transportation and disposal of MSW (Fig. 2).

Since most airlines and airports are highly organized economic units, where most business processes are already automated and robotic, it is proposed to organize business processes for handling MSW at the same high level, relying on rational environmental management and embedding these processes in a circular economy. At the same time, for example, Moscow airports can be used as experimental sites for the implementation of sustainability principles, where the total consumption of natural resources («Na» in Formula 1) will decrease due to an increase in the share of MSW sent for processing and the involvement of primary natural resources in secondary turnover.

Such a stable organization of business processes does not require large expenses from airlines and airports, since the regional operator is responsible for the general handling of MSW, with which the airline or airport enters into a contract. The airline is required to organize separate collection of MSW on board, and from the airport—separate collection of MSW on the ground and transportation of MSW from the aircraft to the site of accumulation of separate waste.

It is obvious that the organization of separate collection of MSW on board the aircraft will entail the development of appropriate regulations that take into account flight safety. However, standard sets for feeding passengers on board with a known composition of generated MSW, as well as a directive duty for cabin crews of separate collection aircraft will make such business processes effective from the point of view of a circular economy. In such business processes, a more detailed separation of MSW is possible, which will make it possible to send a larger volume of the total waste accumulation for recycling and reuse.

In the issue of separate collection of MSW at the airport itself, it is necessary to take into account the highly organized structure of the airport as an organization

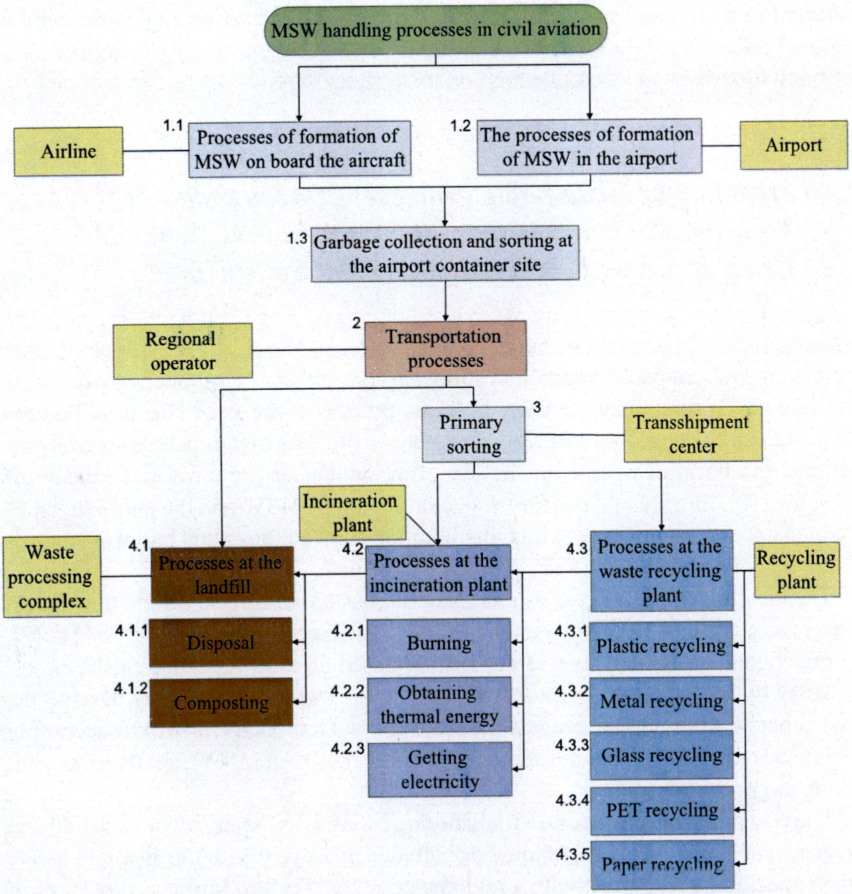

Fig. 2 A typical scheme of business processes of accumulation, transportation and disposal of MSW in civil aviation

that will allow for the collection of MSW with high detail. Such conditions significantly increase the possibilities of separate collection of MSW in civil aviation, for example, in comparison with housing and communal services, where residents of apartment buildings should have a certain culture of separate collection, which is quite problematic.

In accordance with the Decree of the Government of the Russian Federation of November 12, 2016 N 1156 «On the management of solid municipal waste and amendments to the Decree of the Government of the Russian Federation of August 25, 2008 N 641» [Resolution of the Government of the Russian Federation…, 2016] business processes of transportation of MSW at the airport can be organized in two ways. In the first case, the airline can independently conclude an agreement with a regional operator for the export of solid waste generated in the aircraft, and in the second case, the airline can delegate this obligation to the airport, which is reflected in the contract. Accordingly, the airport enters into an agreement with a regional operator for the export, including its own MSW, providing container sites for undivided and separate collection on its territory.

2.2 Modeling of Sustainable Processes of Accumulation, Transportation and Disposal of MSW in Civil Aviation Using Business Process Management System/Tool

As an example, the possible sustainable organization of business processes of separate collection and removal of municipal solid waste at Moscow Domodedovo airport is considered. To model and analyze business processes, we used Business Process Management System/Tool «Business Engineer» [9]. The digital process model was built on the basis of data from the electronic model of the territorial scheme of Moscow [10], in which places for the accumulation of MSW and the entire logistics chain of export, sorting, processing, incineration at incinerators and burial at landfills of the non-recyclable part of MSW were established.

The creation of the process model made it possible to carry out Functional Cost Analysis, as well as ABC-analysis according to the Pareto principle (Fig. 3) [11, 12].

Functional Cost Analysis was carried out based on the calculation of the export of MSW by the regional operator from the airport territory once a day. During the day, about 50 thousand passengers passed through Domodedovo Airport according to statistics [6], who left behind about 3,000 cubic meters of MSW directly on aircraft and in airport buildings.

The cost of business processes for handling MSW is automatically calculated and presented in Table 1. This cost should be included in the tariffs for the export of MSW when concluding a contract with a regional operator. The key business processes of ABC-analysis of category «A», which are formed according to the Pareto principle, are highlighted in red. Thanks to such economic tools, it is possible to evaluate the most profitable directions of sustainable transformation of business processes.

3 Practical Recommendations on the Organization of Sustainable Business Processes for the Export of MSW at Air Enterprises

The proposed analytical model makes it economically feasible to transform the business processes of handling MSW into a closed-cycle economy model, taking into account the principles of ESG. The basis of such transformation will be a gradual transition to separate collection of MSW and their separate processing. Given the high organization of business processes in civil aviation, such a transition can be

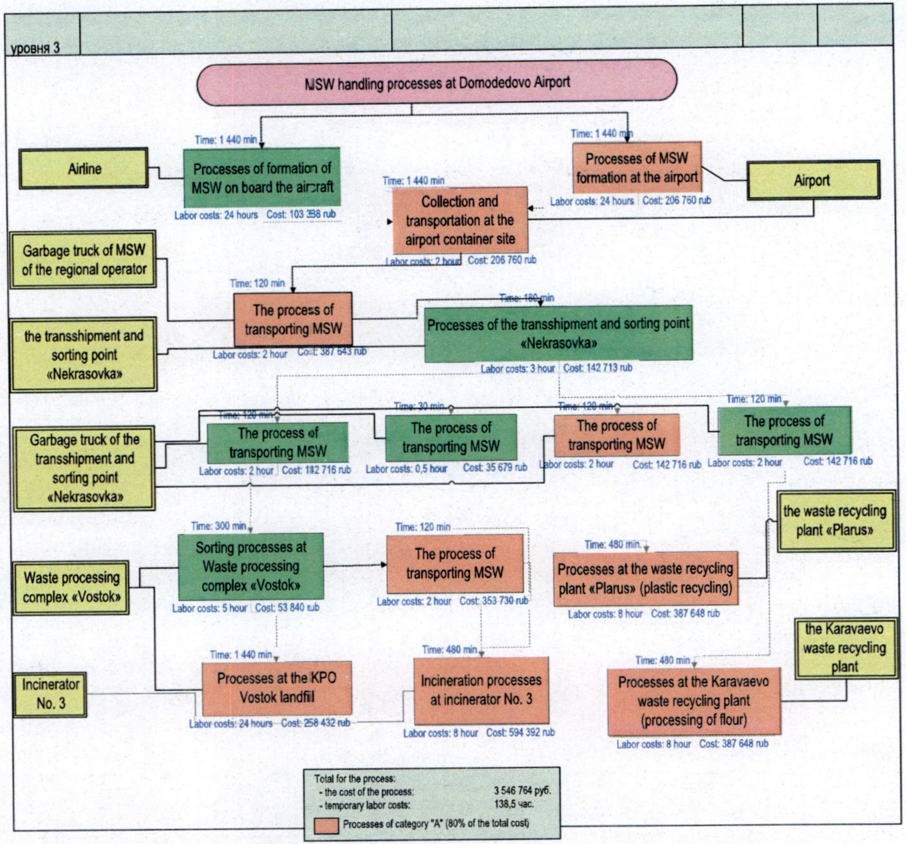

Fig. 3 Functional cost analysis and ABC-analysis based on the Pareto principle of business processes for the removal of solid municipal waste from an airline

Table 1 Calculation of the cost of the process and its time labor costs

№	Code	Business processes	Execution time, min	The cost of temporary resources, rub	The cost of one execution, rub	Number of executions	The cost, rub	Temporary labor costs, hour
1	B3.1.1	Handling of MSW at the airport		3,546,764	3 546,764		3,546,764	138.50
2	B3.1.1.1	Collection and transportation at the container site	1440	206,760	206,760	1	206,760	24.00
3	B3.1.1.2	MSW formation processes on board	1440	103,368	103,368	1	103,368	24.00
4	B3.1.1.3	Processes of MSW formation at the airport	1440	206,760	206,760		206,760	24.00
5	B3.1.1.4	Sorting point processes	180	142,713	142,713		142,713	3.00
6	B3.1.1.5	Processes at the Vostok landfill	1440	258,432	258,432		258,432	24.00
7	B3.1.1.6	Combustion processes at Plant No. 3	480	594,392	594,392		594,392	8.00
8	B3.1.1.7	Processes at a plastic recycling plant	480	387,648	387,648		387,648	8,00
9	B3.1.1.8	Sorting processes at KPO «Vostok»	300	53,840	53,840		53,840	5.00

(continued)

Table 1 (continued)

№	Code	Business processes	Execution time, min	The cost of temporary resources, rub	The cost of one execution, rub	Number of executions	The cost, rub	Temporary labor costs, hour
10	B3.1.1.9	The process of transporting MSW	120	387,646	387,646	1	387,646	2.00
11	B3.1.1.10	Processes at the paper processing plant	480	387,648	387,648		387,648	8.00
12	B3.1.1.11	The process of transporting MSW	120	142,716	142,716		142,716	2.00
13	B3.1.1.12	The process of transporting MSW	30	35,679	35,679		35,679	0.50
14	B3.1.1.13	The process of transporting MSW	120	142,716	142,716		142,716	2.00
15	B3.1.1.14	The process of transporting MSW	120	142,716	142,716		142,716	2.00
16	B3.1.1.15	The process of transporting MSW	120	353,730	353,730		353,730	2.00

carried out much easier than in the housing and utilities sector, where the quality of separate collection of MSW largely depends on the culture of residents.

The practical value of using such an analytical model of business processes for airlines and air companies lies in the possibility of concluding economically sound contracts for the export of MSW with a regional operator, in which the economic mechanisms of tariff formation become transparent. For example, the tariffs for aircraft maintenance at airports can include tariffs for aircraft cleaning and for the removal of MSW, taking into account the principles of sustainability and transition to a circular economy.

The transition to a rational organization of business processes for handling MSW at the airport can be presented as an airline development project, which it is desirable to manage based on a «green» international standard, for example, GPM Global P5.

References

1. Doe J (2021) Resolution adopted by the General Assembly on 6 July 2017. https://ggim.un.org/documents/a_res_71_313.pdf. Accessed 12 Nov 2021
2. Bobylev SN (2020) Cirkulyarnaya ekonomika i ee indikatory dlya Rossii (Circular economy and its indicators for Russia). The world of the New Economy. 14(2):63–72 (In Russian)
3. Kovkov DV, Koryagin ND, Eroshkin SY, Kameneva NA, Zaitsev EG, Sukhorukov TA, Sukhorukov AI (2017) Spatial model of electron-ionic concentrations distribution of in low-temperature air plasmoid over strong radiation contamination of soils and territories. In: Abstracts of the International Congress on Soils of Urban, Industrial, Traffic, Mining and Military Areas, Moscow, 22–26 May
4. Sukhorukov A, Koryagin N, Sulyagina J, Ulitskaya N, Eroshkin S (2019) Digital transformation of airline management as the basis of innovative development. In: Abstracts of the International Scientific Siberian Transport Forum, Novosibirsk, 22–27 May
5. Doe J (2020) Reviewed by Gordon Scott "Environmental, Social, and Governance (ESG) Criteria". https://www.investopedia.com/terms/e/environmental-social-and-governance-esg-criteria.asp. Accessed 12 Nov 2020
6. Doe J (2021) Statistics. Passenger traffic in Russia. https://www.avia-adv.ru/placement/airports/passenger-traffic.htm#placement_1. Accessed 10 Oct 2021
7. Doe J (2021) Federal Classification Catalog of waste. https://rpn.gov.ru/fkko/. Accessed 10 October 2021
8. Sukhorukov AI, Zakharova EA (2021) Aktual'nye podhody k razvitiyu ekonomiki zamknutogo cikla v sfere obrashcheniya s tverdymi kommunal'nymi othodami (Actual approaches to the development of a closed-cycle economy in the field of solid municipal waste management). Bulletin of the Plekhanov Russian University of Economics 5(119):33–44 (In Russian)
9. Koryagin, N.D., Sukhorukov A.I., Bolshedvorskaya, L.G (2016) Processnoe upravlenie na osnove programmnoj sistemy «Biznes-inzhener» (Process management based on the software system "Business Engineer") Moscow: Publishing House of the Zhukovsky Academy. 86 (In Russian).
10. Doe J (2020) Appendix No. 1 to the order of the Department of Housing and Communal Services of the City of Moscow No. 01-01-14-590/19 dated 26 December 2019. Territorial scheme of waste management. https://www.mos.ru/upload/documents/files/5871/02_TSOO-Moskva-2019-12-26-v19.pdf. Accessed 10 Feb 2020
11. Pawelek B, Pociecha J, Baryla M (2017) ABC analysis in corporate bankruptcy prediction. In: Abstracts of the IFCS Conference, Dusseldorf, 24–25 October

12. Sato Y, Kaufman JJ (2005) Value analysis tear-down: a new process for product development and innovation. Industrial Press Inc.
13. Doe J (2021) Decree of the Government of the Russian Federation No. 1156 of November 12, 2016 "On the Management of Solid Municipal Waste and Amendments to the Decree of the Government of the Russian Federation No. 641 of August 25, 2008" (with amendments and additions). https://base.garant.ru/71540160/. Accessed 10 October 2021

Problems of Introducing Digital Products into the Technological Processes of Organizing Air Transportation

Anastasia Stepanenko, Elena Stepanenko, and Leila Nikiforova

Abstract This article considers the aspects of the introduction of digital products in the technological processes of organizing air transportation, as well as the specifics of the processes of organizing air transportation. The aspects of software solutions which are necessary for the effective operation in the air transportation management system are shown. By establishing mathematical relations, the values of the criterion of the effectiveness of digital products in the overall system of transport organization are identified.

Keywords Digitalization · Transportation organization · Introduction of software products · Typical solutions · Regulated processes · Time intervals for the operation · Organization of transportation processes · Efficiency of digital product · Digitalization of the transportation organization system · Efficiency criteria

1 Introduction

The global trend of simplifying processes and accelerating operations dictates certain development paths for the air transport industry. The trend of digitalization of both stages [1] and specific processes in the organization of air transportation has been observed over the past 10 years [2]. Digitalization is a prerequisite for global social transformations that determine the path of business development, and provides unprecedented opportunities for creating value based on its virtualization [3].

A. Stepanenko · E. Stepanenko (✉) · L. Nikiforova
Moscow State Technical University of Civil Aviation, Moscow, Russia
e-mail: e.stepanenko@mstuca.aero

O. A. Gorbachev et al. (eds.), *Proceedings of 10th International Conference on Recent Advances in Civil Aviation*, Lecture Notes in Mechanical Engineering, https://doi.org/10.1007/978-981-19-3788-0_40

2 Materials and Methods

The study of reducing the time spent on performing operations was chosen not only because of the industry specifics of organizing air transportation processes, but also because of the global trend of accelerating production through digitalizing systems. When considering the specifics of organizing work in the field of air transportation, it should be noted that reducing the time for servicing aircraft, organizing the passage of inspection procedures by passengers, handling baggage, mail and cargo, has a direct impact on the profit of airlines. Let us assume that the impact is quite significant, since one of the key factors in the efficiency of the airline's work is the minimum amount of downtime and delays. This factor is formed based on the need to pay for additional time for servicing aircraft at airports. Therefore, it is worth noting that the time for aircraft ground handling for the work of ground services, as well as the time spent by the transport management service, is the most important factor in the effective operation of the airline. In this regard, the formation of a criterion reflecting the change in time intervals is relevant for the air transport industry in the field of introducing digital products.

In this paper, the formation of the criterion is substantiated by the built mathematical dependencies.

3 Literature Review

The development and transformation of transport networks and large transportation points require the introduction of digital products in the organization of transportation processes for the systematization of work and centralized management. There are barriers to the digitalization within the industry. Such as the perceived high cost and complexity of digital solutions [4, 5]. However, speaking about the air transport industry, it should be noted that the implementation of software is the most important factor in the development of the industry. And the failure to use information products makes the functioning of airlines impossible.

The alignment of business goals and areas of digital solutions is an important factor in gaining success of digitalization [6], which indicates the relevance of the topic under consideration.

When organizing air transportation, it is necessary to synchronize more than 20 parallel processes, therefore, simplification of individual parts of work, as well as issues of a unified coordination system, are one of the central tasks for air transport enterprises. Integration of software products allows reducing the time for one operation, accelerating the operation of the entire transportation organization system. The paper develops a criterion that allows tracking the time saving effect, as well as influences the acceleration of the entire transportation organization system. Taking into account the foregoing, an approach to the implementation of digital products can be singled out, which makes it possible to transfer part of the operations to the digital space.

Currently, there are both standard solutions in the field of software products for airlines and specialized developments. Standard solutions for organizing necessary processes are used, for example, for notification of passengers, processing of data and issuance of boarding passes. Specialized developments are mainly aimed at ensuring the coordination of airport services involved in the processes of transportation organization [7]. Existing digital products provide an opportunity to consider an approach to retrofit current digital equipment. Such an approach allows reducing costs for digitalization [8]. However, when improving the existing digital solutions, it is necessary to consider not only the issue of coordination among various software products, but also the functionality of each system. Originally missing functions, implemented as "add-ons" within the program, can change the quality of the main functionality of the system, and, consequently, reduce the efficiency of its work.

The processes of organizing air transportation can be divided into regulated and non-regulated processes.

Regulated processes are a set of operations arranged in a strict order in accordance with regulatory documents. For the air transport industry, a large percentage of operations involve just such processes. The regulations are determined by ICAO documents [9], Federal Aviation Regulations and internal airline documents, for example, Flight Crew Operation Manual [10]. In the presence of legally enshrined processes, we can talk about a standard technology for performing operations, and, consequently, about a simpler approach to digitalization [8, 11]. Taking into account the need to comply with work regulations, the implementation of a software product becomes an urgent solution not only to track the time required for such work, but also to reduce the percentage of errors inevitably made in the presence of manual labor.

Non-regulated processes represent a small part of the work; these are areas of work in the field of organizing transportation, which are not subject to regulations and temporary rationing.

Any process can be divided into stages and operations, respectively, there are three options for using digital solutions within one process: digitalization of the operation, digitalization of the stage, entire process digitalization. Taking into account the specifics of the work, a full transition to software management is not always relevant, therefore, when introducing a software product, many factors reflecting the specifics of the industry should be kept in mind. When implementing a fully automated process, there are high risks of failures, and in the case of air transport, failures in the organization of processes directly affect the safety level, and therefore cannot be tolerated. It is known that even the most reliable software product has an error rate; moreover, an error can occur systematically, destabilizing the entire system of processes. And in the conditions of the need to continuously perform operations, which is characteristic for the organization of work on air transport, a systematic error can lead to negative consequences.

Based on the foregoing, it can be concluded that the implementation of software products for complete digitalization can be relevant only for individual operations and process stages. If a system involves many processes, executed sequentially and in parallel, the implemented solutions should be universal and be able to reduce errors arising from software incompatibility. It is worth paying attention to the stage of digitalization of the process. Most decision areas fall under the categories of data collection and visualization (37%), as well as data analysis and decision-making (42%). This may reflect the fact that many SME manufacturers are still in the early stages of digitalization [12]. Collecting and visualizing data can be an important first step toward greater digitalization of processes..

4 Discussion

The universal nature of the solution will not only preserve the integrity of the system operation, but also ensure that there are no problems of "embedding" the solution into an existing software product during its implementation. When using universal blocks, it is not required to allocate time for personnel training, and therefore, the implementation time, up to the possibility of the full-fledged operation of the solution, will be minimal. Dividing software products into universal blocks makes it possible to rebuild them within the system, depending on changes in the external environment. Rebuilding a system consisting of similar blocks will require less time. However, when considering the software products that are currently used in the organization of air transportation, there arises an issue of a large difference in their functionality.

When forming "universal blocks", operations should be grouped taking into account the possibility of connecting to various specialized software products.

Figure 1 shows an example of the formation of universal blocks for building a general digital system for organizing transportation, as well as an example of functional links between blocks of one layout—"Execution of Documents"—with the rest of the universal blocks.

Speaking about the universal nature of a digital solution, it should be noted that software must be based on a single code. Rearrangement of universal blocks, as well as coordination and synchronization of operations with each other requires a well-orchestrated work of the system. In the case of using digital products for the air transport industry, continuity should be also taken into account. In the organization of air transportation, the absence of breaks in the performance of operations is of particular importance. When applying the critical path system, which is the basis for the organization of transport processes in the air transport industry, there are critical points in the execution of operations. Therefore, when building blocks of a universal digital solution, the necessary time constraints for each element of the system should be taken into account. In this case, the software will fulfill one of the necessary requirements for organizing air transportation.

Speaking about creating software for organizing processes in the air transport industry, it should be noted that there is a need for a separate approach to digitalization of each specific group of works within the framework of a unified information system. When establishing the configuration of universal blocks, one should be guided by

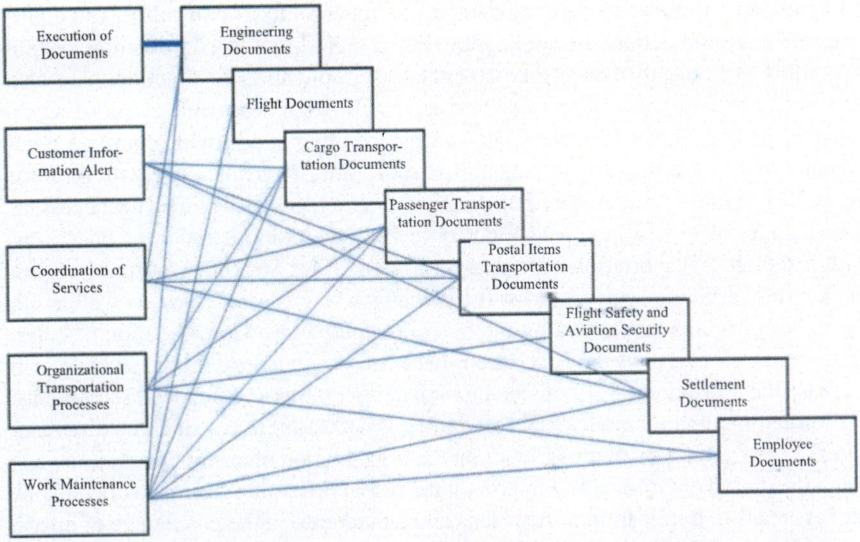

Fig. 1 An example of the formation of universal blocks to create a general digital support for the transportation organization system

the functional relationships of the processes, as well as the time intervals for their execution. Taking into account the characteristics of each operation is the only way to form an effective coordination system within the implemented software product.

Existing software products have overlapping functionality and the need to bundle data from one system to another; when information is transferred, system errors may appear, which jeopardizes the continuity of processes. Thus, the introduction of a common software system, consisting of universal blocks, will reduce the percentage of errors that occur during coordination between several software products.

When switching to a single digital system, it is necessary to formulate criteria for the effectiveness of digitalization of processes, to take into account all the features of operations performed in the organization of air transportation.

The main criteria for the digitalization of processes are: the universal nature of the solution, which can be used in various processes within the entire transportation organization system, the presence of regulations or work standards; increase in speed indicators and other characteristics of stages and operations. Adhering to the general criterion, it is possible to digitalize the processes of organizing air transportation, avoiding most of the errors, without reducing the efficiency of the system. Considering the determined criteria of process digitalization, it should be noted that the use of standardized processes, and in the case of considering the organization of air transportation, most of the processes are strictly regulated in time and operations, is an extremely favorable environment for the introduction of digital products.

The organizational processes of passenger and cargo transportation, as well as technical works are carried out both sequentially and in parallel, therefore, must have a common coordination system. At the same time, many implemented software solutions, even those working at one stage, are not commonly controlled, and therefore duplicate the actions of employees (Fig. 2). A common coordination system will allow not only to reduce the percentage of duplication of functions, but also to simplify management processes. That is, it will make it possible to coordinate a group of processes that are coordinated with each other and provide common functionality in the system of organizing air transportation. This, of course, will improve the quality of management, speed up decision-making, and also reduce the necessary functionality of employees, whose duties include the administration of operations within the air transportation organization system. This statement is true both for the aircraft servicing sector and for the passenger service processes, as well as for the group of works performed by the handling companies. Consolidation of enterprises of the air transport industry, through the use of a universal digital product, will increase the transparency of both technological operations and financial settlements.

Before implementing universal software, it is necessary to consider the disadvantages of the existing structure of work on the organization of air transportation when using the digital environment. First of all, the use of different systems to operate one process leads to duplication of functions and an increase in the percentage of errors.

Duplication of operations in different software products leads to a decrease in the efficiency of digitalization of the entire system, so the main goal of digitalization—simplifying of operations, stages, processes—is not fulfilled. Consequently, when

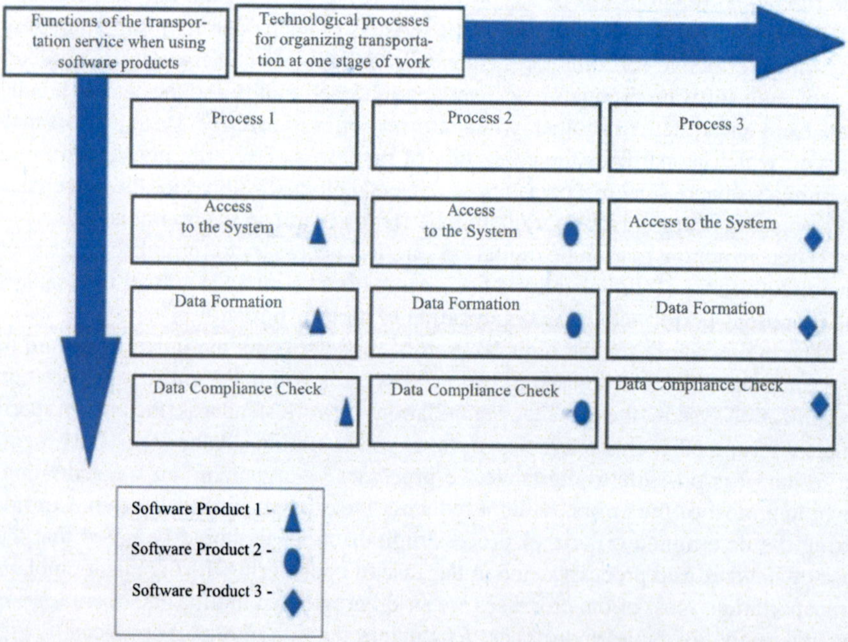

Fig. 2 Duplication of the functions of the transport management service when using various software products at one stage

creating a digitalization strategy for the organization of air transportation and when taking tactical measures, duplication of operations should be monitored. Identifying and eliminating similar operations in different software products can reduce the time for processing one operation to a minimum. Reducing the time it takes to find information will reduce the overall time spent on a process operation. It should be noted that the execution of long-term documents, for example, was identified as a major benefit in a recent study.

To create an effective system for the operation of software products within the air transportation organization system, it is necessary to establish an implementation criterion that will reflect the degree of influence of a digital product on changes in the time intervals for performing operations.

Calculations of the change in time intervals in case of digitalization of operations in transportation organization processes are provided below. To simplify the calculations, we will accept the division of all functional operations into 4 groups according to the execution time.

$$Ln = \sum(tx, ty, tz, tm) \tag{1}$$

where:

Ln is a general indicator of the execution time of sequential processes;
tx is time intervals from 2 to 5 min;
ty is time intervals from 5 to 10 min;
tz is time intervals from 10 to 20 min;
tm is time intervals from 20 to 30 min;

Subject to the adoption of the division of all sequential operations for organizing air transportation into 4 main time intervals (Table 1), technological processes can be represented as the sum of time intervals.

To calculate the effectiveness of the implementation of a digital product, the calculated indicator of the time interval when accessing the digital system is taken as 30 s, and the average indicator of reducing the time interval of the operation using the software product—as 0.3.

The indicators are taken on the basis of the average timing of access to the system of an employee of the air transport management service.

Table 1 Time intervals of operations adopted for calculations

Operation time intervals	Examples of operations in the organization of processes in air transport
2–5	Passport control for passengers, check of consignment notes, etc.
5–10	Cargo marking, cargo weighing, etc.
10–20	Boarding the aircraft, loading baggage, etc.
20–30	Routine inspection of the aircraft for breakdowns before departure, transportation of cargo to the aircraft, etc.

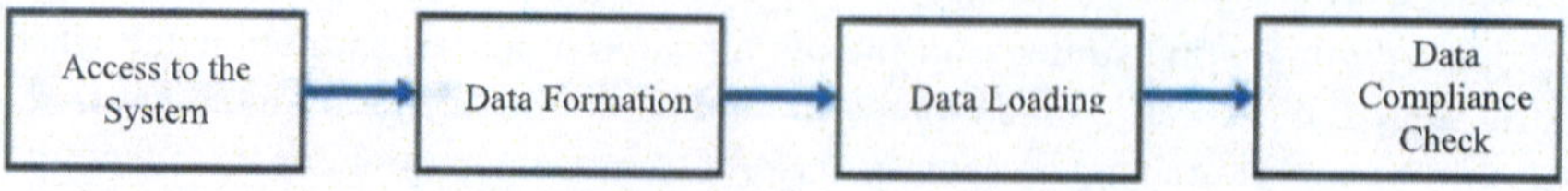

Fig. 3 Standard set of operations for transportation organization processes

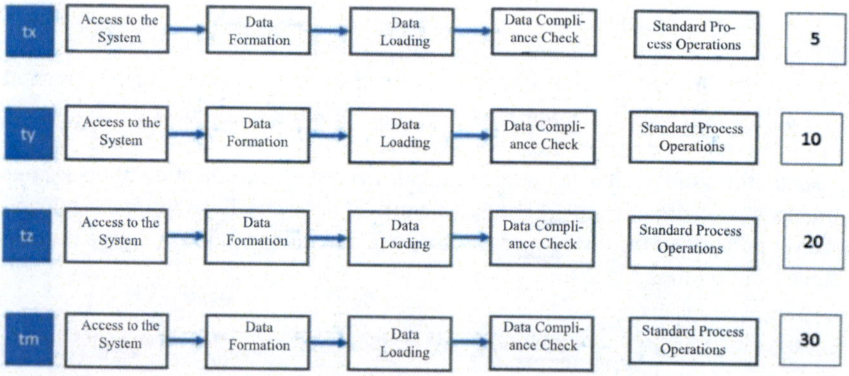

Fig. 4 Standard Set of Operations for Transportation Organization Processes with Indication of Time Intervals

To carry out calculations, a standard set of operations for working with software products for transportation organization processes shall apply (Fig. 3).

When considering a standard set of operations, the calculated time intervals shall be determined based on the maximum knowledge (Fig. 4).

When forming the calculated indicators of the change in intervals, the ratio of changes in the time spent on the operation and the calculated indicators of access to the digital system should be taken into account. To form a database that allows drawing conclusions on the effectiveness of the implementation of information systems in air transportation organization processes, it is necessary to form a mathematical ratio for calculating the time for an operation according to a standard process, taking into account the implementation of the software product:

$$\mathrm{Ln} = \sum((\mathrm{tx} * 0,3) + ng), (\mathrm{ty} * 0,3) + ng)(\mathrm{tz} * 0,3) + ng)(\mathrm{tm} * 0,3) + ng) \tag{2}$$

where:

tx is time intervals from 2 to 5 min;
ty is time intervals from 5 to 10 min;
tz is time intervals from 10 to 20 min;
tm is time intervals from 20 to 30 min;
Ln is regulated time for the process;
g is the time interval required to access the digital system (30 s);

Table 2 Calculated indicators of changes in time intervals

Indicators	0	1	2	3	4	**5**	6	7	8	9	10	11
tx	5	4.8	5.1	5.4	5.7	**6**	6.3	6.6	6.9	7.2	7.5	7.8
ty	10	9.3	9.6	9.9	10.2	**10.5**	10.8	11.1	11.4	11.7	12	12.3
tz	20	18.3	18.6	18.9	19.2	**19.5**	19.8	20.1	20.4	20.7	21	21.3
tm	30	27.3	27.6	27.9	28.2	**28.5**	28.8	29.1	29.4	29.7	30	30.3
Ln	65	59.7	60.9	62.1	63.3	**64.5**	65.7	66.9	68.1	69.3	70.5	71.7

n is the number of accesses to the digital system;

kt is the accepted indicator of reducing the operation interval using a software product (0.3).

The calculated indicators of changes in time intervals were formed taking into account the selected set of operations, and are shown in Table 2.

5 Results

According to the calculated indicators, it can be concluded that there is a proportional increase in the time interval spent on a functional operation at the given indicators. Since the processes are regulated in time, due to the specifics of the organization of air transportation, the excess of time intervals relative to the regulations is unacceptable.

Thus, a criterion must be established for the implementation of digital products in the technological processes of organizing air transportation, such as the number of accesses to the system required to complete one process. The calculated value according to the criterion should not exceed five. If this threshold is met, digitalization of the air transport organization system can reduce the number of failures in parallel and sequential processes without increasing the time intervals for regulated processes.

References

1. Ttsiarovaya-transformatsiya-aviaotrasli-trendy (2021) Digital transformation of the aviation industry: https://www.if24.ru
2. Fedoseeva MS, Tyumenev AV (2019) Digitalization in civil aviation. Theory and practice of project education. 4(12):100–101
3. Chyzhevska L, Shatskova L, Sokolenko L (2021) Digitalization as a vector of information systems development and accounting system modernization. Studia Universitatis "Vasile Goldis" Arad. Economics Series 31(4). 1584-2339; (online) ISSN: 2285-3065 Web: publicatii.uvvg.ro/index.php/studiaeconomia. pp 18–39
4. Galati F, Bigliardi B (2019) Industry 4.0: Emerging themes and future research avenues using a text mining approach. Comput Ind 109:100–113. https://doi.org/10.1016/j.compind.2019.04.018

5. Sevinc A, Gür S, Eren T (2018) Analysis of the difficulties of SMEs in industry 4.0 applications by analytical hierarchy process and analytical network process. Processes 6. https://doi.org/10.3390/pr6120264
6. Ming Heng BJ, Ng AK, Tay RKH (2019) Digitization of work instructions and checklists for improved data management and work productivity. Intell Transp Eng ICITE 2019:79–83. https://doi.org/10.1109/ICITE.2019.8880219
7. Afanasiev VG (1991) Organizaciya mezhdunarodnyh vozdushnyh perevozok. Perevozki passazhirov i bagazha (Organization of International air transportation. Transportation of passengers and luggage). Air transport, Moscow, p 255 (In Russian)
8. Li W, Liu K, Belitski M, Ghobadian A, O'Regan N (2016) e-Leadership through strategic alignment: an empirical study of small- and medium-sized enterprises in the digital age. J Inf Technol 31:185–206. https://doi.org/10.1057/jit.2016
9. Trofimov SV (2000) Legal regulation of the use of air transport. Transport Law 2:30–32
10. Prikaz Mintransa Rossii ot 28 iyunya 2007 g. N 82 (red. ot 15 sentyabrya 2020 g.) "Ob utverzhdenii Federal'nyh aviacionnyh pravil" Obshchie pravila vozdushnyh perevozok passazhirov, bagazha, gruzov i trebovaniya po obsluzhivaniyu passazhirov, gruzootpravitelej, gruzopoluchatelej» (Zaregistrirovano v Minyuste Rossii 27 sentyabrya 2007 g. N 10186.(Order of the Ministry of Transport of Russia dated June 28, 2007 N 82 (edition from September 15, 2020) "On approval of the Federal Aviation Regulations" General rules for the air transportation of passengers, baggage, cargo and requirements for servicing passengers, consignors, consignees "(Registered in the Ministry of Justice of Russia on September 27. 2007 N 10186.) (In Russian)
11. Lins T, Oliveira RAR (2020) Cyber-physical production systems retrofitting in context of industry 4.0. Comput Ind Eng 139. https://doi.org/10.1016/j.cie.2019.106193
12. Zheng T, Ardolino M, Bacchetti A, Perona M (2020). The applications of Industry 4.0 technologies in manufacturing context: a systematic literature review. Int J Prod Res 56:848–861. https://doi.org/10.1080/00207543.2017.1403664

Efficiency Indicators of Airlines' Business Processes

Elena Pronina and **Mihail Rodionov**

Abstract In the article, in accordance with the significantly increased requirements for ensuring the competitiveness and efficiency of airline management, taking into account the specifics of modern Russian air transport anti-crisis risk management, as well as in connection with the transition of many enterprises to process-oriented management, the problem of creating a measuring mechanism for quantitative assessment of the efficiency of business processes of airlines is considered. The authors substantiated the basic set of the system of performance indicators. Taking into account the stochastic nature of business processes, mathematical models are proposed for indicators of reliability, productivity and cost-effectiveness of a business process, as well as approaches to constructing an integral indicator of efficiency.

Keywords Enterprise management efficiency · Business processes · Business processes efficiency · Anti-crisis risk management · Reliability of a business process indicator · Productivity of a business process indicator · Cost-effectiveness of a business process indicator · Business process efficiency integral indicator

1 Introduction

The problem of increasing the efficiency of enterprise management has not lost its relevance for many years. In certain periods of time, different approaches, methodologies, and systems of indicators were used to assess effectiveness, and serious discussions were held on this topic [1].

At the present stage, the task becomes even more acute due to objective reasons: the globalization of markets leads to competition at the international level; consumer demand is becoming more and more sophisticated; and investment resources are not being added—the time for super profits has passed, it is enough to analyze the financial condition of airlines as a result of the pandemic. In addition, there

E. Pronina (✉) · M. Rodionov
Moscow State Technical University of Civil Aviation, Moscow, Russia
e-mail: ev_pronina@mail.ru

O. A. Gorbachev et al. (eds.), *Proceedings of 10th International Conference on Recent Advances in Civil Aviation*, Lecture Notes in Mechanical Engineering, https://doi.org/10.1007/978-981-19-3788-0_41

are subjective, organizational reasons—the existence of duplicate and unnecessary functions and procedures, etc.

In the interests of obtaining the presented scientific results, the authors of the article used a number of methods: content analysis—to study existing approaches to the problem under consideration; event analysis—to study the development of the situation in the air transport industry; systems approach; method of expert assessments, methods of probability theory and mathematical statistics, etc.

2 Investigation

Recently, Russian managers have come to the conclusion that business efficiency is improved if it is managed through business processes. And all modern methods of increasing operational efficiency (in the literature you can find the abbreviation PEX—process excellence), including Lean, 6 Sigma, theory of constraints, reengineering, design thinking, scenario approach, etc., are not applicable to functions isolated from each other, but to single processes [2–4]. Considering international experience, we see examples of outstanding companies (Toyota and others), which, using such approaches, have become industry leaders, introducing innovations without raising prices and reducing costs. However, despite the seeming ease and a huge number of consultants and published literature, only a few manage to repeat this success [5, 6].

Improving the efficiency of enterprise management is a complex strategic task that cannot always be crowned with success, even if all published recommendations are followed.

For Russian airlines, the current situation, in a certain sense, is comparable to the crisis in the air transportation market in the early 1990s after the collapse of the USSR. In the context of a serious aggravation of the international situation, the consequences of the systemic crisis of the 90s, the ongoing global financial and economic recession, the strengthening of anti-Russian economic sanctions, the ongoing coronavirus pandemic, associated infodemics, and others, the role of air transport anti-crisis risk management is significantly increasing.

The content of modern anti-crisis management is forecasting the possibility of crises, their diagnosis, planning, organization and practical implementation of measures to prevent, neutralize, reduce the negative consequences of crises, depending on the conditions of the situation, taking into account the interests of enterprise development [7]. However, in the air transport industry, it is often carried out only sporadically, is not proactive, and is not sufficiently linked to the processes of ensuring the competitiveness and risk management of the enterprise.

One of the directions for solving these issues is the use of the above-mentioned process approach, which, according to the international standard ISO 9001, provides for the implementation of the PDSA concept in business process management. In Russian airlines, the participants in the risk management system are distinguished:

the owner, initiator, performer of risk, controller, methodologist and coordinator, independent expert, guarantor, etc. with the regulatory documents of the enterprise.

The risk management system, providing, among other functions, the adoption of anti-crisis decisions, allows you to vary the processes of rationalizing the company's activities in the interests of ensuring its competitiveness and increasing efficiency. For a comprehensive assessment of the effectiveness of anti-crisis risk management measures, it is necessary to develop an appropriate system of indicators and criteria, as well as mathematical tools, which will have their own specifics in relation to a particular airline. These aspects should be interconnected with a quantitative assessment of the effectiveness of the organization's business processes as a whole.

Bill Hewlett, one of the founders of Hewlett-Packard, said that "you cannot manage what cannot be measured ... but everything that is measurable can be achieved" [8]. In full measure, this statement can be attributed to the management of the efficiency of business processes of aviation enterprises. That is, the following logical chain is built: the system of performance indicators—the assessment of indicators—the control of indicators (if possible, comparison with target values)—management decision-making—improving the efficiency of business processes.

First of all, it is necessary to determine what is meant by the effectiveness of business processes and its management. Let's turn to the primary sources. In [9] the following definition is given: «performance management is the use of information about performance to control the performance, quality, and cost of a process, workflow or business unit for compliance with their specified target levels. Based on this information, areas of improvement are determined that help to achieve the desired efficiency». In the same place [9] it is explained that «the assessment of efficiency is the identification of deviations of the current indicators of the efficiency of the process from the indicators established in accordance with the goals of the organization. Comparisons can be made against standards, targets or actual performance». As you can see, there is no clarity in the definition of performance indicators, or efficiency is understood as operational characteristics such as productivity, quality, and cost of the process [10, 11]. Perhaps the most obvious authors think is the approach to determining efficiency indicators, reflected in [12]: «efficiency indicators characterize the optimality of the use of resources to obtain the required result of the process with the given parameters, that is, when the goal of the process is achieved». Thus, the formation of an integral indicator of the efficiency of a business process can be represented using the following relational model, shown in Fig. 1.

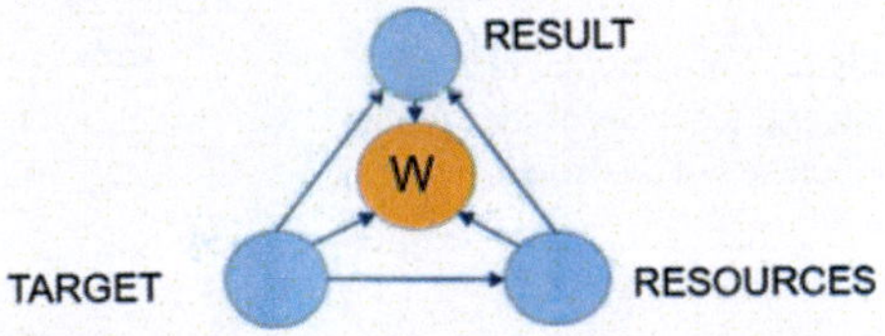

Fig. 1 Relational model of the formation of an integral indicator of the efficiency of a business process

In accordance with this model, a business process is effective if it ensures that the result is consistent with the target, as well as the resources allocated to ensure the achievement of the target. Mathematically, this statement can be reflected as follows [1].

$$\left.\begin{aligned} S_K(\omega,\xi,t) = \frac{U_K(\omega,\xi,t)}{Y_K(\omega,\xi,t)} \rightarrow \min; \\ Y_K(\omega,\xi,t) \geq Y_Ц; \\ \omega \in R[\omega]; t \in [t_0, t_1], \end{aligned}\right\} \quad (1)$$

where $S_K(\omega, \xi, t)$—the cost of resources per unit of the final result of the business process; $Y_K(\omega, \xi, t)$—the final result of the business process; $U_K(\omega, \xi, t)$—the cost of resources to obtain the final result of the business process; $Y_Ц$—target standards; ω—vector of controlled factors (control parameters); ξ— vector of uncontrollable factors (parameters of the external environment); $[t_0, t_1]$—the beginning and end of the period under consideration; $R[\omega]$—restrictions on control parameters.

From the model (1), which reflects the requirements achieving the target of a business process (limitation) with maximum resource savings (objective function), logically follows a basic set of three performance indicators: reliability indicator (whether the target will be achieved), productivity indicator (what result will be obtained), and an indicator of the cost-effectiveness (what price will be achieved both). In this case, the system of indicators of the efficiency of the business process can be represented as follows (Fig. 2), where W—an integral indicator of the efficiency of a business process; W_H, W_P, $W_Э$—indicators of reliability, productivity, and cost-effectiveness of the business process, respectively.

If a business process is considered as a deterministic system, then the values in model (1) are also deterministic, and efficiency modeling is reduced to solving a deterministic mathematical programming problem. In this case, the indicator of the reliability of the business process is not determined, the indicator of productivity is determined by the value Y_K and indicator of cost-effectiveness—the ratio of deterministic values Y_K and U_K.

However, any business process, as you know, is a stochastic system, therefore, the quantities making up (1) are random. In this case, the constraint in model (1) should

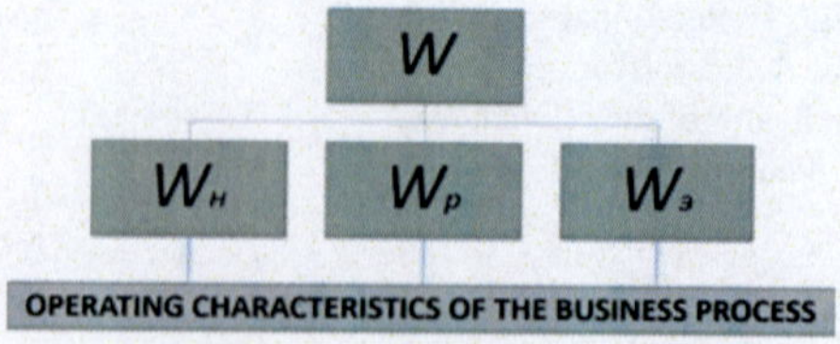

Fig. 2 Formation of an integral indicator of the efficiency of a business process

be represented as

$$P(Y_K \geq Y_Ц) \geq P_Ц, \tag{2}$$

and the objective function is

$$M[S_K] = \frac{M[U_K]}{M[Y_K]}, \tag{3}$$

where $P(Y_K \geq Y_Ц)$—the probability of achieving the target of the business process; $P_Ц$—target standard of this probability; $M[Y_K]$—mathematical expectation of the final result; $M[U_K]$—mathematical expectation of expenditures of resources for achievement Y_K; $M[S_K]$—mathematical expectation of expenditures of resources per unit of the final result Y_K of a business process.

Therefore, the set (4) is a universal set of measures of the efficiency of a business process that allows you to build a stochastic model (5)

$$\{P(Y_K \geq Y_Ц), M[Y_K], M[U_K]\} \tag{4}$$

$$\left.\begin{array}{c} M[S_K] = \dfrac{M[U_K]}{M[Y_K]} \to \min; \\ P(Y_K \geq Y_Ц) \geq P_Ц; \\ S_K \geq 0; U_K \geq 0; Y_K \geq 0; Y_Ц \geq 0 \end{array}\right\} \tag{5}$$

In this case, indicators (4), as we see, do not have the same dimension, they are not reduced to a single measure. Measure $P(Y_K \geq Y_Ц)$ is enclosed within [0, 1], and measures $M[Y_K]$ and $M[U_K]$—within $[0, y_{\max}]$ and $[0, u_{\max}]$ accordingly. Since the system of equations on these intervals should provide $M[Y_K] \to y_{\max}$ and $M[U_K] \to u_{\min}$, then you can build mathematical models of business process efficiency indicators as follows.

$$W_H = P(Y_K \geq Y_Ц); \tag{6}$$

$$W_P = \frac{M[Y_K]}{y_{\max}}; \tag{7}$$

$$W_Э = \frac{u_{\min}}{M[U_K]}, \tag{8}$$

where W_H—the indicator of reliability of the business process; W_P—the indicator of productivity of the business process; $W_Э$—indicator of cost-effectiveness of the business process. Note that the values of all indicators are enclosed within the limits [0, 1].

A more "stringent" measure of the cost-effectiveness of a business process is the cost of resources per unit of the final result, then the indicator of cost-effectiveness can be represented as

$$W_{Э} = \frac{s_{\min}}{M[S_K]} \tag{9}$$

The further task is to construct the integral indicator W. Various options for its solutions are proposed in the literature [13, 14]. For example, an integral indicator is determined using additive or multiplicative models, taking into account the weighting coefficients of the importance of the indicators being reduced. Another approach is to rank the indicators in order of importance. Optimization is carried out according to the most "important" indicator, and then the area of solutions is determined, where this indicator differs from its optimal value by no more than, for example, 10%. In the area of solutions, optimization is performed according to the second indicator, and so on [15].

Let's consider the business process of the airline «air transportation of passengers». To implement the mathematically rigorous approach described above, a significant amount of statistical data is required, which is difficult to present within the limited volume of this article, from the point of view of the completeness of the presentation of the essence of the study. In this regard, a comprehensive assessment of the effectiveness of a given business process for a certain period of time can be carried out according to the following working mathematical models, making certain assumptions:

$$W_H^* = F_T\left(\frac{\mu^*_{Q^Ф} - \mu^*_{Q^{ПЛ}}}{\sqrt{\sigma^{*2}_{\mu^*_{Q^Ф}} + \sigma^{*2}_{\mu^*_{Q^{ПЛ}}}}}\right) \tag{10}$$

$$W_P^* = \frac{\mu^*_{Q^Ф}}{1,3\mu^*_{Q^{ПЛ}}}; \tag{11}$$

$$\sigma^*_{W_i^*} = \sqrt{\frac{W_i^*(1 - W_i^*)}{k-1}}; \tag{12}$$

$$W_i^* = (W_H^* \vee W_P^*); \tag{13}$$

$$W_Э^* = \frac{1}{k-1}\sum_{t=2}^{k} v_{S_Ф}^{Ht}; \tag{14}$$

$$\sigma^*_{W_Э^*} = \frac{\sigma^*_{v^H_{S_Ф}}}{\sqrt{k-1}}; \tag{15}$$

where W_H^*, W_P^*, $W_Э^*$—estimates of the indicators of reliability, productivity, and cost-effectiveness of a business process, accordingly; $\sigma_{W_i^*}^*$, $\sigma_{W_Э^*}^*$—estimates of the standard deviation of the indicator estimates W_H^*, W_P^* and $W_Э^*$, accordingly;

$$\mu_{Q^\Phi}^* = \frac{1}{k}\sum_{t=1}^{k} Q^{\Phi t}; \tag{16}$$

$$\mu_{Q^{ПЛ}}^* = \frac{1}{k}\sum_{t=1}^{k} Q^{ПЛt}; \tag{17}$$

where $\mu_{Q^\Phi}^*$, $\mu_{Q^{ПЛ}}^*$—estimates of the mathematical expectation of the actual and planned passenger turnover, accordingly;

$$\sigma_{\mu_{Q^\Phi}^*}^* = \frac{\sigma_{Q^\Phi}^*}{\sqrt{k}}; \tag{18}$$

$$\sigma_{\mu_{Q^{ПЛ}}^*}^* = \frac{\sigma_{Q^{ПЛ}}^*}{\sqrt{k}}; \tag{19}$$

$$\sigma_{Q^\Phi}^* = \sqrt{\frac{1}{k-1}\sum_{t=1}^{k}\left(Q^{\Phi t} - \mu_{Q^\Phi}^*\right)^2}; \tag{20}$$

$$\sigma_{Q^\Phi}^* = \sqrt{\frac{1}{k-1}\sum_{t=1}^{k}(Q^{\Phi t} - \mu_{Q^\Phi}^*)^2}; \tag{21}$$

where $\sigma_{\mu_{Q^\Phi}^*}^*$, $\sigma_{\mu_{Q^{ПЛ}}^*}^*$—estimates of the standard deviation of the estimates $\mu_{Q^\Phi}^*$ и $\mu_{Q^{ПЛ}}^*$, accordingly; $Q^{\Phi t}$, $Q^{ПЛt}$—actual and planned values of passenger turnover in the t-th year; $\sigma_{Q^\Phi}^*$, $\sigma_{Q^{ПЛ}}^*$—estimates of the standard deviation of the actual and planned passenger turnover, accordingly;

$$v_{S_\Phi}^{Ht} = \frac{v_{S_\Phi}^t}{v_{\max}}; \tag{22}$$

$$v_{S_\Phi}^t = 2 - \frac{S_\Phi^t}{S_\Phi^{t-1}}; \tag{23}$$

$$\sigma_{v_{S_\Phi}^H}^* = \sqrt{\frac{1}{k-2}\sum_{t=2}^{k}\left(v_{S_\Phi}^{Ht} - W_Э^*\right)^2}, \tag{24}$$

where S_{Φ}^{t}—actual cost per passenger-kilometer in year t; $\upsilon_{S_{\Phi}}^{t}$—the rate of decrease in the actual cost per passenger-kilometer in the t-th year; $\upsilon_{S_{\Phi}}^{Ht}$—normalized value $\upsilon_{S_{\Phi}}^{t}$; $\upsilon_{\max}$—maximum value $\upsilon_{S_{\Phi}}^{t}$ for a period of k years; $\sigma_{\upsilon_{S_{\Phi}}^{Ht}}^{*}$—estimation of the standard deviation of the normalized rate of decrease in the actual cost of passenger-kilometers.

In the above models of reliability and productivity indicators of the airline business process «air transportation of passengers», instead of the target standard of the final result, its planned value is taken, since target standards of passenger turnover in the airline are often absent. In this regard, the coefficient 1.3 appeared in the denominator of the performance indicator model (from the condition that in the Russian aviation market, the segment of low-cost transportation, for example, has a growth potential up to a level of at least 30% [16]). The indicator of cost-effectiveness of a business process has also undergone changes due to the absence of a target standard and acquired the meaning of the average normalized rate of decrease in the actual cost of the airline's passenger-kilometer.

3 Results

The assessment of the efficiency of the business process carried out in this way makes it possible to compare the results of the work of various airlines, to make appropriate management decisions, including those of an anti-crisis nature (Table 1).

The results of evaluating the effectiveness of a business process are clearly presented as follows (Fig. 3).

Table 1 Indicators of the efficiency of the business process "air transportation of passengers"

Air company	The indicator of reliability		The indicator of productivity		The indicator of cost-effectiveness	
	W_H^*	$\sigma_{W_H^*}^*$	W_P^*	$\sigma_{W_P^*}^*$	$W_Э^*$	$\sigma_{W_Э^*}^*$
Airline 1	0.832	0.187	0.822	0.191	0.955	0.024
Airline 2	0.821	0.192	0.817	0.193	0.971	0.024
Airline 3	0.695	0.230	0.820	0.192	0.922	0.034
Airline 4	0.610	0.244	0.804	0.198	0.955	0.026
Airline 5	0.722	0.224	0.815	0.194	0.946	0.023
Airline 6	0.764	0.212	0.815	0.194	0.914	0.064

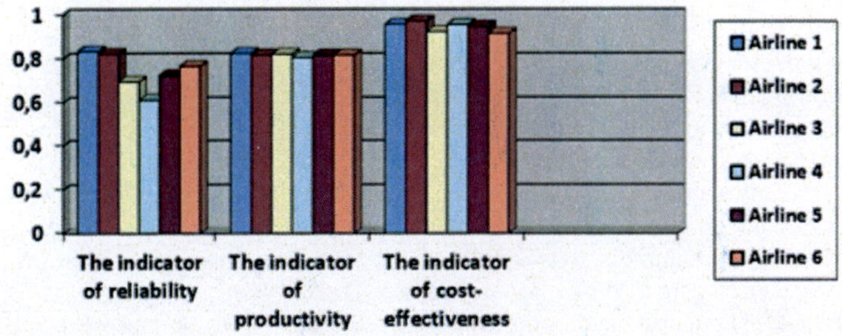

Fig. 3 Indicators of business process efficiency

4 Conclusion

In conclusion, it should be noted that business process performance indicators should not be just measurements—they are the basis for serious discussions and decision-making to improve business processes.

In this regard, it seems relevant to conduct research on the scientific and practical substantiation of the applied methodological approaches to assessing the effectiveness of business processes with the involvement of a significant number of representatives of the expert community. The participation of the latter can be carried out in a wide variety of forms—symposia, forums, conferences, expert councils, etc.

References

1. Pronina EV (2017) Universal'naya sistema pokazatelej dlya ocenki effektivnosti ERP-sistem (A universal system of indicators for evaluating the effectiveness of ERP systems). Ekonomika i predprinimatel'stvo 4(ch. 2):988–992. (In Russian)
2. Matt C, Hess T, Benlian A (2015) Digital transformation strategies. Bus Inf Syst Enf 57:339–343
3. Dumas M, Rosa ML, Mendling J, Reijers HA (2018) Fundamentals of business process management, 2nd edn. Springer, Berlin
4. Drakulevski L, Nakov L (2014) Managing business model as function of organizational dynamism. Management 72:37–44
5. Peters SP, Dijkman RM, Grefen PW (2018) Quantitative effects of advanced resource constructs in business process simulation. In: 22nd International Enterprise Distributed Object Computing Conference (EDOC), IEEE, pp 115–122
6. Peters SP, .Dijkman RM, Grefen PW (2018) Advanced simulation of resource constructs in business process models. In: International Conference on Business Process Management, Springer, pp 159–175
7. Rodionov MA (2020) Information aspects of anti-crisis air transport risk management. Financial Risk Manag Moscow. 2(62):120–128
8. Thompson AA (2007) Strategicheskij menedzhment: koncepcii i situacii dlya analiza (Strategic management: concepts and situations for analysis) (12th edn). Translation from English. Publishing house "Williams", Moscow, 928 p. (In Russian)

9. Belaychuk AA, Yeliferov VG (2020) Svod znanij po upravleniyu biznes-processami (A body of knowledge on business process management): BPM CBOK 3.0, 2-ed. Alpina Publisher, Moscow, 480 p. (In Russian)
10. Clempner J (2014) Verifying soundness of business processes: A decision process. Petri nets approach. Expert Systems with Appl 41(11):5030–5040
11. Heidari F, Loucopoulos P (2014) Quality evaluation framework (QEF): Modeling and evaluating quality of business processes. Int J Account Inf Syst 15(3):193–223
12. Kamennova MS, Krokhin VV, Mashkov IV (2021) Modelirovanie biznes-processov (Business process modeling). Part 2: Yurayt, Moscow, 228 p. (In Russian)
13. Peters SP (2021) Analysis and optimization of resources in business processes. Eindhoven University of Technology, The Netherlands
14. Groger C, Schwarz H, Mitschang B (2014) Prescriptive analytics for recommendation-based business process optimization. In: International Conference on Business Information Systems, Springer, pp 25–37
15. Venttsel YS (2010) Issledovanie operacij: zadachi, principy, metodologiya (Operations research: tasks, principles, methodology). KnoRus, Moscow, 192 p. (In Russian)
16. Godovoy otchet Aeroflota 2020 (Aeroflot annual report 2020). ar2020_rus.pdf (aeroflot.ru). Accessed 12 October 2021 (In Russian)